Probability:

Probability has applications in many areas of modern science, not to mention in our daily life, and its importance as a mathematical discipline cannot be overrated. This engaging book, with its easy to follow writing style, provides a comprehensive, yet concise, introduction to the subject. It covers all of the standard material for undergraduate and first-year-graduate-level courses, as well as many topics that are usually not found in standard texts – such as Bayesian inference, Markov chain Monte Carlo simulation, and Chernoff bounds.

The student-friendly text has the following additional features:

- Is the result of many years of teaching and feedback from students
- Stresses why probability is so relevant and how to apply it
- Offers many real-world examples to support the theory
- Includes more than 750 problems with detailed solutions of the odd-numbered problems
- Gives students confidence in their own problem-solving skills

HENK TIJMS is Professor Emeritus at the Vrije University in Amsterdam. He is the author of several textbooks and numerous papers on applied probability and stochastic optimization. In 2008 Henk received the prestigious INFORMS Expository Writing Award for his publications and books. His activities also include the popularization of probability for high-school students and the general public.

Probability: A Lively Introduction

HENK TIJMS
Vrije Universiteit, Amsterdam

CAMBRIDGE
UNIVERSITY PRESS

University Printing House, Cambridge CB2 8BS, United Kingdom

One Liberty Plaza, 20th Floor, New York, NY 10006, USA

477 Williamstown Road, Port Melbourne, VIC 3207, Australia

314-321, 3rd Floor, Plot 3, Splendor Forum, Jasola District Centre, New Delhi - 110025, India

79 Anson Road, #06-04/06, Singapore 079906

Cambridge University Press is part of the University of Cambridge.

It furthers the University's mission by disseminating knowledge in the pursuit of education, learning and research at the highest international levels of excellence.

www.cambridge.org
Information on this title: www.cambridge.org/9781108407847
DOI: 10.1017/9781108291361

© Henk Tijms 2018

This publication is in copyright. Subject to statutory exception and to the provisions of relevant collective licensing agreements, no reproduction of any part may take place without the written permission of Cambridge University Press.

First published 2018

A catalogue record for this publication is available from the British Library

Library of Congress Cataloging in Publication data
Names: Tijms, H.
Title: Probability: a lively introduction / Henk Tijms, Vrije Universiteit, Amsterdam.
Description: Cambridge : Cambridge University Press, 2017. | Includes index.
Identifiers: LCCN 2017014908 | ISBN 9781108418744 (Hardback : alk. paper) | ISBN 9781108407847 (pbk. : alk. paper)
Subjects: LCSH: Probabilities–Textbooks.
Classification: LCC QA273.2 .T55 2017 | DDC 519.2–dc23 LC record available at https://lccn.loc.gov/2017014908

ISBN 978-1-108-41874-4 Hardback
ISBN 978-1-108-40784-7 Paperback

Cambridge University Press has no responsibility for the persistence or accuracy of URLs for external or third-party internet websites referred to in this publication, and does not guarantee that any content on such websites is, or will remain, accurate or appropriate.

Contents

	Preface	*page* ix
1	**Foundations of Probability Theory**	1
1.1	Probabilistic Foundations	3
1.2	Classical Probability Model	7
1.3	Geometric Probability Model	15
1.4	Compound Chance Experiments	19
1.5	Some Basic Rules	25
1.6	Inclusion–Exclusion Rule	36
2	**Conditional Probability**	42
2.1	Concept of Conditional Probability	42
2.2	Chain Rule for Conditional Probabilities	47
2.3	Law of Conditional Probability	54
2.4	Bayes' Rule in Odds Form	67
2.5	Bayesian Inference – Discrete Case	77
3	**Discrete Random Variables**	85
3.1	Concept of a Random Variable	85
3.2	Expected Value	89
3.3	Expected Value of Sums of Random Variables	99
3.4	Substitution Rule and Variance	106
3.5	Independence of Random Variables	113
3.6	Binomial Distribution	118
3.7	Poisson Distribution	124
3.8	Hypergeometric Distribution	135
3.9	Other Discrete Distributions	140

4	**Continuous Random Variables**	146
4.1	Concept of Probability Density	147
4.2	Expected Value of a Continuous Random Variable	156
4.3	Substitution Rule and the Variance	160
4.4	Uniform and Triangular Distributions	164
4.5	Exponential Distribution	167
4.6	Gamma, Weibull, and Beta Distributions	177
4.7	Normal Distribution	180
4.8	Other Continuous Distributions	193
4.9	Inverse-Transformation Method and Simulation	198
4.10	Failure-Rate Function	202
4.11	Probability Distributions and Entropy	205
5	**Jointly Distributed Random Variables**	209
5.1	Joint Probability Mass Function	209
5.2	Joint Probability Density Function	212
5.3	Marginal Probability Densities	219
5.4	Transformation of Random Variables	228
5.5	Covariance and Correlation Coefficient	233
6	**Multivariate Normal Distribution**	239
6.1	Bivariate Normal Distribution	239
6.2	Multivariate Normal Distribution	248
6.3	Multidimensional Central Limit Theorem	250
6.4	Chi-Square Test	257
7	**Conditioning by Random Variables**	261
7.1	Conditional Distributions	262
7.2	Law of Conditional Probability for Random Variables	269
7.3	Law of Conditional Expectation	276
7.4	Conditional Expectation as a Computational Tool	283
7.5	Bayesian Inference – Continuous Case	294
8	**Generating Functions**	302
8.1	Generating Functions	302
8.2	Branching Processes and Generating Functions	311
8.3	Moment-Generating Functions	313
8.4	Central Limit Theorem Revisited	318
9	**Additional Topics in Probability**	321
9.1	Bounds and Inequalities	321
9.2	Strong Law of Large Numbers	327

9.3	Kelly Betting System	335
9.4	Renewal–Reward Processes	339
10	**Discrete-Time Markov Chains**	348
10.1	Markov Chain Model	349
10.2	Time-Dependent Analysis of Markov Chains	357
10.3	Absorbing Markov Chains	362
10.4	Long-Run Analysis of Markov Chains	373
10.5	Markov Chain Monte Carlo Simulation	386
11	**Continuous-Time Markov Chains**	403
11.1	Markov Chain Model	403
11.2	Time-Dependent Probabilities	414
11.3	Limiting Probabilities	420
	Appendix A: Counting Methods	438
	Appendix B: Basics of Set Theory	443
	Appendix C: Some Basic Results from Calculus	447
	Appendix D: Basics of Monte Carlo Simulation	451
	Answers to Odd-Numbered Problems	463
	Index	532

Preface

Why do so many students find probability difficult? Even the most mathematically competent often find probability a subject that is difficult to use and understand. The difficulty stems from the fact that most problems in probability, even ones that are easy to understand, cannot be solved by using cookbook recipes as is sometimes the case in other areas of mathematics. Instead, each new problem often requires imagination and creative thinking. That is why probability is difficult, but also why it is fun and engaging. Probability is a fascinating subject and I hope, in this book, to share my enthusiasm for the subject.

Probability is best taught to beginning students by using motivating examples and problems, and a solution approach that gives the students confidence to solve problems on their own. The examples and problems should be relevant, clear, and instructive. This book is not written in a theorem–proof style, but proofs flow with the subsequent text and no mathematics is introduced without specific examples and applications to motivate the theory. It distinguishes itself from other introductory probability texts by its emphasis on why probability is so relevant and how to apply it. Every attempt has been made to create a student-friendly book and to help students understand what they are learning, not just to learn it.

This textbook is designed for a first course in probability at an undergraduate level or first-year graduate level. It covers all of the standard material for such courses, but it also contains many topics that are usually not found in introductory probability books – such as stochastic simulation. The emphasis throughout the book is on probability, but attention is also given to statistics. In particular, Bayesian inference is discussed at length and illustrated with several illuminating examples. The book can be used in a variety of disciplines, ranging from applied mathematics and statistics to computer science, operations research, and engineering, and is suitable not only for introductory courses, but also for self-study. The prerequisite knowledge is a basic course in calculus.

Good problems are an essential part of each textbook. The field of probability is well known for being a subject that can best be acquired by the process of learning-by-doing. Much care has been taken to present problems that will enhance the student's understanding of probability. He or she will be asked to think about ideas, rather than simply plugging numbers into formulas. Working through them, it may often be found that probability problems are harder than they first appear. This book has more than 750 carefully designed problems, both "easy ones" and challenging ones. Problems are grouped according to the section they are based on, which should be convenient for both students and instructors. An important feature of this textbook is that it contains detailed solutions to the odd-numbered problems, which helps stimulate active learning and contributes to students' confidence in their own problem-solving skills. It is my belief that there is an enormous increase in content when worked-out solutions to exercises are included. Solutions for the even-numbered problems and further detailed solutions to the odd-numbered problems for instructors can be found on the book's webpage (www.cambridge.org/TijmsProbability). Another added feature is that the student will find many tips on problem-solving strategies.

How to Teach from this Book

This book, which is the result of many years of teaching and feedback from many students, can be used to teach probability courses at several different levels. Since students learn best when they participate actively in the process, the instructor may want to use computer simulation in the course. Appendix D gives the basic ideas of Monte Carlo simulation, together with instructive simulation exercises. Simulation in teaching probability helps to develop and sharpen probabilistic intuition and is popular with many students. Over many years of teaching probability I have also found that the subject of Markov chains is very appealing to students, and the book offers the possibility of a follow-up course on stochastic processes that covers renewal–reward processes, discrete-time Markov chains, and continuous-time Markov chains. In a course on discrete probability, material on discrete-time Markov chains can even be taught right after the notion of conditional probability has been introduced.

1
Foundations of Probability Theory

For centuries, mankind lived with the idea that uncertainty was the domain of the gods and fell beyond the reach of human calculation. Common gambling led to the first steps toward understanding probability theory, and the colorful Italian mathematician and physician Gerolamo Cardano (1501–1575) was the first to attempt a systematic study of the calculus of probabilities. As an ardent gambler himself, Cardano wrote a handbook for fellow gamblers entitled *Liber de Ludo Aleae* (The Book of Games of Chance) about probabilities in games of chance like dice. He originated and introduced the concept of the set of outcomes of an experiment, and for cases in which all outcomes are equally probable, he defined the probability of any one event occurring as the ratio of the number of favorable outcomes to the total number of possible outcomes. This may seem obvious today, but in Cardano's day such an approach marked an enormous leap forward in the development of probability theory.

Nevertheless, many historians mark 1654 as the birth of the study of probability, since in that year questions posed by gamblers led to an exchange of letters between the great French mathematicians Pierre de Fermat (1601–1665) and Blaise Pascal (1623–1662). This famous correspondence laid the groundwork for the birth of the study of probability, especially their question of how two players in a game of chance should divide the stakes if the game ends prematurely. This problem of points, which will be discussed in Chapter 3, was the catalyst that enabled probability theory to develop beyond mere combinatorial enumeration.

In 1657, the Dutch astronomer Christiaan Huygens (1629–1695) learned of the Fermat–Pascal correspondence and shortly thereafter published the book *De Ratiociniis de Ludo Aleae* (On Reasoning in Games of Chance), in which he worked out the concept of expected value and unified various problems that had been solved earlier by Fermat and Pascal. Huygens' work led the field for many years until, in 1713, the Swiss mathematician Jakob Bernoulli

(1654–1705) published *Ars Conjectandi* (The Art of Conjecturing) in which he presented the first general theory for calculating probabilities. Then, in 1812, the great French mathematician Pierre Simon Laplace (1749–1827) published his *Théorie Analytique des Probabilités*. Here, Laplace applied probabilistic ideas to many scientific and practical problems, and his book represents perhaps the single greatest contribution in the history of probability theory. Towards the end of the nineteenth century, mathematicians attempted to construct a solid foundation for the mathematical theory of probability by defining probabilities in terms of relative frequencies. That attempt, however, was marked by much controversy, and the frequency view of probability did not lead to a satisfactory theory.

An acceptable definition of probability should be precise enough for use in mathematics and yet comprehensive enough to be applicable to a wide range of phenomena. It was not until 1933 that the great Russian mathematician Andrey Nikolaevich Kolmogorov (1903–1987) laid a satisfactory mathematical foundation for probability theory by taking a number of axioms as his starting point, as had been done in other fields of mathematics. Axioms state a number of minimal requirements that the mathematical objects in question (such as points and lines in geometry) must satisfy. In the axiomatic approach of Kolmogorov, probability figures as a function on subsets of a so-called sample space, where the sample space represents the set of all possible outcomes of the experiment. He assigned probabilities to these subsets in a consistent way, using the concept of additivity from measure theory. Those axioms are the basis for the mathematical theory of probability, and, as a milestone, are sufficient to logically deduce the law of large numbers. This law confirms our intuition that the probability of an event in a repeatable experiment can be estimated by the relative frequency of its occurrence in many repetitions of the experiment. The law of large numbers is the fundamental link between theory and the real world.

Nowadays, more than ever before, probability theory is indispensable in a wide variety of fields. It is absolutely essential to the field of insurance. Likewise the stock market, "the largest casino in the world," cannot do without it. Call centers and airline companies apply probabilistic methods to determine how many service desks will be needed based on expected demand. In stock control, probability theory is used to find a balance between the cost risks of running out of inventory and the cost of holding too much inventory in a market with uncertain demand. Engineers use probability theory when constructing dikes, to calculate the probability of water levels exceeding their margins. Judges and physicians benefit from a basic knowledge of probability theory, to help make better judicial and medical decisions. In short, probabilistic thinking has become an integral part of modern life.

The purpose of this chapter is to give the student a solid basis for solving probability problems. We first discuss the intuitive and fundamental axioms of probability, and from them derive a number of basic rules for the calculation of probabilities. These rules include the complement rule, the addition rule, and the inclusion–exclusion rule, for which many illustrative examples are provided. These examples, which are instructive and provide insight into the theory, include classical probability problems such as the birthday problem and the hat-check problem.

Traditionally, introductory probability books begin with a comprehensive discussion of set theory and combinatorics before presenting the "real stuff." This will not be done in this book, as it is not necessary and often stifles the student's natural curiosity and fascination with probability. Rather, the appendices at the end of the book provide a self-contained overview of the essentials of set theory and the basic tools from combinatorics, and these tools are not introduced until they are needed.

1.1 Probabilistic Foundations

A probability model is a mathematical representation of a real-world situation or a random experiment. It consists of a complete description of all possible outcomes of the experiment and an assignment of probabilities to these outcomes. The set of all possible outcomes of the experiment is called the *sample space*. A sample space is always such that one and only one of the possible outcomes occurs if the experiment is performed. Here are some examples:

- The experiment is to toss a coin once. The sample space is the set $\{H, T\}$, where H means that the outcome of the toss is a head and T that it is a tail. Each outcome gets assigned a probability of $\frac{1}{2}$ if the coin is fair.
- The experiment is to roll a die once. The sample space is the set $\{1, 2, \ldots, 6\}$, where outcome i means that i dots appear on the up face. Each outcome gets assigned a probability of $\frac{1}{6}$ if the die is fair.
- The experiment is to choose a letter at random from the word "statistics." The sample space is the set $\{s, t, a, i, c\}$. The probabilities $\frac{3}{10}, \frac{3}{10}, \frac{1}{10}, \frac{2}{10}$, and $\frac{1}{10}$ are assigned to the five outcomes s, t, a, i, and c.
- The experiment is to repeatedly roll a fair die until the first six shows up. The sample space is the set $\{1, 2, \ldots\}$ of the positive integers. Outcome k indicates that the first six shows up on the kth roll. The probabilities $\frac{1}{6}, \frac{5}{6} \times \frac{1}{6}, (\frac{5}{6})^2 \times \frac{1}{6}, \ldots$ are assigned to the outcomes $1, 2, 3, \ldots$.

- The experiment is to measure the time until the first emission of a particle from a radioactive source. The sample space is the set $(0, \infty)$ of the positive real numbers, where outcome t indicates that it takes a time t until the first emission of a particle. Taking an appropriate unit of time, the probability $\int_a^b e^{-t}\,dt$ can be assigned to each time interval (a, b) on the basis of physical properties of radioactive material, where $e = 2.71828\ldots$ is the base of the natural logarithm.

In probability applications we are typically interested in particular subsets of the sample space, which in probability language are called *events*. The terms event and subset of outcomes of an experiment are used interchangeably in probability theory. In the second example, the event that an odd number is rolled is the subset $A = \{1, 3, 5\}$ of the sample space. In the fourth example, the event that more than six rolls are needed to get a six is the subset $A = \{7, 8, \ldots\}$ of the sample space. In the fifth example, the event that it takes between 5 and 7 time units until the first emission of a particle is the subset $A = \{t : 5 \leq t \leq 7\}$ of the sample space, where ":" means "such that."

Various choices for the sample space are sometimes possible. In the experiment of tossing a fair coin twice, a possible choice for the sample space is the set $\{HH, HT, TH, TT\}$. Another possible choice is the set $\{0, 1, 2\}$, where the outcome indicates the number of heads obtained. The assignment of probabilities to the elements of the sample space differs for the two choices. In the first choice of the sample space, the four elements are equally likely and each element gets assigned the probability $\frac{1}{4}$. In the second choice of the sample space, the elements are not equally likely and the elements 0, 1, and 2 get assigned the probabilities $\frac{1}{4}, \frac{1}{2}$, and $\frac{1}{4}$. In general, it is preferable to use a sample space with equally likely outcomes whenever possible.

In the first three examples above, the sample space is a finite set. In the fourth example the sample space is a so-called countably infinite set, while in the fifth example the sample space is a so-called uncountable set. Let us briefly explain these basic concepts from set theory, see also Appendix B. The set of natural numbers (positive integers) is an infinite set and is the prototype of a countably infinite set. In general, a nonfinite set is called *countably infinite* if a one-to-one function exists which maps the elements of the set to the set of natural numbers. In other words, every element of the set can be assigned to a unique natural number and conversely each natural number corresponds to a unique element of the set. For example, the set of squared numbers $1, 4, 9, 16, 25, \ldots$ is countably infinite. Not all sets with an infinite number of elements are countably infinite. The set of all points on a line and the set of all real numbers between 0 and 1 are examples of infinite sets that are not countable.

1.1 Probabilistic Foundations

The German mathematician Georg Cantor (1845–1918) proved this result in the nineteenth century. This discovery represented an important milestone in the development of mathematics and logic (the concept of infinity, to which even scholars from ancient Greece had devoted considerable energy, obtained a solid theoretical basis for the first time through Cantor's work). Sets that are neither finite nor countably infinite are called *uncountable*, whereas sets that are either finite or countably infinite are called *countable*.

1.1.1 Axioms of Probability Theory

A probability model consists of a sample space together with the assignment of probability, where probability is a function that assigns numbers between 0 and 1 to subsets of the sample space. The axioms of probability are mathematical rules that the probability function must satisfy. The axioms of probability are essentially the same for a chance experiment with a countable or an uncountable sample space. A distinction must be made, however, between the sorts of subsets to which probabilities can be assigned, whether these subsets occur in countable or uncountable sample spaces. In the case of a countable sample space, probabilities can be assigned to each subset of the sample space. In the case of an uncountable sample space, weird subsets can be constructed to which we cannot associate a probability. Then the probability measure is only defined on a sufficiently rich collection of well-behaved subsets, see Appendix B for more details. These technical matters will not be discussed further in this introductory book. The reader is asked to accept the fact that, for more fundamental mathematical reasons, probabilities can only be assigned to well-behaved subsets when the sample space is uncountable. In the case that the uncountable sample space is the set of real numbers, then essentially only those subsets consisting of a finite interval, the complement of any finite interval, the union of any countable number of finite intervals, or the intersection of any countable number of finite intervals are assigned a probability. These subsets suffice for practical purposes. The probability measure on the sample space is denoted by P. It assigns to each well-behaved subset A a probability $P(A)$ and must satisfy the following properties:

Axiom 1 $P(A) \geq 0$ *for each subset A of the sample space.*

Axiom 2 $P(\Omega) = 1$ *for the sample space Ω.*

Axiom 3 $P(\bigcup_{i=1}^{\infty} A_i) = \sum_{i=1}^{\infty} P(A_i)$ *for every collection of pairwise disjoint subsets $A_1, A_2, \ldots$ of the sample space.*

The *union* $\bigcup_{i=1}^{\infty} A_i$ of the subsets $A_1, A_2, \ldots$ is defined as the set of all outcomes which belong to at least one of the subsets $A_1, A_2, \ldots$. The subsets $A_1, A_2, \ldots$ are said to be *pairwise disjoint* when any two subsets have no element in common. In probability terms, any subset of the sample space is called an *event*. If the outcome of the chance experiment belongs to the subset A, then the event A is said to occur. The events $A_1, A_2, \ldots$ are said to be *mutually exclusive* (or *disjoint*) if the corresponding sets $A_1, A_2, \ldots$ are pairwise disjoint. In other words, events $A_1, A_2, \ldots$ are mutually exclusive if the occurrence of one of these events implies the non-occurrence of the others. For example, suppose that a die is rolled. The outcome is one of the numbers from 1 to 6. Let A be the event that the outcome 1 or 2 occurs and B be the event that the outcome 5 or 6 occurs. Then A and B are mutually exclusive events. As another example, suppose that a coin is tossed until heads appears for the first time. Let A_k be the event that heads appears for the first time at the kth toss. Then the events $A_1, A_2, \ldots$ are mutually exclusive.

The first two axioms simply express a probability as a number between 0 and 1. The crucial third axiom states that, for any infinite sequence of mutually exclusive events, the probability of at least one of these events occurring is the sum of their individual probabilities. This property also holds for any *finite* sequence of mutually exclusive events. Using the concept of the empty set, the proof of this result is almost trivial, see Rule 1.1 in Section 1.5. The countable additivity in Axiom 3 is required to have a unified framework for finite and nonfinite sample spaces. Starting with the three axioms and a few definitions, a powerful and beautiful theory of probability can be developed.

The standard notation for the sample space is the symbol Ω. An outcome of the sample space is denoted by ω. A sample space together with a collection of events and an assignment of probabilities to the events is called a *probability space*. For a countable sample space Ω, it is sufficient to assign a probability $p(\omega)$ to each element $\omega \in \Omega$ such that $p(\omega) \geq 0$ and $\sum_{\omega \in \Omega} p(\omega) = 1$. A probability measure P on Ω is then defined by specifying the probability of each event A as

$$P(A) = \sum_{\omega \in A} p(\omega).$$

In other words, $P(A)$ is the sum of the individual probabilities of the outcomes ω that belong to the set A. It is left to the reader to verify that P satisfies Axioms 1 to 3.

A probability model is constructed with a specific situation or experiment in mind. The assignment of probabilities is part of the translation process from a concrete context into a mathematical model. Probabilities may be assigned to events any way you like, as long as the above axioms are satisfied. To make your choice of the probabilities useful, the assignment should result in a "good" model for the real-world situation. There are two main approaches to assigning probabilities to events. In the relative-frequency approach, probabilities are assigned to the outcomes of a physical experiment having the feature that it can be repeated over and over under identical conditions. Think of spinning a roulette wheel or rolling dice. Then one may speak of *physical probabilities* and such probabilities can be determined experimentally. In the subjective approach, the word probability is roughly synonymous with plausibility and probability is defined as the degree of belief a particular person holds in the occurrence of an event. Think of the chances of your favorite horse winning a race or the chances of your favorite baseball team winning the World Series. Hence judgment is used as the basis for assigning *subjective probabilities*. The use of the subjective approach is usually limited to experiments that are unrepeatable. In this book the emphasis is on physical probabilities, but we will also pay attention to subjective probabilities in Chapters 2 and 7.

1.2 Classical Probability Model

In many experiments with finitely many outcomes $\omega_1, \ldots, \omega_m$, it is natural to assume that all these outcomes are equally likely to occur (this probability model is, in fact, also based on judgment). In such a case, $p(\omega_i) = \frac{1}{m}$ for $i = 1, \ldots, m$ and each event A gets assigned the probability

$$P(A) = \frac{m(A)}{m},$$

where $m(A)$ is the number of outcomes in the set A. This model is sometimes called the *classical probability model* or the *Laplace model*.

Example 1.1 John, Pedro, and Rosita each roll one fair die. How do we calculate the probability that Rosita's score is equal to the sum of the scores of John and Pedro?

Solution. We take the set $\{(i,j,k) : i,j,k = 1,\ldots,6\}$ as sample space for the chance experiment, where the outcome (i,j,k) means that John's score is i dots, Pedro's score is j dots, and Rosita's score is k dots. Each of the 216 possible outcomes is equally probable, and thus gets assigned a probability mass of $\frac{1}{216}$. Rosita's score is equal to the sum of the scores of John and Pedro if and only if one of the 15 outcomes (1,1,2), (1,2,3), (2,1,3), (1,3,4), (3,1,4), (2,2,4), (1,4,5), (4,1,5), (2,3,5), (3,2,5), (1,5,6), (5,1,6), (2,4,6), (4,2,6), (3,3,6) occurs. The probability of this event is thus $\frac{15}{216}$.

Example 1.2 Three players enter a room and are given a red or a blue hat to wear. The color of each hat is determined by a fair coin toss. Players cannot see the color of their own hats, but do see the color of the other two players' hats. The game is won when at least one of the players correctly guesses the color of his own hat and no player gives an incorrect answer. In addition to having the opportunity to guess a color, players may also pass. Communication of any kind between the players is not permissible after they have been given their hats; however, they may agree on a group strategy beforehand. The players decided upon the following strategy. A player who sees that the other two players wear a hat with the same color guesses the opposite color for his/her own hat; otherwise, the player says nothing. What is the probability of winning the game under this strategy?

Solution. This chance experiment can be seen as tossing a fair coin three times. As sample space, we take the set consisting of the eight elements *RRR*, *RRB*, *RBR*, *BRR*, *BBB*, *BBR*, *BRB*, *RBB*, where *R* stands for a red hat and *B* for a blue hat. Each element of the sample space is equally probable and gets assigned a probability of $\frac{1}{8}$. The strategy is winning if one of the six outcomes *RRB*, *RBR*, *BRR*, *BBR*, *BRB*, or *RBB* occurs (verify!). Thus, the probability of winning the game under the chosen strategy is $\frac{3}{4}$.

In Example 1.2 we have encountered a useful problem-solving strategy: see whether the problem can be related to a familiar problem.

As preparation for the next example, consider a task that involves a sequence of r choices. Suppose that n_1 is the number of possible ways the first choice can be made, n_2 is the number of possible ways the second choice can be made after the first choice has been made, and n_3 is the number of possible ways the third choice can be made after the first two choices have been made, etc. Then the total number of possible ways the task can be performed is $n_1 \times n_2 \times \cdots \times n_r$. For example, the total number of possible ways five people can stand in line is $5 \times 4 \times 3 \times 2 \times 1 = 120$. In other words, there are 120 permutations.

Example 1.3 In a Monte Carlo casino the roulette wheel is divided into 37 sections numbered 1 to 36 and 0. What is the probability that all numbers showing up in eight spins of the wheel are different?

Solution. Take as sample space the set of all ordered sequences $(i_1, \ldots, i_8)$, where i_k is the number showing up at the kth spin of the wheel. The sample space has $37 \times 37 \times \cdots \times 37 = 37^8$ equiprobable elements. The number of elements for which all components are different is $37 \times 36 \times \cdots \times 30$. Therefore, the sought probability is

1.2 Classical Probability Model

$$\frac{37 \times 36 \times \cdots \times 30}{37^8} = 0.4432.$$

We reiterate the concept of the binomial coefficient before continuing.

Definition 1.1 *The binomial coefficient $\binom{n}{k}$ denotes the total number of ways to choose k different objects out of n distinguishable objects, with order not mattering.*

In other words, $\binom{n}{k}$ is the total number of combinations of k different objects out of n. The key difference between permutations and combinations is *order*. Combinations are *unordered* selections, permutations are *ordered* arrangements. In Appendix A these important concepts are discussed extensively and illustrated with several combinatorial probability problems. For any integers n and k with $1 \leq k \leq n$, the binomial coefficient can be calculated as

$$\binom{n}{k} = \frac{n!}{k!\,(n-k)!},$$

where $m!$ is shorthand for $1 \times 2 \times \cdots \times m$ with the convention $0! = 1$. Note that $\binom{n}{k} = \binom{n}{n-k}$ with the convention $\binom{n}{0} = 1$. For example, the number of ways to choose three jurors out of five candidates is $\binom{5}{3} = \frac{5!}{3!\,2!} = \frac{120}{6 \times 2} = 10$, with order not mattering.

In no book on introductory probability should problems on tossing coins, rolling dice, and dealing cards be missing. The next examples deal with these sorts of probability problems and use binomial coefficients to solve them.

Example 1.4 A fair coin is tossed 100 times. What is the probability of getting exactly 50 heads?

Solution. Take as sample space all possible sequences of zeros and ones to a length of 100, where a zero stands for tails and a one for heads. The sample space has 2^{100} equiprobable elements. The number of elements having exactly 50 ones is $\binom{100}{50}$. Therefore, the probability of getting exactly 50 heads is

$$\frac{\binom{100}{50}}{2^{100}}.$$

This probability is the ratio of two enormously large numbers and its computation requires special provisions. However, a very accurate approximation to this probability can be given. To this end, consider the general case of $2N$ tosses of a fair coin. Then the probability p_N of getting exactly N heads is

$$p_N = \frac{\binom{2N}{N}}{2^{2N}}.$$

To approximate this probability, we use Stirling's approximation for $n!$. This famous approximation states that

$$n! \approx \sqrt{2\pi n}\left(\frac{n}{e}\right)^n$$

for sufficiently large n, where the mathematical constant e is the Euler number $2.7182818\ldots$. In practice, this approximation is useful for $n \geq 10$. The relative error percentage is about $\frac{100}{12n}\%$. Using Stirling's approximation, we find

$$p_N = \frac{(2N)!}{N!\,N!}\frac{1}{2^{2N}} \approx \frac{\sqrt{2\pi \times 2N}\,(2N/e)^{2N}}{\sqrt{2\pi N}\,(N/e)^N\,\sqrt{2\pi N}\,(N/e)^N}\frac{1}{2^{2N}}.$$

Canceling out common terms in the denominator and numerator, we get

$$p_N \approx \frac{1}{\sqrt{\pi N}}$$

for N sufficiently large. This approximation is not only very accurate, but also gives insight into how the probability p_N depends on N. The approximate value of p_{50} is 0.07979, while the exact value is 0.07959.

Example 1.5 What is the probability that three different face values each appear twice in a roll of six dice?

Solution. Take as sample space the set of all possible sequences of the face values $1, 2, \ldots, 6$ to a length of 6. The sample space has 6^6 equiprobable elements. There are $\binom{6}{3}$ ways to choose the face values for the three pairs of different face values. There are $\binom{6}{2}$ possible combinations of two dice from the six dice for the first pair, $\binom{4}{2}$ possible combinations of two dice from the remaining four dice for the second pair, and then one combination of two dice remains for the third pair. Thus the probability of getting three pairs of different face values in a roll of six dice is

$$\frac{\binom{6}{3} \times \binom{6}{2} \times \binom{4}{2} \times 1}{6^6} = 0.0386.$$

The next example shows that the choice of the sample space is not always unambiguous.

Example 1.6 A bridge hand in which there is no card higher than a nine is called a Yarborough. What is the probability of a Yarborough when you are randomly dealt 13 cards out of a well-shuffled deck of 52 cards?

Solution. The choice of the sample space depends on whether we care about the order in which the cards are dealt from the deck of 52 cards. If we consider the order in which the 13 cards are dealt as being relevant, then we take a sample space made up of *ordered* elements. Imagine that the 52 cards of the deck are numbered $1, 2, \ldots, 52$ and that the 13 cards of the bridge hand are randomly dealt one at a time. The sample space made up of ordered elements is the set of all possible 13-tuples $(i_1, \ldots, i_{13})$, where i_k is the number of the kth card dealt. The sample space has $52 \times 51 \times \cdots \times 40$ equally likely outcomes. The number of cards having a value nine or lower is $8 \times 4 = 32$, and so there are $32 \times 31 \times \cdots \times 20$ outcomes for which there is no card higher than a nine among the 13 cards dealt. Thus, the probability of a Yarborough is

$$\frac{32 \times 31 \times \cdots \times 20}{52 \times 51 \times \cdots \times 40} = 0.000547.$$

Alternatively, this probability can be computed by using a sample space made up of *unordered* elements. The order in which the 13 cards are dealt is not relevant in such a sample space.[1] Each outcome of this sample space is a set of 13 different cards from the deck of 52 cards. In a set we don't care which element is first, only which elements are actually present. The number of ways you can choose a set of 13 different cards from a deck of 52 cards is given by the binomial coefficient $\binom{52}{13}$. Thus, the sample space made up of unordered elements has $\binom{52}{13}$ equally likely outcomes. The number of outcomes with no card above a nine is $\binom{32}{13}$, and so the probability of a Yarborough can also be calculated as

$$\frac{\binom{32}{13}}{\binom{52}{13}} = 0.000547.$$

The odds against a Yarborough are 1,827 to 1. A Yarborough is named after Lord Yarborough (1809–1862). The second Earl of Yarborough would offer his whist-playing friends a wager of £1,000 to 1 against them picking up such a hand. You see that the odds were on his side. There is no record that he ever paid out.

Example 1.7 A well-shuffled standard deck of 52 cards is divided into four piles of 13 cards each. What is the probability that each pile contains exactly one ace?

Solution. A subtle approach is as follows. The deck of cards has 52 positions from top to bottom. Imagine that the first 13 positions are for the first pile, the next 13 positions for the second pile, etc. Take as sample space the collection

---

[1] In combinatorial probability problems it is sometimes possible to calculate a probability both through a sample space with ordered elements and through a sample space with unordered elements. This provides a sanity check against accidental overcounting or undercounting. Many people make mistakes in counting.

of all possible combinations of four positions to be used for the aces. The sample space has $\binom{52}{4}$ equiprobable elements. There are 13^4 combinations of four positions for which one position belongs to the first 13 positions of the deck, one position belongs to the second 13 positions of the deck, one position belongs to the third 13 positions of the deck, and one position belongs to the fourth 13 positions of the deck. Thus, the sought probability is

$$\frac{13^4}{\binom{52}{4}} = 0.1055.$$

Alternatively, the sought probability can be calculated from the formula

$$\frac{\binom{4}{1}\binom{48}{12}\binom{3}{1}\binom{36}{2}\binom{2}{1}\binom{24}{12}}{\binom{52}{13}\binom{39}{13}\binom{26}{13}} = 0.1055.$$

The reader is asked to specify the sample space that is the basis of the alternative calculation.

These examples illustrate the importance of doing probability calculations systematically rather than "intuitively." The key steps in all our calculations are:

- What is the sample space of all possible outcomes of the experiment?
- What are the probabilities assigned to the possible outcomes?
- What outcomes define the event of interest?

Then the probability of the event is calculated by adding up the probabilities of the outcomes composing the event. The beginning student should always be able to go back to these basic steps to avoid potential pitfalls.

The importance of specifying an appropriate sample space is nicely illustrated with the following trick problem. A bag contains 14 red and 7 black cards. Two cards are picked together at random from the bag. Which event is more likely: picking two red cards, or one red and one black card? Many people think that picking two red cards has the largest probability. This is not true, as can be seen by using the following sample space. Think of the red cards labeled as $R_1, \ldots, R_{14}$ and the black cards labeled as $B_1, \ldots, B_7$.[2] Also, it is helpful to imagine that the two cards are picked one at a time (it makes no difference to the probability whether the two cards are picked together or one after the other). Then we can take as sample space the set consisting of the 14×13 outcomes (R_j, R_k), the 14×7 outcomes (R_j, B_k), the 7×14 outcomes

---

[2] In problems involving indistinguishable objects, the calculation of a probability is often made easier by pretending that the objects can be distinguished so that a sample space with equally likely outcomes can be used.

(B_j, R_k), and the 7×6 outcomes (B_j, B_k), where the first component of each outcome refers to the first card picked and the second component to the second card picked. There is a total of 420 equally likely outcomes. The probability of picking two red cards is $\frac{14 \times 13}{420}$ and is less than the probability $\frac{14 \times 7 + 7 \times 14}{420}$ of picking one red and one black card.

The genius Gottfried Wilhelm Leibniz (1646–1716), the inventor of differential and integral calculus along with Sir Isaac Newton (1643–1727), had difficulties in calculating the probability of getting the sum 11 with one throw of two dice. When asked whether 12 points and 11 points are equally likely in a throw of two dice, he erroneously stated: "with two dice, it is equally likely to throw twelve points, than to throw eleven; because one or the other can be done in only one manner." This error would not have been made had an appropriate sample space been used. A mistake such as Leibniz's was not made by the famous astronomer and physicist Galileo Galilei (1564–1642) when he explained to the Grand Duke of Tuscany, his benefactor, why it is that when you toss three dice, the probability of the sum being 10 is greater than the probability of the sum being 9 (the probabilities are $\frac{27}{216}$ and $\frac{25}{216}$, respectively). To get these probabilities, it is convenient to imagine that the three dice are blue, white, and red, and then to use as sample space all ordered triples (b, w, r), where b is the number showing on the blue, w the number showing on the white, and r the number showing on the red die.

In the following problems, you are asked to first specify a sample space before calculating the probabilities. In many cases there is more than one natural candidate for the sample space.

Problems

1.1 Two fair dice are rolled. What is the probability that the sum of the two numbers rolled is odd? What is the probability that the product of the two numbers rolled is odd?

1.2 In a township, there are two plumbers. On a particular day three residents call village plumbers, independently of each other. Each resident randomly chooses one of the two plumbers. What is the probability that all three residents will choose the same plumber?

1.3 Take a random permutation of the letters in the word "randomness." What is the probability that the permutation begins and ends with a vowel? What is the probability that the three vowels are adjacent to each other in the permutation?

1.4 A dog has a litter of four puppies. Can we correctly say that the litter more likely consists of three puppies of one gender and one of the other than that it consists of two puppies of each gender?

1.5 Suppose that you pick m numbers from a set of n different numbers without replacement. What is the probability of getting the largest number in the set?

1.6 Three girls and three boys have a dinner party. They agree that two of them are going to do the dishes. The unlucky persons are determined by drawing lots. What is the probability that two boys will do the dishes?

1.7 A box contains n identical balls, where two balls have a winning number written on them. In a prespecified order, n persons each choose a ball at random from the box. Show that each person has the same probability of getting a winning ball.

1.8 Players A and B play a game of rolling two fair dice. If the largest number rolled is one, two, three, or four, player A wins; otherwise, player B wins. What is the probability of player A winning?

1.9 In the 6/42 lottery, six distinct numbers are picked at random from the numbers $1, \ldots, 42$. What is the probability that number 10 will be picked? What is the probability that each of the six numbers picked is 20 or more?

1.10 Five people are sitting at a table in a restaurant. Two of them order coffee and the other three order tea. The waiter forgets who ordered what and puts the drinks in a random order for the five persons. What is the probability that each person gets the correct drink?

1.11 Two black socks, two brown socks, and one white sock lie mixed up in a drawer. You grab two socks without looking. What is the probability that you have grabbed two black socks or two brown socks?

1.12 Two letters have fallen out of the word "Cincinnati" at random places. What is the probability that these two letters are the same?

1.13 You choose at random two cards from a standard deck of 52 cards. What is the probability of getting a ten and a heart?

1.14 You choose at random a letter from the word "chance" and from the word "choice." What is the probability that the two letters are the same?

1.15 You roll a fair die 10 times. Can you explain why the probability of rolling a total of s points is the same as the probability of rolling a total of $70 - s$ points for $s = 10, \ldots, 34$?

1.16 You roll 12 fair dice. What is the probability that each number will appear exactly twice?

1.17 A building has three floors. Each floor has four apartments and each apartment has one resident. Three residents meet each other coincidentally in the entrance of the building. What is the probability that these three residents live on different floors?

1.18 Together with two friends you are in a group of 10 people. The group is randomly split up into two groups of 5 people each. What is the probability that your two friends and you are in the same group together?

1.19 You have four mathematics books, three physics books, and two chemistry books. The books are put in random order on a bookshelf. What is the probability of having the books ordered per subject on the bookshelf?

1.20 Three friends go to the cinema together on a weekly basis. Before buying their tickets, all three friends toss a fair coin into the air once. If one of the three gets a different outcome than the other two, then that one pays for all three tickets; otherwise, everyone pays his own way. What is the probability that one of the three friends will have to pay for all three tickets?

1.21 You choose a letter at random from the word "Mississippi" 11 times without replacement. What is the probability that you can form the word Mississippi with the 11 chosen letters? *Hint*: It may be helpful to number the letters $1, 2, \ldots, 11$.

1.22 What is the probability that you hold one pair in a five-card poker hand and what is the probability that you hold two pairs?

1.23 Somebody is looking for a top-floor apartment. She hears about two vacant apartments in a building with 7 floors and 8 apartments per floor. What is the probability that there is a vacant apartment on the top floor?

1.24 John and Paul take turns picking a ball at random from a bag containing four red and seven white balls. The balls are drawn out of the bag without replacement. What is the probability that John is the first person to pick a red ball when Paul is the first person to start? *Hint*: Think of the order in which the balls are drawn out of the bag as a permutation of $1, 2, \ldots, 11$.

1.25 You pick a number from 1 to n twice at random. What is the probability that the sum of the two numbers picked is r for $r \leq n+1$? Use this result to find the probability of getting a sum s when rolling two fair dice.

1.3 Geometric Probability Model

Equally likely outcomes are only possible for a finite sample space (why?). The continuous analogue of the probability model having a finite number of equally likely outcomes is the geometric probability model with an uncountable sample space. In applications of this model, a random point is chosen in a certain bounded region. In mathematical words, choosing a *random point* means that equal probabilities are assigned to sets of equal lengths, areas, or volumes. The next examples illustrate the geometric probability model.

Example 1.8 Joe is driving around, looking for a parking spot. He plans to stay in the area for exactly 2 hours and 15 minutes. He comes across a free parking spot that only allows free parking for 2 hours. The parking enforcement officer comes around once every 2 hours. At each inspection the officer marks cars that were not there at his last visit and tickets earlier-marked cars. Joe arrives at a random moment between two visits of the officer, and Joe has no idea how long ago the officer's last visit was. Joe decides to take a gamble and parks in the free spot. What is the probability that Joe will receive a ticket?

Solution. Define a probability model with the interval $(0, 120)$ as sample space. The outcome x means that it takes x minutes until the next visit of the officer, measured from the moment that Joe arrives. To each subinterval, we assign as probability the length of the interval divided by 120. Let A be the event that Joe receives a ticket. The event A occurs if and only if the outcome x is the interval $(105, 120)$. Thus $P(A) = \frac{15}{120}$, so the probability that Joe will receive a ticket is $\frac{1}{8}$.

Example 1.9 You randomly throw a dart at a circular dartboard with radius R. It is assumed that the dart is infinitely sharp and lands at a random point on the dartboard. How do you calculate the probability of the dart hitting the bullseye, having radius b?

Solution. The sample space of this experiment is the set of pairs of real numbers (x, y) with $x^2 + y^2 \leq R^2$, where (x, y) indicates the point at which the dart hits the dartboard. This sample space is uncountable. We first make the following observation. The probability that the dart lands exactly at a *prespecified* point is zero. It only makes sense to speak of the probability of the dart hitting a given region of the dartboard. This observation expresses a fundamental difference between a probability model with a countable sample space and a probability model with an uncountable sample space. The assumption of the dart hitting the dartboard at a random point is translated by assigning the probability

$$P(A) = \frac{\text{the area of the region } A}{\pi R^2}$$

to each subset A of the sample space. Therefore, the probability of the dart hitting the bullseye is $\pi b^2/(\pi R^2) = b^2/R^2$.

1.3 Geometric Probability Model

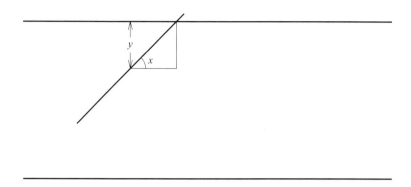

Figure 1.1. The landing of Buffon's needle.

Example 1.10 A floor is ruled with equally spaced parallel lines a distance D apart. A needle of length L is dropped at random on the floor. It is assumed that $L \leq D$. What is the probability that the needle will intersect one of the lines? This problem is known as Buffon's needle problem.

Solution. This geometric probability problem can be translated into the picking of a random point inside a certain region. Let y be the distance from the center of the needle to the closest line and x be the angle at which the needle falls, where x is measured against a line parallel to the lines on the floor; see Figure 1.1. The sample space of the experiment can be taken as the rectangle R consisting of the points (x, y), with $0 \leq x \leq \pi$ and $0 \leq y \leq \frac{1}{2}D$. Dropping the needle at random on the floor can be seen to be equivalent to choosing a random point inside the rectangle R. The needle will land on a line only if the hypotenuse of the right-angled triangle in Figure 1.1 is less than half of the length L of the needle. That is, we get an intersection if and only if $\frac{y}{\sin(x)} < \frac{1}{2}L$. Thus, the probability that the needle will intersect one of the lines equals the probability that a point (x, y) chosen at random in the rectangle R satisfies $y < \frac{1}{2}L\sin(x)$. In other words, the area under the curve $y = \frac{1}{2}L\sin(x)$ divided by the total area of the rectangle R gives the probability of an intersection. This ratio is

$$\frac{\int_0^\pi \frac{1}{2}L\sin(x)\,dx}{\frac{1}{2}\pi D} = \frac{-L\cos(x)}{\pi D}\Big|_0^\pi$$

and so

$$P(\text{needle intersects one of the lines}) = \frac{2L}{\pi D}.$$

Problems

1.26 A stick of length one is broken into two pieces at a random point. What is the probability that the length of the longer piece will be at least three times the length of the shorter piece?

1.27 Franc-carreau was a popular game in eighteenth-century France. In this game, a coin is tossed on a chessboard. The player wins if the coin does not fall on one of the lines of the board. Suppose now that a round coin with a diameter of d is blindly tossed on a large table. The surface of the table is divided into squares whose sides measure a in length, where $a > d$. Define an appropriate probability space and calculate the probability of the coin falling entirely within the confines of a square. *Hint*: Consider the position of the center of the coin.

1.28 Each morning during the week you arrive at a train station at a random moment between 7 a.m. and 8 a.m. Fast trains in the direction of your destination arrive every 15 minutes starting at 7 a.m., whereas slow trains in the direction of your destination arrive every 30 minutes starting at 7:05 a.m. You take the first train that arrives. What is the probability that you will take a slow train on any given morning?

1.29 Consider again Problem 1.28. Suppose that your friend arrives at the train station at a random moment between 7.30 a.m. and 8 a.m. and also takes the first train in the direction of your destination. What is the probability that he gets on the same train as you?

1.30 Two people have agreed to meet at a central point in the city. Independently of one another, each person is to appear at a random moment between 12 p.m. and 1 p.m. What is the probability that the two persons will meet within 10 minutes of one another?

1.31 The numbers B and C are chosen at random between -1 and 1, independently of each other. What is the probability that the quadratic equation $x^2 + Bx + C = 0$ has real roots? Also, derive a general expression for this probability when B and C are chosen at random from the interval $(-q, q)$ for any $q > 0$.

1.32 The Manhattan distance from a point (x, y) in the plane to the origin $(0, 0)$ is defined as $|x| + |y|$. You choose a random point inside the unit square $\{(x, y) : 0 \leq x, y \leq 1\}$. What is the probability that the Manhattan distance between this point and the point $(0, 0)$ is no more than a for $0 \leq a \leq 2$? Does the answer change when the point is randomly chosen in the square $\{(x, y) : -1 \leq x, y \leq 1\}$?

1.33 Consider the following variant of Buffon's needle problem from Example 1.10. A rectangular card with side lengths a and b is dropped at random on the floor. It is assumed that the length $\sqrt{a^2 + b^2}$ of the diagonal of the card is smaller than the distance D between the parallel lines on the floor. Show that the probability of the card intersecting one of the lines is given by $\frac{2(a+b)}{\pi D}$.

1.34 A random point is chosen inside a triangle with height h and base of length b. What is the probability that the perpendicular distance from the point to the base is larger than a given value d with $0 < d < h$? What is the probability that the randomly chosen point and the base of the triangle will form a triangle with an obtuse angle when the original triangle is equilateral?

1.35 You choose a number v at random from $(0, 1)$ and next a number w at random from $(0, 1 - v)$. What is the probability that a triangle can be formed with side lengths v, w, and $1 - v - w$? Next answer the following question. What is the probability that a triangle can be formed with three pieces of a broken stick if the stick is first broken at random into two pieces and next the longer piece is randomly broken into two pieces. *Hint*: Use the fact that the sum of any two side lengths must be greater than the third side length and represent (v, w) as $(x, y(1-v))$, where (x, y) is a random point inside the unit square.

1.36 A stick is broken at two places. The break points are chosen at random on the stick, independently of each other. What is the probability that a triangle can be formed with the three pieces of the broken stick?

1.37 Choose a random point inside a circle with radius r and construct the (unique) chord with the chosen point as its midpoint. What is the probability that the chord is longer than a side of an equilateral triangle inscribed in the circle?

1.4 Compound Chance Experiments

A chance experiment is called a *compound* experiment if it consists of several elementary chance experiments. The question arises as to how, in general, we define a probability space for a compound experiment in which the elementary experiments are physically independent of each other. By physically independent, we mean that the outcomes from any one of the elementary experiments have no influence on the functioning or outcomes of any of the other elementary experiments. We first answer the question for the case of a finite number of physically independent elementary experiments $\varepsilon_1, \ldots, \varepsilon_n$. Assume that each experiment ε_k has a countable sample space Ω_k on which the probability measure P_k is defined, such that the probability $p_k(\omega_k)$ is assigned to each element $\omega_k \in \Omega_k$. The sample space of the compound experiment is then given by the set Ω consisting of all $\omega = (\omega_1, \ldots, \omega_n)$, where $\omega_k \in \Omega_k$ for $k = 1, \ldots, n$. A natural choice for the probability measure P on Ω arises by assigning the probability $p(\omega)$ to each element $\omega = (\omega_1, \ldots, \omega_n) \in \Omega$ by using the *product rule*:

$$p(\omega) = p_1(\omega_1) \times p_2(\omega_2) \times \cdots \times p_n(\omega_n).$$

This choice for the probability measure is not only intuitively the obvious one, but it can also be proved that it is the only probability measure satisfying the property $P(AB) = P(A)P(B)$ when the elementary experiments that generate event A are physically independent of those elementary experiments that give rise to event B. This important result of the uniqueness of the probability measure satisfying this property justifies the use of the product rule for compound chance experiments.

Example 1.11 In the "Reynard the Fox" café, it normally costs $3.50 to buy a pint of beer. On Thursday nights, however, customers pay $0.25, $1.00, or $2.50 for the first pint. In order to determine how much they will pay, customers must throw a dart at a dartboard that rotates at high speed. The dartboard is divided into eight segments of equal size. Two of the segments read $0.25, four of the segments read $1, and two more of the segments read $2.50. You pay whatever you hit. Two friends, John and Peter, each throw a dart at the board and hope for the best. What is the probability that the two friends will have to pay no more than $2 between them for their first pint?

Solution. This problem can be modeled as a compound experiment that consists of two subexperiments. The physically independent subexperiments are the throws of John and Peter. The elementary outcomes of each subexperiment are L, M, and H, where L stands for hitting a low-priced segment, M stands for hitting a medium-priced segment, and H stands for hitting a high-priced segment. Assuming that the dart hits the dartboard at a random point, the probabilities $p(L) = \frac{2}{8}$, $p(M) = \frac{4}{8}$, and $p(H) = \frac{2}{8}$ are assigned to the outcomes L, M, and H. The sample space of the compound experiment consists of the nine outcomes (L, L), (L, M), (M, L), (L, H), (H, L), (M, M), (M, H), (H, M), (H, H). The probability $\frac{1}{4} \times \frac{1}{4} = \frac{1}{16}$ is assigned to each of the outcomes (L, L), (L, H), (H, L), (H, H), the probability $\frac{1}{2} \times \frac{1}{2} = \frac{1}{4}$ to the outcome (M, M), and the probability $\frac{1}{2} \times \frac{1}{4} = \frac{1}{8}$ to each of the outcomes (L, M), (M, L), (H, M), (M, H). The two friends will have to pay no more than $2 between them for their first pint if one of the four outcomes (L, L), (L, M), (M, L), (M, M) occurs. The probability of this event is $\frac{1}{16} + \frac{1}{8} + \frac{1}{8} + \frac{1}{4} = \frac{9}{16}$.

1.4 Compound Chance Experiments

The next example deals with a compound chance experiment having a countably infinite sample space.

Example 1.12 Two desperados play Russian roulette in which they take turns pulling the trigger of a six-cylinder revolver loaded with one bullet. After each pull of the trigger, the magazine is spun to randomly select a new cylinder to fire. What is the probability that the desperado who begins will be the one to shoot himself dead?

Solution. The set $\Omega = \{F, NF, NNF, \ldots\}$ can be taken as sample space for the compound experiment, where the outcome F means that the first attempt is fatal and the outcome $N \cdots NF$ of length n means that no fatal shot is fired on the first $n-1$ attempts but the nth attempt is fatal. There is a remaining possibility: no fatal shot is ever fired. However, it will be seen that the probabilities of the other outcomes sum to 1 and so this possibility must have probability 0. The results of pulling the trigger are independent of one another and with each pull of the trigger there is a probability of $\frac{1}{6}$ that the fatal shot will be fired. Then it is reasonable to assign the probability $p_1 = \frac{1}{6}$ to the outcome F and the probability $p_n = \frac{5}{6} \times \cdots \times \frac{5}{6} \times \frac{1}{6} = \left(\frac{5}{6}\right)^{n-1} \frac{1}{6}$ to the outcome $N \cdots NF$ of length n. Note that $\sum_{k=1}^{\infty} p_k = 1$, as follows from a basic result saying that the geometric series $\sum_{j=0}^{\infty} x^j$ sums to $\frac{1}{1-x}$ for $|x| < 1$. Since $\sum_{k=1}^{\infty} p_k = 1$, the fatal shot will ultimately be fired with probability 1. Define A as the event that the fatal shot is fired by the desperado who begins. Then $P(A) = \sum_{n=0}^{\infty} p_{2n+1}$. This gives

$$P(A) = \sum_{n=0}^{\infty} \left(\frac{5}{6}\right)^{2n} \frac{1}{6} = \frac{1}{6} \sum_{k=0}^{\infty} \left(\frac{25}{36}\right)^k = \frac{1}{6} \left(\frac{1}{1-\frac{25}{36}}\right) = \frac{6}{11}.$$

Problems

1.38 In a tennis tournament between three players A, B, and C, each player plays the others once. The strengths of the players are as follows: $P(A$ beats $B) = 0.5$, $P(A$ beats $C) = 0.7$, and $P(B$ beats $C) = 0.4$. Assuming independence of the match results, calculate the probability that player A wins at least as many games as any other player.

1.39 An electronic system has four components labeled 1, 2, 3, and 4. The system has to be used during a given time period. The probability that component i will fail during that time period is f_i for $i = 1, \ldots, 4$. Failures of the components are physically independent of each other. A system failure occurs if component 1 fails or if at least two of the other three components fail. What is the probability that the system will fail?

1.40 Bill and Mark take turns picking a ball at random from a bag containing four red and seven white balls. The balls are drawn out of the bag with replacement. What is the probability that Bill is the first person to pick a red ball when Mark is the first person to start?

1.41 Two dice are tossed until the dice sum 7 or 8 appears. What is the probability of getting a total of 8 before a total of 7?

1.42 Three desperados A, B, and C play Russian roulette in which they take turns pulling the trigger of a six-cylinder revolver loaded with one bullet. Each time, the magazine is spun to randomly select a new cylinder to fire as long as the deadly shot has not already occurred. The desperados shoot according to the order $A, B, C, A, B, C, \ldots$. Determine, for each of the three desperados, the probability that this desperado will be the one to shoot himself dead.

1.43 Two persons each roll a pair of dice, independently of each other. What is the probability that the sums rolled are different?

1.4.1 A Coin-Tossing Experiment[3]

When a compound chance experiment consists of an infinite number of independent elementary chance experiments, it has an uncountable sample space and the choice of an appropriate probability measure is less obvious. We illustrate how we deal with such experiments by way of an illustration of a compound experiment consisting of an infinite number of tosses of a fair coin. The sample space of this experiment is taken as the set of all infinite sequences $\omega = (\omega_1, \omega_2, \ldots)$, where ω_i is equal to H when the ith coin toss comes up heads, and is equal to T otherwise. This sample space is uncountable, see also Appendix B. In order to be able to define a probability measure on the uncountable sample space, we must begin by restricting our attention to a class of appropriately chosen subsets. The so-called cylinder sets form the basis of this class of subsets. In the case of our chance experiment, a *cylinder set* is the set of all outcomes ω for which the first n elements $\omega_1, \ldots, \omega_n$ have specified values for a given finite value of n. A natural choice for the probability measure on the sample space is to assign the probability $P^{(\infty)}(A) = \left(\frac{1}{2}\right)^n$ to each cylinder set A with n specified elements. In this way, the event that heads first occurs at the kth toss can be represented by the cylinder set A_k whose elements ω have the finite beginning $T, T, \ldots, T, H$, and thus the event can be assigned a probability of $\left(\frac{1}{2}\right)^k$. The set $\bigcup_{k=1}^{\infty} A_k$ represents the event that at some point heads

---

[3] This section can be skipped without loss of continuity.

1.4 Compound Chance Experiments

occurs. The probability measure on the class of cylinder sets can be extended to one defined on a sufficiently general class of subsets capable of representing all possible events of this chance experiment. This extension requires measure theory, however, and is beyond the scope of this book.

It is intuitively clear that the fraction of coin tosses in which heads occurs converges to $\frac{1}{2}$ when the number of tosses increases without limit. We can state this claim more rigorously with the help of the probability measure $P^{(\infty)}$. To do so, we adopt the notation $K_n(\omega)$ to represent the number of heads occurring in the first n elements of ω. Furthermore, let C be the set of all outcomes ω for which $\lim_{n\to\infty} K_n(\omega)/n = \frac{1}{2}$. For very many sequences ω, the number $K_n(\omega)/n$ does not converge to $\frac{1}{2}$ as $n \to \infty$ (e.g., this is the case for any sequence ω with finitely many H's). However, "nature" chooses a sequence from the set C according to $P^{(\infty)}$. More precisely, the probability measure $P^{(\infty)}$ assigns a probability of 1 to the set C, or, in mathematical notation,

$$P^{(\infty)}\left(\left\{\omega : \lim_{n\to\infty} \frac{K_n(\omega)}{n} = \frac{1}{2}\right\}\right) = 1.$$

This type of convergence is called *convergence with probability one*, or *almost-sure convergence*. The terminology of an "almost-sure" event means that realizations not in this event are theoretically possible but will not happen in reality. The convergence result is a special case of the *strong law of large numbers*. This law is of enormous importance: it provides a direct link between theory and practice. It was a milestone in probability theory when, around 1930, A. N. Kolmogorov proved this law from the simple axioms of probability. In Chapter 9 we come back to the strong law of large numbers.

1.4.2 Continuity Property of Probability and Borel–Cantelli

The Kolmogorov axioms guarantee that the probability measure has a nice continuity property. Let us first introduce some terminology for an infinite sequence of sets $E_1, E_2, \ldots$. This sequence is said to be *increasing* if the set $E_n \subseteq E_{n+1}$ for all $n \geq 1$ and is said to be *decreasing* if $E_{n+1} \subseteq E_n$ for all $n \geq 1$, where the notation $A \subseteq B$ means that every element of the set A is also an element of the set B. A natural definition of the set $\lim_{n\to\infty} E_n$ is given by

$$\lim_{n\to\infty} E_n = \begin{cases} \bigcup_{i=1}^{\infty} E_i & \text{if the sequence } E_n \text{ is increasing,} \\ \bigcap_{i=1}^{\infty} E_i & \text{if the sequence } E_n \text{ is decreasing,} \end{cases}$$

where $\bigcup_{i=1}^{\infty} E_i$ is the set of all outcomes that belong to at least one of the sets E_i and $\bigcap_{i=1}^{\infty} E_i$ is the set of all outcomes that belong to each of the sets E_i.

Continuity of probability. *If the infinite sequence of sets $E_1, E_2, \ldots$ is either increasing or decreasing, then*

$$\lim_{n\to\infty} P(E_n) = P\left(\lim_{n\to\infty} E_n\right).$$

The proof is instructive and is given only for the increasing case. Define $F_1 = E_1$ and let the set F_{n+1} consist of the points of E_{n+1} that are not in E_n for $n \geq 1$. It is readily seen that the sets $F_1, F_2, \ldots$ are pairwise disjoint and satisfy $\bigcup_{i=1}^{n} F_i = \bigcup_{i=1}^{n} E_i$ for all $n \geq 1$ and $\bigcup_{i=1}^{\infty} F_i = \bigcup_{i=1}^{\infty} E_i$. By Axiom 3, $P(\bigcup_{i=1}^{\infty} F_i) = \sum_{i=1}^{\infty} P(F_i)$ and $P(\bigcup_{i=1}^{n} F_i) = \sum_{i=1}^{n} P(F_i)$ for all n (see also Rule 1.1 in Section 1.5). Thus,

$$P\left(\lim_{n\to\infty} E_n\right) = P\left(\bigcup_{i=1}^{\infty} E_i\right) = P\left(\bigcup_{i=1}^{\infty} F_i\right) = \sum_{i=1}^{\infty} P(F_i) = \lim_{n\to\infty} \sum_{i=1}^{n} P(F_i)$$

$$= \lim_{n\to\infty} P\left(\bigcup_{i=1}^{n} F_i\right) = \lim_{n\to\infty} P\left(\bigcup_{i=1}^{n} E_i\right) = \lim_{n\to\infty} P(E_n),$$

using the fact that $\bigcup_{i=1}^{n} E_i = E_n$ when the sets E_i are increasing.

The continuity property of probability implies the Borel–Cantelli lemma on which the proof of the strong law of large numbers in Chapter 9 is based.

First Borel–Cantelli lemma. *Suppose that $A_1, A_2, \ldots$ is an infinite sequence of subsets of the sample space Ω such that $\sum_{k=1}^{\infty} P(A_k) < \infty$. Then, $P(\{\omega \in \Omega : \omega \in A_k \text{ for infinitely many } k\}) = 0$.*

To prove the lemma, let the set $C = \{\omega \in \Omega : \omega \in A_k \text{ for infinitely many } k\}$. Define $B_n = \bigcup_{k=n}^{\infty} A_k$ for $n \geq 1$, then $B_1, B_2, \ldots$ is a decreasing sequence of sets. Note that $\omega \in C$ if and only if $\omega \in B_n$ for all $n \geq 1$. This implies that set C equals the intersection of all sets B_n. Thus, by the continuity of probability, $P(C) = \lim_{n\to\infty} P(B_n)$. This gives $P(C) = \lim_{n\to\infty} P(\bigcup_{k=n}^{\infty} A_k)$. Next we use Boole's inequality, stating that $P(\bigcup_{i=1}^{\infty} E_i) \leq \sum_{i=1}^{\infty} P(E_i)$ for any sequence of sets E_i. In Problem 1.44 you are asked to verify this intuitively obvious result. Thus, $P(C) \leq \lim_{n\to\infty} \sum_{k=n}^{\infty} P(A_k)$. This limit is zero, since $\sum_{k=1}^{\infty} P(A_k) < \infty$. This completes the proof that $P(C) = 0$.

Problems

1.44 Use the Kolmogorov axioms to prove the following inequalities:

(a) $P(A) \leq P(B)$ when $A \subseteq B$. *Hint*: Let $C = B \backslash A$ be the set of all outcomes in B that are not in A and note that B is the union of the disjoint sets A and C.

(b) $P\left(\bigcup_{k=1}^{\infty} A_k\right) \leq \sum_{k=1}^{\infty} P(A_k)$ for any sequence of subsets $A_1, A_2, \ldots$. This is *Boole's inequality*, also known as the *union bound*. *Hint*: Define the pairwise disjoint sets B_k by $B_1 = A_1$ and $B_k = A_k \setminus (A_1 \cup \cdots \cup A_{k-1})$ for $k \geq 2$.

1.45 In the coin-tossing experiment of repeatedly tossing a fair coin, a run of length r is said to have occurred if heads have just been tossed r times in a row. Prove that a run of length r is certain to occur somewhere if a fair coin is tossed indefinitely often. *Hint*: Analyze the coin-tossing experiment as an infinite sequence of physically independent r-experiments. The r-experiment consists of r consecutive tosses of the coin. Define B_n as the event that no run of length r occurs in any of the first n of the r-experiments and evaluate $P(\bigcap_{n=1}^{\infty} B_n)$.

1.46 You toss a fair coin until you obtain 10 heads in a row and then you stop. What is the probability of seeing at least 10 consecutive tails in the sequence prior to stopping? *Hint*: Use the result of Problem 1.45.

1.5 Some Basic Rules

The axioms of probability theory directly imply a number of basic rules that are useful for calculating probabilities. Before stating these results, let us repeat some notation from set theory. Suppose that A and B are events for a chance experiment with sample space Ω. The event that at least one of the events A or B occurs is called the *union of A and B*. It consists of all outcomes which are either in A or B or both. The union of A and B is denoted by $A \cup B$. The event that both events A and B occur is called the *intersection of A and B*. It consists of all outcomes that are both in A and in B. The intersection of events A and B is denoted by AB. Another notation for the intersection is $A \cap B$. The notation AB for the intersection of events A and B will be used throughout this book. The notation for the union and intersection of two events extends to any finite or infinite sequence of events. For any event A, the *complementary event* A^c is defined as the event that consists of all outcomes not in A. For example, suppose that the chance experiment consists of rolling a die. Then the sample space is $\Omega = \{1, 2, \ldots, 6\}$, where outcome i means that number i was rolled. If A is the event that the number rolled is odd and B is the event that the number rolled is larger than 3, then $A \cup B = \{1, 3, 4, 5, 6\}$, $AB = \{5\}$, and $A^c = \{2, 4, 6\}$.

Rule 1.1 *For any finite sequence of events $A_1, \ldots, A_n$,*

$$P\left(\bigcup_{i=1}^{n} A_i\right) = \sum_{i=1}^{n} P(A_i) \text{ when the events are mutually exclusive.}$$

Rule 1.2 For any event A,
$$P(A) = 1 - P(A^c).$$
Rule 1.3 For any two events A and B,
$$P(A \cup B) = P(A) + P(B) - P(AB).$$
Rule 1.4 For any finite sequence of events $A_1, \ldots, A_n$,
$$P\left(\bigcup_{i=1}^{n} A_i\right) = \sum_{i=1}^{n} P(A_i) - \sum_{\substack{i,j: \\ i<j}} P(A_i A_j) + \sum_{\substack{i,j,k: \\ i<j<k}} P(A_i A_j A_k) - \cdots$$
$$+ (-1)^{n-1} P(A_1 A_2 \cdots A_n).$$

The proofs of these rules are simple and instructive. They nicely demonstrate how useful propositions can be obtained from "minimal" axioms.

To prove Rule 1.1, denote by $\emptyset$ the empty set of outcomes. We first show that
$$P(\emptyset) = 0.$$
Using Axiom 3 with $A_i = \emptyset$ for all i gives $P(\emptyset) = \sum_{i=1}^{\infty} a_i$, where $a_i = P(\emptyset)$ for all i. This implies that $P(\emptyset) = 0$. Let $A_1, \ldots, A_n$ be any finite sequence of pairwise disjoint sets. Augment this sequence with $A_{n+1} = \emptyset, A_{n+2} = \emptyset, \ldots$. Then, by Axiom 3,
$$P\left(\bigcup_{i=1}^{n} A_i\right) = P\left(\bigcup_{i=1}^{\infty} A_i\right) = \sum_{i=1}^{\infty} P(A_i) = \sum_{i=1}^{n} P(A_i).$$

The proof of Rule 1.2 is as follows. The set $A \cup A^c$ is by definition equal to the sample space. Thus, by Axiom 2, $P(A \cup A^c) = 1$. The sets A and A^c are disjoint. It now follows from Rule 1.1 that $P(A \cup A^c) = P(A) + P(A^c)$. This gives the complement rule $P(A) = 1 - P(A^c)$.

To prove Rule 1.3, denote by A_1 the set of outcomes that belong to A but not to B. Let B_1 be the set of outcomes that are in B but not in A and let $C = AB$ be the set of outcomes that are both in A and B. The sets A_1, B_1, and C are pairwise disjoint. Moreover,
$$A \cup B = A_1 \cup B_1 \cup C, \ A = A_1 \cup C, \ \text{and} \ B = B_1 \cup C.$$
Applying Rule 1.1 gives
$$P(A \cup B) = P(A_1) + P(B_1) + P(C).$$
Also, $P(A) = P(A_1) + P(C)$ and $P(B) = P(B_1) + P(C)$. By substituting the latter two relations into the expression for $P(A \cup B)$ and noting that $C = AB$, we find

1.5 Some Basic Rules

$$P(A \cup B) = P(A) - P(C) + P(B) - P(C) + P(C)$$
$$= P(A) + P(B) - P(AB).$$

Rule 1.4 will be proved for the special case that the sample space is countable.[4] In this case $P(A) = \sum_{\omega \in A} p(\omega)$, where $p(\omega)$ is the probability assigned to the individual element ω of the sample space. Fix ω. If $\omega \notin \cup_{i=1}^{n} A_i$, then ω does not belong to any of the sets A_i and $p(\omega)$ does not contribute to either the left-hand side or the right-hand side of the expression in Rule 1.4. Assume now that $\omega \in \cup_{i=1}^{n} A_i$. Then there is at least one set A_i to which ω belongs. Let s be the number of sets A_i to which ω belongs. On the left-hand side of the expression in Rule 1.4, $p(\omega)$ contributes only once. In the first term of the right-hand side of this expression, $p(\omega)$ contributes s times, in the second term $\binom{s}{2}$ times, in the third term $\binom{s}{3}$ times, and so on. Thus, the coefficient of $p(\omega)$ on the right-hand side is

$$s - \binom{s}{2} + \binom{s}{3} - \cdots + (-1)^{s-1} \binom{s}{s}.$$

Showing that this coefficient is 1 gives Rule 1.4. Noting that $\binom{s}{1} = s$ and $\binom{s}{0} = 1$, we can write the expression $s - \binom{s}{2} + \binom{s}{3} - \cdots + (-1)^{s-1} \binom{s}{s}$ as

$$1 - \left[\binom{s}{0} - \binom{s}{1} + \binom{s}{2} - \binom{s}{3} + \cdots + (-1)^s \binom{s}{s}\right] = 1 - (-1+1)^s = 1,$$

using Newton's binomium $(a+b)^s = \sum_{k=0}^{s} \binom{s}{k} a^k b^{s-k}$. This completes the proof.

Next we give several illustrative applications of the above properties. We first illustrate Rule 1.2, which is known as the *complement rule*. This rule states that the probability of an event occurring is one minus the probability that it does not occur. It is often easier to compute the complementary probability than the probability itself. The complement rule is an extremely useful problem-solving tool in probability. The practical importance of the complement rule can hardly be overestimated. We first illustrate this rule with an example based on a problem that was posed by the nobleman and gambler Chevalier de Méré to the famous French mathematician Blaise Pascal in 1654. De Méré knew that it was advantageous to wager that a six would be rolled at least one time in four rolls of one die, but his experience as a gambler taught him that it was not advantageous to wager on a double six being rolled at least one time in 24 rolls of a pair of dice. However, his mathematical skills were not sufficient to understand why this should be so. De Méré had the idea that four rolls of one die could be translated into $4 \times 6 = 24$ rolls of two dice, because two dice can come up in six times as many ways as one die. He posed the problem to Pascal, who solved it.

---

[4] A proof for the general case is outlined in Problem 3.31 in Chapter 3.

Example 1.13 How many rolls of a fair die are required to have at least a 50% chance of rolling at least one six? How many rolls of two fair dice are required to have at least a 50% chance of rolling at least one double six?

Solution. Let us first consider the experiment of rolling a single die r times, where r is fixed in advance. The sample space of this experiment consists of all elements $(i_1, i_2, \ldots, i_r)$, where i_k denotes the outcome of the kth roll of the die. The sample space has $6 \times 6 \times \cdots \times 6 = 6^r$ equally likely elements. Let A be the event that at least one six is obtained in r rolls of the die. To compute $P(A)$, it is easiest to compute the probability of the complementary event A^c that no six is obtained in r rolls of the die. The set A^c consists of $5 \times 5 \times \cdots \times 5 = 5^r$ elements. Thus, $P(A^c) = 5^r/6^r$ and so

$$P(A) = 1 - \frac{5^r}{6^r}.$$

This probability has the value 0.4213 for $r = 3$ and the value 0.5177 for $r = 4$. Therefore four rolls of the die are required.

The following sample space is taken for the experiment of rolling two dice r times. Imagining that one die is blue and the other is red, the sample space consists of all elements $((i_1,j_1), (i_2,j_2), \ldots, (i_r,j_r))$, where i_k and j_k denote the outcomes of the blue and the red die on the kth roll of the two dice. The sample space consists of $36 \times 36 \times \cdots \times 36 = 36^r$ elements. All elements are equally likely. Let A be the event that at least one double six is obtained in the r rolls of the two dice. The complementary event A^c of rolling no double six in r rolls of the two dice can occur in $35 \times 35 \times \cdots \times 35 = 35^r$ ways. Therefore, $P(A^c) = 35^r/36^r$ and so

$$P(A) = 1 - \frac{35^r}{36^r}.$$

This probability has the value 0.4914 for $r = 24$ and the value 0.5055 for $r = 25$. Therefore 25 rolls of the two dice are required. De Méré must have been an assiduous player in order to have established empirically that the probability of rolling at least one double six in 24 rolls of a pair of dice lies just under one-half.

The complement rule is also the key to the solution of the famous birthday problem.

Example 1.14 What is the probability that in a class of n (≤ 365) children (no twins), two or more children have the same birthday? Assume that the year has 365 days (ignore February 29) and that all possible birthdays are equally likely. What is the smallest value of n such that this probability is at least 50%?

Solution. Take as the sample space all ordered n-tuples of numbers selected from the integers $1, 2, \ldots, 365$, where the first number of the tuple is the birthday of the first child, the second number is the birthday of the second child, etc. The sample space has $365 \times 365 \times \cdots \times 365 = 365^n$ equally likely elements. Let A be the event that at least two children have the same birthday. The number of elements for which all n children have a different birthday is $365 \times 364 \times \cdots \times (365 - n + 1)$. Then, by the complement rule,

$$P(A) = 1 - \frac{365 \times 364 \times \cdots \times (365 - n + 1)}{365^n}.$$

The smallest value of n for which this probability is 50% or more is $n = 23$. The probability has the value 0.5073 for $n = 23$.[5] The finding that a small value of $n = 23$ suffices is no longer surprising when you realize that there are $\binom{23}{2} = 253$ combinations of two children in a class of 23 children, where any combination leads to a match with a probability of $\frac{1}{365}$.

The psychology of probability intuition is an interesting feature of the birthday problem. Many people think that more than 23 people are needed for a birthday match and the number 183 is commonly suggested. A similar misconception can be seen in the words of a lottery official regarding his lottery, in which a four-digit number was drawn daily from the 10,000 number sequence $0000, 0001, \ldots, 9999$. On the second anniversary of the lottery, the official deemed it highly improbable that any of the 10,000 possible numbers had been drawn two or more times in the last 625 drawings. The lottery official was wildly off the mark: the probability that some number will not be drawn two or more times in 625 drawings is inconceivably small and of the order of magnitude 10^{-9}. This probability can be calculated by looking at the problem as a "birthday problem" with 10,000 possible birthdays and a group of 625 people. The birthday problem nicely illustrates that coincidences may not be so unusual after all, see also Problems 1.54 and 1.55. Nearly all coincidences can be explained by simple probability rules. What looks unexpected usually turns out to be expected.

Rule 1.3 is often called the *addition rule* for probabilities. When $P(A)$ and $P(B)$ are added for non-disjoint sets A and B, the probability of the intersection

---

[5] In reality, birthdays are not uniformly distributed throughout the year, but follow a seasonal pattern. The probability of a match only becomes larger for any deviation from equal birth frequencies, as can be understood intuitively by imagining a group of people coming from a planet on which people are always born on the same day. However, for birth frequency variation as occurring in reality, the match probability is very insensitive to deviations from uniform birth rates and the group size of 23 for a fifty-fifty match probability does not change, see T. S. Nunnikhoven, "A birthday problem for nonuniform birth frequencies," *The American Statistician* **46** (1992): 270–274.

of A and B is counted twice and must therefore be subtracted in order to get $P(A \cup B)$. The following three examples illustrate this.

Example 1.15 A single card is randomly drawn from a thoroughly shuffled deck of 52 cards. What is the probability that the card drawn will be either a heart or an ace or both?

Solution. For the sample space of this chance experiment, we take the set consisting of the 52 elements

$$\spadesuit A, \ldots, \spadesuit 2, \quad \heartsuit A, \ldots, \heartsuit 2, \quad \clubsuit A, \ldots, \clubsuit 2, \quad \diamondsuit A, \ldots, \diamondsuit 2,$$

where, for example, the outcome $\clubsuit 7$ means that the seven of clubs is drawn. All possible outcomes are equally likely, and thus each outcome gets assigned the same probability $\frac{1}{52}$. Let A be the event that the card drawn is a heart and B the event that the card drawn is an ace. These two events are not mutually exclusive. We are looking for the probability $P(A \cup B)$ that at least one of the events A and B occurs. This probability can be calculated by applying Rule 1.3:

$$P(A \cup B) = P(A) + P(B) - P(AB).$$

In this case, $P(AB)$ stands for the probability that the card drawn is the ace of hearts. The events A and B correspond to sets that contain 13 and 4 elements, respectively, and thus have respective probabilities $\frac{13}{52}$ and $\frac{4}{52}$. The event AB corresponds to a set that is a singleton and thus has probability $\frac{1}{52}$. Therefore, the probability that the card drawn is either a heart or an ace or both is equal to

$$P(A \cup B) = \frac{13}{52} + \frac{4}{52} - \frac{1}{52} = \frac{16}{52}.$$

Example 1.16 Suppose that you are playing blackjack against the dealer. What is the probability that neither of you are dealt a blackjack when getting two cards each from a thoroughly shuffled standard deck of 52 cards? A blackjack consists of one ace together with one card from the 16 cards formed by the tens, jacks, queens, and kings.

Solution. The sample space consists of all combinations of an unordered pair of cards for the player and an unordered pair of cards for the dealer. It has $\binom{52}{2}\binom{50}{2}$ equiprobable elements. Let A be the event that neither the player nor the dealer gets a blackjack. The sought probability $P(A)$ is easiest computed by using the complement rule. Let A_1 be the event that the player gets a blackjack and A_2 be the event that the dealer gets a blackjack. Then $P(A) = 1 - P(A_1 \cup A_2)$. By the addition rule, $P(A_1 \cup A_2) = P(A_1) + P(A_2) - P(A_1 A_2)$. Thus

$$P(A) = 1 - P(A_1) - P(A_2) + P(A_1 A_2).$$

By simple counting arguments,

$$P(A_1) = P(A_2) = \frac{\binom{4}{1}\binom{16}{1}\binom{50}{2}}{\binom{52}{2}\binom{50}{2}} \quad \text{and} \quad P(A_1 A_2) = \frac{\binom{4}{1}\binom{16}{1}\binom{3}{1}\binom{15}{1}}{\binom{52}{2}\binom{50}{2}}.$$

This leads to the value 0.9052 for the sought probability $P(A)$.

Example 1.17 The eight soccer teams that have reached the quarterfinals of the Champions League are made up of two teams from each of the four countries England, Germany, Italy, and Spain. The four matches to be played in the quarterfinals are determined by drawing lots.

(a) What is the probability that the two teams from the same country play against each other in each of the four matches?
(b) What is the probability that there is a match between the two teams from England or the two teams from Germany?

Solution. Number the eight teams $1, \ldots, 8$ and take as sample space the ordered set of all possible permutations $(t_1, \ldots, t_8)$ of the integers $1, \ldots, 8$, where t_k is the team that is drawn from the bowl as kth team. Any permutation $(t_1, \ldots, t_8)$ corresponds to the four matches (t_1, t_2), (t_3, t_4), (t_5, t_6), (t_7, t_8). The sample space has $8!$ equally likely elements.

(a) Let A be the event that the two teams from the same country play against each other in each of the four matches. There are $8 \times 1 \times 6 \times 1 \times 4 \times 1 \times 2 \times 1 = 384$ outcomes that make up the event A. Therefore

$$P(A) = \frac{384}{8!} = 0.0095.$$

(b) Let E be the event that there is a match between the two teams from England and let G be the event that there is a match between the two teams from Germany. The sought probability is $P(E \cup G)$ and satisfies

$$P(E \cup G) = P(E) + P(G) - P(EG).$$

Each of the events E and G corresponds to $4 \times 2 \times 6! = 5{,}760$ elements in the sample space. To see this, note that there are four possibilities for the match in which the two British teams can be paired with each other, there are two possibilities for the order in which these two teams can be drawn from the bowl, and there are $6!$ possibilities for the order in which the other six teams can be drawn from the bowl. By the same argument, the event EG corresponds to $4 \times 3 \times 2 \times 2 \times 4! = 1{,}152$ elements in the sample space. Thus

$$P(E) = P(G) = \frac{5{,}760}{8!} \quad \text{and} \quad P(EG) = \frac{1{,}152}{8!}.$$

This gives
$$P(E \cup G) = \frac{5{,}760}{8!} + \frac{5{,}760}{8!} - \frac{1{,}152}{8!} = 0.2571.$$

Problems

1.47 Of those visiting a particular car dealer selling second-hand cars and Japanese cars, 55% buy no car, 25% buy a second-hand car, and 30% buy a Japanese car. What is the probability that a visit leads to buying a second-hand Japanese car?

1.48 Suppose that 50% of households in a certain city subscribe to the morning newspaper, 70% of households subscribe to the afternoon newspaper, and 80% of households subscribe to at least one of the two newspapers. What proportion of the households subscribe to both newspapers?

1.49 A transport firm has two vehicles, a truck and a van. The truck is used 75% of the time. Both vehicles are used 30% of the time and neither of the vehicles is used for 10% of the time. What is the probability that the van is used on any given day? What is the probability that only the van is used on any given day?

1.50 The event A has probability $\frac{2}{3}$ and there is a probability of $\frac{3}{4}$ that at least one of the events A and B occurs. What are the smallest and largest possible values for the probability of event B?

1.51 For any two events A and B, prove that the probability of exactly one of them occurring is $P(A) + P(B) - 2P(AB)$. *Hint*: Consider $P\big((A \cap B^c) \cup (B \cap A^c)\big)$.

1.52 Let $A_1, \ldots, A_n$ be any finite sequence of events. Use induction to verify the bounds $\sum_{i=1}^{n} P(A_i) - \sum_{i=1}^{n-1} \sum_{j=i+1}^{n} P(A_i A_j) \le P(\bigcup_{i=1}^{n} A_i) \le \sum_{i=1}^{n} P(A_i)$.

1.53 In a class of 30 children, each child writes down a randomly chosen number from $1, 2, \ldots, 250$. What is the probability that two or more children have chosen the same number?

1.54 Suppose that n independent repetitions are done of an experiment that has m equally likely outcomes $O_1, \ldots, O_m$. What is the probability that some outcome occurs two or more times? Show that this probability can be approximated by $1 - e^{-\frac{1}{2}n(n-1)/m}$ when m is much larger than n. What is the probability that the particular outcome O_1 occurs at least once? Verify that for a fifty-fifty probability the required number of trials is about $1.177\sqrt{m} + 0.5$ for the first probability and about $0.6931m$ for the second probability when m is large. *Hint*: Use the approximation $1 - x \approx e^{-x}$ for x close to zero.

1.5 Some Basic Rules

1.55 Twenty-five people each choose two distinct numbers at random from the numbers $1, 2, \ldots, 25$, independently of each other. What is the probability that at least two people will choose the same two numbers?

1.56 On Wednesday, June 21, 1995, a remarkable thing occurred in the German 6/49 lottery, in which six different numbers are drawn from the numbers $1, \ldots, 49$. On the day in question, the mid-week drawing produced this six-number result: 15-25-27-30-42-48. These were the same numbers as had been drawn previously on Saturday, December 20, 1986, and it was the first time in the 3,016 drawings of the German lottery until June 21, 1995 that the same sequence had been drawn twice. Is this an incredible occurrence, given that in the German lottery there are nearly 14 million possible combinations of the six numbers in question?

1.57 Canadian lottery officials had no knowledge of the birthday problem and its treacherous variants when they put this idea into play. They purchased 500 Oldsmobile cars from nonclaimed prize monies, to be raffled off as bonus prizes among their 2.4 million registered subscribers. A computer chose the winners by selecting 500 subscriber numbers from a pool of 2.4 million registered numbers without regard for whether or not a given number had already appeared. The unsorted list of the 500 winning numbers was published and to the astonishment of lottery officials, one subscriber put in a claim for two cars. How surprising is it that this happened?

1.58 A bowl contains 10 red and 10 blue balls. Three times you pick at random two balls out of the bowl without replacement. What is the probability that at least one pick has two balls of the same color?

1.59 In the Massachusetts Numbers Game, a four-digit number is drawn from the numbers $0000, 0001, \ldots, 9999$ every evening (except Sundays). Let's assume that the same lottery takes place in ten other states each evening.

   (a) What is the probability that the same number will be drawn in two or more states next Tuesday evening?

   (b) What is the probability that on some evening in the coming 300 drawings, the same number will be drawn in two or more states?

1.60 The playlist of your iPhone has 15 songs by each of 30 artists. After each song the playlist is shuffled and the next song is selected at random from the 450 songs. What is the probability that you will hear the same artist more than once in any 10-song block?

1.61 A producer of a new brand of breakfast cereal offers a baseball card in each cereal box purchased. Ten different baseball cards are distributed equally among the cereal boxes. What is probability of getting the cards of your two favorite teams when you buy five boxes?

1.62 What is the probability of getting at least one ace in a poker hand of five cards dealt from an ordinary deck of 52 cards? What is the probability of getting five cards of the same suit in a poker hand of five cards?

1.63 A fair die is rolled six times. What is the probability that each of the numbers 5 and 6 appears at least once?

1.64 In the casino game of Chuck-a-Luck, three dice are contained within an hourglass-shaped, rotating cage. You bet on one of the six numbers $1, \ldots, 6$ and the cage is rotated. You lose money only if your number does not come up on any of the three dice. What is the probability that your number will come up?

1.65 What is the probability that in a class of 23 children *exactly* two children have birthdays on the same day? *Hint*: Number the $\binom{23}{2} = 253$ possible combinations of two children as $1, \ldots, 253$ and define A_i as the event that the two children from combination i are the only two children having the same birthday.

1.66 You roll a fair die six times in a row. What is the probability that all of the six face values will appear? What is the probability that one or more sixes will appear? What is the probability that the largest number rolled is r?

1.67 For the upcoming drawing of the Bingo Lottery, five extra prizes have been added to the pot. Each prize consists of an all-expenses-paid vacation trip. Each prize winner may choose from among three possible destinations $A, B,$ and C. The three destinations are equally popular. The prize winners choose their destinations independently of each other. Calculate the probability that at least one of the destinations A and B will be chosen. Also, calculate the probability that not all of the three destinations will be chosen.

1.68 In the 6/42 lottery, six distinct numbers are picked at random from the numbers $1, \ldots, 42$. What is the probability that at least two of the numbers 7, 14, 21, 28, 35, and 42 will be picked?

1.69 You draw at random five cards from a standard deck of 52 cards. What is the probability of having an ace among the five cards together with either a king or a queen or both?

1.70 John and Paul play the following game. They each roll one fair die. John wins the game if his score is larger than Paul's score or if the product of the two scores is an odd number. Is this a fair game?

1.71 From an ordinary deck of 52 cards, cards are randomly drawn one by one and without replacement. What is the probability that an ace will appear before any of the cards 2 through 10? What is the probability that two aces will appear before any of the cards 2 through 10?

1.72 A random experiment has the possible outcomes $O_1, O_2, \ldots$ with respective probabilities $p_1, p_2, \ldots$. Independent repetitions of the experiment are performed. Define an appropriate probability space to verify that the probability of the outcome O_1 occurring before the outcome O_2 is given by $p_1/(p_1 + p_2)$. Also, verify the more general result that the probability of the rth appearance of outcome O_1 occurring before the sth appearance of outcome O_2 is given by

$$\sum_{k=0}^{s-1} \binom{r+k-1}{r-1} \left(\frac{p_1}{p_1+p_2}\right)^r \left(\frac{p_1}{p_1+p_2}\right)^k \quad \text{for any } r, s \geq 1.$$

Hint: Define $A_{n,k}$ as the event that the rth appearance of outcome O_1 occurs at the nth trial and that outcome O_2 appears exactly k times in the first $n-1$ trials. Use the identity $\sum_{l=0}^{\infty} (l+1) \cdots (l+m) x^l = m! / (1-x)^{m+1}$ for $m \geq 1$ and $|x| < 1$.

1.73 Consider a random permutation of the numbers $1, 2, \ldots, n$ and answer the following questions.

(a) What is the probability that a particular number r belongs to a cycle of a given length k?[6]

(b) What is the probability that two particular numbers r and s belong to the same cycle?

1.74 One hundred prisoners are offered a one-time chance to prevent being exiled to a feared prison camp. The names of the 100 prisoners are placed in 100 boxes, one name to a box, and the boxes are lined up on a table in a room. One by one, the prisoners are led into the room. Each prisoner may inspect up to 50 of the boxes to try to find his name. The prisoners must leave the room behind exactly as they found it, and they cannot communicate after having left the room. Unless they all find their own names, the whole group will be exiled to the feared prison camp. If each prisoner inspects 50 boxes at random, the probability of avoiding exile for the group is $\left(\frac{1}{2}\right)^{100}$ and is thus practically zero. However, the prisoners are allowed to plot a strategy in advance. They decide on a random labeling of the boxes with their own names and to stick to this labeling. Upon entering the room, each prisoner goes to the box labeled with his name. If he finds another prisoner's name in the box, he then looks into the box labeled with that prisoner's name. This process repeats until the

---

[6] A permutation cycle is a subset of a permutation whose elements trade places with one another. For example, take the permutation $(2, 4, 3, 1)$ of $(1, 2, 3, 4)$, then the subset $\{1, 4, 2\}$ is a cycle of length 3 with $1 \to 4 \to 2 \to 1$ and the subset $\{3\}$ is a cycle of length 1 with $3 \to 3$.

prisoner either finds his own name or has inspected 50 boxes. What is the probability that the group will not be exiled? *Hint*: Consider the contents of the boxes as a random permutation of the integers $1, 2, \ldots, 2n$ with $n = 50$ and verify that the probability of having no cycle of length $n + 1$ or more is equal to $1 - \sum_{k=n+1}^{2n} 1/k$.

1.75 In a television show the remaining couple is shown three closed doors behind which is hidden a new car, the car key, and a goat. One member of the couple gets the task to find the car and the other member gets the task to find the car key. The couple wins the car only if both partners succeed in their respective tasks. Each member of the couple is allowed to open two doors, where the second person cannot see what is behind the doors opened by the first person. The couple may, however, agree on a strategy beforehand. What is the best strategy?

1.76 Two-up is a traditional Australian gambling game. A spinner repeatedly throws two pennies up in the air. "Heads" appear if both pennies land with the head side facing up, "tails" appear if both pennies land with the tails side facing up, and "odds" appear if one penny lands with the head side up and the other lands with the tails side up. Consider the following game variant. The spinner wins if three "heads" appear before any appearance of "tails" and also before any intervening sequence of five consecutive "odds." A win pays 8.5 for 1. What is the win probability of the spinner? Is the game favorable for the bettor?

1.6 Inclusion–Exclusion Rule

Many probability problems can be reduced to finding the probability of the occurrence of at least one event from a sequence of appropriately defined events. How to do this is explained in Rule 1.4, which is known as the *inclusion–exclusion rule*. It states that the probability of the union of n events equals the sum of the probabilities of these events taken one at a time, minus the sum of the probabilities of these events taken two at a time, plus the sum of the probabilities of these events taken three at a time, and so on. Many probability problems of a combinatorial nature can be solved by using the inclusion–exclusion rule. Probabilities of intersections of events are usually easier to calculate than probabilities of unions of events. The inclusion–exclusion rule provides us with a way to convert between the two.

Example 1.17 (continued) What is the probability that in none of the four matches do two teams from the same country play against each other?

1.6 Inclusion–Exclusion Rule

Solution. Label the four countries as $i = 1, \ldots, 4$. Let A_i be the event that the two teams from country i play against each other. The sought probability is $1 - P(A_1 \cup A_2 \cup A_3 \cup A_4)$. In the solution of Example 1.17, we already found that $P(A_i) = 5{,}760/8!$ for any i, $P(A_i A_j) = 1{,}152/8!$ for any $i < j$, and $P(A_1 A_2 A_3 A_4) = 384/8!$. If three matches are between the two teams from the same country, then the fourth match is also between the two teams from the same country. Thus $P(A_i A_j A_k) = P(A_1 A_2 A_3 A_4)$ for any $i < j < k$. By the inclusion–exclusion rule, $P(A_1 \cup A_2 \cup A_3 \cup A_4)$ is given by

$$\binom{4}{1}\frac{5{,}760}{8!} - \binom{4}{2}\frac{1{,}152}{8!} + \binom{4}{3}\frac{384}{8!} - \binom{4}{4}\frac{384}{8!} = 0.4286$$

and so the sought probability is $1 - 0.4286 = 0.5714$.

The next example is a particular instance of the balls and bins model in which balls are placed at random into bins. Many probability problems are related to this model, see also Problem 1.87.

Example 1.18 An airport bus deposits 25 passengers at 7 stops. Each passenger is as likely to get off at any stop as at any other, and the passengers act independently of one another. The bus makes a stop only if someone wants to get off. What is the probability that somebody gets off at each stop?

Solution. Take as sample space the set of all possible ordered sequences $(i_1, \ldots, i_{25})$, where i_k is the stop at which the kth passenger gets off. Each sequence is equally likely. Let A_i be the event that nobody gets off at the ith stop. The sought probability is $1 - P(A_1 \cup A_2 \cup \cdots \cup A_7)$. Noting that $P(A_{i_1} \cdots A_{i_k}) = \frac{(7-k)^{25}}{7^{25}}$ and using the fact that the total number of ways one can choose k different events from the events $A_1, \ldots, A_7$ is $\binom{7}{k}$, it follows from the inclusion–exclusion formula that

$$P(A_1 \cup A_2 \cup \cdots \cup A_7) = \sum_{k=1}^{7} (-1)^{k+1} \binom{7}{k} \frac{(7-k)^{25}}{7^{25}} = 0.1438.$$

Thus the sought probability is $1 - 0.1438 = 0.8562$.

To conclude this section, we illustrate the inclusion–exclusion rule with an instance of the classic matching problem.

Example 1.19 Suppose that n people enter a restaurant and check their coats. As they are leaving, the lazy coat-checker returns the coats in random order. What is the probability that at least one person gets the correct coat?

Solution. For the formulation of the sample space for this chance experiment, it is convenient to give label i to the coat of person i for $i = 1, \ldots, n$. Then we take the set of all possible orderings $(e_1, \ldots, e_n)$ of the integers $1, \ldots, n$ as our sample space. In the outcome $\omega = (e_1, \ldots, e_n)$, person i receives the coat with label e_i for $i = 1, \ldots, n$. The total number of possible outcomes is $n \times (n-1) \times \cdots \times 1 = n!$. Since the coats are returned in random order, all possible orderings are equally likely and thus each outcome $(e_1, \ldots, e_n)$ gets assigned the same probability $\frac{1}{n!}$. For any i, let A_i be the event that person i receives his or her own coat. The probability that at least one person gets the correct coat is given by $P(A_1 \cup A_2 \cup \cdots \cup A_n)$. The probabilities in the inclusion–exclusion formula are easy to calculate. The total number of orderings $(e_1, \ldots, e_n)$ with $e_i = i$ is equal to $(n-1)!$, and so $P(A_i) = \frac{(n-1)!}{n!}$ for all i. The number of orderings $(e_1, \ldots, e_n)$ with $e_i = i$ and $e_j = j$ is $(n-2)!$ for $i \neq j$ and so $P(A_i A_j) = \frac{(n-2)!}{n!}$ for all i and j with $i \neq j$. In general, $P(A_{i_1} \cdots A_{i_k}) = \frac{(n-k)!}{n!}$ for all $i_1 \neq i_2 \neq \ldots \neq i_k$. Thus, the sought probability $P(A_1 \cup A_2 \cup \cdots \cup A_n)$ is given by

$$\binom{n}{1}\frac{(n-1)!}{n!} - \binom{n}{2}\frac{(n-2)!}{n!} + \binom{n}{3}\frac{(n-3)!}{n!} - \cdots + (-1)^{n-1}\binom{n}{n}\frac{1}{n!}.$$

Using the formula $\binom{n}{k} = \frac{n!}{k!(n-k)!}$, we get the sought formula

$$P(A_1 \cup A_2 \cup \cdots \cup A_n) = 1 - \sum_{k=0}^{n} \frac{(-1)^k}{k!}.$$

A surprising conclusion can be drawn from this result. A basic result from calculus is that $\sum_{k=0}^{\infty} (-1)^k/k! = e^{-1}$ with $e = 2.71828\ldots$, see Appendix C. Thus, for sufficiently large n, the probability that at least one person will receive his or her own coat is approximately equal to $1 - e^{-1} = 0.63212\ldots$, regardless of how large n is. This approximation is accurate to seven or more decimal places for $n \geq 10$.

Having obtained the probability of at least one person getting the correct coat, it is not difficult to argue that

$$P(\text{exactly } j \text{ persons will receive their own coats}) = \frac{1}{j!} \sum_{k=0}^{n-j} \frac{(-1)^k}{k!}$$

for $j = 0, 1, \ldots, n$. To verify this, denote by N_m the number of permutations of the integers $1, \ldots, m$ so that no integer remains in its original position. As shown above, the probability that a random permutation of $1, \ldots, m$ has no integer in its original position is given by $\sum_{k=0}^{m} (-1)^k/k!$. Therefore,

$$\frac{N_m}{m!} = \sum_{k=0}^{m} \frac{(-1)^k}{k!}.$$

1.6 Inclusion–Exclusion Rule

By the definition of N_m, the number of permutations of the integers $1, \ldots, n$ so that exactly j integers remain in their original positions is $\binom{n}{j} N_{n-j}$. Thus the probability that exactly j persons will receive their own coats is $\binom{n}{j} N_{n-j}/n! = \frac{1}{j!} N_{n-j}/(n-j)!$. Using the expression for $N_m/m!$, we get the formula for the probability of exactly j persons receiving their own coats. This probability tends to the so-called Poisson probability $e^{-1}/j!$ as the number of people becomes large. The Poisson approximation is very accurate. Numerical investigations show that the Poisson probabilities agree with the exact probabilities in four or more decimal places for $n \geq 10$.

The matching problem appears in many disguises. A nice real-life application of the problem is the following. In 2008, a celebrated self-proclaimed psychic medium participated in a one-million-dollar paranormal challenge to test for psychic ability. A child was chosen by the medium as one whom he could connect with telepathically. The medium was shown the 10 target toys that he would have to pick up during the test and then he was placed in an isolated room. A toy was chosen at random and given to the child. The medium's task was to connect to the child and state which toy he was playing with. This was repeated ten times, and each time the child got a different toy. At the end of the trials the medium's answers were checked against the toys the child actually played with. If the medium had six or more correct guesses, then he would win one million dollars. The medium correctly guessed one toy choice out of 10, thus failing the test and losing the challenge. By chance alone, a correct guess of 1 out of 10 random choices was to be expected, while the probability of winning the challenge was just 6.0×10^{-4}.

Problems

1.77 A certain person is taking part in a blind taste test of ten different wines. The person has been made aware of the names of the ten wine producers beforehand, but does not know what order the wines will be served in. Each wine producer may be named only once. After the tasting session is over, it turns out that he has correctly identified five of the ten wines. Do you think he is a connoisseur?

1.78 You belong to a class of 15 students. The teacher hands 15 graded assignments back to the students in random order. What is the probability that you are the only person who receives his own paper?

1.79 Five friends enter a restaurant on a rainy evening and check their coats and umbrellas. As they are leaving, the lazy coat-checker returns both the coats and the umbrellas in random order. What is the probability that at least one person gets both the correct coat and the correct umbrella?

1.80 A blind taste test of ten different wines takes place. The participants are told the names of the ten wines beforehand. The wines are presented in random order and a participant may name each wine only once. There are three Italian wines among the ten wines. A person who has no knowledge about wine participates in the test. What is the probability that none of the three Italian wines is correctly guessed by this person?

1.81 Three people each choose five distinct numbers at random from the numbers $1, 2, \ldots, 25$, independently of each other. What is the probability that there is some number that is chosen by each of the three people?

1.82 What is the probability that a hand of 13 cards from an ordinary deck of 52 cards contains four cards of the same rank, such as four queens?

1.83 What is the probability that in a 13-card bridge hand at least one suit will be missing? What is the probability that in a 5-card poker hand at least one suit will be missing?

1.84 Suppose 8 Germans and 4 Italians are randomly assigned to 6 double rooms. What is the probability that no room will have two people of different nationalities?

1.85 Consider a communication network with four nodes n_1, n_2, n_3, and n_4, and six directed links $l_1 = (n_1, n_2)$, $l_2 = (n_1, n_3)$, $l_3 = (n_2, n_3)$, $l_4 = (n_3, n_2)$, $l_5 = (n_2, n_4)$, and $l_6 = (n_3, n_4)$. A message has to be sent from the source node n_1 to the destination node n_4. The network is unreliable. The probability that the link l_i is functioning is p_i for $i = 1, \ldots, 6$. The links behave physically independently of each other. A path from node n_1 to node n_4 is only functioning if each of its links is functioning. What is the probability that there is some functioning path from node n_1 to node n_4? What is this probability when $p_i = p$ for all i?

1.86 What is the probability that in 30 lottery drawings of six distinct numbers from the numbers 1 to 45, not all of the 45 numbers will come up?

1.87 The *coupon-collector problem* is as follows. Each time you buy a certain product (chewing gum, for example) you receive a coupon (a baseball card, for example), which is equally likely to be any one of n types.

(a) What is the probability of collecting a complete set of coupons in no more than r purchases?

(b) Answer the following questions. How many rolls of a fair die are required so that each number will appear one or more times with a probability of at least 50%? How many people are required in order to have represented all of the 365 birthdays of the year (ignore

1.6 Inclusion–Exclusion Rule 41

February 29) with a probability of at least 50%? How often should a ball be randomly dropped into one of 100 bins so that there is no empty bin with a probability of at least 50%?

1.88 In a group of n boys and n girls, each boy chooses at random a girl and each girl chooses at random a boy. The choices of the boys and girls are independent of each other. If a boy and a girl have chosen each other, they form a couple. What is the probability that no couple will be formed?

1.89 What is the probability that everybody in a randomly formed group of n people shares his or her birthday with someone else in the group? How large should n be so that this probability is at least 50%?

1.90 Three people each write down the numbers $1, \ldots, 10$ in a random order. What is the probability that in some position the three random sequences have the same number?

1.91 Ten persons each choose one name at random from the names of the other nine persons, independently of each other. What is the probability that at least two persons choose each other's names?

1.92 There are n people in a room and each of them is armed and angry. At a chime of the clock, everyone spins around and shoots at random one other person. What is the probability that there is no survivor after the first round?

1.93 Five balls are randomly chosen, with replacement, from a bowl that contains 3 red balls, 5 white balls, and 7 blue balls. What is the probability of getting at least one ball of each color? What is this probability if the balls are picked without replacement?

1.94 Suppose n married couples are invited to a bridge party. Bridge partners are chosen at random, without regard to gender. What is the probability that no one will be paired with his or her spouse?

1.95 An ordinary deck of 52 cards is thoroughly shuffled. The dealer turns over the cards one at a time, counting as he goes "ace, two, three, ..., king, ace, two, ...," and so on, so that the dealer ends up calling out the 13 ranks four times each. A match occurs if the rank the dealer is calling corresponds to the card he turns over. A "supermatch" occurs if all four cards of some same rank are matched (four aces, for example). What is the probability that a "supermatch" will occur?

2
Conditional Probability

The concept of conditional probability lies at the heart of probability theory. It is an intuitive concept. To illustrate this, most people reason as follows to find the probability of getting two aces when two cards are selected at random from an ordinary deck of 52 cards. The probability of getting an ace on the first card is $\frac{4}{52}$. Given that one ace is gone from the deck, the probability of getting an ace on the second card is $\frac{3}{51}$. The sought probability is therefore $\frac{4}{52} \times \frac{3}{51}$. Letting A_1 be the event that the first card is an ace and A_2 the event that the second card is an ace, one intuitively applies the fundamental formula $P(A_1 A_2) = P(A_1) P(A_2 \mid A_1)$, where $P(A_2 \mid A_1)$ is the notation for the conditional probability that the second card will be an ace given that the first card was an ace.

The purpose of this chapter is to present the basics of conditional probability. You will learn about the multiplication rule for probabilities and the law of conditional probabilities. These results are extremely useful in problem solving. Much attention will be given to Bayes' rule for revising conditional probabilities in light of new information. This rule is inextricably bound up with conditional probabilities. The odds form of Bayes' rule is particularly useful and will be illustrated with several examples. Following on from Bayes' rule, we explain Bayesian inference for discrete models and give several statistical applications.

2.1 Concept of Conditional Probability

The starting point for the definition of conditional probability is a chance experiment for which a sample space and a probability measure P are defined. Let A be an event of the experiment. The probability $P(A)$ reflects our knowledge of the occurrence of event A *before* the experiment takes place.

2.1 Concept of Conditional Probability

Therefore, the probability $P(A)$ is sometimes referred to as the *a priori* probability of A or the *unconditional* probability of A. Suppose we are now told that an event B has occurred in the experiment, but we still do not know the precise outcome in the set B. In light of this added information, the set B replaces the sample space as the set of possible outcomes and consequently the probability of the occurrence of event A changes. A conditional probability now reflects our knowledge of the occurrence of the event A given that event B has occurred. The notation for this new probability is $P(A \mid B)$.

Definition 2.1 *For any two events A and B with $P(B) > 0$, the conditional probability $P(A \mid B)$ is defined as*

$$P(A \mid B) = \frac{P(AB)}{P(B)}.$$

Here AB stands for the occurrence of both event A and event B. This is not an arbitrary definition. It can be intuitively reasoned through a comparable property of the relative frequency. Let's define the relative frequency $f_n(E)$ of the occurrence of event E as $\frac{n(E)}{n}$, where $n(E)$ represents the number of times that E occurs in n repetitions of the experiment. If, in n independent repetitions of the experiment, event B occurs r times simultaneously with event A and s times without event A, then we can say that $f_n(AB) = \frac{r}{n}$ and $f_n(B) = \frac{r+s}{n}$. Moreover, defining $f_n(A \mid B)$ as the relative frequency of event A in those repetitions of the experiment in which event B has occurred, we have that $f_n(A \mid B) = \frac{r}{r+s}$. Dividing $f_n(AB)$ by $f_n(B)$ gives $\frac{f_n(AB)}{f_n(B)} = \frac{r}{r+s}$. As a result, we get

$$f_n(A \mid B) = \frac{f_n(AB)}{f_n(B)}.$$

This relationship accounts for the definition of the conditional probability $P(A \mid B)$. The relative frequency interpretation also tells us how a conditional probability must be estimated in a simulation study.

Example 2.1 Someone has rolled two dice out of your sight. You ask this person to answer "yes or no" to the question of whether there is a six among the two rolls. He truthfully answers "yes." What is the probability that two sixes have been rolled?

Solution. Let A be the event that two sixes show up and B the event that at least one six shows up. It is helpful to imagine that one of the dice is red and the other is blue. The sample space of this experiment is the set $\{(i,j) \mid i,j = 1,\ldots,6\}$, where i and j denote the outcomes of the red and the blue die. A probability of $\frac{1}{36}$ is assigned to each element of the sample space. The event A is given by the set $A = \{(6,6)\}$ and the event B by the set $B = \{(1,6), \ldots, (5,6), (6,6), (6,5), \ldots, (6,1)\}$. The probability of event B occurring is $\frac{11}{36}$ and the probability of both event A and event B occurring is $\frac{1}{36}$. Given that you know that event B has occurred, the probability that event A has also occurred is $P(A \mid B)$. Applying the definition of conditional probability gives

$$P(A \mid B) = \frac{P(AB)}{P(B)} = \frac{1/36}{11/36}.$$

Thus, the sought probability is $\frac{1}{11}$ (not $\frac{1}{6}$).

The above example illustrates once again how careful you have to be when you are interpreting the information a problem is conveying. The wording of the problem is crucial: you know that one of the dice turned up a six but you do not know which one. In the case where one of the dice had dropped on the floor and you had seen the outcome six for that die, the probability of the other die turning up a six would have been $\frac{1}{6}$. An intuitive explanation of the difference between the two probabilities is the fact that in the second scenario you have the additional information of which one of the two dice (the red or the blue die) shows a six.

Example 2.2 John, Pedro, and Rosita are experienced darts players. The probability of John hitting the bullseye in a single throw is $\frac{1}{3}$. This hitting probability is $\frac{1}{5}$ for Pedro and $\frac{1}{4}$ for Rosita. The three players each throw one dart simultaneously. Two of the darts hit the bullseye and one of the darts misses the bullseye. What is the probability that John is the one who missed?

Solution. The sample space of the chance experiment consists of the eight elements (H, H, H), (H, H, M), (H, M, H), (H, M, M), (M, H, H), (M, H, M), (M, M, H), (M, M, M), where H stands for "hit" and M stands for "miss." The first component of each element of the sample space refers to John's throw, the second component refers to Pedro's throw, and the third component refers to Rosita's throw. By the independence of the outcomes of the individual throws, we assign the probability $\frac{1}{3} \times \frac{1}{5} \times \frac{1}{4} = \frac{1}{60}$ to the outcome (H, H, H), the probability $\frac{1}{3} \times \frac{1}{5} \times \frac{3}{4} = \frac{3}{60}$ to the outcome (H, H, M), the probability $\frac{1}{3} \times \frac{4}{5} \times \frac{1}{4} = \frac{4}{60}$ to the outcome (H, M, H), the probability $\frac{1}{3} \times \frac{4}{5} \times \frac{3}{4} = \frac{12}{60}$ to the outcome (H, M, M), the probability $\frac{2}{3} \times \frac{1}{5} \times \frac{1}{4} = \frac{2}{60}$ to the outcome (M, H, H), the probability $\frac{2}{3} \times \frac{1}{5} \times \frac{3}{4} = \frac{6}{60}$ to the outcome (M, H, M), the probability $\frac{2}{3} \times \frac{4}{5} \times \frac{1}{4} = \frac{8}{60}$ to the outcome (M, M, H), and the probability $\frac{2}{3} \times \frac{4}{5} \times \frac{3}{4} = \frac{24}{60}$ to the outcome (M, M, M). Let A be the event that John misses and let B be the event that exactly two of the darts hit the target. The event AB occurs if the outcome (M, H, H) occurs and the event B occurs if one of the outcomes $(H, H, M), (H, M, H), (M, H, H)$ occurs, and so $P(AB) = \frac{2}{60}$ and $P(B) = \frac{3}{60} + \frac{4}{60} + \frac{2}{60} = \frac{9}{60}$. Thus, the conditional probability that John is the one who missed given two hits is

$$P(A \mid B) = \frac{P(AB)}{P(B)} = \frac{2}{9}.$$

Problems

2.1 Two fair dice are rolled. What is the conditional probability that the sum of the two dice is 8, given that the two dice show different numbers?

2.2 You toss a nickel, a dime, and a quarter, independently of each other. What is the conditional probability that the quarter shows up heads given that the coins showing up heads represent an amount of at least 15 cents?

2.3 Every evening, two weather stations issue a weather forecast for the next day. The weather forecasts of the two stations are independent of each other. On average, the weather forecast of station 1 is correct in 90% of cases, irrespective of the weather type. This percentage is 80% for station 2. On a given day, station 1 predicts sunny weather for the next day, whereas station 2 predicts rain. What is the probability that the weather forecast of station 1 will be correct?

2.4 A professor has given two subsequent tests to her students. 50% of the students passed both tests and 80% of the students passed the first test. What percentage of those who passed the first test also passed the second test?

2.5 In a certain village, 30% of households have a cat and 25% of households have a dog. Of the households with a cat, 20% also have a dog. What percentage of households with a dog also have a cat?

2.6 The experiment is to toss a fair coin once and to roll a fair die once. Let A be the event that the coin lands "heads" and B the event that the die lands "six." Given that at least one of the two events has occurred, what is the probability that both events have occurred and what is the probability that event A has occurred?

2.7 You simultaneously grab two balls at random from an urn containing two red balls, one blue ball, and one green ball. What is the probability that you have grabbed two non-red balls given that you have grabbed at least one non-red ball? What is the probability that you have grabbed two non-red balls given that you have grabbed the green ball? Can you give an intuitive explanation of why the second probability is larger than the first one?

2.8 The following game is played in a particular carnival tent. The carnival master has two covered beakers, each containing one die. He shakes the beakers thoroughly, removes the lids, and peers inside. You have agreed that whenever at least one of the two dice shows an even number of points, you will bet with even odds that the other die will also show an even number of points. Is this a fair bet?

2.9 Suppose a bridge player's hand of 13 cards contains an ace. What is the probability that the player has at least one more ace? What is the answer to this question if you know that the player had the ace of hearts? Can you explain why the second probability is larger than the first one?

2.10 A hand of 13 cards is dealt from a standard deck of 52 cards. What is the probability that it contains more aces than tens? How does this probability change when you have the information that the hand contains at least one ace?

2.11 A fair die is rolled three times. What is the probability that each number rolled is higher than all those that were rolled earlier? *Hint*: Condition on the event that each roll gives a different outcome.

2.12 Let A and B be two events with $0 < P(A) < 1$ and $0 < P(B) < 1$.

(a) Suppose that $P(A \mid B) > P(B \mid A)$. Verify that $P(A) > P(B)$.
(b) Suppose that $P(A \mid B) > P(A)$. Verify the inequalities $P(B \mid A) > P(B)$ and $P(B^c \mid A) \leq P(B^c)$, where B^c is the complement of event B.
(c) What is $P(A \mid B)$ if A and B are disjoint? What is $P(A \mid B)$ if B is a subset of A?

2.13 In a high-school class, 35% of the students take Spanish as a foreign language, 15% take French as a foreign language, and 40% take at least one of these languages. What is the probability that a randomly chosen student takes French given that the student takes Spanish?

2.14 On a summer camp, 48% of the children are enrolled in swimming and 40% in tennis. One-third of the children signed up for swimming are also enrolled in tennis. What is the probability that a randomly chosen child is enrolled in swimming given that the child signed up for tennis?

2.15 A population of voters contains 45% Democrats and 55% Republicans. It is reported that 70% of the Democrats and 50% of the Republicans favor a particular election issue.

(a) What is the probability that a randomly chosen voter is a Democrat and in favor of the election issue, and what is the probability that a randomly chosen voter is a Republican and in favor of the election issue?

(b) What is the probability that a randomly chosen voter is in favor of the election issue?

(c) What is the probability that a randomly chosen voter is a Democrat given that the voter is in favor of the election issue?

2.16 Suppose that 50% of households in a certain city subscribe to the morning newspaper, 70% of households subscribe to the afternoon newspaper, and 40% of households subscribe to both newspapers. What is the probability that a randomly selected household does not subscribe to the morning newspaper given that the household subscribes to the afternoon newspaper?

2.2 Chain Rule for Conditional Probabilities

Interchanging the roles of A and B in the definition $P(A \mid B) = \frac{P(AB)}{P(B)}$ and noting that $P(BA) = P(AB)$, we have $P(B \mid A) = \frac{P(AB)}{P(A)}$. This equivalent representation of the definition of conditional probability can be written in the intuitively appealing form

$$P(AB) = P(A)P(B \mid A).$$

In words, the probability that events A and B both occur is equal to the probability that event A occurs multiplied by the probability that event B occurs given that A has occurred. This phrasing lines up naturally with the intuitive way people think about probabilities.

The formula $P(AB) = P(A)P(B \mid A)$ is often referred to as the *multiplication rule* for probabilities. This rule is a very useful tool for assigning or calculating probabilities. In many cases, the rule is used in attributing probabilities to elements of the sample space. In illustration of this, consider the experiment in which two marbles are randomly chosen without replacement from a receptacle holding seven red and three white marbles. One possible choice for the sample space of this experiment is the set consisting of four elements (R, R), (R, W), (W, W), and (W, R), where R stands for red and W for white. The first component of each element indicates the color of the first marble chosen and the second component the color of the second marble chosen. On grounds of the reasoning that $P(1^{st}$ marble is red$) = \frac{7}{10}$ and $P(2^{nd}$ marble is white $\mid 1^{st}$ marble is red$) = \frac{3}{9}$, we attribute the probability $P(R, W) = \frac{7}{10} \times \frac{3}{9} = \frac{7}{30}$ to the element (R, W). In the same way we attribute the probabilities $P(R, R) = \frac{7}{10} \times \frac{6}{9} = \frac{7}{15}$, $P(W, W) = \frac{3}{10} \times \frac{2}{9} = \frac{1}{15}$, and $P(W, R) = \frac{3}{10} \times \frac{7}{9} = \frac{7}{30}$ to the remaining elements. It is common practice for this type of problem to assign probabilities to the elements of the sample space as a product of probabilities, one marginal and the others conditional.

A repeated application of the basic formula $P(AB) = P(A)P(B \mid A)$ leads to the so-called chain rule for conditional probabilities:

Rule 2.1 (chain rule) *For any sequence of events $A_1, A_2, \ldots, A_n$,*

$P(A_1 A_2 \cdots A_n)$
$= P(A_1) \times P(A_2 \mid A_1) \times P(A_3 \mid A_1 A_2) \times \cdots \times P(A_n \mid A_1 A_2 \cdots A_{n-1}).$

The analysis of a complex probability experiment is often simplified by imagining that the experiment is performed as a multi-stage experiment in which each stage is a simple probability experiment and then applying the chain rule. This is a very useful problem-solving strategy. Probability problems that can be solved by counting arguments are often more easily tackled by using the chain rule for conditional probabilities. The chain rule decomposes the original problem into a series of simpler problems. To illustrate this, consider question (**a**) from Example 1.17. For $i = 1, \ldots, 4$, let A_i be the event that two teams from the same country play against each other in the ith match. The four matches are sequentially determined by lot. The probability asked for in question (**a**) is then given by $P(A_1 A_2 A_3 A_4)$. Obviously, $P(A_1) = \frac{1}{7}$, $P(A_2 \mid A_1) = \frac{1}{5}$, $P(A_3 \mid A_1 A_2) = \frac{1}{3}$, and $P(A_4 \mid A_1 A_2 A_3) = 1$, so the sought probability $P(A_1 A_2 A_3 A_4)$ equals $\frac{1}{7} \times \frac{1}{5} \times \frac{1}{3} \times 1 = 0.0095$. As another illustration, what is the probability that it takes 10 or more cards before the first ace appears if cards are randomly drawn one by one from an ordinary deck of 52 cards? To answer this question by counting arguments, we take as sample space the set of all possible permutations of the integers $1, 2, \ldots, 52$. Each of the 52! elements is equally likely. Let A be the event that it takes 10 or more cards before the first ace appears. The set A contains $48 \times 47 \times \cdots \times 40 \times 41!$ elements of the sample space and so $P(A) = 48 \times 47 \times \cdots \times 40 \times 41!/52! = 0.4559$. To answer the question with conditional probabilities, define A_i as the event that the ith card drawn is not an ace. Applying Rule 2.1, we find $P(A_1 A_2 \cdots A_9) = \frac{48}{52} \times \frac{47}{51} \times \cdots \times \frac{40}{44} = 0.4559$.

The following important remark is made. In using conditional probabilities, you usually perform the probability calculations without explicitly specifying a sample space; an assignment of probabilities to properly chosen events suffices. Solving a probability problem by counting arguments always requires the specification of a sample space.

Example 2.3 There are four British teams among the eight teams that have reached the quarterfinals of the soccer Champions League. What is the probability that the four British teams will avoid each other in the quarterfinal draw if the eight teams are paired randomly?

Solution. The following probability model can be used to answer the question. A bowl contains four red and four blue balls. Four times you pick at random two balls from the bowl without replacement. What is the probability that you pick each time a red and a blue ball? Defining A_i as the event that the ith pick gives a red and a blue ball, this probability is $P(A_1A_2A_3A_4)$. Then $P(A_1) = \frac{8 \times 4}{8 \times 7} = \frac{4}{7}$ (or, alternatively, $P(A_1) = \binom{4}{1}\binom{4}{1}/\binom{8}{2} = \frac{4}{7}$), $P(A_2 \mid A_1) = \frac{6 \times 3}{6 \times 5} = \frac{3}{5}$, $P(A_3 \mid A_1A_2) = \frac{4 \times 2}{4 \times 3} = \frac{2}{3}$, and $P(A_4 \mid A_1A_2A_3) = 1$. Thus, by applying the chain rule,

$$P(A_1A_2A_3A_4) = P(A_1)P(A_2 \mid A_1)P(A_3 \mid A_1A_2)P(A_4 \mid A_1A_2A_3)$$
$$= \frac{4}{7} \times \frac{3}{5} \times \frac{2}{3} \times 1 = \frac{8}{35}.$$

In other words, the probability that the four British teams will avoid each other in the draw is $\frac{8}{35}$.

Example 2.4 Two players A and B take turns rolling a fair die. Player A rolls the die first and wins on a "six." If A fails in rolling a "six," then player B rolls the die and wins on a "five" or "six." If B fails, then A rolls again and wins on a "four," "five," or "six," and so on. What is the probability that player A will be the winner?

Solution. Let E_i be the event that player A wins on the ith turn for $i = 1, 3,$ and 5. The events E_1, E_3, and E_5 are mutually exclusive and so $P(A \text{ wins}) = P(E_1) + P(E_3) + P(E_5)$. Using the chain rule for conditional probabilities,

$$P(E_1) = \frac{1}{6}, \quad P(E_3) = \frac{5}{6} \times \frac{4}{6} \times \frac{3}{6}, \quad \text{and} \quad P(E_5) = \frac{5}{6} \times \frac{4}{6} \times \frac{3}{6} \times \frac{2}{6} \times \frac{5}{6}.$$

This gives

$$P(A \text{ wins}) = \frac{1}{6} + \frac{5}{6} \times \frac{4}{6} \times \frac{3}{6} + \frac{5}{6} \times \frac{4}{6} \times \frac{3}{6} \times \frac{2}{6} \times \frac{5}{6} = \frac{169}{324}.$$

The game is not fair. Player A has a slight edge over player B.

Example 2.5 A coin-tossing game is played with 10 players including the two friends John and Pete. Each player has the same starting capital of d dollars. There are nine rounds of the game. In each round, two persons play against each other. A fair coin is tossed and, depending on the outcome of the toss, one player pays one dollar to the other player. The two players continue tossing the coin until one of them has won all the other player's money. The winner of the first round has doubled his starting capital and next plays against another player until one has won all the other's money, and so on. Suppose that John is in the first game and Pete plays the last game against the survivor of the first nine players. Who has the best chance of being the ultimate winner – John or Pete?

Solution. To solve this problem, consider first a game in which the player is equally likely to either win or lose one dollar on each gamble. If the player starts with a dollars and plays the game repeatedly until he either goes broke or increases his bankroll to $a + b$ dollars, then the probability of reaching his desired goal and thus not going broke is $\frac{a}{a+b}$. This intuitively obvious result is a special case of the gambler's ruin formula that will be proved in the next section. Next we turn to the stated problem. The answer to this problem is that each of the 10 players has the same probability $\frac{1}{10}$ of being the ultimate winner. This answer might surprise you. Let us first argue that the probability of John being the ultimate winner is $\frac{1}{10}$. Define A_i as the event that John enters the ith round and wins that round. The gambler's ruin formula gives $P(A_1) = \frac{d}{d+d} = \frac{1}{2}$, $P(A_2 \mid A_1) = \frac{2d}{2d+d} = \frac{2}{3}$, $P(A_3 \mid A_1, A_2) = \frac{3d}{3d+d} = \frac{3}{4}$, and so on. Hence, by the chain rule for conditional probabilities,

$$P(A_1 A_2 \cdots A_9) = \frac{1}{2} \times \frac{2}{3} \times \frac{3}{4} \times \cdots \times \frac{9}{10} = \frac{1}{10},$$

verifying that John will be the ultimate winner with probability $\frac{1}{10}$. By the same argument, the player meeting the winner of the first round in the second round will be the ultimate winner with probability $\frac{1}{3} \times \frac{3}{4} \times \cdots \times \frac{9}{10} = \frac{1}{10}$. Following this reasoning through, we find the same win probability of $\frac{1}{10}$ for each of the 10 participants of the game.

Problems

2.17 You and your friend have lunch with three business partners. After lunch each one puts a business card in a bowl. Two of the five cards are randomly drawn from the bowl to determine who pays for the lunch. What is the probability that you and your friend do not have to pay?

2.18 You are dealt a hand of four cards from a well-shuffled deck of 52 cards. What is the probability that you receive the four cards J, Q, K, A in any order, with suit irrelevant? *Hint*: Let A_i be the event that the ith card you receive is a picture card that you have not received before.

2.19 A fair die is rolled six times. What is the probability of one or more sixes and no one occurring?

2.20 There are two Spanish teams and two German teams among the eight teams that have reached the quarterfinals of the soccer Champions League. What is the probability that none of the Spanish teams will be paired with a German team when the teams are paired randomly?

2.2 Chain Rule for Conditional Probabilities

2.21 A jar contains three white and two black balls. Each time, you pick at random one ball from the jar. If it is a white ball, a black ball is inserted instead; otherwise, a white ball is inserted instead. You continue until all balls in the jar are black. What is the probability that you need three picks to achieve this? What is the probability that you need five picks?

2.22 Use conditional probabilities to solve Problems 1.6, 1.9–1.11, and 1.23.

2.23 You have received a tip that the management of a theater will give a free ticket to the first person in line having the same birthday as someone before him or her in line. Assuming that people enter the line one at a time and you can join the line any time, what position in the line maximizes your probability of getting the free ticket?

2.24 You travel from Amsterdam to Sidney with a change of airplane in Dubai and Singapore. You have one piece of luggage. At each stop your luggage is transferred from one airplane to the other. At the airport in Amsterdam there is a probability of 5% that your luggage is not placed in the right plane. This probability is 3% at the airport in Dubai and 2% at the airport in Singapore. What is the probability that your luggage does not reach Sidney with you? If your luggage does not reach Sidney with you, what is the probability that it was lost at Dubai airport?

2.25 Seven individuals have reserved tickets at the opera. The seats they have been assigned are all in the same row of seven seats. The row of seats is accessible from either end. Assume that the seven individuals arrive in a random order, one by one. What is the probability of all seven individuals taking their seats without having to squeeze past an already seated individual? *Hint*: Consider the reverse process of people leaving in random order.

2.26 In a poker game with three players A, B, and C, the dealer is chosen by the following procedure. In the order $A, B, C, A, B, C, \ldots$, a card from a well-shuffled deck is dealt to each player until someone gets an ace. This first player receiving an ace gets to start the game as dealer. Do you think that everyone has an equal chance of becoming dealer?

2.27 Suppose there are n women–men couples at a party. What is the probability that there are at least two couples such that the two women have the same birthday and the two men have the same birthday? *Hint*: Let A_i be the event that there is no match between the ith couple and any of the couples $1, \ldots, i - 1$.

2.28 What is the probability that, in the next m draws of the 5/39 lottery, there will be three consecutive draws with the feature that the same five numbers appear in two or more of these three draws? In each draw five distinct numbers are drawn at random from $1, 2, \ldots, 39$. *Hint*: Define A_i

as the event that each of the three draws i, $i+1$, and $i+2$ has a different set of five numbers.

2.29 A bowl contains r red and b blue balls. The balls are removed from the bowl, one at a time and at random. What is the probability that the first red ball is removed on the kth pick? Can you tell without doing any calculations what the probability is that there are still blue balls in the bowl when the last red ball is picked?

2.30 In a group of 25 people, a person tells a rumor to a second person, who in turn tells it to a third person, and so on. Each person tells the rumor to just one of the people chosen at random, excluding the person from whom he/she heard the rumor. The rumor is told 10 times. What is the probability that the rumor will not be repeated to any one person once more? What is the probability that the rumor will not return to the originator?

2.2.1 Independent Events

In the special case of $P(A \mid B) = P(A)$, the occurrence of event A is not contingent on the occurrence or non-occurrence of event B. Event A is then said to be independent of event B. In other words, if A is independent of B, then learning that event B has occurred does not change the probability that event A occurs. Since $P(A \mid B) = \frac{P(AB)}{P(B)}$, it follows that A is independent of B if the equation $P(AB) = P(A)P(B)$ holds true. This equation is symmetric in A and B: if A is independent of B, then B is also independent of A. Summarizing,

Definition 2.2 *Two events A and B are said to be independent if*

$$P(AB) = P(A)P(B).$$

The reader should be aware that independent events and disjoint events are completely different things. If events A and B are disjoint, you calculate the probability of the union $A \cup B$ by *adding* the probabilities of A and B. For independent events A and B, you calculate the probability of the intersection AB by *multiplying* the probabilities of A and B. Beginning students sometimes think that independent events are disjoint. This is not true. If two events A and B are disjoint, then $A \cap B = \emptyset$ and so $P(AB) = 0$, while $P(AB) = P(A)P(B)$ for independent events A and B. This shows that independent events with nonzero probability are never disjoint.

Example 2.6 Suppose that a red and a blue die are thrown. Let A be the event that the number shown by the red die is 4, B be the event that the sum of the dice is 7, and C be the event that the sum of the dice is 8. Show that the events

2.2 Chain Rule for Conditional Probabilities

A and B are independent, whereas the events A and C are dependent. Can you explain this result intuitively?

Solution. The experiment has 36 possible outcomes (i,j), where i is the number shown by the red die and j the number shown by the blue die. All 36 possible outcomes are equally likely. Simply, by counting, $P(A) = \frac{6}{36}$, $P(B) = \frac{6}{36}$, and $P(AB) = \frac{1}{36}$. Since $P(AB) = P(A)P(B)$, the events A and B are independent. An intuitive explanation of the independence of A and B is the observation that the conditional probability of a sum 7 given the outcome of the red die is the same for each possible outcome of the red die. The events A and C are dependent, as follows by noting that $P(C) = \frac{5}{36}$ and $P(AC) = \frac{1}{36}$. An intuitive explanation is that the sum of the dice can never be 8 if the red die shows a 1.

In practical problems we rarely need to check independence in such detail as in Example 2.6, but independence of events can usually be concluded directly from the physical setup of the underlying chance experiment. Independent events A and B typically arise in a compound experiment consisting of physically independent subexperiments, where one subexperiment alone determines whether event A occurs and another subexperiment alone determines whether event B occurs.

Remark 2.1 In the case that events A, B, and C are pairwise independent, it is not necessarily true that $P(ABC) = P(A)P(B)P(C)$. This can be shown using Example 2.6. In addition to the events A and B from Example 2.6, let C be the event that the number shown by the blue die is even. It is readily verified that the events A, B, and C are pairwise independent. However, $P(ABC)(=0)$ is not equal to $P(A)P(B)P(C)(=\frac{1}{72})$. In general, a collection of events $A_1, A_2, \ldots, A_n$ are called *independent* if

$$P\left(\bigcap_{i \in I} A_i\right) = \prod_{i \in I} P(A_i)$$

for each subset I of indices. An infinite collection of events $A_1, A_2, \ldots$ is said to be independent if each finite subcollection is independent.

A useful result for independent events $A_1, \ldots, A_n$ is the inequality

$$P(\text{none of the events } A_1, \ldots, A_n \text{ will occur}) \leq e^{-\sum_{i=1}^{n} P(A_i)}.$$

This useful bound is easily obtained by noting that the probability is given by $P(A_1^c \cdots A_n^c)$, where A_i^c is the complement of A_i. It is readily verified that the independence of the A_i implies the independence of the A_i^c (see Problem 2.33), and so $P(A_1^c \cdots A_n^c) = \prod_{i=1}^{n} P(A_i^c)$. Noting that $P(A_i^c) = 1 - P(A_i)$ and using the inequality $1 - x \leq e^{-x}$ for $0 \leq x \leq 1$, the bound follows next.

Problems

2.31 Suppose that a red and a blue die are thrown. Let A be the event that the number shown by the red die is even, and B the event that the sum of the dice is odd. Do you think the events A and B are independent?

2.32 Fifty different numbers are arranged in a matrix with 5 rows and 10 columns. You pick at random one number from the matrix. Let A be the event that the number comes from an odd-numbered row and B be the event that the number comes from the first five columns. Are the events A and B independent?

2.33 Let A and B be independent events. Denote by the events A^c and B^c the complements of the events A and B. Verify that the events A and B^c are independent. Conclude from this result that the events A^c and B^c are also independent.

2.34 Let A and B be two events such that $P(A \mid B) = P(A \mid B^c)$ and $0 < P(B) < 1$. Prove that A and B are independent events.

2.35 Suppose $A_1, \ldots, A_n$ are independent events. Prove that the probability that at least one of the events $A_1, \ldots, A_n$ occurs is $1 - \left[(1 - P(A_1)) \cdots (1 - P(A_n))\right]$.

2.36 Suppose that the events A, B, and C are independent, where $P(A) = \frac{1}{2}$, $P(B) = \frac{1}{3}$, and $P(C) = \frac{1}{4}$. What is $P(A \cup B \cup C)$?

2.37 Suppose that $A_1, A_2, \ldots$ is a sequence of *independent* events with the property that $\sum_{n=1}^{\infty} P(A_n) = \infty$. Define A as the event that A_n occurs for infinitely many values of n. Prove that $P(A) = 1$. This result is known as the *second Borel–Cantelli lemma*. *Hint*: Argue that $P(A) = \lim_{n \to \infty} P\left(\bigcup_{k=n}^{\infty} A_k\right)$ and use the inequality $1 - x \leq e^{-x}$ for $0 \leq x \leq 1$.

2.3 Law of Conditional Probability

The law of conditional probability is a natural partitioning rule. To introduce this rule, let us consider the following experiment. A fair die is rolled to yield a number between 1 and 6, and then a fair coin is tossed that many times. Many people would reason as follows to find the probability that heads will not appear. The probability of rolling the number k is $\frac{1}{6}$ for $k = 1, \ldots, 6$. If the number k is rolled, then the probability of no heads is $(\frac{1}{2})^k$. Therefore, the sought probability is

$$\frac{1}{2} \times \frac{1}{6} + \frac{1}{4} \times \frac{1}{6} + \frac{1}{8} \times \frac{1}{6} + \frac{1}{16} \times \frac{1}{6} + \frac{1}{32} \times \frac{1}{6} + \frac{1}{64} \times \frac{1}{6} = 0.1641.$$

This intuitive reasoning is justified by the next rule:

2.3 Law of Conditional Probability

Rule 2.2 *Let A be an event that can only occur if one of the mutually exclusive events $B_1, \ldots, B_n$ occurs. Then,*

$$P(A) = P(A \mid B_1)P(B_1) + P(A \mid B_2)P(B_2) + \cdots + P(A \mid B_n)P(B_n).$$

This rule is called the *law of conditional probability*. In words, a "difficult" unconditional probability $P(A)$ is obtained by averaging "easier" conditional probabilities $P(A \mid B_i)$ over the probabilities $P(B_i)$ for $i = 1, \ldots, n$. Probabilities are usually easier to compute when you have more information. The proof of Rule 2.2 is simple and instructive. In terms of sets, the subset A of the sample space is contained in the union $B_1 \cup \cdots \cup B_n$ of the subsets $B_1, \ldots, B_n$. This implies

$$A = AB_1 \cup AB_2 \cup \cdots \cup AB_n,$$

where the set AB_i is the intersection of the sets A and B_i. The assumption that the sets $B_1, \ldots, B_n$ are pairwise disjoint implies that the sets $AB_1, \ldots, AB_n$ are also pairwise disjoint. By Rule 1.1 in Chapter 1, we then have

$$P(A) = P(AB_1) + P(AB_2) + \cdots + P(AB_n).$$

This relationship and the definition $P(A \mid B) = P(AB)/P(B)$ lead to the law of conditional probability. Obviously, this law is also applicable when the sample space is divided by a countably infinite number of disjoint subsets $B_1, B_2, \ldots$ instead of by a finite number.

It is often the case that the unconditional probability $P(A)$ of an event A is found most easily by conditioning on appropriately chosen events $B_1, \ldots, B_n$ such that the B_i are mutually exclusive and event A can only occur when one of the events B_i occurs. This is one of the most useful problem-solving strategies in probability. A nice illustration of the law of conditional probability is provided by the next example.

Example 2.7 In a television game show, you can win a small prize, a medium prize, and a large prize. The large prize is a sports car. The three prizes are "locked up" in different boxes. There are five keys randomly arranged in front of you. Three of the keys can open only one lock and each of these keys fits a different box. Another key is a dud that does not open any of the locks. The final key is the "master key" that opens all three locks. You have a chance to choose up to two keys. For that purpose, you are asked two quiz questions. For each correct answer, you can select one key. The probability of correctly answering any given quiz question is 0.5. Any key you have gained is tried on all three boxes. What is the probability of winning the sports car?

Solution. Let A be the event that the sports car is won. To find $P(A)$, condition on the first step of the experiment and let B_i be the event that you have correctly answered i questions, $i = 0, 1, 2$. By the law of conditional probability, $P(A) = P(A \mid B_0)P(B_0) + P(A \mid B_1)P(B_1) + P(A \mid B_2)P(B_2)$. It is immediate that $P(B_0) = \frac{1}{4}$, $P(B_1) = \frac{2}{4}$, and $P(B_2) = \frac{1}{4}$. Obviously, $P(A \mid B_0) = 0$. Since two of the five keys fit the box with the sports car, we have $P(A \mid B_1) = \frac{2}{5}$. To get $P(A \mid B_2)$, it is easier to compute the complementary probability. The probability that neither of two randomly chosen keys fits the box with the sports car can be calculated as $\frac{3}{5} \times \frac{2}{4} = \frac{3}{10}$ (or, alternatively, as $\binom{3}{2}/\binom{5}{2} = \frac{3}{10}$). This gives $P(A \mid B_2) = \frac{7}{10}$. Thus, the probability of winning the sports car is

$$P(A) = 0 \times \frac{1}{4} + \frac{2}{5} \times \frac{2}{4} + \frac{7}{10} \times \frac{1}{4} = \frac{3}{8}.$$

All aspects of conditional probability are addressed in the next example.

Example 2.8 Two boxes are placed in front of you. One box contains nine $1 bills and one $5 bill, while the other box contains two $1 bills and one $100 bill. You choose at random one of the two boxes and then two bills are randomly picked out of the chosen box. It appears that these two bills are $1 bills. Next you get the opportunity to pick a third bill out of one of the two boxes. Should you stick to the chosen box or should you switch to the other box when you want to maximize the probability of picking the $100 bill?

Solution. Let A be the event that you have chosen the box with the $100 bill and B the event that you have chosen the other box. Also, let E be the event that the two bills taken out from the chosen box are $1 bills. The probability of picking the $100 bill as third bill is $P(A \mid E) \times 1$ if you stick to the box chosen and is $P(B \mid E) \times \frac{1}{3}$ if you switch to the other box. Since $P(B \mid E) = 1 - P(A \mid E)$, it suffices to find $P(A \mid E)$. To do so, note that

$$P(A \mid E) = \frac{P(AE)}{P(E)} = \frac{P(E \mid A)P(A)}{P(E)}.$$

Since $P(A) = P(B) = \frac{1}{2}$, we have

$$P(E) = P(E \mid A) \times \frac{1}{2} + P(E \mid B) \times \frac{1}{2},$$

by the law of conditional probability. Obviously,

$$P(E \mid A) = \frac{2}{3} \times \frac{1}{2} = \frac{1}{3} \quad \text{and} \quad P(E \mid B) = \frac{9}{10} \times \frac{8}{9} = \frac{4}{5}.$$

2.3 Law of Conditional Probability

This leads to $P(E) = \frac{1}{3} \times \frac{1}{2} + \frac{4}{5} \times \frac{1}{2} = \frac{17}{30}$. Next we get

$$P(A \mid E) = \frac{1/3 \times 1/2}{17/30} = \frac{5}{17} \text{ and } P(B \mid E) = 1 - \frac{5}{17} = \frac{12}{17}.$$

Putting the pieces together, it follows that the probability of picking the $100 bill as third bill is $\frac{5}{17} \times 1 = \frac{5}{17}$ if you stick to the chosen box and $\frac{12}{17} \times \frac{1}{3} = \frac{4}{17}$ if you switch to the other box. You had better stick to the chosen box. A surprising finding!

The next example deals with a famous problem known as the *gambler's ruin problem*. This problem can be seen as a random walk with two absorbing barriers.

Example 2.9 John and Pete enter a coin-tossing game, where John starts with a dollars and Pete with b dollars. They repeatedly flip a coin until one of them has won all of the money. If the coin lands heads, John gets one dollar from Pete; otherwise, John loses one dollar to Pete. The coin lands heads with probability p and tails with probability $q = 1 - p$, where $0 < p < 1$. What is the probability that John will win all of the money?

Solution. Let $P(a, b)$ denote the probability of John winning all of the money when John starts with a dollars and Pete with b dollars. The *gambler's ruin formula* states that

$$P(a,b) = \begin{cases} \frac{1-(q/p)^a}{1-(q/p)^{a+b}} & \text{if } p \neq q \\ \frac{a}{a+b} & \text{if } p = q. \end{cases}$$

To prove this formula, let A be the event that John wins all of the money. A recurrence equation for $P(A) = P(a, b)$ is obtained by conditioning on the outcome of the first flip. Let E be the event that the first flip lands heads and let F be the event that the first flip lands tails. By the law of conditional probability,

$$P(A) = P(A \mid E)P(E) + P(A \mid F)P(F).$$

If the first flip lands heads, you get the changed situation that John has $a + 1$ dollars and Pete has $b - 1$ dollars. This gives $P(A \mid E) = P(a + 1, b - 1)$. Similarly, $P(A \mid F) = P(a - 1, b + 1)$. Since $P(E) = p$ and $P(F) = q$, we obtain the recurrence equation

$$P(a,b) = P(a+1, b-1) \times p + P(a-1, b+1) \times q.$$

Copying this equation with the starting capitals a and b replaced by i and $a + b - i$ gives a system of equations for $P_i = P(i, a + b - i)$. These equations are

$$P_i = pP_{i+1} + qP_{i-1} \quad \text{for } i = 1, \ldots, a+b-1$$

with the boundary conditions $P_0 = 0$ and $P_{a+b} = 1$. The solution can be found through the standard theory for linear difference equations. However, this theory is not needed to find the solution. Noting that $P_i = (p+q)P_i$, the difference equations can be rewritten as $P_{i+1} - P_i = \frac{q}{p}(P_i - P_{i-1})$. Successive substitution of this equation into itself gives

$$P_{i+1} - P_i = \left(\frac{q}{p}\right)^i (P_1 - P_0) \quad \text{for } i = 1, \ldots, a+b-1.$$

Summing this equation over $i = 1, \ldots, k-1$ and noting that $P_0 = 0$, we get the relation $P_k - P_1 = P_1 \sum_{i=1}^{k-1} \left(\frac{q}{p}\right)^i$ and so

$$P_k = P_1 \sum_{i=0}^{k-1} \left(\frac{q}{p}\right)^i \quad \text{for } 1 \leq k \leq a+b.$$

Using the identity $(1-a)\sum_{i=0}^{m} a^i = 1 - a^{m+1}$, it follows that

$$P_k = \frac{1-(q/p)^k}{1-q/p} P_1 \text{ if } p \neq q \text{ and } P_k = kP_1 \text{ if } p = q = \frac{1}{2}.$$

The boundary condition $P_{a+b} = 1$ leads to $P_1 = (1-q/p)/(1-(q/p)^{a+b})$ if $p \neq q$ and $P_1 = 1/(a+b)$ if $p = q = \frac{1}{2}$. Hence,

$$P_k = \frac{1-(q/p)^k}{1-(q/p)^{a+b}} \quad \text{for } k = 1, \ldots, a+b$$

if $p \neq q$ and $P_k = \frac{k}{a+b}$ if $p = q = \frac{1}{2}$, proving the gambler's ruin formula.

To illustrate the gambler's ruin formula, suppose that you go to the casino with $50, and it is your goal to multiply your capital to $250. You decide to play European roulette and to stake each time the same amount of s dollars on red for each spin of the wheel. You get back twice your stake with probability $\frac{18}{37}$ and lose your stake with probability $\frac{19}{37}$. What is the probability of reaching your goal for $s = 10, 25$, and 50? To find the probabilities, we take $p = \frac{18}{37}$ and $q = \frac{19}{37}$ in the gambler's ruin formula and calculate $P(a,b)$ for $a = 5$ and $b = 20$ when $s = 10$, for $a = 2$ and $b = 8$ when $s = 25$, and for $a = 1$ and $b = 4$ when $s = 50$. This results in the probabilities 0.1084, 0.1592, and 0.1790 for $s = 10, 25$, and 50. It is intuitively obvious that the probability gets larger for higher stakes, since your bankroll is exposed to the house advantage of the casino for a shorter time when you play more boldly.

The gambler's ruin formula was derived by parameterizing the starting state and then conditioning on what happens in the first step of the process. This powerful approach can also be used to analyze success runs.

2.3 Law of Conditional Probability

Example 2.10 A sequence of independent trials is performed. Each trial results in a success with probability p and in a failure with probability $1 - p$. What is the probability of getting a run of r or more successes in n trials?

Solution. By a "run" we mean a sequence of consecutive trials that all result in the same outcome. Let us parameterize with respect to the number of trials and define p_k as the probability of getting a run of r or more successes in k trials. Fix $k \geq r$ and let A be the event that a run of r or more consecutive successes occurs in k trials. To get a recursion scheme for $P(A) = p_k$, define B_r as the conditioning event that each of the first r trials results in a success and B_{j-1} as the conditioning event that each of the first $j-1$ trials results in a success but the jth trial in a failure, where $1 \leq j \leq r$. Then $P(B_r) = p^r$ and $P(B_{j-1}) = p^{j-1}(1-p)$ for $1 \leq j \leq r$. The events B_i are disjoint and the event A can only occur if one of the events B_i occurs. Thus $P(A) = \sum_{j=1}^{r+1} P(A \mid B_{j-1}) P(B_{j-1})$, by the law of conditional probability. Noting that $P(A \mid B_r) = 1$ and $P(A \mid B_{j-1}) = p_{k-j}$, we then obtain the recursion scheme

$$p_k = p^r + \sum_{j=1}^{r} p^{j-1}(1-p) p_{k-j} \quad \text{for } k \geq r,$$

where $p_k = 0$ for $k < r$. Starting with $p_r = p^r$, we can compute the p_k recursively until the sought probability p_n is obtained.

The probability of getting a run of either at least r successes or at least r failures, or both, in n trials can be calculated by an extension of the recursion scheme given in Example 2.10. You are asked to derive this extension in Problem 2.69. In Table 2.1 we give the probability of getting a run of either at least r heads or at least r tails, or both, in n tosses of a fair coin for several values of r and n. The numbers in the table may help correct common mistaken notions about the lengths of runs in coin tossing. Most people grossly underestimate the lengths of longest runs. For example, in a series of 200 tosses

Table 2.1. *Probability of a run of either r heads or r tails in n tosses*

n/r	3	4	5	6	7	8	9	10
10	0.826	0.465	0.217	0.094	0.039	0.016	0.006	0.002
25	0.993	0.848	0.550	0.300	0.151	0.073	0.035	0.017
50	1.000	0.981	0.821	0.544	0.309	0.162	0.082	0.041
75	1.000	0.998	0.929	0.703	0.438	0.242	0.126	0.064
100	1.000	1.000	0.972	0.807	0.542	0.315	0.169	0.087
150	1.000	1.000	0.996	0.918	0.697	0.440	0.247	0.131
200	1.000	1.000	0.999	0.965	0.799	0.542	0.318	0.172

with a fair coin, you should expect the longest run of heads to be about seven in a row and the longest run of either heads or tails to be about eight in a row. A rule of thumb is that, in n tosses of a coin with probability p of heads, the probability mass of the longest run of heads is strongly concentrated around the integer nearest to $\log_{1/p}(n(1-p))$, while the probability mass of the longest run of either heads or tails is strongly concentrated around the integer nearest to $\log_{1/p}(n(1-p)) + 1$ provided that $n(1-p)$ is sufficiently large.[1] The theoretical findings for runs in coin tossing put winning streaks in baseball into perspective. In the 2015 baseball season the New York Mets won 11 games in a row, which got a lot of attention. While this event is remarkable in isolation, you can expect the event to happen at a certain moment. In the baseball season each team plays 162 games. The probability is 23.06% that among seven seasons of 162 games there will be some season in which a particular team will win 11 or more games in a row when this team wins with a probability of 50% any single game. The most likely outcome for the longest winning streak of the team in a typical 162-game season is six games. An 11-game winning streak is the most likely outcome in a season when the team has a winning percentage of 70%.

To conclude this section, we use conditional probabilities to solve a real-life problem.

Example 2.11 Two candidates A and B remain in the finale of a television game show. At this point, each candidate must spin a wheel of fortune. The numbers $1, 2, \ldots, 100$ are listed on the wheel and when the wheel has stopped spinning, a pointer stops randomly on one of the numbers. Each player has a choice of spinning the wheel one or two times, whereby a second spin must immediately follow the first. The goal is to reach a total closest to but not exceeding 100 points. A player whose total exceeds 100 gets a final score of zero. The winner is the player who gets the highest score. Should both players have the same final score, then the player to spin the wheel first is the winner. Player A has to spin first. What is the optimal strategy for player A?

Solution. Let us take a general setup and assume that the numbers $1, 2, \ldots, R$ are listed on the wheel. The optimal strategy of the second player B is obvious. This player stops after the first spin only if the score is larger than the final score of player A. To find the optimal strategy of player A, let $S(a)$ be the conditional win probability of player A if A stops after the first spin given that this spin results in a score of a points, and let $C(a)$ be the conditional win probability of

---

[1] This result was obtained in M. F. Schilling, "The surprising predictability of long runs," *Mathematics Magazine* **85** (2012): 141–149.

player A if A continues after the first spin given that this spin results in a score of a points. To find $S(a)$, we condition on the outcome of the first spin of player B. Given that player B has scored b points in the first spin with $b \leq a$, then player A is the final winner if player B scores in the second spin either at most $a - b$ points or more than $R - b$ points. By the law of conditional probability,

$$S(a) = \sum_{b=1}^{a} \left(\frac{a-b}{R} + \frac{b}{R}\right) \times \frac{1}{R} = \frac{a^2}{R^2} \quad \text{for } 1 \leq a \leq R.$$

The expression for $C(a)$ follows directly from $S(a)$. By conditioning on the outcome of the second spin of player A, we have

$$C(a) = \sum_{k=1}^{R-a} S(a+k) \times \frac{1}{R} = \sum_{k=1}^{R-a} \frac{(a+k)^2}{R^3} \quad \text{for } 1 \leq a \leq R.$$

Taking $R = 100$, numerical calculations give that $C(a) > S(a)$ for $1 \leq a \leq 53$ and $S(a) > C(a)$ for $53 < a \leq 100$. Hence, player A stops after the first spin if this spin gives a score larger than 53; otherwise, A continues. Then, by the law of conditional probability, the probability of player A winning is

$$\sum_{a=1}^{53} C(a) \times \frac{1}{100} + \sum_{a=54}^{100} S(a) \times \frac{1}{100} = 0.4596.$$

Problems

2.38 You have two identical boxes in front of you. One of the boxes contains 10 balls numbered 1 to 10 and the other box contains 25 balls numbered 1 to 25. You choose at random one of the boxes and pick a ball at random from the chosen box. What is the probability of picking the ball with number 7 written on it?

2.39 A drunkard removes two randomly chosen letters of the message HAPPY HOUR that is attached to a billboard in a pub. His drunk friend puts the two letters back in a random order. What is the probability that HAPPY HOUR appears again?

2.40 On the television show "Deal or No Deal," you are faced with 26 briefcases in which various amounts of money have been placed, including the amounts $1,000,000 and $750,000. You first choose one case. This case is "yours" and is kept out of play until the very end of the game. Then you play the game and in each round you open a number of the other cases. What is the probability that the cases with $1,000,000 and $750,000 will still be in the game when you are going to open 20 cases?

2.41 In the UK National Lottery, six different numbers are drawn from a bucket with 59 balls numbered 1 to 59. To win the jackpot, you have to match all six numbers drawn. The lottery has added a new prize. If you match exactly two numbers, you win a lucky-dip ticket. The ticket is used on the next draw. If you then match exactly two numbers on the lucky-dip ticket, you get a new lucky-dip ticket, and so on. What is the probability of ever winning the jackpot when you buy one ticket only once?

2.42 Suppose that Joe arrives home on time with probability 0.8. If Joe does not arrive home on time, the probability that his dinner is burnt is 0.5; otherwise, his dinner is burnt with probability 0.15. What is the probability that Joe's dinner will be burnt on any given day? What is the probability that Joe arrived home on time given that his dinner is burnt?

2.43 You owe $20,000 to a loan shark and you will run into deep trouble if the loan shark is not repaid before midnight. You decide to go to a newly opened casino having a special roulette game with the three roulette bets: a 9-numbers bet, a 12-numbers bet, and an 18-numbers bet. For a k-numbers bet, you get back $\frac{36}{k}$ times your stake with probability $\frac{k}{37}$ and lose your stake with probability $1 - \frac{k}{37}$. Your initial bankroll is $10,000. In a desperate attempt to raise your bankroll to $20,000, you stake $5,000 on a 12-numbers bet. If you win, you are done; otherwise, you stake the remaining $5,000 on a 9-numbers bet. Verify that under this betting strategy the probability of reaching your goal of $20,000 is slightly higher than the probability of reaching your goal under the bold strategy of putting $10,000 only once on an 18-numbers bet.

2.44 The game "craps" is played as follows. A player rolls two fair six-sided dice. If their sum is 7 or 11, the player wins right away. If their sum is 2, 3, or 12, the player loses right away. If their sum is any other number, this number becomes the "point." The player then repeatedly rolls two dice until their sum equals the point, in which case the player wins, or equals 7, in which case the player loses. What is the probability of the player winning?

2.45 You are fighting a dragon with three heads. Each time you swing at the dragon with your sword, there is a 70% chance of knocking off one head and a 30% chance of missing. If you miss, an additional head grows immediately before you can swing again at the dragon. You win if you have knocked off all of the dragon's heads, but you must run for your life if the dragon has seven heads. What is your chance of winning?

2.46 A lottery organization distributes 100,000 tickets every week. At one end of the ticket, there is a visible printed number consisting of six digits, say

070469. At the other end of the ticket, another six-digit number is printed, but this number is hidden by a layer of scratch-away silver paint. The ticket holder scratches the paint away to reveal the underlying number. If the number is the same as the number at the other end of the ticket, it is a winning ticket. The two six-digit numbers on each of the one million tickets printed each week are randomly generated in such a way that no two tickets are printed with the same visible numbers or the same hidden numbers. Assume that in a particular week only half of the tickets printed are sold. What is the probability of exactly r winners in that week for $r = 0, 1, \ldots$? *Hint*: Use the results for the matching problem in Example 1.19.

2.47 The points $1, 2, \ldots, 12$ are around a ring like a clock. A transition from a point is to either of the two adjacent points with equal probabilities. What is the probability of visiting all points before returning to the starting point?

2.48 A thoroughly shuffled deck of cards has r red and b black cards. The top card is removed from the deck and then a card is picked at random from the deck. Use the law of conditional probability to find the probability that the second card picked is red. Can you explain the answer intuitively?

2.49 John and Pete each toss a fair coin, independently of each other. John tosses the coin until two heads have been obtained and Pete tosses the coin until three heads have been obtained. What is the probability that John will need more tosses than Pete?

2.50 Twenty-five persons attended a "reverse raffle," in which everyone bought a number. Numbered balls were then drawn out of a bin, one at a time and at random. The last ball in the bin would be the winner. But when the organizers got down to the last ball, they discovered that three numbered balls had been unintentionally overlooked. They added those balls to the bin and continued the draw. Was the raffle still fair? Use the law of conditional probability to answer this question.

2.51 Two players A and B enter a coin-tossing game, where player A starts with $a = 4$ dollars and player B with $b = 5$ dollars. They repeatedly flip a fair coin until one of them has won all of the money. If the coin lands heads, then player A gets one dollar from player B; otherwise, player A loses one dollar to player B. After the game has been played, you are informed that player A is the winner. What is the probability distribution of the lowest value of player A's bankroll during the game?

2.52 The upcoming Tour de France bicycle tournament will take place from July 1 through July 23. One hundred eighty cyclists will participate in the event. What is the probability that two or more participating cyclists will have birthdays on the same day during the tournament? *Hint*: Let B_i be the event that there are i cyclists having their birthday during the tournament.

2.53 A drunkard leaves the pub at 22:30 hours. Twenty steps to the left of the pub is a police station and ten steps to the right is his home. Every 30 seconds the drunkard makes a step either to the left or to the right. Any step is with equal probabilities to the left or to the right, independently of the other steps. The drunkard is thrown in a cell if he reaches the police station. What is the probability that he is not locked up in a police cell and reaches his home no later than midnight?

2.54 A tennis tournament is arranged for 8 players. It is organized as a knockout tournament. First, the 8 players are randomly allocated over four groups of two players each. In the semifinals, the winners of groups 1 and 2 meet each other and the winners of groups 3 and 4. In any match, either player has a probability 0.5 of winning. John and Pete are among the 8 players. What is the probability that they meet each other in the semifinals? What is the probability that they meet each other in the final?

2.55 There are two bags in front of you. One bag has three white balls and one red ball. The other has one white ball and three red balls. You choose one of the bags at random and pick randomly one ball out of this bag. You notice the ball picked is red. You then put it back and pick another ball out of the same bag at random. What is the probability that the second ball picked is red?

2.56 Dave and Eric alternately roll two dice. The first player who fails to surpass the sum of the two dice in the previous roll loses. What is the probability that Dave will win the game when Dave starts the game? *Hint*: Parameterize and define d_s as the probability that Dave will win the game when the game begins with Dave rolling the dice and Dave has to roll more than s points in his first roll.

2.57 Consider again Example 1.18. What is the probability that there will be exactly j stops at which nobody gets off. More generally, what is the probability of exactly j empty bins when m balls are sequentially placed into one of b bins with $m \geq b$, where for each ball a bin is selected at random?

2.58 A jar contains five blue and five red balls. You roll a fair die. Next you simultaneously draw at random as many balls from the jar as the score of the die. What is the probability that each of the balls drawn is blue? What is the inverse probability that the score of the die is r, given that each of the balls drawn is blue?

2.59 Two fair dice are rolled twice. What is the probability that both rolls show the same combination of two numbers?

2.60 In some basketball competition, 14 teams participate. The first six teams in the final ranking are automatically placed for the playoffs. There are no ties in the final ranking. The other eight teams participate in a lottery for the remaining two spots in the playoffs. The team placed 7th in the competition is given 8 tickets in the lottery, the team placed 8th is given 7 tickets, and so on, until the team placed 14th is given 1 ticket. A ticket is picked at random and the first-place draft is awarded to the team holding that ticket. That winner's remaining tickets are removed and a second ticket is selected at random for second place in the draft. What is the probability that the team placed jth in the competition will win the first place in the draft and what is the probability that this team will win the second place in the draft for $j = 7, \ldots, 14$?

2.61 Imagine a coin is tossed infinitely often. The coin lands on heads with probability p and on tails with probability $1-p$. It is assumed that $p \leq \frac{1}{2}$. Use the gambler's ruin formula to explain that after the first toss, the number of heads will ever be equal to the number of tails with probability $2p$.

2.62 A drunkard is wandering back and forth on a road. At each step he moves two units distance to the north with a probability of $\frac{1}{2}$, or one unit to the south with a probability of $\frac{1}{2}$. Show that with probability $\frac{1}{2}(\sqrt{5}-1)$ the drunkard will ever visit the point which is one unit distance south from his starting point. Can you explain why this probability also gives the probability that the number of heads will ever exceed twice the number of tails if a fair coin is tossed over and over?

2.63 Your friend has chosen at random a card from a standard deck of 52 cards, but keeps this card concealed. You have to guess what card it is. Before doing so, you can ask your friend either the question whether the chosen card is red or the question whether the card is the ace of spades. Your friend will answer truthfully. What question would you ask?

2.64 Player 1 tosses $m+1$ times a fair coin and player 2 tosses m times a fair coin. Player 1 wins the game if player 1 tosses more heads than player 2; otherwise, player 2 wins. What is the probability that player 1 will win the game?

2.65 You repeatedly roll two fair dice. What is the probability of two consecutive totals of 7 appearing before a roll with double sixes?

2.66 Consider again Example 2.11. It is now assumed that lots are drawn to determine the winner when there is a tie. Find the optimal strategy for player A and the corresponding win probability.

2.67 A carnival booth offers the following game of chance. Under each of six inverted cups is a colored ball, in some random order. The six balls are colored red, blue, yellow, orange, green, and purple. You get six tokens.

All you have to do is guess the color of the ball under each of the cups, where you handle one cup at a time. Every time you guess, you risk a token. If your guess is wrong, you lose the token. Each time you guess correctly, the ball is uncovered and you keep the token. Use conditional probabilities to obtain the probability of guessing all six balls before running out of tokens. *Hint*: Define $p(i,t)$ as your success probability once you have reached cup i with t tokens left and derive a recursion equation for $p(i,t)$.

2.68 A fair die is rolled repeatedly. Let p_n be the probability that the sum of scores will ever be n. Use the law of conditional probability to find a recursion equation for p_n. Verify numerically that p_n tends to $\frac{1}{3.5} = 0.2857$ as n gets large. Can you give an intuitive explanation of this result?

2.69 Use the recursive approach given in Example 2.10 to solve the following two longest-run problems.

(a) A sequence of independent trials is performed. Each trial results in a success with probability p and in a failure with probability $1-p$. What is the probability of getting a run of either at least r successes or at least r failures, or both, in N trials?

(b) A bowl contains r red balls and b blue balls. Balls are randomly picked out of the bowl, one by one and without replacement. What is the probability that the longest run of red balls will be L or more?

2.70 Each of seven dwarfs has his own bed in a common dormitory. Every night, they retire to bed one at a time, always in the same sequential order. On a particular evening, the youngest dwarf, who always retires first, has had too much to drink. He randomly chooses one of the seven beds to fall asleep on. As each of the other dwarfs retires, he chooses his own bed if it is not occupied, and otherwise randomly chooses another unoccupied bed. What is the probability that the kth dwarf can sleep in his own bed for $k = 1, \ldots, 7$? *Hint*: Use a general setup and define $p(k,n)$ as the probability that the kth dwarf will not sleep in his own bed for the situation of n dwarfs with the first dwarf being drunk. Use a conditioning argument to obtain a recursion equation for $p(k,n)$.

2.71 (difficult) Consider the three-player variant of Example 2.11. What is the optimal strategy for the first player A and what is the overall win probability of player A? Next use the probability distribution function of the final score of player A to find the overall win probabilities of the players B and C.

2.4 Bayes' Rule in Odds Form

A rule for revising (conditional) probabilities in light of new information is Bayes' rule. This rule, which is nothing other than logical thinking, is the queen of conditional probability. Bayes' rule can be stated in various forms, but the simplest is the so-called "odds form." The essence of Bayesian thinking is best understood by considering Bayes' rule for the situation where there is a question of a hypothesis being either true or false. An example of such a situation is a court case where the defendant is either guilty or not guilty. Let H represent the event that the hypothesis is true, and $\overline{H}$ the complementary event that the hypothesis is false. Before examining the evidence, a Bayesian analysis begins with assigning prior probabilities $P(H)$ and $P(\overline{H}) = 1 - P(H)$ to the mutually exclusive events H and $\overline{H}$. Prior probabilities are often based on subjective assessments. How do the prior probabilities change once evidence in the form of the knowledge that the event E has occurred becomes available? In our example of the court case, event E could be the evidence that the accused has the same blood type as the perpetrator's, whose blood has been found at the scene of the crime. The updated value of the probability that the hypothesis is true given that event E has occurred is denoted by $P(H \mid E)$. The calculation of the posterior probability $P(H \mid E)$ requires the rule of Bayes.

The rule of Bayes can be expressed in several ways. A convenient form uses odds. Odds are often used to represent probabilities. Gamblers usually think in terms of "odds" instead of probabilities. The *odds* of an event A are defined by

$$o_A = \frac{P(A)}{1 - P(A)}.$$

For example, an event A with $P(A) = \frac{2}{3}$ has odds 2. It is said that the odds are 2 to 1 (written 2:1) in favor of the event A. Conversely, odds o_A of an event A correspond to the probability

$$P(A) = \frac{o_A}{1 + o_A}.$$

The odds form of Bayes' rule is stated in the next rule.

Rule 2.3 *The posterior probability $P(H \mid E)$ satisfies*

$$\frac{P(H \mid E)}{P(\overline{H} \mid E)} = \frac{P(H)}{P(\overline{H})} \times \frac{P(E \mid H)}{P(E \mid \overline{H})}.$$

In words, Bayes' rule in odds form states that

$$\text{posterior odds} = \text{prior odds} \times \text{likelihood ratio}.$$

This insightful formula follows from first principles. Using the definitions $P(AB) = P(A)P(B \mid A)$ and $P(BA) = P(B)P(A \mid B)$, we get the basic form of the rule of Bayes:

$$P(A \mid B) = \frac{P(A) \times P(B \mid A)}{P(B)}.$$

This is one of the most beautiful formulas in probability theory. Next, Bayes' rule in odds form follows by noting that

$$P(H \mid E) = \frac{P(H)P(E \mid H)}{P(E)} \text{ and } P(\overline{H} \mid E) = \frac{P(\overline{H})P(E \mid \overline{H})}{P(E)}.$$

The factor $\frac{P(H)}{P(\overline{H})}$ represents the prior odds in favor of the hypothesis H before the evidence has been presented. The ratio of $P(E \mid H)$ and $P(E \mid \overline{H})$ is called the *likelihood ratio* or the *Bayes factor*. Bayes' rule updates the prior odds of the hypothesis H by multiplying them by the likelihood ratio and thus measures how much new evidence should alter a belief in a hypothesis. The likelihood ratio is the probability of finding the evidence if the hypothesis is true divided by the probability of finding the evidence if the hypothesis is not true. If the likelihood ratio is greater than one then the evidence supports the hypothesis, if the likelihood ratio is less than one it supports the negation of the hypothesis, and if the likelihood ratio equals one then the evidence favors neither. The likelihood ratio is a measure of the probative value of the evidence. In practical situations such as in judicial decision making, the likelihood ratio is typically determined by an expert. However, it is not the expert's task to tell the court what the prior odds are. The prior probability $P(H)$ represents the personal opinion of the court before the evidence is taken into account.

With two pieces of evidence E_1 and E_2 that are sequentially obtained, Bayes' rule can be applied iteratively. You could use the first piece of evidence to calculate initial posterior odds, and then use that posterior odds as new prior odds to calculate second posterior odds given the second piece of evidence. As an illustration, suppose that a closed box contains either three red balls or two red balls together with one blue ball. Initially you believe that the two possibilities are equally likely. Then a randomly chosen ball is removed from the box. It appears to be a red ball. What is your new belief in the hypothesis H that the box originally contained three red balls? The posterior odds are calculated as $\frac{1/2}{1/2} \times \frac{1}{2/3} = \frac{3}{2}$. Therefore, the probability $\frac{3/2}{1+3/2} = \frac{3}{5}$ represents your new belief in the hypothesis H. Next, a second ball is randomly taken out of the box. Again, it appears to be a red ball. Then the new posterior odds are calculated as $\frac{3/5}{2/5} \times \frac{1}{1/2} = 3$ and your belief in the hypothesis H is revised according to the probability $\frac{3}{1+3} = \frac{3}{4}$.

2.4 Bayes' Rule in Odds Form

In applying Bayes' rule in odds form, the main step is identifying the hypothesis H and the evidence E. Bayes' rule forces you to make transparent any implicit assumption you are making.

Before we give several applications of Bayes' rule, the following remark is made. In both legal and medical cases, the conditional probabilities $P(H \mid E)$ and $P(E \mid H)$ are sometimes confused with each other. As a classic example, suppose that at the scene of a crime blood is found that must belong to the offender. The blood type is found in approximately one in every hundred thousand people. In the police database a person is found who matches the blood type of the offender, but there is no other evidence against this person. The prosecutor argues: "the probability that the suspect would have the particular blood type is 0.001% if he were innocent. The suspect has the particular blood type and therefore the probability is only 0.001% that he is not guilty." Letting H be the event that the suspect is innocent and E be the event that the blood of the suspect matches the blood type of the offender, the prosecutor confuses the probability $P(H \mid E)$ – the relevant probability – with the probability $P(E \mid H)$. A famous example of the prosecutor's fallacy is the court case of People vs. Collins in Los Angeles in 1964. In this case, a couple matching the description of a couple that had committed an armed robbery was arrested. Based on expert testimony, the district attorney claimed that the frequency of couples matching the description was roughly 1 in 12 million. Although this was the estimate for the probability that an innocent couple matches the description, the district attorney treated this estimate as if it was the probability that a couple matching the description is innocent and incorrectly concluded that the couple was guilty beyond reasonable doubt. The prosecutor's fallacy had dramatic consequences in the case of Regina vs. Sally Clark in the UK in 1999. Sally Clark was convicted for murder because of the cot deaths of two of her newborn children within a period of one year. A revision of her process benefited from Bayesian arguments and led to her release in 2001.

Example 2.12 It is believed by a team of divers that a sought-after wreck will be in a certain sea area with probability $p = 0.4$. A search in that area will detect the wreck with probability $d = 0.9$ if it is there. What is the revised probability of the wreck being in the area when the area is searched and no wreck is found?

Solution. Let the hypothesis H be the event that the wreck is in the area in question and the evidence E be the event that the wreck has not been detected in that area. Before the outcome of the search, the events H and $\overline{H}$ have the subjective probabilities

$$P(H) = 0.4 \text{ and } P(\overline{H}) = 0.6.$$

Further, we have

$$P(E \mid H) = 0.1 \text{ and } P(E \mid \overline{H}) = 1.$$

Then, by Bayes' rule in odds form,

$$\frac{P(H \mid E)}{P(\overline{H} \mid E)} = \frac{0.4}{0.6} \times \frac{0.1}{1} = \frac{1}{15}.$$

The updated value of the subjective probability that the wreck is in the area is equal to $\frac{1/15}{1+1/15} = \frac{1}{16}$.

The above example is not far from reality. Bayesian analysis has guided successful searches in several real-life situations. In 1966, a mid-air collision over the Mediterranean between a KC-135 tanker and a B-52 bomber caused four hydrogen bombs to fall near the Spanish town of Palomares. Three of the bombs fell on land, but the fourth fell into the Mediterranean and was lost. In May 1968, a 3500-ton nuclear attack submarine, the USS Scorpion, vanished in the Atlantic Ocean with a crew of 99. No GPS was available for the search efforts at that time. In both cases the lost object at sea was recovered thanks to Bayesian search theory. In 2009, Air France flight 447 en route to Paris from Rio de Janeiro vanished during stormy weather over a remote part of the Atlantic Ocean. A Bayesian search procedure helped locate the flight's black box in 2011.[2]

Example 2.13 Suppose that there is a DNA test that determines with 100% accuracy whether or not a particular gene for a certain disease is present. A woman would like to do the DNA test, but wants to have the option of holding out hope that the gene is not present in her DNA even if it is determined that the gene for the illness is, indeed, present. She makes the following arrangement with her doctor. After the test, the doctor will toss a fair coin into the air, and will tell the woman the test results only if those results are negative and the coin has turned up heads. In every other case, the doctor will not tell her the test results. Suppose that there is a 1 in 100 probability that the woman does have the gene for the disease in question before she is tested. What is the revised value of this probability if the woman's doctor does not inform her of the test results?

Solution. Let the hypothesis H be the event that the woman has the particular gene and the evidence E be the event that she is not told the test results.

---

[2] Lawrence D. Stone et al., "Search for the wreckage of Air France flight AF 447," *Statistical Science* **59** (2014): 69–80.

The priors are $P(H) = 0.01$ and $P(\overline{H}) = 0.99$. Further, we have $P(E \mid H) = 1$ and $P(E \mid \overline{H}) = 0.5$. Then, by applying Bayes' rule in odds form, we get

$$\frac{P(H \mid E)}{P(\overline{H} \mid E)} = \frac{0.01}{0.99} \times \frac{1}{0.5} = \frac{1}{49.5}.$$

In the perception of the woman, the updated value of probability that she has the particular gene is equal to $\frac{1/49.5}{1+49/5} = \frac{2}{101}$ when the test is done under the deal with her doctor and she is not told the test results.

Example 2.14 A murder is committed. The perpetrator is either one or the other of the two persons X and Y. Both persons are on the run from the authorities, and after an initial investigation, both fugitives appear equally likely to be the perpetrator. Further investigation reveals that the actual perpetrator has blood type A. Ten percent of the population belongs to the group having this blood type. Additional inquiry reveals that person X has blood type A, but offers no information concerning the blood type of person Y. In light of this new information, what is the probability that person X is the perpetrator?

Solution. In answering this question, use H to denote the event that person X is the perpetrator. Let E represent the new evidence that person X has blood type A. Before the appearance of evidence E, the events H and $\overline{H}$ have probabilities $P(H) = P(\overline{H}) = \frac{1}{2}$. Further, we have that $P(E \mid H) = 1$ and $P(E \mid \overline{H}) = \frac{1}{10}$. By Bayes' rule in odds form,

$$\frac{P(H \mid E)}{P(\overline{H} \mid E)} = \frac{1/2}{1/2} \times \frac{1}{1/10} = 10.$$

The odds in favor, then, are 10 to 1 that person X is the perpetrator given that this person has blood type A. Otherwise stated, $P(H \mid E) = \frac{10}{11}$. The probability of Y being the perpetrator is $1 - \frac{10}{11} = \frac{1}{11}$ and not, as may be thought, $\frac{1}{10} \times \frac{1}{2} = \frac{1}{20}$. The error in this reasoning is that the probability of person Y having blood type A is not $\frac{1}{10}$ because Y is not a randomly chosen person; rather, Y is first of all a person having a 50% probability of being the perpetrator, whether or not he is found at a later time to have blood type A. The Bayesian analysis sharpens our intuition in a natural way.

Another nice illustration of Bayes' rule in odds form is provided by legal arguments used in the discussion of the O. J. Simpson trial.[3] In the analysis of this example, we will use an extension of Bayes' rule in odds form:

---

[3] This example is based on the article J. F. Merz and J. P. Caulkins, "Propensity to abuse–propensity to murder?," *Chance* **8** (1995): 14.

$$\frac{P(H \mid E_1 E_2)}{P(\overline{H} \mid E_1 E_2)} = \frac{P(H \mid E_1)}{P(\overline{H} \mid E_1)} \times \frac{P(E_2 \mid H E_1)}{P(E_2 \mid \overline{H} E_1)}.$$

To prove this extension of Bayes' rule, it suffices to verify the basic relation

$$P(A \mid BC) = \frac{P(A \mid C) P(B \mid AC)}{P(B \mid C)}$$

for any events A, B, and C. Then apply this basic relation with $(A = H, B = E_1, C = E_2)$ and with $(A = \overline{H}, B = E_1, C = E_2)$. By taking the quotient of $P(H \mid E_1 E_2)$ and $P(\overline{H} \mid E_1 E_2)$, the desired result follows next. The formula for $P(A \mid BC)$ is obtained as follows:

$$P(A \mid BC) = \frac{P(ABC)}{P(BC)} = \frac{P(AC) P(B \mid AC)}{P(BC)} = \frac{P(C) P(A \mid C) P(B \mid AC)}{P(C) P(B \mid C)}$$
$$= \frac{P(A \mid C) P(B \mid AC)}{P(B \mid C)}.$$

Example 2.15 Nicole Brown was murdered at her home in Los Angeles on the night of June 12, 1994. The prime suspect was her husband O. J. Simpson, at the time a well-known celebrity famous both as a television actor and a retired professional football star. This murder led to one of the most heavily publicized murder trials in the United States during the last century. The fact that the murder suspect had previously physically abused his wife played an important role in the trial. The celebrity lawyer Alan Dershowitz, a member of the team of lawyers defending the accused, tried to belittle the relevance of this fact by stating that only 0.1% of the men who physically abuse their wives actually end up murdering them. Was the fact that O. J. Simpson had previously physically abused his wife irrelevant to the case?

Solution. The answer to the question is no. In this particular court case it is important to make use of the crucial fact that Nicole Brown was murdered. The question, therefore, is not what the probability is that abuse leads to murder, but the probability that the husband is guilty in light of the fact that he had previously abused his wife. This probability can be estimated with the help of Bayes' formula and a few facts based on crime statistics. Define the following:

G = the event that the husband is guilty of the murder of his wife
M = the event that the wife has been murdered
E = the event that the husband has physically abused his wife in the past.

The probability in question is the conditional probability $P(G \mid EM)$. By the extension of Bayes' rule in odds form,

$$\frac{P(G \mid ME)}{P(\overline{G} \mid ME)} = \frac{P(G \mid M)}{P(\overline{G} \mid M)} \times \frac{P(E \mid GM)}{P(E \mid \overline{G}M)},$$

where $\overline{G}$ represents the event that the husband is not guilty of the murder of his wife. How do we estimate the conditional probabilities on the right-hand side of this formula? In 1992, 4,936 women were murdered in the United States, of which roughly 1,430 were murdered by their (ex-)husbands or boyfriends. This results in an estimate of $\frac{1,430}{4,936} = 0.29$ for the prior probability $P(G \mid M)$ and an estimate of 0.71 for the prior probability $P(\overline{G} \mid M)$. Furthermore, it is also known that roughly 5% of married women in the United States have at some point been physically abused by their husbands. If we assume that a woman who has been murdered by someone other than her husband had the same chance of being abused by her husband as a randomly selected woman, then the probability $P(E \mid \overline{G}M)$ is equal to 5%. Our estimate of the probability $P(E \mid GM)$ is based on the reported remarks made by Simpson's defense attorney, Alan Dershowitz, in a newspaper article. In this article, Dershowitz admitted that a substantial percentage of the husbands who murder their wives have, previous to the murders, also physically abused their wives. Given this statement, the probability $P(E \mid GM)$ will be taken to be 0.5. By substituting the various estimated values for the probabilities into Bayes' formula in odds form, we find that

$$\frac{P(G \mid ME)}{P(\overline{G} \mid ME)} = \frac{0.29}{0.71} \times \frac{0.5}{0.05} = 4.08.$$

By $P(\overline{G} \mid ME) = 1 - P(G \mid ME)$, we can convert the odds to a probability. This results in $P(G \mid ME) = 0.81$. In words, there is an estimated probability of 81% that the husband is the murderer of his wife in light of the knowledge that he had previously physically abused her. The fact that O. J. Simpson had physically abused his wife in the past was therefore certainly very relevant to the case.

The next example involves a subtle application of Bayes' rule in odds form.

Example 2.16 A diamond merchant has lost a case containing a very expensive diamond somewhere in a large city in an isolated area. The case has been found again but the diamond has vanished. However, the empty case contains the DNA of the person who took the diamond. The city has 150,000 inhabitants, and each is considered a suspect in the diamond theft. An expert declares that the probability of a randomly chosen person matching the DNA profile is 10^{-6}. The police search a database with 5,120 DNA profiles and find one person matching the DNA from the case. Apart from the DNA evidence, there is no additional background evidence related to the suspect. On the basis of the extreme infrequency of the DNA profile and the fact that the population of potential finders of the diamond is only 150,000 people, the prosecutor jumps to the conclusion that the odds of the suspect not being the thief are practically nil and calls for a tough sentence. What do you think of this conclusion?

Solution. The conclusion made by the prosecutor could not be more wrong. The prosecutor argues: "The probability that a person chosen at random would match the DNA profile found on the diamond case is negligible and the number of inhabitants of the city is not very large. The suspect matches this DNA profile, thus it is nearly one hundred percent certain that he is the perpetrator." This is a textbook example of the faulty use of probabilities. The probability that the suspect is innocent of the crime is generally quite different from the probability that a randomly chosen person matches the DNA profile in question. What we are actually looking for is the probability that among all persons matching the DNA profile in question, the arrested person is the perpetrator. Counsel for defense could reason as follows to estimate this probability: "We know that the suspect has matching DNA, but among the other $150{,}000 - 5{,}120 = 144{,}880$ individuals the expected number of people matching the DNA profile is $144{,}880 \times 10^{-6} = 0.14488$. So the probability that the suspect is guilty is $1/(1 + 0.14488) = 0.8735$. It is not beyond reasonable doubt that the suspect is guilty and thus the suspect must be released."

The reasoning of the defense is on the whole correct and can be supported by Bayes' rule. Let us perform the analysis under the assumption that each of the 150,000 inhabitants of the city is equally likely to be the finder of the diamond and that the finder keeps the diamond with probability p_0 when the finder is a person in the database and with probability p_1 otherwise. Let

$$r = \frac{p_0}{p_1}.$$

It will appear that only the value of the ratio r is required and thus not the values of p_0 and p_1. The following formula will be derived:

$$P(\text{suspect is the thief}) = \frac{r}{r + (n - n_0)\lambda},$$

where $n = 150{,}000$, $n_0 = 5{,}120$, and $\lambda = 10^{-6}$ denotes the probability that a randomly chosen person matches the DNA profile in question. This posterior probability has the value 0.8735 for $r = 1$ and the value 0.9325 for $r = 2$. This result confirms that the counsel for defense is correct in stating that it is not beyond reasonable doubt that the suspect is guilty, so the suspect must be released if there is no additional proof. The derivation of the above formula uses again the extension of Bayes' rule and proceeds as follows. Define the following events:

H = the hypothesis that the thief is a person from the database
E_1 = the evidence that the thief has the observed DNA profile
E_2 = the evidence that only one person in the database has the observed DNA profile.

The subtlety of the Bayesian analysis lies in the specification of H, E_1, and E_2. With this specification, the probability that the suspect is the thief is given by $P(H \mid E_1 E_2)$ (verify!). Since $P(H)$ is just the conditional probability that the diamond has been found and kept by a person from the database given that the diamond was not brought back, we have

$$P(H) = \frac{(n_0/n)p_0}{(n_0/n)p_0 + ((n-n_0)/n)p_1} \text{ and } P(\overline{H}) = \frac{((n-n_0)/n)p_1}{(n_0/n)p_0 + ((n-n_0)/n)p_1}.$$

This shows that the prior odds of H are given by

$$\frac{P(H)}{P(\overline{H})} = r \frac{n_0}{n - n_0}.$$

Using the fact that $P(E_1 \mid H) = P(E_1 \mid \overline{H}) = \lambda$, it follows from Bayes' rule in odds form that

$$\frac{P(H \mid E_1)}{P(\overline{H} \mid E_1)} = \frac{P(H)}{P(\overline{H})}.$$

In other words, the evidence E_1 does not affect the ratio of the prior probabilities $P(H)$ and $P(\overline{H})$. Next we use the evidence E_2. Noting that

$$P(E_2 \mid HE_1) = (1-\lambda)^{n_0 - 1} \text{ and } P(E_2 \mid \overline{H}E_1) = n_0 \lambda (1-\lambda)^{n_0 - 1},$$

a second application of Bayes' rule in odds form gives that

$$\frac{P(H \mid E_1 E_2)}{P(\overline{H} \mid E_1 E_2)} = \frac{P(H \mid E_1)}{P(\overline{H} \mid E_1)} \times \frac{P(E_2 \mid HE_1)}{P(E_2 \mid \overline{H}E_1)} = r \frac{n_0}{n - n_0} \times \frac{1}{n_0 \lambda}.$$

This leads to $P(H \mid E_1 E_2) = r/(r + (n-n_0)\lambda)$, as was to be shown.

Problems

2.72 In a binary transmission channel, a 1 is transmitted with probability 0.8 and a 0 with probability 0.2. The conditional probability of receiving a 1 given that a 1 was sent is 0.95, the conditional probability of receiving a 0 when a 0 was sent is 0.99. What is the probability that a 1 was sent when receiving a 1? Use Bayes' rule in odds form to answer this question.

2.73 An oil explorer performs a seismic test to determine whether oil is likely to be found in a certain area. The probability that the test indicates the presence of oil is 90% if oil is indeed present in the test area, while the probability of a false positive is 15% if no oil is present in the test area. Before the test is done, the explorer believes that the probability of the presence of oil in the test area is 40%. Use Bayes' rule in odds form to revise the value of the probability of oil being present in the test area given that the test gives a positive signal.

2.74 In a certain region, it rains on average once every 10 days during the summer. Rain is predicted on average for 85% of the days when rainfall actually occurs, while rain is predicted on average for 25% of the days when it does not rain. Assume that rain is predicted for tomorrow. What is the probability of rainfall actually occurring on that day?

2.75 You have five coins colored red, blue, white, green, and yellow. Apart from the variation in color, the coins look identical. One of the coins is unfair and when tossed it comes up heads with probability $\frac{3}{4}$; the other four are fair coins. You have no further information about the coins, apart from having observed that the blue coin, tossed three times, came up heads on all three tosses. On the grounds of this observation, you indicate that the blue coin is the unfair one. What is the probability of your being correct in this assumption?

2.76 The final match of world championship soccer is to be played between England and the Netherlands. The star player for the Dutch team, Dennis Nightmare, has been injured. The probability of his being fit enough to play in the final is being estimated at 75%. Pre-game predictions have estimated that, without Nightmare, the probability of a Dutch win is 30% and with Nightmare, 50%. Later, you hear that the Dutch team has won the match. Without having any other information about the events that occurred, what would you say was the probability that Dennis Nightmare played in the final?

2.77 A family is chosen at random from all two-child families without twins. Each child is a boy or a girl with equal probability. What is the probability that the chosen family has two boys if the family has a boy among the two children? What is the answer to this question if, among the two children, the family has a boy born on a Tuesday? What is the probability that the chosen family has two boys if, among the two children, the family has a boy born on one of the first k days of the week? In each of the three cases, it is assumed that the information was obtained by asking the father "Yes or no, do you have a son/ a son born on Tuesday/ a son born on the first k days of the week?" and getting a truthful answer. *Note*: The wording of the boy–girl problem is of utmost importance. If the formulation of the boy–girl problem is ambiguous and leaves room for different interpretations, then there are several answers that are right for different background assumptions.

2.78 On the island of liars, each inhabitant lies with probability $\frac{2}{3}$. You overhear an inhabitant making a statement. Next you ask another inhabitant whether the inhabitant you overheard spoke truthfully. What is the probability that the inhabitant you overheard indeed spoke truthfully, given that the other inhabitant says so?

2.79 Suppose that a person has confessed to a crime under a certain amount of police pressure. Besides this confession, there is no other proof that the person is guilty. Use Bayes' rule in odds form to verify that the confession only adds to the evidence of guilt if the confession is more likely to come from the guilty than from the innocent. Do you think that in real life a hardened person who is guilty is more likely to confess than an unstable person who is innocent?

2.80 An opaque bowl contains one ball. The ball is equally likely to be red or blue. A red ball is added to the bowl. Then a ball is randomly picked from the bowl. The ball that has been picked from the bowl turns out to be red. Use Bayes' rule in odds form to calculate the probability that the bowl originally contained a red ball.

2.81 A doctor discovers a lump in a woman's breast during a routine physical examination. The lump could be a cancer. Without performing any further tests, the probability that the woman has breast cancer is 0.01. A further test can be done. On average, this test is able to establish correctly whether a tumor is benign or cancerous 90% of the time. A positive test result indicates that a tumor is cancerous. What is the probability that the woman has breast cancer if the test result is positive?

2.82 On average, one in every 125 births produces a set of fraternal twins and one in every 300 births produces a set of identical twins (identical twins are always of the same sex but fraternal twins are random). Elvis had a twin brother, Jesse Garon, who died at birth. What is the probability that Elvis was an identical twin?

2.83 A box contains 10,000 coins. One of the coins has heads on both sides but all the other coins are fair coins. You choose at random one of the coins. Use Bayes' rule in odds form to find the probability that you have chosen the two-headed coin given that the first n tosses have all resulted in heads. What are the numerical values of the posterior probability for $n = 10$, 15, and 25?

2.5 Bayesian Inference – Discrete Case

Bayesian inference is historically the original approach to statistics, predating what is nowadays called classical statistics by a century. The foundation of

Bayesian inference was laid by the English clergyman Thomas Bayes (1702–1761), but the Bayesian interpretation of probability was developed mainly by Pierre Simon Laplace (1749–1827). Astronomers have contributed much to the Bayesian approach. Astronomers cannot do experiments on the universe and thus have to make probabilistic inferences from evidence left behind. This is very much the same situation as in forensic science, where Bayesian inference also plays a very important role. In Bayesian inference one typically deals with nonrepeatable chance experiments.

Fundamentally, the distinction between Bayesian inference and frequentist inference in classical statistics concerns the interpretation of probability. To illustrate, suppose that an astronomer wants to make a statement about the photon flux of a nonvariable star. In the frequentist view, the true flux is a single fixed value and a statistical estimate of this unknown value can only be based on many measurements of the flux. In a Bayesian view, one can meaningfully treat the flux of the star as a random variable and talk about the probability that the true value of the flux of the star is between specific bounds. That probability is subjective and represents one's knowledge of the value based on prior information and available data. The fundamental difference between the classical and the Bayesian approach can also be illustrated nicely with the next example. Imagine a multiple-choice exam consisting of 50 questions, each of which has three possible answers. A student receives a pass grade if he/she correctly answers more than half of the questions. Take the case of a student who manages to answer 26 of the 50 questions correctly and claims not to have studied, but rather to have obtained 26 correct answers merely by guessing. The classical approach in statistics is to calculate the probability of correctly answering 26 *or more* of the 50 questions by luck alone. This excess probability is equal to 0.0049 and is called the *p*-value in classical statistics. On the basis of the small *p*-value one might conclude that the student is bluffing and in fact did prepare for the exam. However, the *p*-value does not give the probability that the student went unprepared to the exam, but only gives the probability of getting 26 or more correct answers when the student did not prepare for the exam. In reality, one might have information about the earlier achievements of the student in question. The Bayesian approach does use this information and then calculates the probability that the student did not prepare for the exam given that he/she answered 26 of the 50 questions correctly. This is the probability we are actually looking for! The Bayesian approach requires that we first specify a prior distribution for the various ways the student may have prepared for the exam. This distribution concerns the situation *before* the exam and may be based on information of the student's earlier academic performance in homework or previous exams.

2.5 Bayesian Inference – Discrete Case

The Bayesian approach to statistical inference treats population parameters not as fixed, unknown constants but as random variables – subject to change as additional data arise. Probability distributions have to be assigned to these parameters by the investigator before observing data. These distributions are called *prior distributions* and are inherently subjective. They represent the investigator's uncertainty about the true value of the unknown parameters. Assigning probabilities by degree of belief is consistent with the idea of a fixed but unknown value of the parameter. Treating the unknown parameter as a random variable does not mean that we believe the parameter is random. Rather, it expresses the state of our knowledge about the parameter. In the Bayesian approach we revise our beliefs about the population parameters by learning from data that are obtained after having specified the prior distribution. The revised distribution of the unknown parameter is called the *posterior distribution*. It is obtained from Bayes' rule and reflects our new beliefs about the population parameters. An illuminating illustration of thinking about the parameter of interest as a random state of nature is provided by the following example. Suppose that you have three symmetric dice in your pocket. Dice 1 has hidden the numbers 1, 1, 2, 2, 2, 2 on its six faces, dice 2 the numbers 1, 1, 1, 2, 2, 2, and dice 3 the numbers 1, 1, 1, 1, 2, 2. You randomly select one die from your pocket without looking. Somebody else rolls the die once and truthfully informs you that a 1 has shown up. What is the probability that you have chosen die i for $i = 1, 2,$ and 3? Before the die is rolled, the state of nature of the chosen die is described by the prior distribution $\left(\frac{1}{3}, \frac{1}{3}, \frac{1}{3}\right)$. After the die has been rolled, the state of nature of the chosen die is described by the posterior distribution $\left(\frac{2}{9}, \frac{3}{9}, \frac{4}{9}\right)$, as the reader may verify by using Bayes' rule in odds form (or the Bayesian formula given in Example 2.17 below).

As said above, Bayesian analysis has its origin in the work of Reverend Thomas Bayes, in whose *Essay toward solving a problem in the doctrine of chance*, published posthumously in 1763, an early attempt is made to establish what we now refer to as Bayes' rule. Although the method of Bayes was enthusiastically taken up by Pierre Simon Laplace and other leading probabilists in the nineteenth century, it was not used much until the late 1980s and early 1990s, when powerful computers became widely accessible and new computational methods were developed. Nowadays, Bayesian methods are used widely to address pressing questions in diverse application areas such as astrophysics, neurobiology, weather forecasting, prediction of election results, spam filtering, and criminal justice. In many cases the Bayesian approach produces answers that are more logical and easier to understand than those produced by the frequentist approach of classical statistics.

Example 2.17 A new treatment is tried out for a disease for which the historical success rate of the standard treatment is 35%. The discrete uniform probability distribution on $0, 0.01, \ldots, 0.99, 1$ is taken as prior for the success probability of the new treatment. The experimental design is to make exactly 10 observations by treating 10 patients. The experimental study yields seven successes and three failures. What is the posterior probability that the new treatment is more effective than the standard treatment?[4]

Solution. Model the unknown success probability of the new treatment by the random variable Θ. Assume that our state of knowledge about the unknown parameter is expressed by the "non-informative" prior distribution $p_0(\theta) = \frac{1}{101}$ for $\theta = 0, 0.01, \ldots, 0.99, 1$. To update the prior distribution given the observed data, we need the so-called likelihood function $L(\text{data} \mid \theta)$. This function is defined as the probability of getting the data given that the success probability has value θ. In the particular situation of seven successes in the treatment of 10 patients,

$$L(\text{data} \mid \theta) = \binom{10}{7}\theta^7(1-\theta)^3 \quad \text{for } \theta = 0, 0.01, \ldots, 0.99, 1.$$

To find the posterior probability $p(\theta) = P(\Theta = \theta \mid \text{data})$, we use

$$P(\Theta = \theta \mid \text{data}) = \frac{P(\text{data} \mid \theta)p_0(\theta)}{P(\text{data})}.$$

By the law of conditional probability, $P(\text{data}) = \sum_\theta P(\text{data} \mid \theta)p_0(\theta)$. Thus, using the notation $L(\text{data} \mid \theta)$ for $P(\text{data} \mid \theta)$, the posterior distribution is given by the Bayesian formula

$$p(\theta) = \frac{L(\text{data} \mid \theta)p_0(\theta)}{\sum_{\theta'} L(\text{data} \mid \theta')p_0(\theta')} \quad \text{for } \theta = 0, 0.01, \ldots, 0.99, 1.$$

The *posterior* $p(\theta)$ is proportional to *prior* $p_0(\theta) \times$ *likelihood* $L(\text{data} \mid \theta)$. Letting $\theta_i = \frac{i}{100}$ and inserting the formulas for $L(\text{data} \mid \theta)$ and $p_0(\theta)$, we get

$$p(\theta_i) = \frac{\theta_i^7(1-\theta_i)^3}{\sum_{k=0}^{100} \theta_k^7(1-\theta_k)^3} \quad \text{for } i = 0, 1, \ldots, 100.$$

In particular, the posterior probability of the new treatment being more effective than the standard treatment is given by

$$\sum_{i=36}^{100} p(\theta_i) = 0.9866.$$

---

[4] This example is based on the paper by D. A. Berry, "Bayesian clinical trials," *Nature Reviews Drug Discovery* **5** (2006): 27–36.

2.5 Bayesian Inference – Discrete Case

The posterior probability of the new treatment not being more effective than the standard treatment is 0.0134. In this particular example, the value 0.0134 of the posterior probability is not very different from the value 0.0260 of the excess probability of obtaining seven or more successes in 10 trials under the null hypothesis that the new treatment causes no difference and thus has success probability 0.35. In classical statistics, this so-called *p*-value would have been calculated and the hypothesis that the new treatment causes no difference would also have been rejected.

In general, *p*-values overestimate the evidence against the null hypothesis. To illustrate this, consider the following experiment. Suppose that there is reason to believe that a coin might be slightly biased toward heads. To test this, you decide to throw the coin 1,000 times. Before performing the experiment, you express your uncertainty about the unbiasedness of the coin by assuming that the probability of getting heads in a single toss of the coin can take on the values $\theta = 0.50, 0.51$, and 0.52 with respective prior probabilities $p_0(\theta) = \frac{1}{2}$, $\frac{1}{3}$, and $\frac{1}{6}$. Next the experiment is performed and 541 heads are obtained in 1,000 tosses of the coin. The likelihood of getting 541 heads in 1,000 tosses is $L(541 \mid \theta) = \binom{1,000}{541} \theta^{541}(1-\theta)^{459}$ for $\theta = 0.50, 0.51$, and 0.52. This leads to the posterior probability $p(\theta \mid 541) = 0.1282$ for $\theta = 0.50$. In other words, your posterior belief that the coin is fair equals 0.1282. In classical statistics, one would compute the probability of getting 541 or more heads in 1,000 tosses of the coin under the hypothesis that the coin is fair. This excess probability is equal to 0.0052. Many classical statisticians would consider this small *p*-value as significant evidence that the coin is biased toward heads. However, your subjective Bayesian probability of 0.1282 for the hypothesis of a fair coin is not strong enough evidence for such a conclusion. The difference in the conclusions can be explained as follows. The *p*-value is based on the set of all possible observations that cast as much or more doubt on the hypothesis as the actual observations do. It is not possible to base the *p*-value only on the actual data, because it frequently happens that all individual outcomes have such small probabilities that every outcome would look significant. The inclusion of unobserved data means that the resulting *p*-value may greatly exaggerate the strength of evidence against the hypothesis.[5] The *p*-value, which is the probability of getting data at least as extreme as the observed data when the null hypothesis is true, is

---

[5] That's why physicists required an extremely small *p*-value of about 1 in 3.5 million (the 5-sigma rule) before declaring the "discovery" of the long-sought Higgs boson. The evidence of the efficacy of the famous Salk vaccine against polio was also based on an extremely small *p*-value, see Problem 3.65. It is true that *p*-values are overused and often misused, but they were the right tool in the search for the Higgs boson and in showing the efficacy of the Salk vaccine.

often interpreted incorrectly as the probability that the null hypothesis is true. The Bayesian posterior probability does have the desired probabilistic interpretation.

In the Bayesian approach, only the actual data obtained matter. However, in order to give a Bayesian conclusion about the truth of the hypothesis using only the actual data, you must first choose a prior distribution for the truth of the hypothesis. Assuming a "non-informative" prior such as the uniform distribution, the results of the trials carry essentially all the influence in the posterior distribution. In practice, it happens in many cases that the prior can be based on earlier experimental studies or scientific theory.

The Bayesian approach has the appealing property that you can continuously update your beliefs as information accrues. In many situations data arrive sequentially and updating one's beliefs is a natural process. Think of updating sequential polling results to predict the outcome in elections. In Problem 2.86 dealing with Example 2.17, you are asked to update your beliefs after each trial, using the posterior obtained after the preceding trials as prior. This gives the same result as when the update is based on the aggregate observations. The Bayesian approach gives the user more flexibility in the experimental design. Whereas the frequentist approach requires trials to reach a prespecified sample size before stopping, the Bayesian approach allows the user to stop trials early when adding more observations will not appreciably change the conclusion. Bayesian inference is immune to the selection bias associated with early termination of an experiment.

Example 2.18 Two candidates A and B are contesting the election of governor in a given state. The candidate who wins the popular vote becomes governor. Suppose that, based on earlier polls, the prior distribution of the fraction of the voting population in favor of candidate A is $p_0(\theta) = \frac{\theta - 0.29}{4.41}$ for $\theta = 0.30, \ldots, 0.50$ and $p_0(\theta) = \frac{0.71 - \theta}{4.41}$ for $\theta = 0.51, \ldots, 0.70$. A random sample of the voting population is undertaken to find out the current preference of the voters. The sample size of the poll is 1,000. It turns out that 517 of the polled voters favor candidate A. What is the new value of the probability that candidate A will win the election?

Solution. The prior distribution is a discrete triangular distribution with expected value 0.50. In particular, the prior probability of candidate A getting the majority of the votes at the election is $p_0(0.51) + \cdots + p_0(0.70) = 0.476$. The poll gives that 517 of the 1,000 polled voters favor candidate A. In light of this new information, the updated probability of candidate A getting the majority of the votes at the time of election is $p(0.51) + \cdots + p(0.70)$, where $p(\theta)$ is the posterior probability that the fraction of the voting population

2.5 Bayesian Inference — Discrete Case

in favor of candidate A equals θ. Using the likelihood $L(\text{data} \mid \theta) = \binom{1{,}000}{517}\theta^{517}(1-\theta)^{483}$, we get the posterior distribution

$$p(\theta) = \frac{\binom{1{,}000}{517}\theta^{517}(1-\theta)^{483}p_0(\theta)}{\sum_{a=30}^{70}\binom{1{,}000}{517}\left(\frac{a}{100}\right)^{517}\left(1-\frac{a}{100}\right)^{483}p_0\left(\frac{a}{100}\right)}.$$

Performing the numerical calculations, we find that the posterior probability of candidate A getting the majority of the votes at the election equals $p(0.51) + \cdots + p(0.70) = 0.763$. Let's see what the posterior probability of candidate A winning the election is when the prior $p_0(\theta) = \frac{c}{1-\theta^2}$ for $\theta = 0.30, 0.31, \ldots, 0.70$ is used, where the normalizing constant $c = 0.0174484$. Note that the shape of this prior distribution having expected value 0.520 is much different from the shape of the triangular prior distribution. We then find the value 0.786 for the posterior probability of candidate A winning the election.

Problems

2.84 Your friend has fabricated a loaded die. In doing so, he has chosen at random one of the values 0.1, 0.2, 0.3, or 0.4 and has loaded the die in such a way that any roll of the die results in the outcome 6 with a probability that is equal to the randomly chosen value. He does not tell you what the chosen value is. You ask him to roll the die 300 times and to inform you how often the outcome 6 has appeared. You are informed that the outcome 6 has appeared 75 times. What is the posterior distribution of the probability that a single roll of the die gives a 6?

2.85 You wonder who is the better player of the tennis players Alassi and Bicker. Your prior assigns equal probabilities to the possible values 0.4, 0.5, and 0.6 for the probability that Alassi wins a given match. Then you learn about a tournament at which a best-of-five series of matches is played between Alassi and Bicker over a number of days. In such an encounter the first player to win three matches is the overall winner. It turns out that Alassi wins the best-of-five contest. How should you update your prior? *Note*: This problem is taken from J. Albert, "Teaching Bayes' rule: a data-oriented approach," *The American Statistician* **51** (1997): 247–253.

2.86 Consider again Example 2.17. Assume that the 10 observations for the new treatment are *SSFSSFSSSF*, where *S* stands for success and *F* for failure. Update the posterior probability of the new treatment being more effective than the standard treatment after each observation in order to see how this probability changes when additional information becomes available.

2.87 A student has passed a final exam by supplying correct answers for 26 out of 50 multiple-choice questions. For each question, there was a choice of three possible answers, of which only one was correct. The student claims not to have learned anything in the course and not to have studied for the exam, and says that his correct answers are the product of guesswork. Do you believe him when you have the following prior information? There are three ways the student may have done the exam: totally unprepared, half prepared, and well prepared. The three possibilities have the respective prior probabilities 0.2, 0.3, and 0.5. Any of the 50 questions is correctly answered by the student with probability $\frac{1}{3}$ if he is totally unprepared, with probability 0.45 if he is half prepared, and with probability 0.8 if he is well prepared.

2.88 Your friend is a basketball player. To find out how good he is in free throws, you ask him to shoot 10 throws. You assume the three possible values 0.25, 0.50, and 0.75 for the success probability of the free shots of your friend. Before the 10 throws are shot, you believe that these three values have the respective probabilities 0.2, 0.6, and 0.2. What is the posterior distribution of the success probability given that your friend scores 7 times out of the 10 throws?

3
Discrete Random Variables

In performing a chance experiment, one is often not interested in the particular outcome that occurs but in a specific numerical value associated with that outcome. Any function that assigns a real number to each outcome in the sample space of the experiment is called a random variable. Intuitively, a random variable takes on its value by chance. The observed value, or realization, of a random variable is completely determined by the realized outcome of the chance experiment and consequently probabilities can be assigned to the possible values of the random variable.

The first purpose of this chapter is to familiarize you with the concept of a random variable and with characteristics such as the expected value and the variance of a random variable. In addition, we give rules for the expected value and the variance of a sum of random variables, including the square-root rule. These rules are easiest explained and understood in the context of discrete random variables. Such random variables can take on only a finite or countably infinite number of values (the so-called continuous random variables that can take on a continuum of values are treated in the next chapter). The second purpose of the chapter is to introduce you to important discrete random variables such as the binomial, Poisson, hypergeometric, geometric, and negative binomial random variables among others. A comprehensive discussion of these random variables is given, together with appealing examples. Much attention is given to the Poisson distribution, which is the most important discrete distribution. Unlike most other introductory texts, we consider at length the practically useful Poisson heuristic for weakly dependent trials.

3.1 Concept of a Random Variable

The concept of a random variable is always difficult for the beginner. Intuitively, a random variable is a function that takes on its value by chance.

A random variable is not a variable in the traditional sense of the word and actually it is a little misleading to call it a variable. Formally, a *random variable* is defined as a real-valued function on the sample space of a chance experiment. A random variable X assigns a numerical value $X(\omega)$ to each element ω of the sample space. For example, suppose that the random variable X is defined as the smallest of the two numbers that will result from the experiment of rolling a fair die twice. Then the random variable X assigns the numerical value $X(\omega) = \min(i,j)$ to the outcome $\omega = (i,j)$ of the chance experiment, where i is the number obtained in the first roll and j is the number obtained in the second roll. As said before, a random variable X takes on its value by chance. A random variable gets its value only *after* the underlying chance experiment has been performed. *Before* the experiment is performed, you can only speak of the probability that the random variable will take on a specific value (or a value in a specific interval for the case that the range of the random variable is some continuum of real numbers). It is common to use uppercase letters such as X, Y, and Z to denote random variables, and lowercase letters x, y, and z to denote their possible numerical values.

Illustrative examples of random variables are:

- The number of winners in a football pool next week.
- The number of major hurricanes that will hit the United States next year.
- The number of claims that will be submitted to an insurance company next week.
- The amount of rainfall that the city of London will receive next year.
- The lifetime of a newly bought battery.
- The duration of your next mobile phone call.

The first three examples are examples of discrete random variables taking on a discrete number of values and the other three examples describe continuous random variables taking on a continuum of values.

In this chapter we consider only discrete random variables. A random variable X is said to be *discrete* if its set of possible values is finite or countably infinite. The set of possible values of X is called the *range* of X and is denoted by I. The probabilities associated with these possible values are determined by the probability measure P on the sample space of the chance experiment. The *probability mass function* of a discrete random variable X is defined by $P(X = x)$ for $x \in I$, where the notation $P(X = x)$ is shorthand for

$$P(X = x) = P(\{\omega : X(\omega) = x\}).$$

In words, $P(X = x)$ is the probability mass assigned by the probability measure P to the set of all outcomes ω for which $X(\omega) = x$.

As an example, suppose that you are going to roll a fair die twice. The sample space of this experiment consists of 36 equally likely outcomes (i, j), where i is the number obtained in the first roll and j is the number obtained in the second roll. Let the random variable X be defined as the smallest of the two numbers that will result from this experiment. The random variable X assigns the numerical value $\min(i, j)$ to the outcome (i, j) of the sample space. Thus, the range of X is the set $I = \{1, 2, \ldots, 6\}$. The random variable X takes on the value 1 if one of the 11 outcomes $(1, 1), (1, 2), \ldots, (1, 6), (2, 1), (3, 1), \ldots, (6, 1)$ occurs and so $P(X = 1) = \frac{11}{36}$. In the same way, $P(X = 2) = \frac{9}{36}$, $P(X = 3) = \frac{7}{36}$, $P(X = 4) = \frac{5}{36}$, $P(X = 5) = \frac{3}{36}$, and $P(X = 6) = \frac{1}{36}$. As another example, suppose that the random variable S is defined as the sum of the two rolls. Then S assigns the numerical value $i + j$ to the outcome (i, j) of the sample space and so the range of S is the set $I = \{2, 3, \ldots, 12\}$. It is left to the reader to verify that the probability mass function of S is given by $P(S = 2) = P(S = 12) = \frac{1}{36}$, $P(S = 3) = P(S = 11) = \frac{2}{36}$, $P(S = 4) = P(S = 10) = \frac{3}{36}$, $P(S = 5) = P(S = 9) = \frac{4}{36}$, $P(S = 6) = P(S = 8) = \frac{5}{36}$, and $P(S = 7) = \frac{6}{36}$.

Example 3.1 In your pocket you have three dimes (coins of 10 cents) and two quarters (coins of 25 cents). You grab at random two coins from your pocket. What is the probability mass function of the amount of money you will grab?

Solution. The sample space of the chance experiment is chosen as $\Omega = \{(D, D), (D, Q), (Q, D), (Q, Q)\}$. The outcome (D, D) occurs if the first coin taken is a dime and the second one is also a dime, the outcome (D, Q) occurs if the first coin taken is a dime and the second one is a quarter, etc. The probability $\frac{3}{5} \times \frac{2}{4} = \frac{3}{10}$ is assigned to the outcome (D, D), the probability $\frac{3}{5} \times \frac{2}{4} = \frac{3}{10}$ to the outcome (D, Q), the probability $\frac{2}{5} \times \frac{3}{4} = \frac{3}{10}$ to the outcome (Q, D), and the probability $\frac{2}{5} \times \frac{1}{4} = \frac{1}{10}$ to the outcome (Q, Q). Let the random variable X denote the total number of cents you will grab. The random variable X has 20, 35, and 50 as possible values. The random variable X takes on the value 20 if the outcome (D, D) occurs, the value 35 if either the outcome (D, Q) or (Q, D) occurs, and the value 50 if the outcome (Q, Q) occurs. Thus, the probability mass function of X is

$$P(X = 20) = \frac{3}{10}, \ P(X = 35) = \frac{3}{10} + \frac{3}{10} = \frac{3}{5}, \text{ and } P(X = 50) = \frac{1}{10}.$$

Example 3.2 You are going to remove cards one at a time from a well-shuffled standard deck of 52 cards until you get an ace. Let the random variable X be the number of cards that must be removed. What is the probability mass function of X?

Solution. The range of the random variable X is the set $\{1, 2, \ldots, 49\}$. Obviously, $P(X = 1) = \frac{4}{52}$. Using the chain rule for conditional probabilities, we find for $k = 2, \ldots, 49$:

$$P(X = k) = \frac{48}{52} \times \cdots \times \frac{48 - (k-2)}{52 - (k-2)} \times \frac{4}{52 - (k-1)}.$$

Alternatively, $P(X = k)$ can be calculated as $\left[\binom{48}{k-1}/\binom{52}{k-1}\right] \times \frac{4}{52-(k-1)}$. This representation is based on $P(A \mid B) = P(A)P(B \mid A)$ and can be explained as the probability that the first $k - 1$ picks are non-aces multiplied by the conditional probability that the kth pick is an ace given that the first $k - 1$ picks are non-aces.

The last example gives rise to the following important observation. Often, an explicit listing of the sample space is not necessary to assign a probability distribution to a random variable. Usually, the probability distribution of a random variable is modeled without worrying about the assignment of probability to an underlying sample space. In most problems, you will perform probability calculations without explicitly specifying a sample space; an assignment of probabilities to properly chosen events usually suffices.

Problems

3.1 As an added incentive for students to do homework for their probability course, the professor randomly picks two students at each class meeting. These two students are asked to explain a homework problem. On a particular day, five students from the class of 20 students have not done the homework. Let the random variable X be the number of students picked who have not done the homework. What is the probability mass function of X?

3.2 Imagine that people enter a room one by one and announce their birthdays. Let the random variable X be the number of people required to have a matching birthday. What is the probability mass function of X?

3.3 A bag contains three coins. One coin is two-headed and the other two are normal. A coin is chosen at random from the bag and is tossed twice. Let the random variable X denote the number of heads that will appear. What is the probability mass function of X?

3.4 In a charity lottery, 1,000 tickets numbered $000, 001, \ldots, 999$ are sold. Each contestant buys only one ticket. The prize winners of the lottery are determined by drawing one number at random from the numbers

000, 001, ..., 999. You are a prize winner when the number on your ticket is the same as the number drawn or is a random permutation of the number drawn. What is the probability mass function of the number of prize winners?

3.5 Accidentally, two depleted batteries get into a set of five batteries. To remove the two depleted batteries, the batteries are tested one by one in random order. Let the random variable X denote the number of batteries that must be tested to find the two depleted batteries. What is the probability mass function of X?

3.6 You are allowed to toss a fair coin either five times or until tails appears, whichever occurs first. You receive starting capital of $10 when the first toss results in heads; otherwise, you receive nothing. Each subsequent toss that results in heads doubles your current capital. Let the random variable X be the end capital you will get. What is the probability mass function of X?

3.7 A box contains seven envelopes. Two envelopes contain $10 each, one contains $5, and the other four are empty. Envelopes are drawn at random from the box, one by one and without replacement. You may keep the money from the envelopes that are drawn before an empty envelope is drawn. Let the random variable X be the amount of money you will get. What is the probability mass function of X?

3.2 Expected Value

The most important characteristic of a random variable is its *expected value*. The concept of expected value goes back to Christiaan Huygens (1629–1695) and Blaise Pascal (1623–1662). In his influential text *De Ratiociniis de Ludo Aleae* (On Reasoning in Games of Chance), published in 1657, Huygens worked out the concept of expected value based on an embryonic idea of Pascal. The concept of expected value was introduced in the context of a game of chance. To gain a good understanding of precisely what the concept is, it helps to consider a repeatable game of chance. Imagine a casino game where the player has a 0.70 probability of losing 1 dollar and probabilities 0.25 and 0.05 of winning 2 and 3 dollars, respectively. A player who plays this game a large number of times reasons intuitively as follows in order to determine the average win per game in n games. In approximately $0.70n$ repetitions of the game, the player loses 1 dollar per game and in approximately $0.25n$ and $0.05n$ repetitions of the game, the player wins 2 and 3 dollars, respectively. This means that the total win in dollars is approximately equal to

$$(0.70n) \times (-1) + (0.25n) \times 2 + (0.05n) \times 3 = -(0.05)n,$$

or the average win per game is approximately -0.05 dollars (meaning that the average "win" is actually a loss). If we define the random variable X as the win achieved in just a single repetition of the game, then the number -0.05 is said to be the expected value of X. In the casino game, the expected value of X is calculated as

$$-1 \times P(X = -1) + 2 \times P(X = 2) + 3 \times P(X = 3) = -0.05.$$

The expected value of the discrete random variable X is a weighted mean of the values the random variable can take, the weights being furnished by the probability mass function of the random variable.

We now state the general definition of the expected value of a discrete random variable.

Definition 3.1 *The expected value of the discrete random variable X having I as its set of possible values is defined by*

$$E(X) = \sum_{x \in I} x P(X = x),$$

provided that the sum is well-defined.

The expected value of a random variable X is also known as the *mean*, or *first moment* and *average value* of X. The nomenclature of the expected value may be misleading. The expected value is in general not a typical value that the random variable can take. Before we give several examples, we make the following remarks. Definition 3.1 is only meaningful if the sum is well-defined. The sum is always well-defined if the range I is finite. However, the sum over countably infinite many terms is not always well-defined when both positive and negative terms are involved. For example, the infinite series $1 - 1 + 1 - 1 + \cdots$ has sum 0 when you sum the terms according to $(1 - 1) + (1 - 1) + \cdots$, whereas you get sum 1 when you sum the terms according to $1 + (-1 + 1) + (-1 + 1) + (-1 + 1) + \cdots$. Such abnormalities cannot happen when all terms in the infinite summation are nonnegative. The sum of infinitely many *nonnegative* terms is always well-defined by allowing ∞ as a possible value for the sum. For a sequence $a_1, a_2, \ldots$ consisting of both positive and negative terms, the infinite series $\sum_{k=1}^{\infty} a_k$ is well-defined with a finite sum if and only if $\sum_{k=1}^{\infty} |a_k| < \infty$. This is a basic result from calculus. Such a series is said to be *absolutely convergent*. For an absolutely convergent series $\sum_{k=1}^{\infty} a_k$, the finite sum is uniquely determined and does not depend on the order in which the individual terms are added.

3.2 Expected Value

For a discrete random variable X with range I, it is said that the expected value $E(X)$ *exists* if $\sum_{x \in I} |x| P(X = x) < \infty$ or if X is a nonnegative random variable, where the value $E(X) = \infty$ is allowed when X is nonnegative. It is important to emphasize that the expected value $E(X)$ is finite if and only if $\sum_{k \in I} |k| P(X = k) < \infty$. An example of a random variable X for which $E(X)$ does not exist is the random variable X with probability mass function $P(X = 0) = 0$ and $P(X = k) = \frac{3}{\pi^2 k^2}$ for $k = \pm 1, \pm 2, \ldots$. In this example, $\sum_{k=-\infty}^{\infty} |k| P(X = k) = \infty$, by the fact that $\sum_{k=1}^{\infty} \frac{1}{k} = \infty$. However, the expected value of the nonnegative random variable with probability mass function $\frac{6}{\pi^2 k^2}$ for $k = 1, 2, \ldots$ does exist (and is equal to ∞).

Example 3.1 (continued) What is the expected value of the random variable X to be defined as the number of cents you will grab from your pocket?

Solution. Since the probability mass function of X is $P(X = 20) = \frac{3}{10}$, $P(X = 35) = \frac{3}{5}$, and $P(X = 50) = \frac{1}{10}$, the expected value of X is

$$E(X) = 20 \times \frac{3}{10} + 35 \times \frac{3}{5} + 50 \times \frac{1}{10} = 32 \text{ cents.}$$

Example 3.2 (continued) What is the expected value of the random variable X to be defined as the number of cards that must be picked until you get the first ace?

Solution. Using the result found earlier for $P(X = k)$, we get

$$E(X) = 1 \times \frac{4}{52} + \sum_{k=2}^{49} k \frac{48}{52} \times \cdots \times \frac{48 - (k-2)}{52 - (k-2)} \times \frac{4}{52 - (k-1)} = 10.6.$$

Example 3.3 Joe and his friend make a guess every week whether the Dow Jones index will have risen at the end of the week or not. Both put $10 in the pot. Joe observes that his friend is just guessing and is making his choice by the toss of a fair coin. Joe asks his friend if he could contribute $20 to the pot and submit his guess together with that of his brother. The friend agrees. In each week, however, Joe's brother submits a prediction opposite to that of Joe. The person having a correct prediction wins the entire pot. If more than one person has a correct prediction, the pot is split evenly. How favorable is the game to Joe and his brother?

Solution. Let the random variable X denote the payoff to Joe and his brother in any given week. Either Joe or his brother will have a correct prediction. If Joe's friend is wrong he wins nothing, and if he is correct he shares the $30 pot with either Joe or his brother. Thus X takes on the values 30 and 15 with equal chances. This gives $E(X) = \frac{1}{2} \times 30 + \frac{1}{2} \times 15 = 22.5$ dollars. Joe and his brother have an expected profit of $2.5 every week.

Example 3.4 Three friends go to the cinema together every week. Each week, in order to decide which friend will pay for the other two, they all toss a fair coin into the air simultaneously. They continue to toss coins until one of the three gets a different outcome from the other two. What is the expected value of the number of trials required?

Solution. Let the random variable X denote the number of trials until one of the three friends gets a different outcome from the other two. The probability that any given trial does not lead to three equal outcomes is $1 - \frac{1}{8} - \frac{1}{8} = \frac{3}{4}$. Thus,

$$P(X = j) = (1-p)^{j-1}p \quad \text{for } j = 1, 2, \ldots$$

with $p = \frac{3}{4}$. The expected value of X is

$$E(X) = \sum_{j=1}^{\infty} j(1-p)^{j-1}p = \frac{p}{[1-(1-p)]^2} = \frac{1}{p},$$

using the fact that $\sum_{j=1}^{\infty} jx^{j-1} = 1/(1-x)^2$ for all $0 < x < 1$, see Appendix C. Thus, the expected value of the number of trials required is $\frac{4}{3}$.

The concept of expected value is at the heart of the sequential decision problem that is discussed in the next example.

Example 3.5 Eleven closed boxes are put in random order in front of you. One of these boxes contains a devil's penny and the other 10 boxes contain money. You know which dollar amounts $a_1, \ldots, a_{10}$ are in the 10 boxes. You may open as many boxes as you wish, but they must be opened one by one. You can keep the money from the boxes you have opened, as long as you have not opened the box with the devil's penny. Once you open this box, the game is over and you lose all the money gathered so far. What is a good stopping rule when you want to maximize the expected value of your gain? What is the expected value of the gain of the game when $a_i = 1$ for all i?

Solution. The *one-stage-look-ahead rule* is the key. This is an intuitively appealing rule that looks only one step ahead, as the name says. The rule prescribes stopping in the states in which it is at least as good to stop now as to continue one more step and then stop. For the situation that you have collected so far a dollars and $k+1$ boxes are still closed, including the box with the devil's penny, define the random variable $X_k(a)$ as the amount by which your capital gathered so far would change when you would decide to open one more box. The one-stage-look-ahead rule prescribes continuing if the expected value of $X_k(a)$ is positive and stopping otherwise. Let $A = a_1 + a_2 + \cdots + a_{10}$ be the original amount of dollars in the 10 boxes. To calculate $E[X_k(a)]$, you need not know how the remaining amount of $A - a$ dollars is distributed over the k remaining closed boxes not containing the devil's penny. To see this, imagine that the dollar amounts $b_1, \ldots, b_k$ (say) are in these k closed boxes. Then $b_1 + \cdots + b_k = A - a$ and

$$E[X_k(a)] = \frac{1}{k+1}b_1 + \cdots + \frac{1}{k+1}b_k - \frac{1}{k+1}a$$
$$= \frac{1}{k+1}(A-a) - \frac{1}{k+1}a.$$

Therefore, $E[X_k(a)] > 0$ only if $a < \frac{1}{2}A$, regardless of the value of k. This suggests stopping as soon as you have collected $\frac{1}{2}A$ dollars or more and continuing otherwise.[1] It is noteworthy that this stopping rule depends only on the aggregated amount of dollars in the boxes and not on the number of boxes. To test your understanding of the solution, what would be the optimal stopping rule when there is a 12th box with a second devil's penny?

For the special case that $a_i = 1$ for all i, the stopping rule prescribes stopping as soon as you have opened $\frac{10}{2} = 5$ boxes without having opened the box with the devil's penny. Let the random variable X be the gain of the game for this special case. The random variable X can take on only the values 0 and 5. If you open 5 randomly chosen boxes out of the 11 boxes, then the probability of not opening the box with the devil's penny is

$$\frac{10}{11} \times \frac{9}{10} \times \frac{8}{9} \times \frac{7}{8} \times \frac{6}{7} = \frac{6}{11}.$$

Therefore, $P(X = 5) = \frac{6}{11}$ and $P(X = 0) = 1 - \frac{6}{11} = \frac{5}{11}$. This gives

$$E(X) = 0 \times \frac{5}{11} + 5 \times \frac{6}{11} = \frac{30}{11}.$$

In general, one should resort to computer simulation to find the expected value of the gain of the game for specific values of the a_i. Simulation gives an expected value of about 15.45 for optimal play in the following instance of the devil's penny problem. A card game is played with 11 cards: an ace, two, three, ..., nine, ten, and a joker. These cards are thoroughly shuffled and

---

[1] Using advanced optimization theory, it can be shown that this stopping rule is indeed optimal for the criterion of the expected gain. In general, the one-stage-look-ahead rule is optimal if the problem has the property that continuing in a state in which this rule calls for stopping results again in a state in which the rule calls for stopping. To put it differently, the set of states in which stopping is better than continuing for one step and then stopping must be a closed set from which no escape is possible.

laid face down. You can flip over as many cards as you wish, but they must be flipped over one by one. Once you flip over the joker card, the game is over and you win nothing. If you stop before the joker card is flipped over, you win as many dollars as the sum of the values of the cards you have flipped over (the ace counts for $1). This is the devil's penny problem with $a_i = i$ for all i.

Expected value is a useful concept in casino games, lotteries, and sports betting. The expected value is what the player will win or lose on average if the player were to make the same bet over and over again. This result is known as the strong law of large numbers. This law will be made more precise in Section 9.2.

The next example gives a practical application of the concept of expected value. The example is about group testing for disease and has its historical significance in World War II. The young men who were drafted were subject to a blood test in the fight against syphilis, at a time when mass-produced penicillin was not available. The Harvard economist Robert Dorfman devised an elegant method to cut down on the number of blood tests needed. The idea was to pool the blood samples of soldiers together and test the pooled sample.

Example 3.6 A large group of people are undergoing a blood test for a particular disease. The probability that a randomly chosen person has the disease in question is equal to $p = 0.005$. In order to save on costs, it is decided to split the group into smaller groups each consisting of r people. The blood samples of the r people are then mixed and tested all at once. The mixture will only test negative if all the individual samples are negative. If the test result is negative, then one test will have been sufficient for that whole group. Otherwise, r extra tests will be necessary in order to test each of the r people individually. What value of r minimizes the average number of tests per person?

Solution. Let the random variable X be the number of tests needed for a group of r people. The possible values of the random variable X are 1 and $r+1$. The random variable X takes on the value 1 if and only if all the r samples are negative. An individual sample will test negative with probability $1 - p$. Assuming independence between the test outcomes of individual people, it follows that $P(X = 1)$ is equal to $(1-p) \times (1-p) \times \cdots \times (1-p) = (1-p)^r$. Since $P(X = r+1) = 1 - P(X = 1)$, the probability $P(X = r+1)$ is equal to $1 - (1-p)^r$. Thus, the expected value of X is

$$E(X) = 1 \times (1-p)^r + (r+1) \times (1 - (1-p)^r) = 1 + r\left(1 - (1-p)^r\right).$$

Therefore, by combining blood samples from r people into one mixture, the average number of tests per person is $\frac{1}{r}[1 + r(1 - (1-p)^r)]$. For the case $p = 0.005$, this expression is minimal for $r = 15$ and its minimal value is 0.1391. Thus, group testing cuts down the number of tests performed by about 86% when $p = 0.005$ when the test concerns many people. Also, an interesting quantitative result can be given. For p close to zero, the optimal value of the group size and the minimal value of the average number of tests per person are approximately equal to $\frac{1}{\sqrt{p}}$ and $2\sqrt{p}$. This result follows easily by noting that $1 - (1-p)^r \approx rp$ for p close to zero and next minimizing the function $\frac{1}{x}(1 + px^2)$ with respect to x (verify!).

Problems

3.8 There are 10 cards face down numbered 1 through 10. You pick at random one card. Your payoff is $0.50 if the number on the card is less than 5 and is the dollar value on the card otherwise. What is the expected value of your payoff?

3.9 Four professors give one course each. The courses have 15, 20, 70, and 125 students. No student takes more than one course. Let the random variable X be the number of students in a randomly chosen class and Y be the number of students in the class of a randomly chosen student. Without doing any calculation, can you explain why $E(X)$ is larger than $E(Y)$? What are $E(X)$ and $E(Y)$?

3.10 A firm exports to a variety of countries. An order for $10,000 has been received from a politically unstable country. The export firm estimates that there is a probability of 75% that the full amount of $10,000 will be paid, a probability of 15% that only $5,000 will be paid, and a probability of 10% that nothing will be paid. Therefore, the firm has decided to purchase export credit insurance. The insurance company has asked a price of $2,000 to insure the order. The insurance guarantees compensation for any debt experienced on the order. What is the expected value of the cost of risk reduction?

3.11 A bowl contains ten white and two red balls. In one shot you pick m balls at random from the bowl, where m can be chosen at your discretion. If all the balls picked are white, you win m dollars; otherwise, you win nothing. What is the expected value of your winnings and what is the maximizing value of m?

3.12 Six cards are taken from a deck: two kings and four aces. The six cards are thoroughly shuffled, after which the two top cards are revealed.

If both cards are aces, you win $1.25; otherwise, you lose $1. Is this a fair bet? A gambling game is said to be *fair* when the expected value of your net winnings is zero.

3.13 Consider again Example 3.2. Let the random variable X_j be the number of picks until an ace is obtained for the jth time. What is the expected value of X_j for $j = 2, 3,$ and 4? Can you explain the answer intuitively?

3.14 In the game "Unders and Overs," two dice are rolled and you can bet whether the total of the two dice will be under 7, over 7, or equal to 7. The gambling table is divided into three sections marked "Under 7," "7," and "Over 7." The payoff odds for a bet on "Under 7" are 1 to 1, for a bet on "Over 7" are 1 to 1, and for a bet on "7" are 4 to 1 (payoffs of r to 1 mean that you get $r + 1$ dollars back for each dollar bet if you win; otherwise, you get nothing back). Each player can put chips on one or more sections of the gambling table. Your strategy is to bet one chip on "Under 7" and one chip on "7" each time. What is the expected value of your net winnings?

3.15 In European roulette, players bet on the outcome of a turning wheel, which is fitted with 37 spokes numbered from 0 to 36. Of the spokes numbered from 1 to 36, 18 are red and 18 are black. The 0 represents a win for the casino. A bet on the color red has a 1 to 1 payout. There is a stake limit of $1,000. A player decides to make no more than 11 bets under a doubling strategy and to stop as soon as he has won a bet. The player begins by staking one dollar on red. If he loses, he doubles his stake on red, and continues doubling until red wins. If the player has lost 10 bets in a row, the maximum amount of $1,000 is staked in the 11th bet. What is the expected value of the total amount staked by the player and what is the expected value of the amount won by the player?

3.16 A fair coin is tossed repeatedly until heads appears for the first time or the coin has been tossed m times, whichever occurs first. The payoff is two dollars if heads turns up in the first toss, four dollars if heads turns up for the first time in the second toss, etc. In general, the payoff is 2^k dollars if heads turns up for the first time in the kth toss, for $1 \leq k \leq m$. The payoff is zero if tails is tossed m times in a row. What is the expected value of the payoff? What should be the stake to make the game a fair bet? *Note*: This game is known as the St. Petersburg game when $m = \infty$.

3.17 You throw darts at a circular target on which two concentric circles of radius 1 cm and 3 cm are drawn. The target itself has a radius of 5 cm. You receive 15 points for hitting the target inside the smaller circle, 8 points for hitting the middle annular region, and 5 points for hitting the outer annular region. The probability of hitting the target at all is 0.75.

If the dart hits the target, then the hitting point is a random point on the target. Let the random variable X be the number of points scored on a single throw of the dart. What is the expected value of X?

3.18 The following game is played in a particular carnival tent. You pay one dollar to draw blindly three balls from a box without replacement. The box contains 10 balls and four of those balls are gold colored. You get back your original one-dollar stake if you draw exactly two gold-colored balls, while you win 10 dollars and get back your original one-dollar stake if you draw three gold-colored balls; otherwise, you get nothing back. What is the expected value of your net winnings?

3.19 You spin a game-board spinner with 1,000 equal sections numbered 1 to 1,000. After your first spin, you have to decide whether to spin the spinner for a second time. Your payoff is the total score of your spins, as long as this score does not exceed 1,000; otherwise, your payoff is zero. What strategy maximizes the expected value of your payoff?

3.20 A casino owner has given you an advance of a fiches to play a particular casino game. Your win probability for each play of the game is p, where $p < \frac{1}{2}$. The stake for each play is one fiche. If you win, you get two fiches back; otherwise, you lose the stake. The fiches are given to you as a reward for services rendered to the casino and under the following agreement. You must play until you have lost all fiches or have won b fiches on top of the advance of a fiches, where b must be set beforehand. In the case that you lose all fiches, you are owed nothing by the casino; otherwise, you return the advance of a fiches and keep the b fiches won. You decide to choose the level b in such a way that the expected value of the number of fiches you may keep is maximal.

(a) Show that for large a, the maximizing value of b satisfies $b \approx \frac{1}{\ln(q/p)}$ with $q = 1 - p$, regardless of how large a is. *Hint*: Use the fact that the gambler's ruin formula from Example 2.9 can be rewritten as $P(a, b) = \frac{(q/p)^{-a} - 1}{(q/p)^{-a} - (q/p)^b}$ and note that $(q/p)^{-a}$ can be neglected for large a when $p < \frac{1}{2}$.

(b) Verify that the corresponding approximations for the expected value of the number of fiches you may keep and for the probability of winning the b fiches are $\frac{b}{e}$ and $\frac{1}{e}$ for large a.

Note: This problem is based on M. Orkin and R. Kakigi, "What is the worth of free casino credit?," *The American Mathematical Monthly* **102** (1995): 3–8.

3.21 The following dice game is offered to you. You may simultaneously roll one red die and three blue dice. The stake is $1. If none of the blue dice matches the red die, you lose your stake; otherwise, you get paid $k+1$ dollars if exactly k of the blue dice match the red die. In the case that exactly one blue die matches the red die, you get paid an additional $0.50 if the other two blue dice show the same number. What is the expected payoff of the game?

3.22 You play a game in which four fair dice are rolled. The stake is $1. The payoff is $100 if all four dice show the same number and $10 if two dice show the same even number and the other two dice show the same odd number. Is this a favorable game? What about the following game with a stake of $2? A fair coin is tossed no more than five times. The game ends if the coin comes up tails or five straight heads appear, whichever happens first. You get a payoff of $1 each time heads appears, plus a bonus of $25 if five heads appear in a row.

3.23 A stick of length 1 is broken at random into two pieces. You bet on the ratio of the length of the longer piece to the length of the smaller piece. You receive $k if the ratio is between k and $k+1$ for some k with $1 \leq k \leq m-1$, while you receive $m if the ratio is larger than m. Here, m is a given positive integer. What should your stake be to make this a fair bet?

3.24 In the dice game of Pig, you repeatedly roll a single die in any round. Upon rolling a 1, your turn is over and nothing is added to your score. Otherwise, you can stop whenever you want and then the total number of points rolled is added to your score. The goal is to maximize the expected number of points gained in one round. Under the hold-at-20, rule you stop when you have rolled 20 points or more in the round. Explain why the hold-at-20 rule is optimal.

3.25 You play a dice game in which you repeatedly roll a single die. If the total number of points rolled exceeds 10, the game is over and you get no reward. Otherwise, you can stop whenever you want and then your dollar reward is the accumulated number of points rolled. What is an optimal stopping rule when you want to maximize the expected value of your reward?

3.26 A game machine can be used to drop balls into bins. The balls are dropped one at a time and any ball will land at random into one of 25 bins. You can stop dropping balls whenever you wish. At the end of the game, you win $1 for every bin with exactly one ball and you lose $0.50 for every bin with two or more balls. Empty bins do not count. How does the one-stage-look-ahead rule work for this problem? Also answer this question when you lose $\frac{1}{2}k$ dollars rather than half a dollar for every bin containing $k \geq 2$ balls.

3.27 A bag contains w white and r red balls. You pick balls out of the bag at random, one at a time and without replacement. You win one dollar each time you pick a white ball. If you pick a red ball, the game is over and you lose all the money gathered so far. You can stop picking balls whenever you wish. What is an optimal stopping rule when you want to maximize your expected gain?

3.28 Consider the following dice game in which you repeatedly roll two dice. The dollar reward for each roll is the dice total, provided that the two numbers shown are different; otherwise, the game is over and you lose the total reward gathered so far. You can stop rolling the dice whenever you wish. What is an optimal stopping rule when you want to maximize the expected value of your total reward?

3.29 Let the random variable X be nonnegative and integer valued. Verify that

$$E(X) = \sum_{k=0}^{\infty} P(X > k).$$

Apply this useful formula to calculate the expected value of the largest of 10 randomly chosen numbers from 1 to 100.

3.3 Expected Value of Sums of Random Variables

Let X and Y be two random variables that are defined on the same sample space with probability measure P. As an example, take X as the smallest of the two numbers rolled and Y as the sum of the two numbers rolled when the experiment is to roll two dice. Since X and Y are defined on the same sample space, the sum $X + Y$ is also a random variable. It assigns to each element ω of the sample space the numerical value $X(\omega) + Y(\omega)$. The following basic rule is of utmost importance.

Rule 3.1 *For any two random variables X and Y,*

$$E(X + Y) = E(X) + E(Y),$$

provided that $E(X)$ and $E(Y)$ exist and are finite.

The proof is simple for the discrete case. Define the random variable Z by $Z = X + Y$. Then, for all z,

$$P(Z = z) = \sum_{x,y: x+y=z} P(X = x, Y = y),$$

where $P(X = x, Y = y)$ denotes the probability of the joint event that X takes on the value x and Y the value y. Thus, by $E(Z) = \sum_z z P(Z = z)$, we have $E(Z) = \sum_z z \sum_{x,y: x+y=z} P(X = x, Y = y)$. This gives

$$E(Z) = \sum_z \sum_{x,y: x+y=z} (x+y) P(X = x, Y = y)$$
$$= \sum_{x,y} (x+y) P(X = x, Y = y).$$

Thus, $E(Z) = \sum_{x,y} x P(X = x, Y = y) + \sum_{x,y} y P(X = x, Y = y)$. This expression can be rewritten as

$$E(Z) = \sum_x x \sum_y P(X = x, Y = y) + \sum_y y \sum_x P(X = x, Y = y).$$

For fixed x, the events $\{X = x, Y = y\}$ are disjoint and their union is the event $\{X = x\}$. Thus, by the third axiom of probability theory,

$$\sum_y P(X = x, Y = y) = P(X = x) \quad \text{for any } x.$$

In the same way, $\sum_x P(X = x, Y = y) = P(Y = y)$ for any y. This gives

$$E(Z) = \sum_x x P(X = x) + \sum_y y P(Y = y) = E(X) + E(Y),$$

as was to be proved. It is pointed out that the various manipulations with the summations in the above analysis are justified by the assumption that $\sum_x |x| P(X = x)$ and $\sum_y |y| P(Y = y)$ are finite.

Rule 3.1 applies to any finite number of random variables $X_1, \ldots, X_n$. If $E(X_i)$ exists and is finite for all $i = 1, \ldots, n$, then a repeated application of Rule 3.1 gives

$$E(X_1 + \cdots + X_n) = E(X_1) + \cdots + E(X_n).$$

The result that the expected value of a finite sum of random variables equals the sum of the expected values is extremely useful. It is only required that the relevant expected values exist, but dependencies between the random variables are allowed.

An immediate consequence of Rule 3.1 is that

$$E(X) \geq E(Y) \quad \text{if } X \geq Y,$$

where $X \geq Y$ means that $X(\omega) \geq Y(\omega)$ for all elements ω of the sample space. To see this, note that $X - Y \geq 0$ and thus $E(X - Y) \geq 0$. Also, by the definition of expectation, $E(-Y) = -E(Y)$. Applying Rule 3.1, we now get $E(X - Y) = E(X) - E(Y)$, showing that $E(X) \geq E(Y)$ if $X \geq Y$.

3.3 Expected Value of Sums of Random Variables

Example 3.7 In order to introduce a new brand of breakfast cereal, the producer has introduced a campaign offering a baseball card in each cereal box purchased. There are n different baseball cards. These cards are distributed equally among the cereal boxes, so that if a single box is bought there is a $1/n$ chance it will contain a particular baseball card. As a baseball fan, you want to collect a complete set of baseball cards. How many cereal boxes do you expect to buy in order to get all n baseball cards?

Solution. Define the random variable X as the number of cereal boxes that must be purchased in order to get a complete set of baseball cards. The expected value of X can be calculated by a direct approach using the probability mass function of X. However, a much simpler approach is to define the random variable Y_i as

$Y_i =$ the number of cereal boxes needed in order to go from $i-1$ to i different baseball cards.

Then we can write X as $X = Y_1 + Y_2 + \cdots + Y_n$ and so

$$E(X) = \sum_{i=1}^{n} E(Y_i).$$

In order to calculate $E(Y_i)$, note that $\frac{n-(i-1)}{n}$ gives the probability that the next cereal box purchased will contain a new baseball card when as many as $i-1$ different baseball cards have already been collected. Thus,

$$P(Y_i = k) = \left(1 - \frac{n-(i-1)}{n}\right)^{k-1} \frac{n-(i-1)}{n} \quad \text{for } k = 1, 2, \ldots.$$

Using the fact that $\sum_{k=1}^{\infty} k x^{k-1} = \frac{1}{(1-x)^2}$ for $|x| < 1$, we get $E(Y_i) = \frac{n}{n-i+1}$ for each i. Then, by $E(X) = \sum_{i=1}^{n} E(Y_i)$, it follows that

$$E(X) = n \sum_{k=1}^{n} \frac{1}{k}.$$

Using this formula, we can give an approximation that gives insight into the way that $E(X)$ increases as a function of n. A well-known approximation is

$$\sum_{k=1}^{n} \frac{1}{k} \approx \ln(n) + \gamma + \frac{1}{2n},$$

where $\gamma = 0.57721566\ldots$ is the Euler constant, see Appendix C. Hence we have the insightful approximation

$$E(X) \approx n \ln(n) + \gamma n + \frac{1}{2}.$$

This approximation is very accurate. For example, the approximate and exact values of $E(X)$ are 29.298 and 29.290 when $n = 10$.

The problem in Example 3.7 is known as the *coupon-collector problem*. This problem has many applications. A nice application is the birthday-coverage problem. What is the expected number of people needed so that each of the 365 possible birthdays of the year (ignore February 29) is represented in the group of people? The answer is $365 \sum_{k=1}^{365} \frac{1}{k} = 2{,}364.65$. The calculation of the probability mass function of the number of coupons needed is discussed in Problem 1.87, see also the Markov chain approach in Chapter 10.

3.3.1 Expected Value and Indicator Random Variables

A powerful trick that is often applicable to calculate the expected value of a random variable is to represent the random variable as the sum of so-called *indicator random variables*. These random variables can take on only the values 0 and 1. Just like the complement rule and the law of conditional probability, the trick of indicator random variables is a very useful problem-solving tool in probability. Also, indicator random variables provide a link between probability and expected value: if A is a given event and the indicator variable I_A is to be defined as 1 if A occurs and 0 otherwise, then $E(I_A) = P(A)$. This is seen directly from

$$E(I_A) = 1 \times P(I_A = 1) + 0 \times P(I_A = 0) = P(A).$$

Example 3.8 Suppose that m balls (or tokens) are placed sequentially into one of b bins, where the bin for each ball is selected at random. What is the expected value of the number of empty bins?

Solution. Let the random variable X be the number of empty bins. Then,

$$X = X_1 + \cdots + X_b,$$

where the indicator variable X_i is defined as

$$X_i = \begin{cases} 1 & \text{if the } i\text{th bin is empty,} \\ 0 & \text{otherwise.} \end{cases}$$

The probability that the ith bin will be empty is $\left(\frac{b-1}{b}\right)^m$. Therefore,

$$E(X_i) = 0 \times P(X_i = 0) + 1 \times P(X_i = 1) = \left(1 - \frac{1}{b}\right)^m, \quad i = 1, \ldots, b.$$

3.3 Expected Value of Sums of Random Variables

This gives
$$E(X) = b\left(1 - \frac{1}{b}\right)^m.$$

A useful approximation applies to $E(X)$ for b large. Noting that $\left(1 - \frac{1}{b}\right)^b \approx e^{-1}$ for b large and writing $\left(1 - \frac{1}{b}\right)^m$ as $\left(1 - \frac{1}{b}\right)^{b \times (m/b)}$, it follows that
$$E(X) \approx be^{-m/b} \quad \text{for } b \text{ large}.$$

Example 3.9 What is the expected value of the number of times that, in a thoroughly shuffled deck of 52 cards, two adjacent cards are of the same rank (two aces, two kings, etc.)?

Solution. Let the random variable X_i be equal to 1 if the cards in positions i and $i+1$ are of the same rank and 0 otherwise. Then $P(X_i = 1) = \frac{3}{51}$ and so $E(X_i) = \frac{3}{51}$ for $i = 1, \ldots, 51$. The expected value of the number of times that two adjacent cards are of the same rank is given by $E(X_1 + \cdots + X_{51}) = 51 \times \frac{3}{51} = 3$.

Example 3.10 Suppose that you have a graph with n nodes. Each pair of nodes is connected by an edge with probability p, independently of the occurrence of edges between other pairs of nodes. What is the expected number of isolated nodes?

Solution. Label the nodes $k = 1, \ldots, n$ and let the indicator variable I_k be 1 if node k is isolated and 0 otherwise. Then the number of isolated nodes is $\sum_{k=1}^{n} I_k$. For each k, $P(I_k = 1) = (1-p)^{n-1}$ and so $E(I_k) = (1-p)^{n-1}$. Thus, the expected number of isolated nodes is
$$\sum_{k=1}^{n} E(I_k) = n(1-p)^{n-1}.$$

As a corollary,
$$P(\text{there is at least one isolated node}) \leq n(1-p)^{n-1}.$$

This bound is a direct consequence of the inequality $P(X \geq 1) \leq E(X)$ for any nonnegative, integer-valued random variable X (explain!).

The Linearity Property for General Random Variables

The linearity property of the expected value holds for any type of random variables, including continuous random variables. A continuous random variable such as the decay time of a radioactive particle can take on a continuum of possible values. Continuous random variables are to be discussed in Chapter 4 and subsequent chapters. In Chapter 4, the expected value of a continuous random variable will be defined by an integral rather than by a sum as in the discrete case. The models of discrete and continuous random variables are the most important ones, but are not exhaustive. Also, there are so-called *mixed* random variables, having both a discrete component and a continuous component. Think of your delay in the queue at a supermarket counter or the amount reimbursed on an automobile insurance policy in a given year. These random variables take on either the discrete value zero with positive probability or a value in a continuous interval.

For the case of discrete random variables, a proof of the linearity property of the expected value could be given by using first principles. However, the proof is more tricky for the general case. If the random variables are continuously distributed, we need the concept of joint probability density and using this concept the linearity property can be proved, see Rule 5.6 in Chapter 5. For the general case including the case of mixed random variables, the proof requires some technical machinery from measure theory. Then, we need to proceed from the general definition of integral and expectation in measure theory. This is beyond the scope of this book.

Problems

3.30 A group of m people simultaneously enter an elevator at the ground floor. Each person randomly chooses one of the r floors $1, 2, \ldots, r$ as the exit floor, where the choices of the persons are independent of each other. The elevator only stops on a floor if at least one person wants to exit on that floor. No other people enter the elevator at any of the floors $1, 2, \ldots, r$. What is the expected value of the number of stops the elevator will make?

3.31 Use indicator variables to give an alternative proof of the inclusion–exclusion formula in Rule 1.4. *Hint*: Use the relation $I_{A_1^c \cap \cdots \cap A_n^c} = (1 - I_{A_1}) \cdots (1 - I_{A_n})$, where the indicator variable I_A is equal to 1 if the event A occurs and 0 otherwise.

3.32 What is the expected value of the number of times that two adjacent letters will be the same in a random permutation of the 11 letters of the word "Mississippi?"

3.33 Twelve married couples participate in a tournament. The group of 24 people is randomly split into eight teams of three people each. What is the expected value of the number of teams with a married couple?

3.34 A particle starting at the origin moves every time unit, one step up or down with equal probabilities $\frac{1}{2}$. Verify that the expected number of returns of the random walk to the zero level during the first n time units is approximately equal to $\sqrt{2/\pi}\sqrt{n}$ for n large. On the basis of this result, can you explain intuitively why the time between two successive returns of the random walk to the zero level is infinite? *Hint*: Use Example 1.5 and the fact that $\sum_{j=1}^{m} 1/\sqrt{j} \approx 2\sqrt{m}$ for m large.

3.35 In a bowl, there are $2r$ red balls and $2b$ blue balls. Each time, you simultaneously remove two balls from the bowl until the bowl is empty. What is the expected number of times that you pick two balls of the same color?

3.36 What is the expected number of distinct birthdays within a randomly formed group of 100 persons? What is the expected number of children in a class with r children sharing a birthday with some child in another class with s children? Assume that the year has 365 equally likely birthdays.

3.37 Let S be a given set consisting of n distinct items. A new set T is constructed as follows. In each step, an item is chosen at random from the set S (with replacement) and a copy of the item is added to the set T. What is the expected value of the number of distinct items in the set T after n steps?

3.38 Consider again Problem 1.92. What is the expected value of the number of people who survive the first round?

3.39 A bag contains r red and w white balls. Each time, you take one ball out of the bag at random and without replacement. You stop as soon as all the red balls have been taken out of the bag. What is the expected number of white balls remaining in the bag when you stop?

3.40 Consider again Problem 2.30. What is the expected value of the number of persons having knowledge of the rumor?

3.41 What is the expected number of times that two consecutive numbers will show up in a lottery drawing of six different numbers from the numbers $1, 2, \ldots, 45$?

3.42 You play a sequence of s games, where $s \geq 2$. The outcomes of the various games are independent of each other. The probability that you will win the kth game is $\frac{1}{k}$ for $k = 1, 2, \ldots, s$. You get one dollar each time you win two games in a row. What is the expected value of the total amount you will get?

3.43 You distribute randomly 25 apples over 10 boxes. What is the expected value of the number of boxes that will contain more than 3 apples?

3.44 Verify that the expected number of permutation cycles in a random permutation of the integers $1, \ldots, n$ is about $\ln(n) + \gamma$ for n large, where $\gamma = 0.57722\ldots$ is Euler's constant. *Hint*: For any fixed k with $1 \leq k \leq n$, define $X_i = 1$ if the integer i belongs to a cycle of length k

and $X_i = 0$ otherwise. Use the fact that $\frac{1}{k}\sum_{i=1}^{n} X_i$ is the number of cycles of length k.

3.45 Take a random permutation of the integers $1, 2, \ldots, n$. Let us say that the integers i and j with $i \neq j$ are switched if the integer i occupies the jth position in the random permutation and the integer j the ith position. What is the expected value of the total number of switches?

3.4 Substitution Rule and Variance

Suppose that X is a discrete random variable with a given probability mass function. In many applications, we wish to compute the expected value of some function of X. Note that any function of X, such as X^2 or $\sin(X)$, is also a random variable (why?). Let $g(x)$ be a given real-valued function. Then the probability mass function of the discrete random variable $Y = g(X)$ can be calculated as

$$P(g(X) = y) = \sum_{x:\, g(x)=y} P(X = x)$$

with the convention that an empty sum is zero. However, it is not necessary to use the probability mass function of the random variable $g(X)$ to calculate its expected value. The expected value of $g(X)$ can be calculated directly from the probability mass function of X.

Rule 3.2 *For any function g of the random variable X,*

$$E[g(X)] = \sum_{x \in I} g(x)\, P(X = x),$$

provided that $\sum_{x \in I} |g(x)|\, P(X = x) < \infty$, or the function $g(x)$ is nonnegative.

This rule is called the *substitution rule* and is also known as the *law of the unconscious statistician*. The proof of the rule is simple. Let $Y = g(X)$ and denote by J the range of Y. Then,

$$\sum_{x \in I} g(x) P(X = x) = \sum_{y \in J} \sum_{x:\, g(x)=y} y P(X = x) = \sum_{y \in J} y P\big(g(X) = y\big)$$
$$= \sum_{y \in J} y P(Y = y) = E(Y) = E\big[g(X)\big].$$

Note that the order in which the terms of the series are added does not matter in view of the assumption that the series is absolutely convergent or has only nonnegative terms.

A frequently made mistake of beginning students is to set $E[g(X)]$ equal to $g(E(X))$. In general, $E[g(X)] \neq g(E(X))$! Stated differently, the average value of the input X does not determine in general the average value of the output $g(X)$. As a counterexample, take the random variable X with $P(X = 1) = P(X = -1) = 0.5$ and take the function $g(x) = x^2$. An exception is the case of a linear function $g(x) = ax + b$. Then, by Rule 3.2,

Rule 3.3 *For any constants a and b,*

$$E(aX + b) = aE(X) + b,$$

provided that $E(X)$ is finite.

This rule is valid for any type of random variable X for which $E(X)$ is finite. Another result that is valid for any type of random variable is *Jensen's inequality*:

Rule 3.4 *Suppose that the function $g(x)$ is convex on a line segment containing the range of the random variable X. Then,*

$$E[g(X)] \geq g(E(X)),$$

provided that the expectations are finite.

The proof is simple. For ease, assume that $g(x)$ is differentiable. A differentiable function $g(x)$ is convex only if, for any point x_0, the graph of $g(x)$ lies entirely above the tangent line at the point x_0:

$$g(x) \geq g(x_0) + g'(x_0)(x - x_0) \quad \text{for all } x.$$

Choosing $x = X$ and $x_0 = E(X)$ in this inequality and taking the expectation on both sides of the inequality, the inequality is preserved and Jensen's inequality follows. Here we use Rule 3.3 and the fact that $E[g_1(X)] \geq E[g_2(X)]$ if $g_1(x) \geq g_2(x)$ for all x.

3.4.1 Variance

An important case of a function of X is the random variable $g(X) = (X - \mu)^2$, where $\mu = E(X)$ denotes the expected value of X and is assumed to be finite. The expected value of $(X - \mu)^2$ is called the *variance* of X. It is denoted by

$$\text{var}(X) = E[(X - \mu)^2].$$

It is a measure of the spread of the possible values of X. Often one uses the *standard deviation*, which is defined as the square root of the variance. It is useful to work with the standard deviation, because of the fact that it has the same units (e.g., dollar or centimeter) as $E(X)$. The standard deviation of a random variable X is usually denoted by $\sigma(X)$, and thus is defined by

$$\sigma(X) = \sqrt{\text{var}(X)}.$$

The formula for var(X) allows for another useful representation. Since $(X - \mu)^2 = X^2 - 2\mu X + \mu^2$, it follows from the linearity of the expectation operator and Rule 3.3 that $E[(X - \mu)^2] = E(X^2) - 2\mu E(X) + \mu^2$. Therefore, var($X$) is also given by

$$\text{var}(X) = E(X^2) - \mu^2.$$

Rule 3.3 for the expectation operator has the following analogue for the variance operator:

Rule 3.5 *For any constants a and b,*

$$\text{var}(aX + b) = a^2 \text{var}(X).$$

The proof is simple and is again based on the linearity of expectation. It follows from $E(aX + b)^2 = a^2 E(X^2) + 2ab\mu + b^2$ and $[E(aX + b)]^2 = (a\mu + b)^2 = a^2\mu^2 + 2ab\mu + b^2$ that var($aX + b$) = $a^2 E(X^2) - a^2\mu^2$, showing the desired result. Rule 3.5 implies

$$\sigma(aX + b) = |a|\sigma(X).$$

To illustrate this relation, suppose that the rate of return on stock A is a random variable X taking on the values 30%, 10%, and -10% with respective probabilities 0.25, 0.50, and 0.25. The rate of return on stock B is a random variable Y taking on the values 50%, 10%, and -30% with the same probabilities 0.25, 0.50, and 0.25. Without calculating the actual values of the standard deviation, one can see that the standard deviation of the rate of return on stock B is twice as large as that on stock A. The explanation is that the random variable Y is distributed as $2X - 0.1$.

Example 3.11 What is the standard deviation of the total score of a roll of two dice?

Solution. Let the random variable X denote the total score. Since $E(X) = 3.5 + 3.5 = 7$ and $P(X = k) = P(X = 14 - k) = \frac{k-1}{36}$ for $2 \le k \le 7$, we get

$$\text{var}(X) = \sum_{k=2}^{7} k^2 \frac{k-1}{36} + \sum_{k=8}^{12} k^2 \frac{14-k-1}{36} - 7^2 = \frac{35}{6}.$$

The standard deviation of X is $\sqrt{\text{var}(X)} = 2.415$.

3.4 Substitution Rule and Variance

Example 3.12 Suppose that the random variable X has the so-called Poisson distribution $P(X = k) = e^{-\lambda}\lambda^k/k!$ for $k = 0, 1, \ldots$. What are the expected value and the variance of X?

Solution. It will be shown that the Poisson distribution has the remarkable property that its variance is equal to its expected value. That is,

$$\text{var}(X) = E(X) = \lambda.$$

To verify this, note that $E(X) = \lambda e^{-\lambda} + 2\frac{\lambda^2}{2!}e^{-\lambda} + 3\frac{\lambda^3}{3!}e^{-\lambda} + \cdots$ and so

$$E(X) = \lambda e^{-\lambda}\left(1 + \frac{\lambda}{1!} + \frac{\lambda^2}{2!} + \cdots\right) = \lambda e^{-\lambda}e^{\lambda} = \lambda,$$

where the third equality uses the power-series expansion $e^x = 1 + \frac{x}{1!} + \frac{x^2}{2!} + \cdots$ for every x. Using the identity $k^2 = k(k-1) + k$, we have

$$E(X^2) = \sum_{k=0}^{\infty} k(k-1)e^{-\lambda}\frac{\lambda^k}{k!} + \sum_{k=0}^{\infty} ke^{-\lambda}\frac{\lambda^k}{k!} = \lambda^2\sum_{k=2}^{\infty} e^{-\lambda}\frac{\lambda^{k-2}}{(k-2)!} + E(X)$$

$$= \lambda^2\sum_{n=0}^{\infty} e^{-\lambda}\frac{\lambda^n}{n!} + \lambda = \lambda^2 + \lambda.$$

Next, by $\text{var}(X) = E(X^2) - [E(X)]^2$, we get $\text{var}(X) = \lambda$.

Example 3.8 (continued) What is the variance of the random variable X, denoting the number of empty boxes?

Solution. To find $\sigma^2(X) = E(X^2) - [E(X)]^2$, we again use the representation $X = X_1 + \cdots + X_b$, where $X_i = 1$ if the ith bin is empty and $X_i = 0$ otherwise. By the algebraic formula $(\sum_{i=1}^{n} a_i)^2 = \sum_{i=1}^{n} a_i^2 + 2\sum_{i=1}^{n-1}\sum_{j=i+1}^{n} a_i a_j$ and the linearity of the expectation operator,

$$E(X^2) = \sum_{i=1}^{b} E(X_i^2) + 2\sum_{i=1}^{b-1}\sum_{j=i+1}^{b} E(X_i X_j).$$

Note that $E(X_i^2) = E(X_i)$, since X_i is 0 or 1. We have already found that $E(X_i) = \left(1 - \frac{1}{b}\right)^m$. To find $E(X_i X_j)$ for $i \neq j$, note that $X_i X_j$ equals 1 if $X_i = X_j = 1$ and equals 0 otherwise. Hence, $E(X_i X_j) = P(X_i = 1, X_j = 1)$. For any $i \neq j$,

$$P(X_i = 1, X_j = 1) = \left(\frac{b-2}{b}\right)^m.$$

Thus, $E(X_iX_j) = \left(1 - \frac{2}{b}\right)^m$ for all $i \neq j$. This leads to

$$E(X^2) = b\left(1 - \frac{1}{b}\right)^m + b(b-1)\left(1 - \frac{2}{b}\right)^m.$$

Next it follows from $\sigma^2(X) = E(X^2) - [E(X)]^2$ that

$$\sigma^2(X) = b(b-1)\left(1 - \frac{2}{b}\right)^m + b\left(1 - \frac{1}{b}\right)^m\left[1 - b\left(1 - \frac{1}{b}\right)^m\right].$$

Noting that $\left(1 - \frac{x}{b}\right)^b \approx e^{-x}$ for b large, we find the approximation (verify!)

$$\sigma^2(X) \approx b\left(e^{-m/b} - e^{-2m/b}\right) \quad \text{for } b \text{ large.}$$

The next example deals with the so-called newsboy problem.

Example 3.13 Every morning, rain or shine, young Billy Gates can be found at the entrance to the metro, hawking copies of *The Morningstar*. Demand for newspapers varies from day to day, but Billy's regular early-morning haul yields him 200 copies. He purchases these copies for $1 per paper, and sells them for $1.50 apiece. Billy goes home at the end of the morning, or earlier if he sells out. He can return unsold papers to the distributor for $0.50 apiece. From experience, Billy knows that demand for papers on any given morning is uniformly distributed between 150 and 250, where each of the possible values $150, \ldots, 250$ is equally likely. What are the expected value and the standard deviation of Billy's net earnings on any given morning?

Solution. Denote by the random variable X the number of copies Billy would have sold on a given morning if he had ample supply. The actual number of copies sold by Billy is X if $X \leq 200$ and 200 otherwise. The probability mass function of X is $P(X = k) = \frac{1}{101}$ for $k = 150, \ldots, 250$. Billy's net earnings on any given morning is a random variable $g(X)$, where the function $g(x)$ is given by

$$g(x) = \begin{cases} -200 + 1.5x + 0.5(200 - x), & x \leq 200 \\ -200 + 1.5 \times 200, & x > 200. \end{cases}$$

Applying the substitution rule, we have that $E[g(X)]$ can be calculated as

$$\sum_{k=150}^{250} g(k)P(X = k) = \frac{1}{101}\sum_{k=150}^{200}(-100 + k) + \frac{1}{101}\sum_{k=201}^{250} 100$$

and so

$$E[g(X)] = \frac{3{,}825}{101} + \frac{5{,}000}{101} = 87.3762.$$

3.4 Substitution Rule and Variance

To find the standard deviation of $g(X)$, we apply the formula $\text{var}(Z) = E(Z^2) - [E(Z)]^2$ with $Z = g(X)$. This gives

$$\text{var}[g(X)] = E([g(X)]^2) - (E[g(X)])^2.$$

Letting $h(x) = [g(x)]^2$, then $h(x) = (-100+x)^2$ for $x \leq 200$ and $h(x) = 100^2$ for $x > 200$. Using the substitution rule again, we get

$$E[h(X)] = \sum_{k=150}^{250} h(k)P(X=k) = \frac{1}{101}\sum_{k=150}^{200}(-100+k)^2 + \frac{1}{101}\sum_{k=201}^{250} 100^2.$$

Hence, $E([g(X)]^2) = E[h(X)] = 297{,}925/101 + 500{,}000/101 = 7{,}900.2475$. Therefore, the variance of Billy's net earnings on any given morning is

$$\text{var}[g(X)] = 7{,}900.2475 - (87.3762)^2 = 265.647.$$

Concluding, Billy's net earnings on any given morning has an expected value of 87.38 dollars and a standard deviation of $\sqrt{265.64} = 16.30$ dollars.

Problems

3.46 There are two investment projects A and B each involving an initial investment of $2,000. The possible payoffs of investment A are $1,000, $2,000, and $3,000 with respective probabilities 0.20, 0.40, and 0.40, while the possible payoffs of investment B are $1,600, $2,000, and $2,750 with respective probabilities 0.25, 0.35, and 0.40. Calculate the expected value and the standard deviation of the payoff for each of the two investments.

3.47 Let the random variable X have a discrete uniform distribution on the integers $a, a+1, \ldots, b$ with $a < b$, that is, $P(X = k) = 1/(b-a+1)$ for $a \leq k \leq b$. What is the variance of X?

3.48 Suppose that the random variable X satisfies $P(X = a) = p$ and $P(X = b) = 1 - p$, where $a < b$. Verify that $\text{var}(X)$ is largest for $p = \frac{1}{2}$.

3.49 Consider again Problem 3.33. What is the standard deviation of the number of teams with a married couple?

3.50 In each drawing of the 6/45 lottery, six different integers are randomly chosen from $1, 2, \ldots, 45$. What are the expected value and the standard deviation of the number of integers from $1, 2, \ldots, 45$ that do not show up in 15 drawings?

3.51 The number of paramedical treatments for a particular sports injury is a random variable X with probability mass function $P(X = k) = \frac{11-k}{55}$ for

$1 \leq k \leq 10$. What are the expected value and standard deviation of X? An insurance policy reimburses the costs of the treatments up to a maximum of five treatments. What are the expected value and the standard deviation of the number of reimbursed treatments?

3.52 You have to make a one-time business decision how much stock to order in order to meet a random demand during a single period. The demand is a random variable X with a given probability mass function $\{p_k, k = 0, 1, \ldots\}$. Suppose you decide to order Q units. What is the expected value of the stock left over at the end of the period? What is the expected value of the amount of demand that cannot be satisfied from stock?

3.53 The University of Gotham City renegotiates its maintenance contract with a particular copy-machine distributor on a yearly basis. For the coming year, the distributor has come up with the following offer. For a prepaid cost of $50 per repair call, the university can opt for a fixed number of calls. For each visit beyond that fixed number, the university will pay $100. If the actual number of calls made by a repairman remains below the fixed number, no money will be refunded. Based on previous experience, the university estimates that the number of repairs that will be necessary in the coming year has a Poisson distribution with expected value 150. The university signs a contract with a fixed number of 155 repair calls. What are the expected value and the standard deviation of the maintenance costs in excess of the prepaid costs?

3.54 At the beginning of every month, a pharmacist orders an amount of a certain costly medicine that comes in strips of individually packed tablets. The wholesale price per strip is $100, and the retail price per strip is $400. The medicine has a limited shelf life. Strips not purchased by month's end will have reached their expiration date and must be discarded. When it so happens that demand for the item exceeds the pharmacist's supply, he may place an emergency order for $350 per strip. The monthly demand for this medicine takes on the possible values 3, 4, 5, 6, 7, 8, 9, and 10 with respective probabilities 0.3, 0.1, 0.2, 0.2, 0.05, 0.05, 0.05, and 0.05. The pharmacist decides to order eight strips at the start of each month. What are the expected value and the standard deviation of the net profit made by the pharmacist on this medicine in any given month?

3.55 A contest has two rounds. You are automatically placed for the second round, while n other people are in the first round. Each of these people enters the second round with probability $\frac{1}{n}$, independently of each other. You win the contest with probability $\frac{1}{k+1}$ when k other people join the second round. Use Jensen's inequality to show that you win the contest with a probability of at least $\frac{1}{2}$, regardless of the number of people in the first round. *Hint*: Argue that this probability is given by $E\left(\frac{1}{1+X}\right)$, where X is the number of people surviving the first round.

3.5 Independence of Random Variables

In Chapter 2, we dealt with the concept of independent events. It makes intuitive sense to say that random variables are independent when the underlying events are independent. Let X and Y be two random variables that are defined on the same sample space with probability measure P. The following definition does not require that X and Y are discrete random variables, but applies to the general case of two random variables X and Y.

Definition 3.2 *Random variables X and Y are said to be independent if*

$$P(X \le x, Y \le y) = P(X \le x)P(Y \le y)$$

for any two real numbers x and y, where $P(X \le x, Y \le y)$ represents the probability of occurrence of both event $\{X \le x\}$ and event $\{Y \le y\}$.

In words, the random variables X and Y are independent if the event of the random variable X taking on a value less than or equal to x and the event of the random variable Y taking on a value less than or equal to y are independent for all real numbers x, y. More generally, random variables $X_1, \ldots, X_n$ are said to be independent if

$$P(X_1 \le x_1, \ldots, X_n \le x_n) = P(X_1 \le x_1) \cdots P(X_n \le x_n)$$

for all real numbers $x_1, \ldots, x_n$.[2] This definition implies that X_i and X_j are pairwise independent for any $i \ne j$. This is easily seen. We get $P(X_i \le x_i, X_j \le x_j) = P(X_i \le x_i)P(X_j \le x_j)$ for all x_i, x_j by letting the other $x_k \to \infty$ in $P(X_1 \le x_1, \ldots, X_n \le x_n) = P(X_1 \le x_1) \cdots P(X_n \le x_n)$ and using the fact that $\lim_{x_k \to \infty} P(X_k \le x_k) = 1$.

Using the axioms of probability theory, it can be shown that Definition 3.2 is equivalent to

$$P(X \in A, Y \in B) = P(X \in A)P(Y \in B)$$

for any two sets A and B of real numbers. The proof is omitted. In fact, a rigorous proof for the non-discrete case is beyond the scope of this book and

---

[2] An infinite collection of random variables is said to be independent if every finite subcollection of them is independent. In applications, the independence or otherwise of random variables is usually obvious from the physical construction of the underlying process.

involves measurability constraints on the sets A and B. It is not difficult to verify the following two rules from the alternative definition of independence.

Rule 3.6 *If X and Y are independent random variables, then the random variables $f(X)$ and $g(Y)$ are independent for any two functions f and g.*

In the case that X and Y are discrete random variables, another representation of independence can be given.

Rule 3.7 *Discrete random variables X and Y are independent if and only if*

$$P(X = x, Y = y) = P(X = x)P(Y = y) \quad \text{for all } x, y.$$

A very useful rule applies to the calculation of the expected value of the product of two independent random variables:

Rule 3.8 *If the random variables X and Y are independent, then*

$$E(XY) = E(X)E(Y),$$

assuming that $E(X)$ and $E(Y)$ are finite.

The proof goes as follows for discrete random variables X and Y. Let I and J denote the sets of possible values of the random variables X and Y. Define the random variable Z by $Z = XY$. Then,

$$E(Z) = \sum_z zP(Z = z) = \sum_z z \sum_{x,y:\, xy=z} P(X = x, Y = y)$$

$$= \sum_z \sum_{x,y:\, xy=z} xyP(X = x, Y = y).$$

Hence, using the independence of X and Y,

$$E(Z) = \sum_{x,y} xyP(X = x, Y = y) = \sum_{x,y} xyP(X = x)P(Y = y)$$

$$= \sum_{x \in I} xP(X = x) \sum_{y \in J} yP(Y = y) = E(X)E(Y).$$

The manipulations with the summations are justified by the assumption that the series $\sum_{x \in I} xP(X = x)$ and $\sum_{y \in J} yP(Y = y)$ are absolutely convergent.

The converse of the above result is not true. It is possible that $E(XY) = E(X)E(Y)$, while X and Y are not independent. A simple example is as follows. Suppose that two fair dice are tossed. Denote by the random variable V_1 the number appearing on the first die and by the random variable V_2 the number appearing on the second die. Let $X = V_1 + V_2$ and $Y = V_1 - V_2$. It is obvious that the random variables X and Y are not independent. We leave it to the reader to verify that $E(X) = 7$, $E(Y) = 0$, and $E(XY) = E(V_1^2 - V_2^2) = 0$. Hence, $E(XY) = E(X)E(Y)$.

Rule 3.8 has been proved for the case that X and Y are discrete random variables. The rule is valid for any type of independent random variables X and Y, and the rule can also be extended to the case of finitely many independent random variables. The technical machinery that is required to establish those extended results is beyond the scope of this introductory book.

In Rule 3.1 we proved that the expectation operator has the linearity property. This property holds for the variance operator only under an independence assumption.

Rule 3.9 *If the random variables X and Y are independent, then*

$$\text{var}(X + Y) = \text{var}(X) + \text{var}(Y).$$

The proof is as follows. Putting $\mu_X = E(X)$ and $\mu_Y = E(Y)$, it follows that $\text{var}(X + Y) = E[(X + Y - (\mu_X + \mu_Y))^2] = E[(X - \mu_X + Y - \mu_Y)^2]$ can be worked out as

$$E[(X - \mu_X)^2] + E[(Y - \mu_Y)^2] + 2E[(X - \mu_X)(Y - \mu_Y)]$$
$$= \text{var}(X) + \text{var}(Y) + 2E[(X - \mu_X)]E[(Y - \mu_Y)] = \text{var}(X) + \text{var}(Y),$$

using the linearity property of the expectation operator and using Rule 3.8 together with $E[(X - \mu_X)] = E[(Y - \mu_Y)] = 0$.

Rule 3.9 is valid for any type of independent random variables X and Y. Also, Rule 3.9 can be extended directly to the case of finitely many independent random variables by using the algebraic formula $(a_1 + \cdots + a_n)^2 = \sum_{i=1}^{n} a_i^2 + 2\sum_{i=1}^{n-1}\sum_{j=i+1}^{n} a_i a_j$. Hence, for independent random variables $X_1, X_2, \ldots, X_n$,

$$\text{var}(X_1 + X_2 + \cdots + X_n) = \text{var}(X_1) + \text{var}(X_2) + \cdots + \text{var}(X_n).$$

This result has the following important corollary:

Rule 3.10 *If the random variables $X_1, X_2, \ldots, X_n$ are independent and have the same probability distribution with standard deviation σ, then the standard deviation of the sum $X_1 + X_2 + \cdots + X_n$ is given by*

$$\sigma(X_1 + X_2 + \cdots + X_n) = \sigma\sqrt{n}.$$

In statistics, Rule 3.10 is usually formulated as

$$\sigma\left(\frac{X_1 + X_2 + \cdots + X_n}{n}\right) = \frac{\sigma}{\sqrt{n}}.$$

This alternative formulation follows by noting that $\sigma(aX) = a\sigma(X)$ for any constant $a > 0$. Rule 3.10 is known as the *square-root law*. This law is sometimes called de Moivre's equation. The law was discovered by Abraham de Moivre (1667–1754) in 1730.[3] This law had an immediate impact on the way the mass of golden coins that were struck by the London Mint was controlled. The allowable deviation in the guinea was $\frac{1}{400}$ of its intended weight of 128 grains, which amounts to 0.32 grains. A run of 100 coins was taken periodically from the Mint and the total weight of these coins was compared to the weight of a standard of 100 coins. For almost 600 years the watchdogs allowed a variability of $100 \times 0.32 = 32$ grains in the weight of the 100 coins, but after the discovery of the square-root law in 1730, the allowable variability for 100 coins was changed to $\sqrt{100} \times 0.32 = 3.2$ grains. The gold for the coins was provided by the kings of England. Ignorance of the square-root law may have cost them a fortune.

Convolution Formula

Suppose that X and Y are two discrete random variables each having the set of nonnegative integers as the range of possible values. A useful rule is

Rule 3.11 *If the nonnegative, integer-valued random variables X and Y are independent, then*

$$P(X+Y=k) = \sum_{j=0}^{k} P(X=j)P(Y=k-j) \quad \text{for } k = 0, 1, \ldots.$$

This rule is called the *convolution rule*. The proof is as follows. Fix k. Let A be the event $\{X+Y=k\}$ and B_j be the event $\{X=j\}$. Then $A = \bigcup_{j=0}^{\infty} AB_j$. The events AB_j are disjoint and so $P(A) = \sum_{j=0}^{\infty} P(AB_j)$. Thus,

$$P(X+Y=k) = \sum_{j=0}^{\infty} P(X+Y=k, X=j).$$

Since $P(X+Y=k, X=j) = P(X=j, Y=k-j)$, we have by the independence of X and Y that

$$P(X+Y=k, X=j) = P(X=j)P(Y=k-j).$$

Thus, $P(X+Y=k) = \sum_{j=0}^{\infty} P(X=j)P(Y=k-j)$ for all k. Noting that $P(Y=k-j) = 0$ for $j > k$, we next get the convolution formula.

---

[3] The French-born Abraham de Moivre was the leading probabilist of the eighteenth century and lived most of his life in England. The protestant de Moivre left France in 1688 to escape religious persecution. He was a good friend of Isaac Newton and supported himself by calculating odds for gamblers and insurers and by giving private lessons to students.

3.5 Independence of Random Variables

A very important consequence of Rule 3.11 is the next rule:

Rule 3.12 *Suppose that the random variables X and Y are independent and have the Poisson distributions $P(X = k) = e^{-\lambda}\lambda^k/k!$ and $P(Y = k) = e^{-\mu}\mu^k/k!$ for $k = 0, 1, \ldots$. Then the random variable $X + Y$ has the Poisson distribution $P(X + Y = k) = e^{-(\lambda+\mu)}(\lambda+\mu)^k/k!$ for $k = 0, 1, \ldots$.*

The proof is based on the convolution Rule 3.11. We have

$$P(X+Y=k) = \sum_{j=0}^{k} e^{-\lambda}\frac{\lambda^j}{j!} e^{-\mu}\frac{\mu^{k-j}}{(k-j)!} = \frac{e^{-(\lambda+\mu)}}{k!}\sum_{j=0}^{k}\binom{k}{j}\lambda^j\mu^{k-j}.$$

Next, by applying Newton's binomium $(a+b)^k = \sum_{j=0}^{k}\binom{k}{j}a^j b^{k-j}$, we get the desired result $P(X+Y=k) = e^{-(\lambda+\mu)}\frac{(\lambda+\mu)^k}{k!}$ for $k = 0, 1, \ldots$.

Problems

3.56 A drunkard is standing in the middle of a very large town square. He begins to walk. Each step is a unit distance in one of the four directions east, west, north, and south. All four possible directions are equally probable. The direction of each step is chosen independently of the direction of the other steps. The drunkard takes a total of n steps. Verify that the squared distance between the drunkard's position after n steps and his starting position has expected value n for any value of n. *Hint*: The squared distance can be written as $(\sum_{i=1}^{n} X_i)^2 + (\sum_{i=1}^{n} Y_i)^2$, where the random variables X_i and Y_i denote the changes in the x-coordinate and the y-coordinate of the position of the drunkard caused by his ith step.

3.57 Let the random variable X be defined by $X = YZ$, where Y and Z are independent random variables each taking on the values -1 and 1 with probability 0.5. Verify that X is independent of both Y and Z, but not of $Y + Z$.

3.58 Suppose that $X_1, \ldots, X_n$ are independent random variables each having expected value μ and variance σ^2. Define the sample mean and the sample variance by the random variables $\overline{X}_n = \frac{1}{n}\sum_{k=1}^{n} X_k$ and $S_n^2 = \frac{1}{n}\sum_{k=1}^{n}(X_k - \overline{X}_n)^2$. Show that $E(\overline{X}_n) = \mu$ and $E(S_n^2) = \frac{n-1}{n}\sigma^2$. *Hint*: Writing $X_k - \overline{X}_n$ as $X_k - \mu + \mu - \overline{X}_n$, verify that $\sum_{k=1}^{n}(X_k - \overline{X}_n)^2 = \sum_{k=1}^{n}(X_k - \mu)^2 + n(\overline{X}_n - \mu)^2$.

3.59 Let X_i denote the number of integers smaller than i that precede i in a random permutation of the integers $1, \ldots, 10$. What are the expected value and the variance of the sum $X_2 + \cdots + X_{10}$? *Hint*: Verify that $X_i = \sum_{k=2}^{i} R_k$, where $R_k = 1$ if number $k-1$ precedes number k in the random permutation and $R_k = 0$ otherwise.

3.60 Suppose that X and Y are independent random variables each having the same probability mass function $\{(1-p)^{i-1}p,\ i = 1, 2, \ldots\}$. What is the probability mass function of $X + Y$?

3.61 Let $X_1, X_2, \ldots$ be a sequence of independent random variables each having the same distribution with finite mean. Suppose N is a nonnegative, integer-valued random variable such that the event $\{N \geq n\}$ does not depend on $X_n, X_{n+1}, \ldots$ for any $n \geq 1$. For the case that the X_k are nonnegative, prove *Wald's equation*[4] stating that $E\left(\sum_{k=1}^{N} X_k\right) = E(N)E(X_1)$. *Hint*: Define the indicator variable I_k by $I_k = 1$ if $N \geq k$ and $I_k = 0$ otherwise and note that $\sum_{k=1}^{N} X_k = \sum_{k=1}^{\infty} X_k I_k$.

3.6 Binomial Distribution

Let us first introduce the Bernoulli random variable, being the building block of the binomial random variable. A random variable X is said to have a *Bernoulli distribution* with parameter p if the random variable can only assume the values 1 or 0 with

$$P(X = 1) = p \quad \text{and} \quad P(X = 0) = 1 - p,$$

where $0 < p < 1$. A Bernoulli random variable X can be thought of as the outcome of an experiment that can only result in "success" or "failure." Such an experiment is called a *Bernoulli trial*. Since $E(X) = 0 \times (1-p) + 1 \times p = p$ and $E(X^2) = E(X) = p$, it follows that the expected value and the variance of X are given by

$$E(X) = p \quad \text{and} \quad \text{var}(X) = p(1-p).$$

To introduce the binomial distribution, imagine that n independent repetitions of a Bernoulli trial are performed. Let the random variable X be defined as the total number of successes in the n Bernoulli trials, each having the same probability p of success. Then,

$$P(X = k) = \begin{cases} \binom{n}{k} p^k (1-p)^{n-k} & \text{for } k = 0, 1, \ldots, n, \\ 0 & \text{otherwise.} \end{cases}$$

---

[4] The assumption of nonnegative X_k can be dropped for Wald's equation.

3.6 Binomial Distribution

This probability mass function is said to be the *binomial distribution* with parameters n and p. The derivation of the formula for $P(X = k)$ is simple. The sample space of the compound experiments consists of all possible n-tuples of 0's and 1's, where a "1" means a "success" and a "0" means a "failure." Any specific element containing k ones and $n - k$ zeros gets assigned the probability $p^k(1 - p)^{n-k}$, by the independence of the trials. The number of elements containing k ones and $n - k$ zeros is $\binom{n}{k}$. This completes the explanation of the formula for $P(X = k)$. The expected value and the variance of the random variable X are given by

$$E(X) = np \quad \text{and} \quad \text{var}(X) = np(1 - p).$$

These results can be derived by working out $E(X) = \sum_{k=1}^{n} kP(X = k)$ and $E(X(X - 1)) = \sum_{k=1}^{n} k(k - 1)P(X = k)$. However, this requires quite some algebra. A simpler approach is as follows. Write the random variable X as $X = X_1 + \cdots + X_n$, where $X_1, \ldots, X_n$ are independent random variables each having a Bernoulli distribution with parameter p. Next, using the fact that $E(X_i) = p$ and $\text{var}(X_i) = p(1 - p)$, the results follow from the Rules 3.1 and 3.8.

It is worthwhile pointing out that the binomial distribution has nearly all of its mass within three standard deviations of the expected value when $np(1 - p) \geq 25$. In a bold advertisement campaign, a beer company once made clever use of this fact. The beer company invited 100 beer drinkers for a blind taste test to compare the company's beer with the beer of a key competitor. The live taste test was broadcast at half time of the Superbowl, in front of tens of millions of people watching the playoffs on television. The brilliant move of the marketeers was not to invite the average beer drinker, but beer drinkers of the competing beer. Since many beers taste about the same and the typical beer drinker cannot tell the difference between two such beers in a blind taste test, the marketeers were pretty sure that the test would reveal that 35% or more of the beer drinkers would prefer the other beer over their favorite beer (a value less than 35 for a binomial random variable with $n = 100$ and $p = \frac{1}{2}$ would be more than three standard deviations below the expected value). Such a test result would be very impressive, since it concerns beer drinkers who stated that they swear by the competing beer. In the advertisement campaign the laws of probability indeed acted in favor of the beer company, as anticipated.

A classical problem of historical significance in which the binomial distribution shows up is the *problem of points*. This problem was the subject of the correspondence between the great French mathematicians Blaise Pascal (1623–1662) and Pierre de Fermat (1601–1665). This correspondence laid the groundwork for the birth of the study of probability. It was the catalyst that enabled probability theory to develop beyond mere combinatorial enumeration.[5]

Example 3.14 Two equally strong players A and B play a ball game such that a total of six points is required to win the game. Each inning counts for one point. The pot is 24 ducats and goes to the winner of the game. By some incident, the players cannot finish the game when player A has five points and player B three. How should the pot be divided between the two players?

Solution. This problem had been the subject of debate for quite some time before it was solved by Pascal and Fermat in the seventeenth century. In the fifteenth and sixteenth centuries, the Italian mathematicians Pacioli, Tartaglia, and Cardano proposed different schemes to divide the pot, but these schemes were based on deterministic arguments and were not satisfying. The idea of Pacioli was to divide the pot according to the score 5:3 and to split the pot of 24 ducats into $\frac{5}{8} \times 24 = 15$ ducats for player A and 9 ducats for player B. Tartaglia made the following comment to Pacioli's rule: "his rule seems neither agreeable nor good, since, if one player has, by chance, one point and the other no points, then, by following this rule, the player who has one point should take the whole pot, which obviously does not make sense." Tartaglia proposed another scheme that is based on the difference between the numbers of points already won by the two players. He calculated that the difference $5 - 3 = 2$ in the points already won by the two players is one-third of the required 6 points and proposed giving player A half of the pot plus one-third of half of the pot, which amounts to $12 + \frac{1}{3} \times 12 = 16$ ducats.

In the summer of 1654, Pascal and Fermat addressed the problem of points. They realized that in a fair division of the pot, the amounts of prize money received by the players should be proportional to the respective win probabilities of the players if the game were to be continued. These probabilities depend only on the number of points left to be won. At the point of stopping, player A is 1 point away from the required 6 points and player B 3 points. In the actual game, at most $1 + 3 - 1 = 3$ more innings would be needed to declare a winner. A trick to solve the problem is to imagine that three additional innings were played. The probability of player A being the ultimate winner if the original game was to be continued is the same as the probability

---

[5] Until the middle of the seventeenth century, probability calculations were restricted to counting the number of favorable outcomes and the total number of outcomes in games of chance, using the handbook *Liber de Ludo Aleae* (The Book of Games of Chance) by the colorful Italian mathematician and physician Gerolamo Cardano (1501–1576).

3.6 Binomial Distribution

that A would win one or more innings in three additional innings (explain!). This probability is given by the binomial probability

$$\sum_{k=1}^{3} \binom{3}{k} \left(\frac{1}{2}\right)^k \left(\frac{1}{2}\right)^{3-k} = \frac{7}{8}.$$

Therefore, player A should receive $\frac{7}{8} \times 24 = 21$ ducats and player B 3 ducats.

Example 3.15 Chuck-a-Luck is a carnival game of chance and is played with three dice. To play this game, the player chooses one number from the numbers $1, \ldots, 6$. The three dice are then rolled. If the player's number does not come up at all, the player loses 10 dollars. If the chosen number comes up one, two, or three times, the player wins $10, $20, or $30, respectively. What is the expected win for the house per wager?

Solution. This game seems at first glance to be favorable for the player. This is actually not the case, even if the expected value of the number of times the chosen number comes up is equal to $\frac{1}{2}$. Let the random variable X denote the number of times the player's number comes up. The random variable X can be seen as the number of successes in $n = 3$ independent Bernoulli trials with a success probability of $p = \frac{1}{6}$ and thus has a binomial distribution. This gives

$$P(X = k) = \binom{3}{k} \left(\frac{1}{6}\right)^k \left(\frac{5}{6}\right)^{3-k} \quad \text{for } k = 0, 1, 2, 3.$$

So, the expected win (in dollars) for the house per wager is

$$10 \times \frac{125}{216} - 10 \times \frac{75}{216} - 20 \times \frac{15}{216} - 30 \times \frac{1}{216} = 0.787.$$

A healthy house edge of 7.87%!

Example 3.16 A military early-warning installation is constructed in a desert. The installation consists of five main detectors and a number of reserve detectors. If fewer than five detectors are working, the installation ceases to function. Every two months an inspection of the installation is mounted and at that time all detectors are replaced by new ones. There is a probability of 0.05 that any given detector will cease to function between two successive inspections. The detectors function independently of one another. How many reserve detectors are needed to ensure a probability of less than 0.1% that the system will cease to function between two successive inspections?

Solution. Suppose that r reserve detectors are installed. Let the random variable X denote the number of detectors that will cease to function between two consecutive inspections. Then the random variable X has a binomial distribution with parameters $n = 5 + r$ and $p = 0.05$. The probability that the system will cease to function between two inspections is

$$P(X > r) = \sum_{k=r+1}^{5+r} \binom{5+r}{k} 0.05^k \, 0.95^{5+r-k}.$$

This probability has the value 0.0038 for $r = 2$ and the value 0.00037 for $r = 3$. Therefore, three reserve detectors should be installed.

Problems

3.62 Daily Airlines flies every day from Amsterdam to London. The price for a ticket on this popular route is $75. The aircraft has a capacity of 150 passengers. Demand for tickets is greater than capacity, and tickets are sold out well in advance of flight departures. The airline company sells 160 tickets for each flight, to protect itself against no-show passengers. The probability of a passenger being a no-show is $q = 0.1$. No-show passengers are refunded half the price of their tickets. Passengers that do show up and are not able to board the flight due to the overbooking are refunded the full amount of their tickets plus an extra $425 compensation. What is the probability that more passengers will turn up for a flight than the aircraft has the seating capacity for? What are the expected value and standard deviation of the daily return for the airline?

3.63 What are the chances of getting at least one six in one throw of 6 dice, at least two sixes in one throw of 12 dice, and at least three sixes in one throw of 18 dice? Which game do you think is more likely to win? This problem is known as the Newton–Pepys problem and was brought to the attention of Isaac Newton by Samuel Pepys, who was President of the Royal Society of London and apparently a gambling man.

3.64 The Yankees and the Mets are playing a best-four-of-seven series. The winner takes all the prize money of one million dollars. The two teams are evenly matched and the outcomes of the games are independent of each other. Unexpectedly, the competition must be suspended when the Yankees lead two games to one. How should the prize money be divided between the two teams?

3.65 The Salk vaccine against polio was tested in 1954 in a carefully designed field experiment. Approximately 400,000 children took part in this experiment. Using a randomization procedure, the children were randomly divided into two groups of equal size, a treatment group and a control group. The vaccine was given only to the children in the treatment group; the control-group children received placebo injections. The children did not know which of the two groups they had been placed into. The diagnosticians also lacked this information (double-blind experiment). Fifty-seven children in the treatment group went on to contract polio, while 142 children in the control group contracted the illness. How likely would such a difference in outcomes be when the assignment to the treatment or control group had absolutely no effect on the outcomes?

3.66 A game of chance played historically by Canadian Indians involved throwing eight flat beans into the air and seeing how they fell. The beans were symmetrical and were painted white on one side and black on the other. The bean thrower would win one point if an odd number of beans came up white, two points if either zero or eight white beans came up, and would lose one point for any other configurations. Does the bean thrower have the advantage in this game?

3.67 In order to qualify for a certain program, you are put through n tests. You pass any test with probability p, independently of the other tests. If you have passed k of the n tests, then you are admitted to the program with probability $\frac{k}{n}$ for $k = 0, 1, \ldots, n$. What is the unconditional probability that you will be admitted to the program?

3.68 The final match of the soccer World Cup is a draw at the end of extra time. The teams then proceed to penalty kicks. Each team attempts five shots. Suppose that each shot results in a goal with probability 0.7, independently of the other shots. What is the probability that the match is still tied after one round of five shots by each team?

3.69 You flip 100 coins. Those that land heads are set aside. You then flip the coins that have landed tails and again set aside all those that land heads. Finally, you flip a third time all coins that landed tails twice and again set aside all those that land heads. What is the probability mass function of the number of coins that are set aside?

3.70 In an ESP experiment, a medium has to guess the correct symbol on each of 250 Zener cards. Each card has one of the five possible Zener symbols on it and each of the symbols is equally likely to appear. The medium will get $100,000 if he gives 82 or more correct answers. Can you give a quick assessment of the probability that the medium has to be paid out?

3.71 On bridge night, the cards are dealt ten times. Only twice you do receive cards with an ace. From the beginning, you had your doubts as to whether the cards were being shuffled thoroughly. Are these doubts confirmed?

3.72 An experiment has r possible outcomes $O_1, O_2, \ldots, O_r$ with respective probabilities of $p_1, p_2, \ldots, p_r$. Suppose that n independent repetitions of the experiment are performed. Let the random variable X_i be the number of times that the outcome O_i occurs. Argue that

$$P(X_1 = x_1, X_2 = x_2, \ldots, X_r = x_r) = \frac{n!}{x_1! x_2! \cdots x_r!} p_1^{x_1} p_2^{x_2} \cdots p_r^{x_r}$$

for all nonnegative integers $x_1, x_2, \ldots, x_r$ with $x_1 + x_2 + \cdots + x_r = n$. This distribution is called the *multinomial distribution*.

3.73 A particular game is played with five poker dice. Each die displays an ace, king, queen, jack, ten, and nine. Players may bet on two of the six images displayed. When the dice are thrown and the bet-on images turn up, the player receives three times the amount wagered. In all other cases, the amount of the wager is forfeited. Is this game advantageous for the player?

3.7 Poisson Distribution

A random variable X is said to have a *Poisson distribution* with parameter $\lambda > 0$ if

$$P(X = k) = \begin{cases} e^{-\lambda} \frac{\lambda^k}{k!} & \text{for } k = 0, 1, \ldots, \\ 0 & \text{otherwise,} \end{cases}$$

where $e = 2.71828\ldots$ is the base of the natural logarithm. In Example 3.12 it was verified that

$$E(X) = \lambda \quad \text{and} \quad \text{var}(X) = \lambda.$$

The Poisson distribution is named after the French mathematician Siméon-Denis Poisson (1781–1840). In his 1837 book *Recherches sur la Probabilité des Jugements en Matière Criminelle et en Matière Civile* (Research on the Probability of Criminal and Civil Verdicts), he indirectly introduced a probability distribution that would later develop into one of the most important distributions in probability theory. Poisson himself did not recognize the huge practical importance of the distribution that would later be named after him, and he dedicated just one page to this distribution in his book. It was L. von Bortkiewicz, in 1898, who first discerned and explained the importance of the Poisson distribution in his book *Das Gesetz der Kleinen Zahlen* (The Law of Small Numbers). One unforgettable example from this book applies the Poisson model to the number of Prussian cavalry deaths attributed to fatal horse kicks in each of the 14 cavalry corpses over the 20 years 1875 to 1894, see also Problem 6.18 in Chapter 6.

3.7 Poisson Distribution

The Poisson model was of great importance in the analysis of the distribution of hits of flying bombs (V-1 and V-2 missiles) in London during World War II. The British authorities were anxious to know if these weapons could be aimed accurately at a particular target, or whether they were landing at random. If the missiles were in fact only randomly targeted, the British could simply disperse important installations to decrease the likelihood of their being hit. An area of 36 km² in South London was divided into 576 regions, 250 m wide by 250 m long, and the number of hits in each region was determined. The 576 regions were struck by 535 bombs, and so the average number of hits per region was 0.9288. There were 229 regions with zero hits, 211 regions with one hit, 93 regions with two hits, 35 regions with three hits, 7 regions with four hits, 1 region with five hits, and 0 regions with six or more hits. The analysis showed that the observed relative frequencies of the number of hits were each very close to the corresponding probabilities given by a Poisson distribution with expected value 0.9288. This implied that the distribution of hits in the area was much like the distribution of hits when each of the many flying bombs was to fall on any of the equally sized regions with the same small probability, independently of the other flying bombs. It will be seen below that this characterizes the Poisson distribution. The analysis convinced the British military that the bombs struck at random and had no advanced aiming ability.

The Poisson distribution can be seen as a limiting case of the binomial distribution with parameters n and p, when n is very large and p is very small. That is, in a very large number of independent repetitions of a chance experiment having a very small probability of success, the total number of successes is approximately Poisson distributed. This insight is essential in order to apply the Poisson distribution in practical situations. To give a precise mathematical formulation of the limiting result, let X represent a binomially distributed random variable with the parameters n and p. Assume now that n becomes *very large* and p becomes *very small*, while np is kept equal to the constant λ. The following is then true:

$$\lim_{n\to\infty, p\to 0} P(X=k) = e^{-\lambda}\frac{\lambda^k}{k!} \quad \text{for } k=0,1,\ldots.$$

The proof is as follows. Since $p = \frac{\lambda}{n}$, $P(X=k) = \binom{n}{k}\left(\frac{\lambda}{n}\right)^k\left(1-\frac{\lambda}{n}\right)^{n-k}$ and so

$$P(X=k) = \frac{\lambda^k}{k!}\left(1-\frac{\lambda}{n}\right)^n \left[\frac{n!}{n^k(n-k)!}\right]\left(1-\frac{\lambda}{n}\right)^{-k}.$$

Now let's look at the different terms separately. Take a fixed value for k, where $0 \le k \le n$. The term $\frac{n!}{n^k(n-k)!}$ can be written as

$$\frac{n(n-1)\cdots(n-k+1)}{n^k} = \left(1 - \frac{1}{n}\right)\cdots\left(1 - \frac{k-1}{n}\right).$$

With a *fixed* k, this term approaches 1 as $n \to \infty$, as does the term $(1 - \lambda/n)^{-k}$. A well-known result is that $(1 + b/n)^n$ tends to e^b as $n \to \infty$ for every real number b, see Appendix C. This results in

$$\lim_{n \to \infty} \left(1 - \frac{\lambda}{n}\right)^n = e^{-\lambda},$$

which completes the physical explanation of the Poisson distribution.

The Poisson distribution is useful in studying rare events. It can be used to model many practical phenomena. The explanation is that in many practical situations you can speak of many independent repetitions of a Bernoulli trial with a very small "success" probability. Examples include the number of credit cards that are stolen yearly in a certain area, the number of damage claims filed yearly with an insurance company, the yearly number of impacts of meteorites, and so on.

Example 3.17 The low earth orbit contains many pieces of space debris. It is estimated that an orbiting space station will be hit by space debris beyond a critical size and speed on average once in 400 years. What is the probability that a newly launched space station will not be penetrated in the first 20 years?

Solution. The event that the newly launched space station will be penetrated in the first 20 years is a rare event. Therefore, the Poisson model is an appropriate model to describe this event. The probability that a newly launched space station will not be penetrated in the first 20 years can be estimated by $e^{-\frac{1}{400} \times 20} = 0.9512$.

The Poisson model is characterized by the pleasant fact that one does not need to know the precise number of trials and the precise value of the probability of success; it is enough to know what the product of these two values is. The value λ of this product is usually known in practical applications and determines uniquely the Poisson distribution. What's very useful is the fact that the standard deviation of the Poisson distribution is the square root of the expected value of the distribution. A useful rule of thumb is that the distribution has nearly all of its mass within three standard deviations from the expected value when $\lambda \geq 25$. The probability $P(|X - \lambda| > 3\sqrt{\lambda})$ is on the order of 10^{-3} or less when $\lambda \geq 25$. In other words, it would be exceptional for a Poisson-distributed random variable to take on a value that differs more than $3\sqrt{\lambda}$ from the expected value λ. This is a very useful fact for statistical applications of

3.7 Poisson Distribution

the Poisson model. In order to illustrate this, suppose you read in a newspaper that, based on an average of 1,000 traffic deaths per year in previous years, the number of traffic deaths for last year rose 12%. How can you evaluate this? The number of traffic deaths over a period of one year can be modeled as a Poisson-distributed random variable with expected value 1,000 (why is this model reasonable?). An increase of 12% on an average of 1,000 is an increase of 120, or rather an increase of $120/\sqrt{1,000} = 3.8$ standard deviations above the expected value of 1,000. This can be considered exceptional. In this way, we find justification for the conclusion that the increase in the number of traffic deaths is not coincidental. What would your conclusions have been if, based on an average of 100 traffic deaths per year, a total of 112 traffic deaths occurred in the past year? You see that the Poisson model is a practically useful model to make quick statistical assessments.

Example 3.18 The Pegasus Insurance Company has introduced a policy that covers certain forms of personal injury with a standard payment of $100,000. The yearly premium for the policy is $25. On average, 100 claims per year lead to payment. There are more than one million policyholders. What is the probability that more than 15 million dollars will have to be paid out in the space of a year?

Solution. In fact, every policyholder conducts a personal experiment in probability after purchasing this policy, which can be considered to be "successful" if the policyholder files a rightful claim during the ensuing year. Let the random variable X denote the total number of claims that will be approved for payment during the year of coverage. In view of the many policyholders, there is a large number of independent probability experiments each having a very small probability of success. This means that the probability distribution of the random variable X can be modeled by a Poisson distribution with parameter $\lambda = 100$. The probability of having to pay out more than 15 million dollars is given by $P(X > 150)$. Since $E(X) = 100$ and $\sigma(X) = 10$, a value of 150 claims lies five standard deviations above the expected value. Thus, without doing any further calculations, we can draw the conclusion that the probability of paying out more than 15 million dollars in the space of a year must be extremely small. The precise value of the probability $P(X > 150)$ is $1 - \sum_{k=0}^{150} e^{-100} \frac{100^k}{k!} = 1.23 \times 10^{-6}$. Not a probability the insurance executives need worry about.

Example 3.19 In the kingdom of Lightstone, the game of 6/40 lottery is played. You win the jackpot when the six numbers on your ticket match the six winning lottery numbers from the numbers 1 to 40. At the time of an oil sheik's visit to Lightstone, the jackpot is listed at 12.5 million dollars. The oil sheik decides to take a gamble and orders his retinue to fill in 5 million tickets in his name. These 5 million tickets are not filled in by hand but are random picks generated by the computer of the lottery organization (inevitably, many six-number sequences will be generated more than once[6]). Suppose that the local people have purchased 2 million tickets and that these tickets are also random picks. Each ticket costs $1. What is the probability that the sheik will be among the jackpot winners and what is the probability that the sheik will be the only winner? What is the probability that the sheik's initial outlay will be won back from the jackpot?

Solution. The probability that a particular ticket is a winning ticket for the jackpot is $1/\binom{40}{6} = 1/3{,}838{,}380$. Very many tickets are filled in by the sheik and the local people. Therefore, the Poisson model is applicable. Let the random variable X denote the number of winning tickets for the jackpot among the 5 million tickets of the sheik and the random variable Y be the number of winning tickets for the jackpot among the 2 million tickets of the locals. The random variables X and Y can be modeled as Poisson random variables with respective parameters

$$\lambda = \frac{5{,}000{,}000}{3{,}838{,}380} \quad \text{and} \quad \mu = \frac{2{,}000{,}000}{3{,}838{,}380}.$$

The random variables X and Y are independent. The probability that the sheik will be among the jackpot winners is

$$P(X \geq 1) = 1 - e^{-\lambda} = 0.7282.$$

The probability that the sheik is the only jackpot winner is

$$P(X \geq 1, Y = 0) = P(X \geq 1)P(Y = 0) = (1 - e^{-\lambda})e^{-\mu} = 0.4325.$$

To answer the last question, let $p(r, s)$ denote the probability that the sheik has r winning tickets for the jackpot and the locals have s winning tickets for the jackpot. Then,

$$p(r, s) = e^{-\lambda}\frac{\lambda^r}{r!} \times e^{-\mu}\frac{\mu^s}{s!} \quad \text{for } r, s = 0, 1, \ldots.$$

Let $A = \{(r, s) : \frac{r}{r+s} \times 12.5 \geq 5\}$. Then, the probability that the sheik's initial outlay will be won back from the jackpot is $\sum_{(r,s) \in A} p(r, s) = 0.6935$.

---

[6] The expected number of different combinations among 5 million random picks is 2,795,050, as follows by using the result of Example 3.8 with $m = 5 \times 10^6$ and $b = \binom{40}{6}$.

A scenario as described in Example 3.19 actually happened in the Irish National Lottery. In the original 6/36 format of this lottery, the chances of matching all six numbers drawn is 1 in 1,947,792. At the lottery's initial cost of £0.50 per ticket, all possible combinations could be purchased for £973,896. When the May 1992 bank holiday draw reached a jackpot prize of £1,700,000, a Dublin syndicate successfully pulled off a brute-force attack and managed to buy over 80% of all possible lottery combinations. The consortium won the jackpot but had to share it with two other players who also guessed the six winning numbers correctly. Nevertheless, the consortium made a profit of over £300,000, because they also won numerous smaller prizes matching four or five numbers. To prevent such a brute-force attack from happening again, the Irish National Lottery became a 6/42 lottery.

The Poisson distribution has a remarkable splitting property. Where Rule 3.12 shows that the merging of two independent Poisson random variables gives again a Poisson random variable, the next rule shows that the splitting of a Poisson random variable gives two *independent* Poisson random variables. This result is of great importance.

Rule 3.13 *Suppose the number of events that occur in a given time interval is Poisson distributed with expected value λ. Each event is marked with probability p, independently of the other events. Let the random variable X be the number of marked events in the given time interval and Y be the number of unmarked events. Then,*

(a) *X and Y are Poisson distributed with expected values λp and $\lambda(1-p)$.*
(b) *The Poisson random variables X and Y are independent.*

The proof is as follows. Let the random variable N be the number of events occurring in the given time interval. Noting that $P(X=j, Y=k) = P(X=j, Y=k, N=j+k)$ and using the formula $P(AB) = P(A \mid B)P(B)$, we get

$$P(X=j, Y=k) = P(X=j, Y=k \mid N=j+k)P(N=j+k)$$

$$= \binom{j+k}{j} p^j(1-p)^k e^{-\lambda} \frac{\lambda^{j+k}}{(j+k)!}.$$

Since $\binom{j+k}{j} = \frac{(j+k)!}{j!k!}$ and $e^{-\lambda} = e^{-\lambda p}e^{-\lambda(1-p)}$, this result can be restated as

$$P(X=j, Y=k) = e^{-\lambda p}\frac{(\lambda p)^j}{j!} \times e^{-\lambda(1-p)}\frac{(\lambda(1-p))^k}{k!} \quad \text{for all } j, k \geq 0.$$

Next, by the relations $P(X=j) = \sum_{k=0}^{\infty} P(X=j, Y=k)$ and $P(Y=k) = \sum_{j=0}^{\infty} P(X=j, Y=k)$, it follows that

$$P(X=j) = e^{-\lambda p}\frac{(\lambda p)^j}{j!} \quad \text{and} \quad P(Y=k) = e^{-\lambda(1-p)}\frac{(\lambda(1-p))^k}{k!}$$

for all $j, k \geq 0$. Thus, X and Y are Poisson distributed with expected values λp and $\lambda(1-p)$. Moreover,

$$P(X=j, Y=k) = P(X=j)P(Y=k) \quad \text{for all } j, k \geq 0,$$

showing the remarkable result that X and Y are independent.

3.7.1 Poisson Heuristic[7]

The Poisson distribution is derived for the situation of many independent trials, each having a small probability of success. In case the independence assumption is not satisfied, but there is a "weak" dependence between the trial outcomes, the Poisson model may still be useful as an approximate model. In surprisingly many probability problems, the Poisson heuristic enables us to obtain quick estimates for probabilities that are otherwise difficult to calculate. This approach requires that the problem is reformulated in the framework of a series of weakly dependent trials. The idea of the method will be illustrated with extensions of the birthday problem and the matching problem.

The Almost-Birthday Problem and the Birthday-Coverage Problem

In Example 1.14, we discussed the classical birthday problem. The exact solution to this problem is easily derived. This is different for the almost-birthday problem. In this problem, the question is: What is the probability that, within a randomly formed group of m people, two or more people will have birthdays within one day of each other? The derivation of an exact formula for this probability is far from simple, but the Poisson heuristic is particularly simple to apply. The idea is to consider all possible combinations of two people and to trace whether, in any of those combinations, both people have birthdays within one day of each other. Only when such a combination exists can it be said that two or more people out of the whole group have birthdays on the same day. What you are doing, in fact, is conducting $n = \binom{m}{2}$ trials. A trial is said to be successful when the two people involved have birthdays within one day of each other. Every trial has the same success probability $p = \frac{3}{365}$ (explain!). Let the random variable X be the number of successful trials. The probability that, in a group of m people, two or more people will have birthdays within one day of each other is then equal to $P(X \geq 1)$. Although the outcomes of the trials are dependent on one another, this dependence can be considered to be

---

[7] This section can be skipped at first reading.

weak because of the large number (365) of possible birth dates. It is therefore reasonable to approximate the distribution of X by a Poisson distribution with expected value $\lambda = np = \frac{3}{2}m(m-1)/365$. In particular, $P(X \geq 1) \approx 1 - e^{-\lambda}$. Thus,

P(two or more people have birthdays within one day of each other)

$$\approx 1 - e^{-\frac{3}{2}m(m-1)/365}.$$

More generally, the Poisson heuristic leads to the approximation formula

$$1 - e^{-(\frac{2r+1}{2})m(m-1)/365}$$

for the probability that two or more people in a randomly assembled group of n people have their birthdays within r days of each other. For $r = 1$, the smallest value of m for which this approximate probability is at least 0.5 is 14. The approximate value of the probability that two or more people in a group of $n = 14$ people will have birthdays within one day of each other is $1 - e^{-0.74795} = 0.5267$. The exact value is 0.5375, and is calculated from the following formula for the probability of two or more people in a group of n people having their birthdays within r days from each other:[8]

$$1 - \frac{(365 - 1 - nr)!}{365^{n-1}(365 - (r+1)n)!}.$$

Another interesting problem is the following. How many people are required in order to have all 365 possible birthdays covered with a probability of at least 50%? This problem is called the birthday-coverage problem and can be framed as a coupon-collector problem, where birthdays correspond to coupons and people to purchases. The coupon-collector problem is as follows. Each time you purchase a certain product you receive a coupon, which is equally likely to be any one of n types. What is the probability of collecting a complete set of coupons in exactly r purchases? Imagine that you conduct a trial for each of the n coupons. Trial i is said to be successful if coupon i is *not* among the r purchases. Each trial has the same success probability $p = \left(\frac{n-1}{n}\right)^r$. A complete set of coupons is obtained only if there is no successful trial. Thus, by the Poisson heuristic, the probability of collecting a complete set of coupons in exactly r purchases is approximately

$$e^{-n \times [(n-1)/n]^r}.$$

---

[8] This formula is taken from J. I. Naus, "An extension of the birthday problem," *The American Statistician* **22** (1968): 27–29.

Table 3.1. *Probabilities for the card-matching problem*

k	0	1	2	3	4	5	6	7
exact	0.0162	0.0689	0.1442	0.1982	0.2013	0.1613	0.1052	0.060
approx.	0.0183	0.0733	0.1465	0.1954	0.1954	0.1563	0.1042	0.060

Taking $n = 365$, the smallest value of r for which $e^{-365 \times (364/365)^r} \geq 0.5$ is $r = 2{,}285$. Thus, the minimum number of people required in order to have all 365 possible birthdays covered with a probability of at least 50% is approximately 2,285. The exact answer is 2,287 people, see Problem 1.87.

A Matching Problem with Repetition

An instance of the classical matching problem was discussed in Example 1.19. Let's apply the Poisson heuristic to an instance of the generalized matching problem with repetition. Consider the Las Vegas card game, which goes back to Pierre Rémond de Montmort (1678–1719). An ordinary deck of 52 cards is thoroughly shuffled. The dealer turns over the cards one at a time, counting as he goes "ace, two, three, ..., king, ace, two, ...," and so on, so that the dealer ends up calling out the 13 ranks four times each. A match occurs if the card that comes up matches the rank called out by the dealer as he turns it over. Using the Poisson heuristic, it is easy to calculate an approximation to the probability of one or more matches. Imagine a trial for each card turned over. The trial is said to be successful if the card matches the rank called out by the dealer. There are $n = 52$ trials, and the success probability for each trial is $p = \frac{4}{52}$. The probability distribution of the number of matches is then approximated by a Poisson distribution with an expected value of $\lambda = 52 \times \frac{4}{52} = 4$. In particular, the probability of no match is approximately equal to $e^{-4} = 0.0183$. The exact value of this probability is 0.0162. In Table 3.1, we give the exact and approximate values of the probability of exactly k matches for several values of k.[9] Where the approximate values are obtained with little effort, it is very complex to obtain the exact values.

Problems

3.74 The Brederode Finance Corporation has begun the following advertising campaign in Holland. Each new loan application submitted is

---

[9] The exact values are taken from F. F. Knudsen and I. Skau, "On the asymptotic solution of a card-matching problem," *Mathematics Magazine* **69** (1996): 190–197.

accompanied by a chance to win a prize of $25,000. Every month, 50 zip codes from all Dutch zip codes will be drawn in a lottery. In Holland, each house address has a zip code and there are about 450,000 zip codes. Each serious applicant whose zip code is drawn will receive a $25,000 prize. Brederode Finance Corporation receives 200 serious loan applications each month. Use the Poisson distribution to calculate the probability distribution of the monthly amount that they will have to give away.

3.75 In the famous problem of Chevalier de Méré, players bet first on the probability that a six will turn up at least one time in four rolls of a fair die; subsequently, players bet on the probability that a double six will turn up in 24 rolls of a pair of fair dice. In a generalized version of the de Méré problem, the dice are rolled a total of $4 \times 6^{r-1}$ times; each individual roll consists of r fair dice being rolled simultaneously. A roll of the r dice is called a king's roll if each of the dice shows up a six. Argue that the probability of at least one king's roll occurring in $4 \times 6^{r-1}$ rolls converges to $1 - e^{-2/3} = 0.4866$ if $r \to \infty$.

3.76 Around 500 marriages take place in a certain town each year. What is an appropriate model to describe the probability distribution of the yearly number of marriages having the feature that both partners were born on the same day?

3.77 In a coastal area, the average number of serious hurricanes is 3.1 per year. The numbers of hurricanes in successive years are independent of each other. What is an appropriate probability model to calculate the probability of a total of more than 10 serious hurricanes in the next two years?

3.78 A total of 126 goals were scored by the soccer teams in the 48 games played in the group matches of the 1998 World Cup. It happened that 26 times zero goals were scored by a team, 34 times one goal, 24 times two goals, 8 times three goals, 1 time four goals, 2 times five goals, and 1 time six goals. Can you make plausible that the number of goals scored per team per game can be quite well described by a Poisson distribution with an expected value of 1.3125?

3.79 On average, there are 4.2 fatal shark attacks each year worldwide. What is the probability that there will more than seven fatal shark attacks next year worldwide?

3.80 In 1986 a woman won the New Jersey lottery twice in four months. The event was widely reported as an amazing coincidence with an incredibly small probability of happening. Use the Poisson distribution to explain that somebody winning the top lottery prize twice in a short period is practically a sure thing. Imagine 100 6/42 lotteries all over the world. In each lottery, each of one million people sends in five randomly chosen sequences of six numbers from the numbers 1 to 42 for each drawing. In each lottery there are two drawings every week. What is the probability that in at least one of the lotteries some person will win the jackpot twice in a period of three years?

3.81 The football pool is a betting pool based on predicting the outcome of 13 football matches in the coming week. In the Dutch football pool the average number of weekly winners who have correctly predicted all 13 matches is 0.25. Last week there were 3 winners. Can you explain why this is exceptional?

3.82 Let X be a random variable on the integers $0, 1, \ldots$ and $\lambda > 0$ be a given number. Verify that X is Poisson distributed with expected value λ if and only if $E[\lambda g(X+1) - Xg(X)] = 0$ for any bounded function $g(x)$ on the nonnegative integers. This result is called the *Stein–Chen identity* for the Poisson distribution. *Hint*: To verify the "if" assertion, choose $g(x)$ as $g(r) = 1$ and $g(x) = 0$ for $x \neq r$, where r is a given nonnegative integer.

3.83 Use the Poisson heuristic to approximate the probability that three or more people from a randomly selected group of 25 people will have birthdays on the same day and the probability that three or more persons from the group will have birthdays falling within one day of each other.

3.84 Use the Poisson heuristic to approximate the probability that some suit will be missing in a bridge hand of 13 cards.

3.85 Consider again Problem 1.94. Use the Poisson heuristic to approximate the probability that no couple will be paired as bridge partners.

3.86 What is the probability that no two cards of the same face value (two aces, for example) will succeed one another in a well-shuffled deck of 52 playing cards? Use the Poisson heuristic to verify that this probability is approximately e^{-3}.

3.87 A company has 75 employees in service. The administrator of the company notices, to his astonishment, that there are seven days on which two or more employees have birthdays. Use the Poisson heuristic to argue that this is not so astonishing after all.

3.88 What is the probability that in a random permutation of the integers $1, 2, \ldots, n$, no two integers have interchanged their original positions? Use the Poisson heuristic to show that this probability is approximately equal to $\frac{1}{\sqrt{e}}$ for large n.

3.89 Imagine a family of five brothers and sisters, together with their spouses. The family does a gift exchange every year at Christmas. Each member of the family buys a Christmas gift. The gifts are labeled 1 to 10, where each person knows the label of their gift and the label of the gift of their spouse. Cards with the numbers 1, ..., 10 are put in a hat. The family members consecutively pull a card out of the hat in order to determine the present he or she will receive. If a person draws a card with their own label, or the card with the label of their spouse, all the cards go back in the hat and the drawing is redone. Use the Poisson heuristic to approximate the probability that the drawing need not be redone.

3.90 Consider again Example 3.8. Use the Poisson heuristic to verify that the probability mass function of the number of empty bins can be approximated by a Poisson distribution with expected value $b(1 - \frac{1}{b})^m$ for large b.

3.91 Use the Poisson heuristic to approximate the probability that in a randomly formed group of m people, nobody has a lone birthday. What is the smallest value of m so that this probability is at least 50%?

3.92 Sixteen teams remain in a soccer tournament. A drawing of lots will determine which eight matches will be played. Before the drawing takes place, it is possible to place bets with bookmakers over the outcome of the drawing. You are asked to predict all eight matches, paying no regard to the order of the two teams in each match. Use the Poisson heuristic to approximate the probability distribution of the number of correctly predicted matches.

3.93 In the 6/45 lottery, six different numbers are randomly picked from the numbers 1 to 45. Use the Poisson heuristic to approximate the probability of two or more consecutive numbers in a random pick and the probability of three or more consecutive numbers.

3.8 Hypergeometric Distribution

The *urn* model is at the root of the hypergeometric distribution. In this model, you have an urn that is filled with R red and W white balls. You randomly select n balls out of the urn without replacing any. Define the random variable X as the number of red balls among the selected balls. Then,

$$P(X = r) = \frac{\binom{R}{r}\binom{W}{n-r}}{\binom{R+W}{n}} \quad \text{for } r = 0, 1, \ldots, n.$$

By the convention $\binom{a}{b} = 0$ for $b > a$, we have that $P(X = r) = 0$ for those r with $r > R$ or $n - r > W$. This probability mass function is called the *hypergeometric distribution* with parameters R, W, and n. The explanation of the formula for $P(X = r)$ is as follows. The number of ways in which r red and $n - r$ white balls can be chosen from the urn is $\binom{R}{r}\binom{W}{n-r}$, while the total number of ways in which n balls can be chosen from the urn is $\binom{R+W}{n}$. The ratio of these two expressions gives $P(X = r)$. The expected value and the variance of the random variable X are given by

$$E(X) = n\frac{R}{R+W} \text{ and } \text{var}(X) = n\frac{R}{R+W}\left(1 - \frac{R}{R+W}\right)\frac{R+W-n}{R+W-1}.$$

The simplest way to prove these results is to use indicator variables. Imagine that the balls are to be selected one at a time. Let the random variable X_i be equal to 1 if the ith ball drawn is red and 0 otherwise. Then $X = \sum_{i=1}^{n} X_i$ and $E(X) = \sum_{i=1}^{n} E(X_i)$. The random variables $X_1, \ldots, X_n$ are interchangeable, though they are dependent.[10] Thus, $E(X_i) = E(X_1)$ for all i. Obviously, $P(X_1 = 1) = \frac{R}{R+W}$ and so $E(X_1) = \frac{R}{R+W}$. This gives $E(X) = n\frac{R}{R+W}$. To find var(X), we use the relation

$$E(X^2) = \sum_{i=1}^{n} E(X_i^2) + 2\sum_{i=1}^{n-1}\sum_{j=i+1}^{n} E(X_i X_j)$$

together with the observation that $E(X_i^2) = E(X_1^2)$ for all i and $E(X_i X_j) = E(X_1 X_2)$ for all $i \neq j$. Therefore,

$$E(X^2) = nE(X_1^2) + n(n-1)E(X_1 X_2).$$

Since $E(X_1 X_2) = P(X_1 = 1, X_2 = 1) = P(X_1 = 1)P(X_2 = 1 \mid X_1 = 1)$, it follows that $E(X_1 X_2) = \frac{R}{R+W} \times \frac{R-1}{R+W-1}$. Thus,

$$E(X^2) = n\frac{R}{R+W} + n(n-1)\frac{R}{R+W} \times \frac{R-1}{R+W-1}.$$

Next, it is a matter of some algebra to get the desired expression for var(X).

The hypergeometric model is a versatile model and has many applications, particularly to lotteries. The hypergeometric distribution with parameters R, W, and n can be approximated by the binomial distribution with parameters n and $p = \frac{R}{R+W}$ if $R + W \gg n$. Then it does not make much difference whether the sampling is with or without replacement. Indeed, the above formulas for $E(X)$ and var(X) are very close to those of the binomial distribution with parameters n and $p = \frac{R}{R+W}$ if $R + W \gg n$.

---

[10] In general, random variables $X_1, \ldots, X_n$ are said to be interchangeable if, for every permutation $i_1, \ldots, i_n$ of $1, \ldots, n$, $P(X_{i_1} \leq x_1, \ldots, X_{i_n} \leq x_n) = P(X_1 \leq x_1, \ldots, X_n \leq x_n)$ for all $x_1, \ldots, x_n$.

3.8 Hypergeometric Distribution

Example 3.20 The New York state lottery offers the game called Quick Draw, a variant of Keno. The game can be played in bars, restaurants, bowling areas, and other places. A new game is played every four or five minutes and so a lot of games can be played on a day. In the game of Quick Draw a maximum of 20 individual bets can be made with a single game card. Each individual bet can be for 1, 2, 5, or 10 dollars. A player chooses four numbers from 1 to 80. The lottery then randomly chooses 20 numbers from 1 to 80. The number of matches determines the payoff. The payoffs on an individual bet are 55 times the bet size for four matches, 5 times the bet size for three matches, once the bet size for two matches, and nothing otherwise. In November 1997, the state lottery offered a promotion "Big Dipper Wednesday," where payoffs on the game were doubled on the four Wednesdays in that month. Is this a good deal for the player or just a come-on for a sucker bet?

Solution. Let us first have a look at the game with the ordinary payoffs. The hypergeometric model with $R = 4$, $W = 76$, and $n = 20$ is applicable to this game of Quick Draw. Let the random variable X indicate how many numbers on a single ticket are matched. Then,

$$P(X = k) = \frac{\binom{4}{k}\binom{76}{20-k}}{\binom{80}{20}} \quad \text{for } k = 0, 1, \ldots, 4.$$

This probability has the numerical values 0.308321, 0.432732, 0.212635, 0.043248, and 0.003063 for $k = 0, 1, 2, 3$, and 4. The expected payoff per one-dollar bet is $1 \times 0.212635 + 5 \times 0.043248 + 55 \times 0.003063 = 0.59734$ dollars. In other words, you would expect to lose about 40 cents per dollar bet on average; the usual house edge for this type of game. Now, what about the "Big Dipper Wednesday?" The expected payoff per one-dollar bet for the game with double payoffs becomes twice as much and is 1.1947 dollars. This is more than the amount of the one-dollar bet! The player now enjoys a 19.47% edge over the house in the game with double payoffs. It seems that the New York state lottery made a miscalculation.

The hypergeometric distribution exhibits in various guises, which have at first sight little to do with red and white balls in an urn. Two examples will be given to illustrate this.

Example 3.21 In a close election between two candidates A and B in a small town, the winning margin of candidate A is 1,422 to 1,405 votes. However, 101 votes are found to be illegal and have to be thrown out. It is not said how the illegal votes are divided between the two candidates. Assuming that the illegal votes are not biased in any particular way and the count is otherwise reliable, what is the probability that the removal of the illegal votes will change the result of the election?

Solution. The problem can be translated into the urn model with 1,422 red and 1,405 white balls. If a is the number of illegal votes for candidate A and b the number of illegal votes for candidate B, then candidate A will no longer have most of the votes only if $a - b \geq 17$. Since $a + b = 101$, the inequality $a - b \geq 17$ boils down to $a \geq 59$. The probability that the removal of the illegal votes will change the result of the election is the same as the probability of picking 59 or more red balls from an urn that contains 1,422 red and 1,405 white balls. This probability is given by

$$\sum_{a=59}^{101} \frac{\binom{1{,}422}{a}\binom{1{,}405}{101-a}}{\binom{2{,}827}{101}} = 0.0592.$$

Example 3.22 Two people, perfect strangers to one another, both living in the same city of one million inhabitants, meet each other. Each has approximately 500 acquaintances in the city. Assuming that for each of the two people, the acquaintances represent a random sampling of the city's various population sectors, what is the probability of the two people having an acquaintance in common?

Solution. A bit of imagination shows that this problem can be translated into the urn model with $R = 500$ red and $W = 999{,}498$ white balls, where the red balls represent the 500 acquaintances of the first person. The sought probability is given by the probability that at least one red ball will be drawn when 500 balls are randomly taken out of the urn. This probability is equal to $1 - \binom{999{,}498}{500} / \binom{999{,}998}{500} = 0.2214$. The probability of 22% is surprisingly large. Events are often less "coincidental" than we may tend to think!

Problems

3.94 In the game "Lucky 10," 20 numbers are drawn from the numbers 1 to 80. You tick 10 numbers on the game form. What is the probability of matching r of the 20 numbers drawn for $r = 0, 1, \ldots, 10$?

3.95 The New Amsterdam lottery offers the game Take Five. In this game, players must tick five different numbers from the numbers 1 to 39. The lottery draws five distinct numbers from the numbers 1 to 39. For every one dollar staked, the payoff is $100,000 for five correct numbers, $500 for four correct numbers, and $25 for three correct numbers.

For two correct numbers, the player wins a free game. What is the house percentage for this lottery?

3.96 Ten identical pairs of shoes are jumbled together in one large box. Without looking, someone picks four shoes out of the box. What is the probability that, among the four shoes chosen, there will be both a left and a right shoe?

3.97 A psychologist claims that he can determine from a person's handwriting whether the person is left-handed or not. You do not believe the psychologist and therefore present him with 50 handwriting samples, of which 25 were written by left-handed people and 25 were written by right-handed people. You ask the psychologist to say which 25 were written by left-handed people. Will you change your opinion of him if the psychologist correctly identifies 18 of the 25 left-handers?

3.98 The Massachusetts Cash Winfall lottery was established in 2004 and was ended in 2012. In this lottery, the jackpot was won when the six numbers chosen from 1 through 46 were correctly predicted. The Cash Winfall lottery had the special feature that the jackpot was "rolled down," with the secondary prizes increased when the jackpot rose above $2 million and was not won. Several gambling syndicates made big profits by buying tickets in bulk when the jackpot was approaching $2 million. Suppose that a syndicate has invested $400,000 in buying 200,000 randomly selected $2 tickets when the jackpot is "rolled down." The roll-down prizes are $25,000, $925, and $27.50 for matching five, four, or three numbers, respectively. What is the expected amount of money won by the syndicate? How can you approximate the standard deviation of the amount won?

3.99 For a final exam, your professor gives you a list of 15 items to study. He indicates that he will choose eight for the actual exam. You will be required to answer correctly at least five of those. You decide to study 10 of the 15 items. What is the probability that you will pass the exam?

3.100 In the 6/45 lottery, six different numbers are drawn at random from the numbers 1 to 45. What are the probability mass functions of the largest number drawn and the smallest number drawn?

3.101 You play Bingo together with 35 other people. Each player purchases one card with 24 different numbers that were selected at random out of the numbers 1 to 80. The organizer of the game calls out chosen distinct numbers between 1 and 80, randomly and one at a time. What is the probability that more than 70 numbers must be called out before one of the players has achieved a full card? What is the probability that you will be the first player to achieve a full card while no other player has a full card at the same time as you? What is the probability that you will be among the first players achieving a full card? *Hint*: Let $Q_k = P(X > k)$, where the random variable X counts how many numbers have to be called out before a *particular* player has achieved a full card.

3.102 Suppose that r different numbers are picked at random from the numbers $1, 2, \ldots, s$. Let the random variable X be the sum of the r numbers picked. Show that $E(X) = \frac{1}{2}r(s+1)$ and $\sigma^2(X) = \frac{1}{12}r(s+1)(s-r)$. *Hint*: Proceed as in the derivation of the expected value and the variance of the hypergeometric distribution.

3.103 A bowl contains a red and b white balls. You randomly pick without replacement one ball at a time until you have r red balls. What is the probability mass function of the number of picks needed?

3.104 Suppose you know that the hands of you and your bridge partner contain eight of the 13 spades in the deck. What is the probability of a 3-2 split of the remaining five spades in the bridge hands of your opponents?

3.105 A deck of cards has 8 diamonds and 7 spades. The diamonds are assigned to player A and the spades to player B. The cards are turned up one by one, and the player whose suit is first to be turned up five times wins. What is the probability that player A wins?

3.106 In the 6/49 lottery, six different numbers are drawn at random from $1, 2, \ldots, 49$. What is the probability that the next drawing will have no numbers in common with the last two drawings?

3.107 There is a concert and 2,500 tickets are to be raffled off. You have sent in 100 applications. The total number of applications is 125,000. What are your chances of getting a ticket? Can you explain why this probability is approximately equal to $1 - e^{-2}$?

3.9 Other Discrete Distributions

In this section we give several other discrete distributions that are frequently encountered in probability applications.

3.9.1 Discrete Uniform Distribution

A random variable X is said to have a *discrete uniform distribution* on the integers $a, a+1, \ldots, b$ if

$$P(X = k) = \frac{1}{b - a + 1} \quad \text{for } k = a, a+1, \ldots, b.$$

The random variable X can be thought of as the result of an experiment with finitely many outcomes, each of which is equally likely. Using the fact that $\sum_{k=1}^{n} k = \frac{1}{2}n(n+1)$ and $\sum_{k=1}^{n} k^2 = \frac{1}{6}n(n+1)(2n+1)$ for all $n \geq 1$, it is a matter of some algebra to verify that

$$E(X) = \frac{a+b}{2} \quad \text{and} \quad \text{var}(X) = \frac{(b-a+1)^2 - 1}{12}.$$

Example 3.23 Two friends lose each other while wandering through a crowded amusement park. They cannot communicate by mobile phone. Fortunately, they have foreseen that losing each other could happen and have made a prior agreement. The younger friend stays put at the entrance of a main attraction of the park and the older friend searches the main attractions in random order (the "wait-for-mommy" strategy). The park has 15 main attractions. What are the expected value and the standard deviation of the number of searches before the older friend finds his younger friend?

Solution. Let the random variable X denote the number of searches until the older friend finds his younger friend. The random variable X has a discrete uniform distribution on the integers $1, 2, \ldots, 15$. Therefore, the expected and the standard deviation of X are $E(X) = 8$ and $\sigma(X) = 4.32$.

3.9.2 Geometric Distribution

A random variable X is said to have a *geometric distribution* with parameter p if

$$P(X = k) = \begin{cases} p(1-p)^{k-1} & \text{for } k = 1, 2, \ldots, \\ 0 & \text{otherwise.} \end{cases}$$

The random variable X can be interpreted as the number of trials until the first success occurs in a sequence of independent Bernoulli trials with success probability p. Note that the geometric probability $(1-p)^{k-1}p$ is maximal for $k = 1$. This explains that in a sequence of independent Bernoulli trials, successes are likely to show up "clumped" rather than evenly spaced.

Using the relations $\sum_{k=1}^{\infty} k x^{k-1} = (1-x)^{-2}$ and $\sum_{k=1}^{\infty} k(k-1)x^{k-2} = 2(1-x)^{-3}$ for $|x| < 1$, see Appendix C, it is easily verified that

$$E(X) = \frac{1}{p} \quad \text{and} \quad \text{var}(X) = \frac{1-p}{p^2}.$$

Example 3.24 Each time, you simultaneously toss a fair coin and roll a fair die until heads is tossed or a six is rolled, or both. What is the probability mass function of the number of trials?

Solution. Let the random variable N be the number of trials until heads is tossed or a six is rolled, or both. Then, by the independence of the trials,

$$P(N > k) = \left(\frac{1}{2}\right)^k \times \left(\frac{5}{6}\right)^k = \left(\frac{5}{12}\right)^k \quad \text{for } k = 0, 1, \ldots.$$

This shows that N has a geometric distribution with parameter $1 - \frac{5}{12} = \frac{7}{12}$.

The result of Example 3.24 is a special case of the general result that $\min(X, Y)$ is geometrically distributed with parameter $p_1 + p_2 - p_1 p_2$ when the random variables X and Y are independent and geometrically distributed with parameters p_1 and p_2 (verify!).

3.9.3 Negative Binomial Distribution

A random variable X is said to have a *negative binomial distribution* with parameters r and p if

$$P(X = k) = \begin{cases} \binom{k-1}{r-1} p^r (1-p)^{k-r} & \text{for } k = r, r+1, \ldots, \\ 0 & \text{otherwise.} \end{cases}$$

The random variable X can be interpreted as the number of trials until the rth success occurs in a sequence of independent Bernoulli trials with success probability p. The explanation is as follows. The probability of having the rth success at the kth trial equals the binomial probability $\binom{k-1}{r-1} p^{r-1} (1-p)^{k-1-(r-1)}$ of having $r - 1$ successes among the first $k - 1$ trials multiplied by the probability p of having a success at the kth trial. To obtain $E(X)$ and $\text{var}(X)$, note that X can be written as $X_1 + \cdots + X_r$, where X_i is the number of trials needed in order to go from $i - 1$ to i successes. The random variables $X_1, \ldots, X_r$ are independent and each have a geometric distribution with parameter p. Since $E(X_i) = p$ and $\text{var}(X_i) = (1-p)/p^2$, an application of Rules 3.1 and 3.8 gives

$$E(X) = \frac{r}{p} \quad \text{and} \quad \text{var}(X) = \frac{r(1-p)}{p^2}.$$

Example 3.25 Suppose that a single die is rolled over and over. How many rolls are needed so that the probability of rolling a six three times within this number of rolls is at least 50%?

Solution. Let the random variable X be defined as the number of rolls of the die until a six appears for the third time. Then the random variable X has a negative binomial distribution with parameters $r = 3$ and $p = \frac{1}{6}$. We are asking for the smallest integer k for which $P(X \le k) \ge 0.5$.[11] The smallest integer k for which

$$\sum_{j=3}^{k} \binom{j-1}{2} \left(\frac{1}{6}\right)^3 \left(\frac{5}{6}\right)^{j-3} \ge 0.5$$

is given by $k = 16$. Hence, 16 rolls are needed.

Example 3.26 A particle moves in the positive quadrant of the plane until it is absorbed on either the x-axis or the y-axis, whichever is reached first. The steps of the particle are independent of each other. If the particle is at the point (i, j) with i and j positive integers, it moves to $(i - 1, j)$ with probability p and to $(i, j - 1)$ with probability $1 - p$. Suppose that the particle starts at the point (r, s). What is the probability of absorption at the point $(0, k)$ for $k = 1, \ldots, s$?

Solution. Let us say that a success occurs each time the particle makes a step to the left. Then, the random walk of the particle can be seen as a sequence of independent Bernoulli trials each having success probability p. The sought probability is nothing other than the probability of having the rth success at the $(r + s - k)$th trial. This probability is given by $\binom{r+s-k-1}{r-1} p^r (1-p)^{s-k}$ for $k = 1, \ldots, s$.

Problems

3.108 In the final of the World Series baseball, two teams play a series consisting of at most seven games until one of the two teams has won four games. Two unevenly matched teams are pitted against each other and the probability that the weaker team will win any given game is equal to 0.45. Assuming that the results of the various games are independent of each other, calculate the probability of the weaker team winning the final. What are the expected value and the standard deviation of the number of games the final will take?

3.109 You perform a sequence of independent Bernoulli trials, each having success probability $\frac{3}{4}$. What is the probability of 15 successes occurring before 5 failures?

---

[11] This value of k is the median of the distribution. In general, the median of an integer-valued random variable X is any integer m with $P(X \le m) \ge 0.5$ and $P(X \ge m) \ge 0.5$. The median of a probability distribution need not be unique.

3.110 A red bag contains 15 balls and a blue bag 5 balls. Each time, you pick one of the two bags and remove one ball from the chosen bag. At each pick, the red bag is chosen with probability $\frac{3}{4}$ and the blue bag with probability $\frac{1}{4}$. What is the probability that the red bag will be emptied first? What is the probability that the red bag is emptied while there are still k balls in the blue bag for $k = 1, \ldots, 5$?

3.111 Two players A and B each roll a fair die until player A has rolled a 1 or 2, or player B has rolled a 4, 5, or 6. The first player to roll one of his assigned numbers is the winner, where player A is also the winner if both players roll at the same time one of their assigned numbers. What is the probability of player A winning the game? What is the probability mass function of the length of the game?

3.112 In European roulette, the ball lands on one of the numbers $0, 1, \ldots, 36$ at every spin of the wheel. A gambler offers at even odds the bet that the house number 0 will come up at least once in every 25 spins of the wheel. What is the gambler's expected profit per dollar bet?

3.113 Players A and B toss their own coins at the same time. A toss of the coin of player A results in heads with probability a and a toss of the coin of player B with probability b. The first player to toss heads wins. If they both get heads at the same time, the game ends in a draw. What is the probability of player A winning and what is the probability of a draw? What is the probability mass function of the length of the game?

3.114 A fair coin is tossed until heads appears for the third time. Let the random variable X be the number of tails shown up to that point. What is the probability mass function of X? What are the expected value and the standard deviation of X?

3.115 You are offered the following game. You can repeatedly pick at random a number from 1 to 25. Each pick costs you one dollar. If you decide to stop, you get paid the dollar amount of your last pick. What strategy should you use to maximize your expected net payoff?

3.116 You toss a biased coin with probability p of heads, while your friend tosses at the same time a fair coin. What is the probability distribution of the number of tosses until both coins simultaneously show the same outcome?

3.117 John and Pete had a nice evening at the pub. They decided to play the following game in order to determine who would pay for the beer. Each of them rolls two dice. The game ends if the dice total of John is the same as that of Pete; otherwise, they play another round. Upon ending the game, John pays for the beer if the dice total is odd; otherwise,

Pete pays. What is the probability mass function of the length of the game? What is the probability of John paying for the beer?

3.118 You have a thoroughly shuffled deck of 52 cards. Each time, you choose one card from the deck. The drawn card is put back in the deck and all 52 cards are again thoroughly shuffled. You continue this procedure until you have seen all four different aces. What are the expected value and the standard deviation of the number of times you have to draw a card until you have seen all four different aces?

4
Continuous Random Variables

In many practical applications of probability, physical situations are better described by random variables that can take on a *continuum* of possible values rather than a *discrete* number of values. Examples are the decay time of a radioactive particle, the time until the occurrence of the next earthquake in a certain region, the lifetime of a battery, the annual rainfall in London, electricity consumption in kilowatt hours, and so on. These examples make clear what the fundamental difference is between discrete random variables taking on a discrete number of values and continuous random variables taking on a continuum of values. Whereas a discrete random variable associates *positive* probabilities with its individual values, any individual value has probability *zero* for a continuous random variable. It is only meaningful to speak of the probability of a continuous random variable taking on a value in some interval. Taking the lifetime of a battery as an example, it will be intuitively clear that the probability of this lifetime taking on a specific value becomes zero when a finer and finer unit of time is used. If you can measure the heights of people with infinite precision, the height of a randomly chosen person is a continuous random variable. In reality, heights cannot be measured with infinite precision, but the mathematical analysis of the distribution of people's heights is greatly simplified when using a mathematical model in which the height of a randomly chosen person is modeled as a continuous random variable. Integral calculus is required to formulate the continuous analogue of a probability mass function of a discrete random variable.

The first purpose of this chapter is to familiarize you with the concept of the probability density of a continuous random variable. This is always a difficult concept for the beginning student. However, integral calculus enables us to give an enlightening interpretation of probability density. The second purpose of this chapter is to introduce you to important probability densities such as the uniform, exponential, gamma, Weibull, beta, normal, and lognormal densities among others. In particular, the exponential and normal distributions are treated in depth. Many practical phenomena can be modeled by these distributions, which are of fundamental importance. Many examples are given to illustrate this. Also, much attention is given to the central limit theorem, being the most important theorem of probability theory. At the end of this chapter, we discuss special topics such as the inverse-transformation method for generating a random observation from a continuous random variable, the important concept of failure-rate function, and the principle of maximum entropy in probability.

4.1 Concept of Probability Density

The most simple example of a continuous random variable is the choice of a random number from the interval $(0, 1)$. The probability that the randomly chosen number will take on a prespecified value is zero. It only makes sense to speak of the probability of the randomly chosen number falling in a given subinterval of $(0, 1)$. Choosing a *random number* from the interval $(0, 1)$ means that the number is chosen such that the probability of the number falling in any given subinterval is equal to the length of that subinterval. For example, if a dart is thrown at random into the interval $(0, 1)$, the probability of the dart hitting exactly the point 0.25 is zero, but the probability of the dart landing somewhere in the interval between 0.2 and 0.3 is 0.1 (assuming that the dart has an infinitely thin point). No matter how small Δx is, any subinterval of length Δx has probability Δx of containing the point at which the dart will land. You might say that the probability mass associated with the landing point of the dart is smeared out over the interval $(0, 1)$ in such a way that the density is the same everywhere. Denote by the random variable X the point at which the dart will land. Then, the cumulative probability $P(X \leq a)$ equals a for any $0 \leq a \leq 1$ and can be represented as

$$P(X \leq a) = \int_0^a f(x)\,dx \quad \text{for } 0 \leq a \leq 1,$$

where $f(x)$ is a probability density that is identically equal to 1 on the interval $(0, 1)$. This probability density is called the *uniform density* on $(0, 1)$. Before defining the concept of probability density within a general framework, it is instructive to consider the next example.

Example 4.1 A stick of unit length is broken at a random point into two pieces. What is the probability that the ratio of the length of the shorter piece to that of the longer piece is less than or equal to a for any $0 < a < 1$?

Solution. The sample space of the chance experiment is the interval $(0, 1)$, where the outcome $\omega = u$ means that the point at which the stick is broken is a distance u from the beginning of the stick. Let the random variable X denote the ratio of the length of the shorter piece to that of the longer piece of the broken stick. Fix $0 < a < 1$. The probability that the ratio of the length of the shorter piece to that of the longer piece is less than or equal to a is nothing other than the probability that a random number from the interval $(0,1)$ falls either into $(\frac{1}{1+a}, 1)$ or into $(0, 1 - \frac{1}{1+a})$ (verify!). The latter probability is equal to $2(1 - \frac{1}{1+a}) = \frac{2a}{1+a}$. Thus,

$$P(X \leq a) = \frac{2a}{1+a} \quad \text{for } 0 < a < 1.$$

Obviously, $P(X \leq a) = 0$ for $a \leq 0$ and $P(X \leq a) = 1$ for $a \geq 1$. Denote by $f(a) = \frac{2}{(1+a)^2}$ the derivative of $\frac{2a}{1+a}$ for $0 < a < 1$ and let $f(a) = 0$ outside the interval $(0,1)$. Then, for any a, we can represent $P(X \leq a)$ as

$$P(X \leq a) = \int_{-\infty}^{a} f(x)\, dx.$$

The integral representation for $P(X \leq a)$ is the continuous analogue of the cumulative probability function in the discrete case: if X is a discrete random variable having possible values $a_1, a_2, \ldots$ with associated probabilities $p_1, p_2, \ldots$, then the probability that X takes on a value less than or equal to a is represented by

$$P(X \leq a) = \sum_{i:\, a_i \leq a} p_i \quad \text{for all } a.$$

We now come to the definition of a continuous random variable. Let X be a random variable that is defined on a sample space with probability measure P. It is assumed that the set of possible values of X is a finite or infinite interval on the real line.

Definition 4.1 *The random variable X is said to be (absolutely) continuously distributed if a function $f(x)$ exists such that*

$$P(X \leq a) = \int_{-\infty}^{a} f(x)\, dx \quad \text{for each real number } a,$$

where the function $f(x)$ satisfies

$$f(x) \geq 0 \text{ for all } x \quad \text{and} \quad \int_{-\infty}^{\infty} f(x)\, dx = 1.$$

4.1 Concept of Probability Density

The notation $P(X \leq a)$ stands for the probability that is assigned by the probability measure P to the set of all outcomes ω for which $X(\omega) \leq a$. The function $f(x)$ is called the *probability density function* of X, and is a sort of analogue of the probability mass function of a discrete random variable.

The *cumulative distribution function* of any random variable X, denoted by $F(x)$, describes the probability that the random variable X will take on a value less than or equal to x:

$$F(x) = P(X \leq x).$$

Sometimes one simply speaks of the distribution function. The cumulative distribution function $F(x)$ is an *increasing* function, that is, $F(x_1) \leq F(x_2)$ if $x_1 \leq x_2$. This property is obvious from $P(A) \leq P(B)$ if $A \subseteq B$. Moreover,

$$\lim_{x \to -\infty} F(x) = 0 \text{ and } \lim_{x \to \infty} F(x) = 1,$$

by the continuity of probability. Unlike the cumulative distribution function of a discrete random variable, that of a continuous random variable has no jumps and is continuous everywhere. The cumulative distribution function is an important concept that makes sense for any type of random variable. For discrete random variables the probability mass function can be reconstructed from the sizes of the jumps of the cumulative distribution function, while for continuous random variables the probability density function can be obtained by taking the derivative of the cumulative distribution function.

Before we illustrate the above concepts with two examples, we make the following remark. Beginning students often misinterpret the nonnegative number $f(a)$ as a probability, namely as the probability $P(X = a)$. This interpretation is wrong, as is obvious by the fact that $f(a)$ can be larger than 1. Nevertheless, it is possible to give an intuitive interpretation of the nonnegative number $f(a)$ in terms of probabilities. This will be done in the next section. In that section we also prove the intuitively obvious fact that $P(X = a) = 0$ for all a. This is a characteristic property of continuous random variables.

Example 4.2 Suppose that the lifetime X of a battery has the cumulative distribution function

$$P(X \leq x) = \begin{cases} 0 & \text{for } x < 0, \\ \frac{1}{4}x^2 & \text{for } 0 \leq x \leq 2, \\ 1 & \text{for } x > 2. \end{cases}$$

Does the random variable X satisfy the definition of a continuous random variable?

Solution. The cumulative distribution function $F(x) = P(X \leq x)$ is a continuous function and is differentiable at each point x, except for the two points $x = 0$ and $x = 2$. The derivative of $F(x)$ is continuous at each point at which $F(x)$ is differentiable. Therefore, we define the function $f(x)$ by

$$f(x) = \begin{cases} \frac{1}{2}x & \text{for } 0 < x < 2, \\ 0 & \text{otherwise.} \end{cases}$$

In each of the finite number of points x at which $P(X \leq x)$ has no derivative, it does not matter what value we give to $f(x)$. These values do not affect $\int_{-\infty}^{a} f(x)\,dx$. Usually, we give $f(x)$ the value 0 at any of these exceptional points. We can now conclude from the fundamental theorem of integral calculus that

$$P(X \leq a) = \int_{-\infty}^{a} f(x)\,dx \quad \text{for each number } a.$$

This completes the formal verification that X is a continuous random variable with probability density function $f(x)$.[1]

Example 4.3 A continuous random variable X has a probability density of the form $f(x) = ax + b$ for $0 < x < 1$ and $f(x) = 0$ otherwise. What conditions on the constants a and b must be satisfied? What is the cumulative distribution function of X?

Solution. A function represents a probability density only if it is nonnegative and integrates to 1 over the real line. The requirements for $f(x)$ are $ax + b \geq 0$ for $0 < x < 1$ and $\int_0^1 (ax+b)\,dx = 1$. The first requirement gives $a+b \geq 0$ and $b \geq 0$. The second requirement gives $\frac{1}{2}a + b = 1$. The cumulative distribution function of X is equal to

$$F(x) = \int_0^x (av + b)\,dv = \frac{1}{2}ax^2 + bx \quad \text{for } 0 \leq x \leq 1.$$

Further, $F(x) = 0$ for $x < 0$ and $F(x) = 1$ for $x > 1$.

The probability of any event defined in terms of a continuous random variable X with density function $f(x)$ can be calculated as an integral. First, consider the event $\{X > a\}$. The probability of this event allows for the integral representation

---

[1] In general, let X be a random variable with cumulative distribution function $F(x)$ and V be a finite subset of $R = (-\infty, \infty)$. Suppose that $F(x)$ is continuous on R and differentiable on $R \backslash V$ with a continuous derivative. Then, by the fundamental theorem of integral calculus, X is continuously distributed with density $f(x) = F'(x)$ for $x \in R \backslash V$ and $f(x) = 0$ for $x \in V$.

4.1 Concept of Probability Density

$$P(X > a) = \int_a^\infty f(x)\,dx.$$

To see this, note that the event $\{X > a\}$ is the complement of the event $\{X \le a\}$ and so $P(X > a) = 1 - P(X \le a)$. Since $\int_{-\infty}^\infty f(x)\,dx = 1$, it follows that $P(X > a)$ is equal to $\int_{-\infty}^\infty f(x)\,dx - \int_{-\infty}^a f(x)\,dx = \int_a^\infty f(x)\,dx$. More generally, we can express $P(a < X \le b)$ in terms of the density $f(x)$. For any constants a and b with $a < b$,

$$P(a < X \le b) = \int_a^b f(x)\,dx.$$

To see this, note that the event $\{X \le b\}$ is the union of the two disjoint events $\{a < X \le b\}$ and $\{X \le a\}$. As a result, $P(X \le b) = P(a < X \le b) + P(X \le a)$, or, equivalently, $P(a < X \le b) = P(X \le b) - P(X \le a)$. This gives that $P(a < X \le b)$ is equal to $\int_{-\infty}^b f(x)\,dx - \int_{-\infty}^a f(x)\,dx = \int_a^b f(x)\,dx$.

The integral representation of $P(a < X \le b)$ tells us that this probability is given by the area under the graph of $f(x)$ between the points a and b.

Example 4.4 The maximum outdoor air temperature (in degrees Celsius) in a certain area on any given day in May can be modeled as a continuous random variable X with density function $f(x) = \frac{1}{4,500}(30x - x^2)$ for $0 < x < 30$ and $f(x) = 0$ otherwise. What are the probabilities $P(X \le 10)$, $P(X > 25)$, and $P(15 < X \le 20)$?

Solution. The probabilities $P(X \le 10)$ and $P(X > 25)$ are given by

$$P(X \le 10) = \int_0^{10} \frac{1}{4,500}(30x - x^2)\,dx = \frac{1}{4,500}\left(15x^2 - \frac{1}{3}x^3\right)\Big|_0^{10} = \frac{7}{27}$$

$$P(X > 25) = \int_{25}^{30} \frac{1}{4,500}(30x - x^2)\,dx = \frac{1}{4,500}\left(15x^2 - \frac{1}{3}x^3\right)\Big|_{25}^{30} = \frac{2}{27}.$$

The probability $P(15 < X \le 20)$ is evaluated as

$$P(15 < X \le 20) = \int_{15}^{20} \frac{1}{4,500}(30x - x^2)\,dx = \frac{1}{4,500}\left(15x^2 - \frac{1}{3}x^3\right)\Big|_{15}^{20} = \frac{13}{54}.$$

Problems

4.1 Sizes of insurance claims can be modeled by a continuous random variable X with probability density $f(x) = c(10 - x)$ for $0 < x < 10$ and $f(x) = 0$ otherwise, where c is some constant. What is the value of c? What is the probability that the size of a particular claim is no more than 5 and what is the probability that the size is more than 2?

4.2 Let the random variable X be the portion of a flood insurance claim for flooding damage to a house. The probability density of X has the form $f(x) = c(3x^2 - 8x - 5)$ for $0 < x < 1$. What is the value of the constant c? What is the cumulative distribution function of X?

4.3 The mileage (in thousands of miles) you can get out of a specific type of tire is a continuous random variable X with probability density function $f(x) = \frac{2}{625}xe^{-x^2/625}$ for $x > 0$ and $f(x) = 0$ otherwise. What is the probability that the tire will last at most 15,000 miles? What is the probability that the tire will last more than 30,000 miles? What is the probability that the tire will last more than 20,000 miles but not more than 25,000 miles?

4.4 The lengths of phone calls (in minutes) made by a travel agent can be modeled as a continuous random variable X with probability density $f(x) = 0.25e^{-0.25x}$ for $x > 0$. What is the probability that a particular phone call will take more than 7 minutes?

4.5 A particular pumping engine will only function properly if an essential component functions properly. The time to failure of the component (in thousands of hours) is a random variable X with probability density $f(x) = 0.02xe^{-0.01x^2}$ for $x > 0$. What is the proportion of pumping engines that will not fail before 10,000 hours of use? What is the probability that the engine will survive for another 5,000 hours, given that it has functioned properly during the past 5,000 hours?

4.6 The probability density function $f(x)$ of the electrical resistance (in ohms) of a strain gauge produced by a certain firm can be modeled by $f(x) = \frac{1}{25}(x - 115)$ for $115 < x \le 120$, $f(x) = \frac{1}{25}(125 - x)$ for $120 < x < 125$, and $f(x) = 0$ otherwise. To meet the specification of a customer, the resistance must be in the range (117, 123). What proportion of strain gauges is not acceptable?

4.7 Liquid waste produced by a factory is removed once a week. The weekly volume of waste in thousands of gallons is a continuous random variable X with probability density function $f(x) = 105x^4(1 - x)^2$ for $0 < x < 1$ and $f(x) = 0$ otherwise. How can you choose the capacity of a storage tank so that the probability of overflow during a given week is no more than 5%?

4.8 You have to make a one-time business decision on how much stock to order in order to meet a random demand during a single period. The demand is a continuous random variable X with a given probability density $f(x) = \mu^2 xe^{-\mu x}$ for $x > 0$. Suppose that you decide to order Q units. What is the "stockout" probability that the stock Q will not be enough to meet the demand?

4.1.1 Interpretation of the Probability Density

The use of the word "density" originated with the analogy to the distribution of matter in space. In physics, any finite volume, no matter how small, has a positive mass, but there is no mass at a single point. A similar description applies to continuous random variables, on the grounds of the fact that the probability that a continuous random variable X with density $f(x)$ will take on a value between a and b can be seen as the area under the graph of $f(x)$ between the points a and b. Using the continuity property of probability (see Section 1.4.2), we have

$$P(X = a) = \lim_{n \to \infty} P\left(a - \frac{1}{n} < X \leq a + \frac{1}{n}\right) = \lim_{n \to \infty} \int_{a-\frac{1}{n}}^{a+\frac{1}{n}} f(x)\,dx,$$

and so $P(X = a) = \int_a^a f(x)dx$. This verifies that

$$P(X = a) = 0 \quad \text{for each real number } a.$$

Hence, for a continuous random variable X, it makes no sense to speak of the probability that the random variable X will take on a *prespecified* value. This probability is always zero. It only makes sense to speak of the probability that the continuous random variable X will take on a value in some interval. Incidentally, since $P(X = c) = 0$ for any number c, the probability that X takes on a value in an interval with endpoints a and b is not influenced by whether or not the endpoints are included. In other words, for any two real numbers a and b with $a < b$, we have

$$P(a \leq X \leq b) = P(a < X \leq b) = P(a \leq X < b) = P(a < X < b).$$

The fact that the area under the graph of $f(x)$ can be interpreted as a probability leads to an intuitive interpretation of $f(a)$. Let a be a given continuity point of $f(x)$. Consider now a small interval of length Δa around the point a, say $[a - \frac{1}{2}\Delta a, a + \frac{1}{2}\Delta a]$. Since

$$P\left(a - \frac{1}{2}\Delta a \leq X \leq a + \frac{1}{2}\Delta a\right) = \int_{a-\frac{1}{2}\Delta a}^{a+\frac{1}{2}\Delta a} f(x)\,dx$$

and $\int_{a-\frac{1}{2}\Delta a}^{a+\frac{1}{2}\Delta a} f(x)\,dx \approx f(a)\Delta a$ for Δa small, we have the insightful result

$$P\left(a - \frac{1}{2}\Delta a \leq X \leq a + \frac{1}{2}\Delta a\right) \approx f(a)\Delta a \quad \text{for } \Delta a \text{ small}.$$

In other words, the probability of random variable X taking on a value in a *small* interval around point a is approximately equal to $f(a)\Delta a$ when Δa is the length of the interval. You see that the number $f(a)$ itself is *not* a probability, but it is a relative measure for the likelihood that the random variable X will take on a value in the immediate neighborhood of point a. Stated differently, the probability density function $f(x)$ expresses how densely the probability mass of the random variable X is smeared out in the neighborhood of point x. Hence the name of density function. The probability density function provides the most useful description of a continuous random variable. The graph of the density function provides a good picture of the likelihood of the possible values of the random variable.

4.1.2 Verification of a Probability Density

In general, how can we verify whether a random variable X has a probability density? In concrete situations, we first determine the cumulative distribution function $F(a) = P(X \leq a)$ and next we verify whether $F(a)$ can be written in the form $F(a) = \int_{-\infty}^{a} f(x)\,dx$. A sufficient condition is that $F(x)$ is continuous at every point x and is differentiable except for a finite number of points x. The following two examples are given in illustration of this point.

Example 4.5 In a Vickrey auction, bidders submit written bids without knowing the bid of the other people in the auction. The highest bidder wins but the price paid is the second-highest bid. Suppose that in a Vickrey auction of a collector's item, 10 bids are made. The bids are independent of each other. Each bid size is between 0 and 1 and the probability that any bid size is more than x is $1 - x$ for $0 \leq x \leq 1$. What is the probability density of the price paid for the collector's item?

Solution. Let the random variable X be the price paid for the item. The random variable X is larger than x if and only if two or more bids are larger than x. Therefore, $P(X > x)$ is equal to the binomial probability $\sum_{k=2}^{10} \binom{10}{k} (1-x)^k x^{10-k}$ and so $P(X \leq x) = 1 - x^{10} - 10(1-x)x^9$ for $0 \leq x \leq 1$. Obviously, $P(X \leq x) = 0$ for $x < 0$ and $P(X \leq x) = 1$ for $x > 1$. The cumulative distribution function $P(X \leq x)$ is continuous at every point x. Also, it is differentiable with a continuous derivative except at the two points $x = 0$ and $x = 1$. Therefore, X has a probability density. By differentiation, we get the density function $f(x) = 90(x^8 - x^9)$ for $0 < x < 1$ and $f(x) = 0$ otherwise.

Example 4.6 A random point is picked inside a circular disk with radius r. Let the random variable X denote the distance from the center of the disk to this point. Does the random variable X have a probability density function and, if so, what is its form?

Solution. To answer the question, we first define a sample space with an appropriate probability measure P for the chance experiment. The sample space is taken as the set of all points (x, y) in the two-dimensional plane with $x^2 + y^2 \leq r^2$. Since the point inside the circular disk is chosen at random, we assign to each well-defined subset A of the sample space the probability

$$P(A) = \frac{\text{area of region } A}{\pi r^2}.$$

The cumulative distribution function $P(X \leq x)$ is easily calculated. The event $X \leq a$ occurs if and only if the randomly picked point falls into the disk of radius a with area πa^2. Therefore,

$$P(X \leq a) = \frac{\pi a^2}{\pi r^2} = \frac{a^2}{r^2} \quad \text{for } 0 \leq a \leq r.$$

Obviously, $P(X \leq a) = 0$ for $a < 0$ and $P(X \leq a) = 1$ for $a > r$. The distribution function $P(X \leq x)$ is continuous at every point x. Also, it is differentiable with a continuous derivative except at the points $x = 0$ and $x = r$. Therefore, X has the density function

$$f(x) = \begin{cases} \frac{2x}{r^2} & \text{for } 0 < x < r, \\ 0 & \text{otherwise.} \end{cases}$$

All of the foregoing examples follow the same procedure in order to find the probability density function of a random variable X. The cumulative distribution function $P(X \leq x)$ is determined first and this distribution function is then differentiated to obtain the probability density.

Problems

4.9 Let the radius of a circle be a random variable X with density function $f(x) = 1$ for $0 < x < 1$ and $f(x) = 0$ otherwise. What is the probability density of the area of the circle?

4.10 The density function of the continuous random variable X is $f(x) = \frac{6}{7}(x + \sqrt{x})$ for $0 < x < 1$ and $f(x) = 0$ otherwise. What is the probability density of $\frac{1}{X}$?

4.11 Let X be a positive random variable with probability density function $f(x)$. Define the random variable Y by $Y = X^2$. What is the probability density function of Y? Also, find the density function of the random variable $W = V^2$ if V is a number chosen at random from the interval $(-a, a)$ with $a > 0$.

4.12 A random point Q is chosen inside the unit square. What is the density function of the sum of the coordinates of the point Q? What is the density function of the product of the coordinates of the point Q? Use geometry to find these densities.

4.13 The number X is chosen at random between 0 and 1. Determine the density functions of the random variables $V = X/(1-X)$ and $W = X(1-X)$.

4.14 A stick of unit length is broken at random into two pieces. Let the random variable X be the length of the shorter piece. What is the density function of X? Also, use the cumulative distribution function of X to give an alternative derivation of the density function of the random variable $X/(1-X)$ from Example 4.1.

4.15 A random point is chosen inside the unit square. Let the random variables V and W be defined as the largest and the smallest of the two coordinates of the point. What are the probability densities of the random variables V and W?

4.16 Suppose you decide to take a ride on the ferris wheel at an amusement park. The ferris wheel has a diameter of 30 m. After several turns, the ferris wheel suddenly stops due to a power outage. Let the random variable X be your height above the ground when the ferris wheel stops, where it is assumed that the bottom of the ferris wheel is level with the ground. Verify that the density function of X is given by $\left(15\pi\sqrt{1-(x/15-1)^2}\right)^{-1}$ for $0 < x < 30$. *Hint*: The random variable X can be modeled as $X = 15 + 15\cos(\Theta)$, where Θ is a randomly chosen angle between 0 and π (make a drawing!).

4.2 Expected Value of a Continuous Random Variable

The definition of the expected value of a continuous random variable parallels that of a discrete random variable.

Definition 4.2 *The expected value of a continuous random variable X with probability density function $f(x)$ is defined by*

$$E(X) = \int_{-\infty}^{\infty} xf(x)\,dx,$$

provided that the integral is well-defined.

It is said that $E(X)$ *exists* if the integral $\int_{-\infty}^{\infty} xf(x)\,dx$ is well-defined. For a random variable X that can take on both positive and negative values, the integral is well-defined with a finite value if and only if $\int_{-\infty}^{\infty} |x|f(x)\,dx < \infty$. Such an integral is said to be *absolutely convergent*. For a nonnegative random variable X, the absolute convergence of the integral is not required. Then $E(X)$ always exists, in accordance with the basic convention from integral calculus that an integral with a nonnegative integrand is always well-defined by allowing ∞ as possible value. For example, the expected value of the random variable X with the two-sided density function $\frac{1}{\pi(1+x^2)}$ for $-\infty < x < \infty$ is not defined, because $\int_{-\infty}^{\infty} |x|\frac{1}{\pi(1+x^2)}\,dx = 2\int_0^{\infty} x\frac{1}{\pi(1+x^2)}\,dx = \infty$. However, the expected value of the nonnegative random variable with the one-sided density function $\frac{2}{\pi(1+x^2)}$ for $x > 0$ does exist (and is equal to ∞). It is important to emphasize that the expected value of a random variable X is finite if and only if $\int_{-\infty}^{\infty} |x|f(x)\,dx < \infty$.

The definition of expected value in the continuous case parallels the definition $E(X) = \sum x_i p(x_i)$ for a discrete random variable X with $x_1, x_2, \ldots$ as possible values and $p(x_i) = P(X = x_i)$. For dx small, the quantity $f(x)\,dx$ in a discrete approximation of the continuous case corresponds to $p(x)$ in the discrete case. The summation becomes an integral when dx approaches zero. Results for discrete random variables are typically expressed as sums. The corresponding results for continuous random variables are expressed as integrals.

Example 4.4 (continued) What is the expected value of the random variable X with probability density function $f(x) = \frac{1}{4,500}(30x - x^2)$ for $0 < x < 30$ and $f(x) = 0$ otherwise?

Solution. The expected value of X is calculated as

$$E(X) = \frac{1}{4,500}\int_0^{30} x(30x - x^2)\,dx = \frac{1}{4,500}\left(10x^3 - \frac{1}{4}x^4\right)\bigg|_0^{30} = 15.$$

Example 4.6 (continued) What is the expected value of the random variable X representing the distance from a random point inside the circular disk with radius 1 to the center of the disk?

Solution. The random variable X has the probability density function $f(x) = \frac{2x}{r^2}$ for $0 < x < r$ and $f(x) = 0$ otherwise. This gives

$$E(X) = \int_0^r x\frac{2x}{r^2}\,dx = \frac{2}{3}\frac{x^3}{r^2}\bigg|_0^r = \frac{2}{3}r.$$

Example 4.1 (continued) A stick of unit length is broken at a random point into two pieces. What is the expected value of the ratio of the length of the shorter piece to that of the longer piece? What is the expected value of the ratio of the length of the longer piece to that of the shorter piece?

Solution. Let the random variable X be the ratio of the length of the shorter piece to that of the longer piece and Y be the ratio of the length of the longer piece to that of the shorter piece. In Example 4.1, we showed that $P(X \leq x) = \frac{2x}{x+1}$ for $0 \leq x \leq 1$ with $f(x) = \frac{2}{(x+1)^2}$ as its density function for $0 < x < 1$. Hence,

$$E(X) = \int_0^1 x \frac{2}{(x+1)^2} dx = 2\int_0^1 \frac{1}{x+1} dx - 2\int_0^1 \frac{1}{(x+1)^2} dx$$

$$= 2\ln(x+1)\Big|_0^1 + 2\frac{1}{x+1}\Big|_0^1 = 2\ln(2) - 1.$$

In order to calculate $E(Y)$, note that $Y = \frac{1}{X}$. Therefore,

$$P(Y \leq y) = P\left(X \geq \frac{1}{y}\right) \quad \text{for } y > 1.$$

This leads to $P(Y \leq y) = 1 - \frac{2}{y+1}$ for $y > 1$. Thus, the density function of Y is $\frac{2}{(y+1)^2}$ for $y > 1$ and 0 otherwise. This gives

$$E(Y) = \int_1^\infty y \frac{2}{(y+1)^2} dy = 2\ln(y+1)\Big|_1^\infty + 2\frac{1}{y+1}\Big|_1^\infty = \infty.$$

This finding is in agreement with the expression $2(\frac{1}{2} + \cdots + \frac{1}{m+1})$ for the expected payoff in Problem 3.23, which expected payoff tends to infinity as $m \to \infty$. A little calculus was enough to find the result $E(Y) = \infty$ that otherwise is difficult to obtain from a simulation study. A more relevant measure for the random variable Y is the median. If R is a continuous random variable, then a *median* of R is any number m such that $P(R \leq m) = P(R \geq m) = 0.5$. The median of Y is $m = 3$, as follows by solving the equation $1 - \frac{2}{y+1} = 0.5$.

Problems

4.17 The time (in hundreds of hours) until failure of the power supply to a radar system is a random variable X with probability density function $f(x) = \frac{1}{625}(x - 50)$ for $50 < x \leq 75$, $f(x) = \frac{1}{625}(100 - x)$ for $75 < x < 100$, and $f(x) = 0$ otherwise. What is the expected value of X?

4.18 The time (in milliseconds) for a particular chemical reaction to complete in water is a random variable X with probability density function

$f(x) = \pi\sqrt{2}\cos(\pi x)$ for $0 < x < 0.25$ and $f(x) = 0$ otherwise. What is the expected value of X?

4.19 The javelin thrower Big John throws the javelin more than x meters with probability $P(x)$, where $P(x) = 1$ for $0 \le x < 50$, $P(x) = [1-(x-50)^2]/1{,}200$ for $50 \le x < 80$, $P(x) = (90-x)^2/400$ for $80 \le x < 90$, and $P(x) = 0$ for $x \ge 90$. What is the expected value of the distance thrown in his next shot?

4.20 What is the expected value of the random variable X in Problems 4.2, 4.4, and 4.6?

4.21 A random point is chosen inside a triangle with height h and base length b. What is the expected value of the perpendicular distance of the point to the base?

4.22 Consider again Example 4.5. What is the expected value of the price paid for the collector's item?

4.23 A random point is chosen inside the unit square $\{(x,y) : 0 \le x, y \le 1\}$. What is the expected value of the distance from this point to the point $(0,0)$?

4.24 A random point is chosen inside the unit square. What is the expected value of the distance from this point to the closest side of the unit square?

4.25 Let X be a continuous random variable with probability density function $f(x)$.

(a) Explain why the natural definition of the conditional expected value of X given that $X > a$ is

$$E(X \mid X > a) = \frac{1}{P(X > a)} \int_a^\infty x f(x)\, dx.$$

Also, explain the definition $E(X \mid X \le a) = \frac{1}{P(X \le a)} \int_{-\infty}^a x f(x)\, dx$.

(b) What is $E(X \mid X > a)$ for $0 < a < 1$ when X is a randomly chosen number from $(0,1)$?

(c) What are $E(X \mid X > a)$ and $E(X \mid X \le a)$ for $a > 0$ when X has the exponential density function $f(x) = \lambda e^{-\lambda x}$ for $x > 0$ and $f(x) = 0$ otherwise.

4.26 Let X be a nonnegative continuous random variable with density function $f(x)$. Use an interchange of the order of integration to verify that

$$E(X) = \int_0^\infty P(X > x)\, dx.$$

Use this result to answer the following question. What is the expected value of the smallest of n independent random numbers from $(0,1)$?

4.3 Substitution Rule and the Variance

The substitution rule and the concept of the variance of a random variable were discussed in Sections 3.4 and 3.5 for the case of a discrete random variable. The same results apply to the case of a continuous random variable.

Rule 4.1 *Let X be a continuous random variable with probability density $f(x)$. Then, for any given function $g(x)$, the expected value of the random variable $g(X)$ can be calculated from*

$$E[g(X)] = \int_{-\infty}^{\infty} g(x) f(x)\, dx,$$

provided that $\int_{-\infty}^{\infty} |g(x)| f(x)\, dx < \infty$ or the function $g(x)$ is nonnegative.

We first give the proof for the case that the random variable $g(X)$ is nonnegative. The proof is based on the fact that

$$E(Y) = \int_0^{\infty} P(Y > y)\, dy$$

for any nonnegative random variable Y, see Problem 4.26. Assuming that $g(X)$ is nonnegative, we get

$$E[g(X)] = \int_0^{\infty} P(g(X) > y)\, dy = \int_0^{\infty} dy \int_{x:\, g(x) > y} f(x)\, dx$$

$$= \int_{x:\, g(x) > 0} f(x)\, dx \int_0^{g(x)} dy = \int_{-\infty}^{\infty} g(x) f(x)\, dx.$$

It is noted that the interchange of the order of integration is justified by the fact that the integrand is nonnegative. The proof for the general case follows by writing $g(X)$ as the difference of the two nonnegative random variables $g^+(X) = \max(g(X), 0)$ and $g^-(X) = -\min(g(X), 0)$. The assumption $\int_{-\infty}^{\infty} |g(x)| f(x)\, dx < \infty$ implies that the expected values of $g^+(X)$ and $g^-(X)$ are finite.

We give two illustrative examples of the substitution rule.

Example 4.7 A warranty on an appliance specifies that an amount of $250 will be reimbursed if the appliance fails during the first year, an amount of $125 if it fails during the second year, and nothing if it fails after the second year. The time, measured in years, until the appliance fails has the probability density $f(x) = 0.2e^{-0.2x}$ for $x > 0$. What is the expected value of the warranty reimbursement?

Solution. Let the random variable X be the time until failure of the appliance. Define the function $g(x)$ by $g(x) = 250$ for $0 \leq x \leq 1$, $g(x) = 125$ for

4.3 Substitution Rule and the Variance

$1 < x \le 2$, and $g(x) = 0$ for $x > 2$. Then $g(X)$ is the warranty payment. Its expected value is

$$E[g(X)] = \int_0^1 250 \times \frac{1}{5}e^{-x/5}\, dx + \int_1^2 125 \times \frac{1}{5}e^{-x/5}\, dx$$
$$= 250(1 - e^{-1/5}) + 125(e^{-1/5} - e^{-2/5}) = 63.87 \text{ dollars}.$$

The substitution rule often simplifies the calculation of the expected value. As an illustration, consider again Example 4.1 in which the random variable X is the ratio of the length of the shorter piece to that of the longer piece of a stick of unit length that is broken at random into two pieces. In the foregoing section, we calculated $E(X)$ by determining first the density function of X and then applying the definition of expected value. However, the substitution rule provides a simpler way to calculate $E(X)$. Let the random variable U be distributed as a random point chosen in $(0, 1)$. Then U has the density function $f(u) = 1$ for $0 < u < 1$ and X is distributed as $g(U)$, where the function $g(u)$ is defined by $g(u) = u/(1 - u)$ for $0 < u \le \frac{1}{2}$ and $g(u) = (1 - u)/u$ for $\frac{1}{2} < u < 1$. This gives

$$E(X) = \int_0^{1/2} \frac{u}{1-u}\, du + \int_{1/2}^1 \frac{1-u}{u}\, du = 2\int_{1/2}^1 \frac{1-u}{u}\, du$$
$$= 2\ln(u) - 2u \Big|_{1/2}^1 = 2\ln(2) - 1.$$

Variance of a Continuous Random Variable

Next we deal with the variance and the standard deviation of a continuous random variable. Recall from Chapter 3 that the variance of a random variable X is defined by

$$\text{var}(X) = E[(X - \mu)^2],$$

where $\mu = E(X)$ and $E(X)$ is assumed to be finite. This definition applies to any type of random variable. If X is a continuous random variable with probability density $f(x)$, then the variance of X can be calculated from

$$\text{var}(X) = \int_{-\infty}^{\infty} (x - \mu)^2 f(x)\, dx.$$

Using the alternative representation $\text{var}(X) = E(X^2) - \mu^2$, the variance of X is usually calculated from

$$\text{var}(X) = \int_{-\infty}^{\infty} x^2 f(x)\, dx - \mu^2.$$

Rule 4.2 Let X be a continuous random variable with a finite expected value. Then, for any constants a and b,

$$E(aX + b) = aE(X) + b \quad \text{and} \quad \text{var}(aX + b) = a^2 \text{var}(X).$$

This rule is the continuous analogue of Rules 3.3 and 3.5 for the discrete case and can be verified directly from Rule 4.1. The variance of X does not have the same dimension as the values of X. Therefore, one often uses the standard deviation of the random variable X. This measure is defined by

$$\sigma(X) = \sqrt{\text{var}(X)}.$$

As an illustration, consider again Example 4.6. The variance of the random variable X representing the distance from a random point inside the circular disk with radius 1 to the center of the disk is calculated as

$$\text{var}(X) = \int_0^r x^2 \frac{2x}{r^2} dx - \left(\frac{2}{3}r\right)^2 = \frac{2r^2}{4} - \frac{4}{9}r^2 = \frac{1}{18}r^2.$$

The standard deviation of the distance X is $\sigma(X) = \sqrt{\text{var}(X)} = \sqrt{\frac{1}{18}}r$.

Example 4.8 Let the random variable X be a randomly chosen number from the interval (a, b). What are the expected value and the variance of X?

Solution. The probability that X will fall into a subinterval of width w is $\frac{w}{b-a}$. Hence, $P(X \leq x) = \frac{x-a}{b-a}$ for $a \leq x \leq b$ and so the density function of X is given by the uniform density $f(x) = \frac{1}{b-a}$ for $a < x < b$ and $f(x) = 0$ otherwise. This gives

$$E(X) = \int_a^b x \frac{1}{b-a} dx = \frac{1}{2} \frac{x^2}{b-a} \bigg|_a^b = \frac{1}{2} \frac{b^2 - a^2}{b-a} = \frac{a+b}{2},$$

using the fact that $b^2 - a^2 = (b-a)(b+a)$. Similarly, we find

$$E(X^2) = \int_a^b x^2 \frac{1}{b-a} dx = \frac{1}{3} \frac{x^3}{b-a} \bigg|_a^b = \frac{1}{3} \frac{b^3 - a^3}{b-a} = \frac{a^2 + ab + b^2}{3},$$

using the fact that $b^3 - a^3 = (b^2 + ab + a^2)(b-a)$. Thus,

$$\text{var}(X) = \frac{a^2 + ab + b^2}{3} - \left(\frac{a+b}{2}\right)^2 = \frac{(b-a)^2}{12}.$$

Problems

4.27 Suppose that the random variable X has the probability density $f(x) = 12x^2(1-x)$ for $0 < x < 1$ and $f(x) = 0$ otherwise.
  (a) Use Jensen's inequality from Section 3.4 to verify that $E[(\frac{1}{X})] \geq \frac{5}{3}$.
  (b) What are the expected value and the standard deviation of $\frac{1}{X}$?

4.28 Let the random variables V and W be defined by $V = \sqrt{U}$ and $W = U^2$, where U is a number chosen at random between 0 and 1. What are the expected values and the standard deviations of V and W?

4.29 An insurance policy for water damage pays an amount of damage up to $450. The amount of damage is a random variable X with density function $f(x) = \frac{1}{1,250}$ for $0 < x < 1,250$. The amount of damage exceeding $450 is covered by a supplement policy up to $500. What is the expected value of the amount of damage paid by the supplement policy?

4.30 Consider again Problem 4.7. Assume that the storage tank has a capacity of 0.9 expressed in thousands of gallons. The cost of removing $x > 0$ units of waste at the end of the week is $1.25 + 0.5x$. Additional costs of $5 + 10z$ are incurred when the capacity of the storage tank is not sufficient and an overflow of $z > 0$ units of waste occurs during the week. What is the expected value of the weekly costs?

4.31 A manufacturer has to make a last production run for a product that is near the end of its lifetime. The final demand for the product can be modeled as a continuous random variable X having probability density $f(x) = \frac{1}{2,500}xe^{-x/50}$ for $x > 0$. It is decided to make a last production run of 250 units of the product. The manufacturer earns $2 for each unit of product sold but incurs a cost of $0.50 for each unit of demand occurring when out of stock. What is the expected value of the net profit of the manufacturer? What is the probability that the manufacturer runs out of stock?

4.32 A car owner insures his car, worth $20,000, for one year under a policy with a deductible of $1,000. There is a probability of 0.01 of a total loss of the car during the policy year and a probability of 0.02 of a repairable damage. The cost (in thousands of dollars) of a repairable damage has the probability density $f(x) = \frac{1}{200}(20 - x)$ for $0 < x < 20$. What is the expected value of the insurance payment? Note that this payment is a mixed random variable.

4.33 Choose a point at random in $(0, 1)$, then this point divides the interval $(0, 1)$ into two subintervals. What is the expected length of the subinterval covering a given point s with $0 < s < 1$?

4.34 What is the standard deviation of the random variable X in each of the Problems 4.2, 4.4, and 4.6?

4.35 What are the expected value and the standard deviation of the area of the circle whose radius is a random variable X with density function $f(x) = 1$ for $0 < x < 1$ and $f(x) = 0$ otherwise?

4.36 A point Q is chosen at random inside a sphere with radius r. What are the expected value and the standard deviation of the distance from the center of the sphere to the point Q?

4.37 Let X be a continuous random variable with probability density $f(x)$ and finite expected value $E(X)$.

(a) What constant c minimizes $E[(X - c)^2]$ and what is the minimal value of $E[(X - c)^2]$?

(b) Prove that $E(|X - c|)$ is minimal if c is chosen equal to the median of X.

4.38 Consider again Problem 4.16. Calculate the expected value and the standard deviation of the height above the ground when the ferris wheel stops.

4.39 In an inventory system, a replenishment order is placed when the stock on hand of a certain product drops to the level s, where the reorder point s is a given positive number. The total demand for the product during the lead time of the replenishment order has the probability density $f(x) = \lambda e^{-\lambda x}$ for $x > 0$. What are the expected value and standard deviation of the shortage (if any) when the replenishment order arrives?

4.40 Suppose that the continuous random variable X has the probability density function $f(x) = (\alpha/\beta)(\beta/x)^{\alpha+1}$ for $x > \beta$ and $f(x) = 0$ for $x \leq \beta$ for given values of the parameters $\alpha > 0$ and $\beta > 0$. This density is called the *Pareto* density, which provides a useful probability model for income distributions, among others.

(a) Calculate the expected value, the variance, and the median of X.

(b) Assume that the annual income of the employed, measured in thousands of dollars, in a given country follows a Pareto distribution with $\alpha = 2.25$ and $\beta = 2.5$. What percentage of the working population has an annual income of between 25,000 and 40,000 dollars?

(c) Why do you think the Pareto distribution is a good model for income distributions? *Hint*: Use the probabilistic interpretation of the density $f(x)$.

4.4 Uniform and Triangular Distributions

The uniform and triangular distributions are often used in practice when not much is known about the underlying distribution of the observations. Also, the triangular distribution is related to the uniform distribution.

4.4.1 Uniform Distribution

A continuous random variable X is said to have a *uniform distribution* over the interval (a, b) if X has the probability density function

$$f(x) = \begin{cases} \frac{1}{b-a} & \text{for } a < x < b, \\ 0 & \text{otherwise.} \end{cases}$$

This density has two parameters a and b with $b > a$. Figure 4.1 gives the graph of the uniform density function. The uniform distribution on the interval $(0, 1)$ provides the probability model for random-number generators, which generate random numbers between 0 and 1. The uniform distribution is also used as a model for a quantity that is known to vary randomly between a and b but about which little else is known.

Since the density $f(x)$ of the uniformly distributed random variable X is constant over the interval (a, b), the random variable X is just as likely to be near any value in (a, b) as any other value. This property is expressed by

$$P\left(c - \frac{1}{2}\Delta \leq X \leq c + \frac{1}{2}\Delta\right) = \int_{c-\frac{1}{2}\Delta}^{c+\frac{1}{2}\Delta} \frac{1}{b-a} dx = \frac{\Delta}{b-a},$$

regardless of c provided that the points $c - \frac{1}{2}\Delta$ and $c + \frac{1}{2}\Delta$ belong to the interval (a, b). As shown in Example 4.8, the expected value and the variance of the random variable X are given by

$$E(X) = \frac{1}{2}(a+b) \quad \text{and} \quad \text{var}(X) = \frac{1}{12}(b-a)^2.$$

Also, an explicit expression can be given for the cumulative distribution function $F(x) = \int_{-\infty}^{x} f(y)\,dy$. This function satisfies $F(x) = 0$ for $x < a$, $F(x) = 1$ for $x > b$, and

$$F(x) = \frac{x - a}{b - a} \quad \text{for } a \leq x \leq b.$$

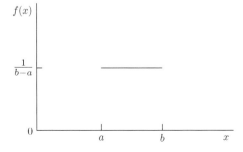

Figure 4.1. Uniform density.

Problems

4.41 The lifetime of a light bulb has a uniform probability density on (2, 12). The light bulb will be replaced upon failure or upon reaching age 10, whichever occurs first. What are the expected value and the standard deviation of the age of the light bulb at the time of replacement?

4.42 A rolling machine produces sheets of steel of different thickness. The thickness of a sheet of steel is uniformly distributed between 120 and 150 mm. Any sheet having a thickness of less than 125 mm must be scrapped. What are the expected value and the standard deviation of a non-scrapped sheet of steel?

4.4.2 Triangular Distribution

A continuous random variable X is said to have a *triangular distribution* over the interval (a, b) if X has the probability density function

$$f(x) = \begin{cases} \frac{2}{b-a} \frac{x-a}{m-a} & \text{for } a < x \leq m \\ \frac{2}{b-a} \frac{b-x}{b-m} & \text{for } m \leq x < b \end{cases}$$

and $f(x) = 0$ otherwise. This density has three parameters a, b, and m with $a < m < b$. Figure 4.2 gives the graph of the triangular density function. The density function increases linearly on the interval $[a, m]$ and decreases linearly on the interval $[m, b]$. The triangular distribution is often used as probability model when little information is available about the quantity of interest but one knows its lowest possible value a, its most likely value m, and its highest possible value b. The expected value and the variance of the random variable X are given by

$$E(X) = \frac{1}{3}(a+b+m) \quad \text{and} \quad \text{var}(X) = \frac{1}{18}(a^2 + b^2 + m^2 - ab - am - bm).$$

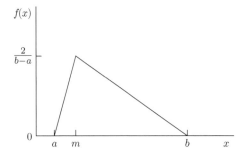

Figure 4.2. Triangular density.

The derivation is a matter of straightforward algebra and is left to the reader. Also, the cumulative distribution function $F(x) = \int_{-\infty}^{x} f(y)\,dy$ allows for an explicit expression. We have $F(x) = 0$ for $x < a$, $F(x) = 1$ for $x > b$, and

$$F(x) = \begin{cases} \dfrac{(x-a)^2}{(b-a)(m-a)} & \text{for } a \leq x < m \\ 1 - \dfrac{(b-x)^2}{(b-a)(b-m)} & \text{for } m \leq x \leq b. \end{cases}$$

The triangular distribution is related to the uniform distribution. In Problem 5.30 in Section 5.3, you are asked to verify that both $\frac{1}{2}(X+Y)$ and $|X-Y|$ have a triangular distribution when the random variables X and Y are independent and uniformly distributed on $(0, 1)$.

4.5 Exponential Distribution

A positive continuous random variable X is said to have an *exponential distribution* with parameter $\lambda > 0$ if X has the density function

$$f(x) = \lambda e^{-\lambda x} \quad \text{for } x > 0$$

and $f(x) = 0$ otherwise. It is left to the reader to verify that the nonnegative function $\lambda e^{-\lambda x}$ integrates to 1 over the interval $(-\infty, \infty)$ and thus is indeed a density function. The parameter λ is a scale parameter. An exponentially distributed random variable X takes on only positive values. Figure 4.3 displays the exponential density function with $\lambda = 1$. The expected value and the variance of the random variable X are given by

$$E(X) = \frac{1}{\lambda} \quad \text{and} \quad \text{var}(X) = \frac{1}{\lambda^2}.$$

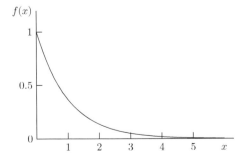

Figure 4.3. Exponential density ($\lambda = 1$).

These formulas are obtained by evaluating the integrals $\int_0^\infty x\lambda e^{-\lambda x}\,dx$ and $\int_0^\infty x^2\lambda e^{-\lambda x}\,dx$. Using partial integration, the reader easily verifies that these integrals are $\frac{1}{\lambda}$ and $\frac{2}{\lambda^2}$, showing that $E(X) = \frac{1}{\lambda}$ and $E(X^2) = \frac{2}{\lambda^2}$.

The cumulative distribution function $F(x) = \int_{-\infty}^x f(y)\,dy$ equals

$$F(x) = \begin{cases} 1 - e^{-\lambda x} & \text{for } x \geq 0 \\ 0 & \text{for } x < 0. \end{cases}$$

Memoryless Property

The exponential distribution is extremely important in probability. It not only models many real-world phenomena, but also allows for tractable mathematical analysis. The reason for its mathematical tractability is the *memoryless property*. This property states that

$$P(X > t + s \mid X > s) = P(X > t) \quad \text{for all } t > 0,$$

regardless of the value of s. In words, imagining that the exponentially distributed random variable X represents the lifetime of an item, the residual life of an item has the *same* exponential distribution as the original lifetime, regardless of how long the item has already been in use. The proof is simple. For any $x \geq 0$, we have that $P(X > x) = e^{-\lambda x}$. Applying the definition $P(A \mid B) = P(AB)/P(B)$ with $A = \{X > t + s\}$ and $B = \{X > s\}$, and noting that $P(X > t + s, X > s) = P(X > t + s)$, we find

$$P(X > t + s \mid X > s) = \frac{P(X > t + s)}{P(X > s)} = \frac{e^{-\lambda(t+s)}}{e^{-\lambda s}} = e^{-\lambda t} = P(X > t),$$

showing the memoryless property. The exponential distribution is the only continuous distribution possessing this property. The technical proof of this uniqueness result is omitted.

The exponential distribution is often used as probability model for the time until a *rare* event occurs. Examples are the time until the next earthquake in a certain region and the decay time of a radioactive particle. A very useful result is that under general conditions, the time until the first occurrence of a rare event is approximately exponentially distributed. To make this result plausible, consider the next example. Maintenance of an operating unit in a reliability system occurs at the scheduled times $\tau, 2\tau, \ldots$, where $\tau > 0$ is fixed. Each maintenance takes a negligible time and the unit is again as good as new after each maintenance. There is a probability $p > 0$ that the unit will fail between two maintenance inspections, where p is very close to zero. Let the random variable X be the time until the first system failure. Then, for any

n, we have $P(X > n\tau) = (1-p)^n$ and so $P(X > n\tau) \approx e^{-np}$, using the fact that $e^{-p} \approx 1 - p$ for p close to zero. Hence, taking $t = n\tau$ and replacing n by t/τ, it follows that

$$P(X > t) \approx e^{-\lambda t} \quad \text{for } t > 0,$$

where $\lambda = p/\tau$ denotes the inverse of the expected time until the first system failure. It is important to note that it suffices to know the expected value of the time until the first system failure in order to approximate the probability distribution of the failure time by an exponential distribution. An application of this very useful result will be given in Example 4.9 below.

Also, the probability of exceeding some *extreme* level is often approximately equal to an exponential tail probability of the form $\alpha e^{-\beta t}$ for constants $\alpha, \beta > 0$. An interesting example concerns the probability that a high tide of h meters or more above sea level will occur in any given year somewhere along the Dutch coastline. This probability is approximately equal to $e^{-2.97h}$ for values of h larger than 1.70 m. This empirical result was used in the design of the Delta works that were built following the 1953 disaster when the sea flooded a number of polders in the Netherlands.

Example 4.9 A reliability system has two identical units, where one unit is in full operation and the other unit is in cold standby. The lifetime of an operating unit has an exponential density with expected value $1/\mu$. Upon failure of the operating unit, the other unit is put into operation provided a standby unit is available. The replacement time of a failed unit is fixed and is equal to $\tau > 0$. A system failure occurs if no standby unit is available at the moment the operating unit fails. It is assumed that the probability $1 - e^{-\mu\tau}$ is close to zero, that is, the probability of an operating unit failing during the replacement time τ is very small. What is an approximation to the probability distribution of the time until the first system failure?

Solution. Let the random variable X denote the time until the first system failure. In Example 7.11 in Section 7.3, it will be shown that

$$E(X) = \frac{2 - e^{-\mu\tau}}{\mu(1 - e^{-\mu\tau})}.$$

Under the assumption that $1 - e^{-\mu\tau}$ is very small, the occurrence of a system failure is a rare event and then the probability distribution of X can be approximated by

$$P(X > t) \approx e^{-t/E(X)} \quad \text{for } t > 0.$$

As an illustration, take $\mu = 1$ and $\tau = 0.02$. Then $E(X) = 51.5017$. The probability $P(X > t)$ has the approximate values $0.8235, 0.6154, 0.3788, 0.1435$, and 0.0206 for $t = 10, 25, 50, 100$, and 200. The approximation is very accurate. Simulation with 100,000 runs for each t gives the simulated values 0.8232, $0.6148, 0.3785, 0.1420$, and 0.0204.

4.5.1 Poisson Process

The Poisson distribution is closely related to the exponential distribution. Both distributions are the fundamentals of the Poisson process. To introduce this process, let $X_1, X_2, \ldots, X_n$ be a sequence of independent random variables each having the same exponential density $\lambda e^{-\lambda x}$. Define the random variable S_n by

$$S_n = X_1 + X_2 + \cdots + X_n.$$

If the X_i can be interpreted as the interoccurrence times of some specific event (for example, failures of a device, or emission of particles from a radioactive source), then S_n represents the time at which the nth event occurs. An important result is that S_n has the probability density function

$$\frac{\lambda^n x^{n-1}}{(n-1)!} e^{-\lambda x} \quad \text{for } x > 0.$$

A proof of this result will be presented in Section 8.3. This density is called the *Erlang density* with parameters n and λ. The corresponding cumulative distribution function is given by

$$P(S_n \leq x) = 1 - \sum_{j=0}^{n-1} e^{-\lambda x} \frac{(\lambda x)^j}{j!} \quad \text{for } x > 0.$$

The easiest way to verify this formula is by differentiating its right-hand side term by term. This is left as an exercise for the reader.

For any $t \geq 0$, define the random variable $N(t)$ as

$$N(t) = \max\{n : X_1 + \cdots + X_n \leq t\}$$

with the convention that $N(t) = 0$ if $X_1 > t$. In words, $N(t)$ is the number of events occurring in $(0, t)$. Note that $N(0) = 0$, by definition. The collection $\{N(t), t \geq 0\}$ of random variables indexed by the time parameter t is called a *Poisson process*. The Poisson process is said to have *rate* λ when the interoccurrence times X_i have the expected value $1/\lambda$. The Poisson process is of utmost importance in applied probability and plays an essential role in

4.5 Exponential Distribution

applications of continuous-time Markov chains, which will be discussed in Chapter 11. We can now state the following important result.

Rule 4.3 *Let $X_1, X_2, \ldots$ be a sequence of independent random variables each having the same exponential density with expected value $1/\lambda$, where the random variable X_i describes the amount of time between the $(i-1)$th and the ith occurrences of some specific event. Let $N(t) = \max\{n : \sum_{j=1}^{n} X_j \leq t\}$. Then, for any $t > 0$,*

$$P(N(t) = k) = e^{-\lambda t} \frac{(\lambda t)^k}{k!} \quad \text{for } k = 0, 1, \ldots.$$

In words, $N(t)$ has a Poisson distribution with expected value λt.

The proof is instructive and goes as follows. Fix $t > 0$. First, we verify that $P(N(t) = 0) = e^{-\lambda t}$. This follows directly from $P(N(t) = 0) = P(X_1 > t)$ and $P(X_1 > t) = e^{-\lambda t}$. Next, observe that n or more events occur in $(0, t]$ if and only if the epoch of the nth occurrence of the event is before or at time t. This gives the important relation

$$P(N(t) \geq n) = P(S_n \leq t).$$

Since $P(N(t) = k) = P(N(t) \geq k) - P(N(t) \geq k+1)$, we have $P(N(t) = k) = P(S_k \leq t) - P(S_{k+1} \leq t)$. Thus, by the above formula for $P(S_n \leq t)$,

$$P(N(t) = k) = 1 - \sum_{j=0}^{k-1} e^{-\lambda t} \frac{(\lambda t)^j}{j!} - \left(1 - \sum_{j=0}^{k} e^{-\lambda t} \frac{(\lambda t)^j}{j!}\right) = e^{-\lambda t} \frac{(\lambda t)^k}{k!},$$

as was to be verified.

On the basis of the memoryless property of the exponential distribution, the result of Rule 4.3 can be extended as follows. For any $s, t > 0$,

$$P(N(s+t) - N(s) = k) = e^{-\lambda t} \frac{(\lambda t)^k}{k!} \quad \text{for } k = 0, 1, \ldots,$$

regardless of s and independently of what happened before time s. That is, the probability distribution of the number of arrivals in a given time interval depends only on the length of the interval. Therefore, it is said that the Poisson process $\{N(t), t \geq 0\}$ has *stationary increments*. This result is based on the fact that the amount of time measured from time point s to the first occurrence of an event after time s has the same exponential distribution as X_1. Intuitively, this fact is clear from the memoryless property of the exponential distribution, see also Problem 9.34 for a formal proof. Also, it will be intuitively clear from the memoryless property of the exponential distribution that the Poisson process $\{N(t), t \geq 0\}$ has *independent increments*, that is, the random variables

$N(t_1)-N(t_0),\ldots,N(t_n)-N(t_{n-1})$ are independent for all $0 \leq t_0 < t_1 \cdots < t_n$. The proof of this property stating that the Poisson process is memoryless will be omitted.

Example 4.10 In a hospital, five babies are born per day on average. It is reasonable to model the times of the arrivals of the babies by a Poisson process. Let the random variable X measure the time from midnight to the first arrival of a baby. What are the expected value and the median of X? What is the probability that more than two babies are born between 12 o'clock midnight and 6 o'clock in the morning?

Solution. Let us take the hour as unit of time. Then the random variable X has the exponential density $\lambda e^{-\lambda t}$ with $\lambda = \frac{5}{24}$. The expected value of X is $1/\lambda = 4.8$ hours. The median is found by solving m from $1 - e^{-\lambda m} = 0.5$ and thus is equal to $\ln(2)/\lambda = 3.33$ hours. The probability that more than two babies are born between 12 o'clock midnight and 6 o'clock in the morning is given by the Poisson probability

$$1 - e^{-6\lambda} - 6\lambda e^{-6\lambda} - (6\lambda)^2 e^{-6\lambda}/2! = 0.1315.$$

In Rule 3.13 we showed that a Poisson-distributed random variable can be split into two independent Poisson random variables. This important property will be used in the next example.

Example 4.11 A piece of radioactive material emits alpha particles according to a Poisson process with a rate of 0.84 particles per second. A counter detects each emitted particle with probability 0.95, independently of any other particle. In a 10-second period, the number of detected particles is 12. What is the probability that more than 15 particles were emitted in that period?

Solution. The number of particles that will be emitted during a 10-second period has a Poisson distribution with expected value $10 \times 0.84 = 8.4$. By Rule 3.13, the number of emitted particles that will be missed by the counter in the 10-second period has a Poisson distribution with expected value $0.05 \times 8.4 = 0.420$, independently of how many particles were detected in that period. Therefore, the sought probability is given by the probability of having more than three emissions of undetected particles in the 10-second period. This probability is $1 - \sum_{j=0}^{3} e^{-0.420}(0.420)^j/j! = 0.00093$.

Example 4.12 In a given city, traffic accidents occur according to a Poisson process with an average of $\lambda = 10$ accidents per week. In a certain week, seven accidents have occurred. What is the probability that exactly one accident has occurred on each day of that week?

Solution. Let the random variable $N(t)$ be the number of accidents occurring in the time interval $(0, t)$, where a day is taken as time unit. Denote by the epoch $t = u - 1$ the beginning of day u for $u = 1, \ldots, 7$. The probability we are looking for is $P(N(u) - N(u-1) = 1 \text{ for } u = 1, \ldots, 7 \mid N(7) = 7)$ with the convention $N(0) = 0$. By the property of independent increments, the random variables $N(1), N(2) - N(1), \ldots, N(7) - N(6)$ are independent and so this probability can be written as

$$\frac{P(N(1) = 1) \times P(N(2) - N(1) = 1) \times \cdots \times P(N(7) - N(6) = 1)}{P(N(7) = 7)}.$$

By the property of stationary increments, the random variables $N(1), N(2) - N(1), \ldots, N(7) - N(6)$ are Poisson distributed with expected value $\lambda/7$. Also, $N(7)$ is Poisson distributed with expected value λ. Thus, the sought probability is given by

$$\frac{e^{-\lambda/7}(\lambda/7) \times e^{-\lambda/7}(\lambda/7) \times \cdots \times e^{-\lambda/7}(\lambda/7)}{e^{-\lambda}\lambda^7/7!} = \frac{7!}{7^7} = 0.0162.$$

In the remark below it will be explained why this probability is so small. Note that the probability $7!/7^7$ is the same as the probability of getting exactly one random number in each of the seven intervals $(0, \frac{1}{7}), (\frac{1}{7}, \frac{2}{7}), \ldots, (\frac{6}{7}, 1)$ when drawing seven independent random numbers from $(0, 1)$.

Remark 4.1 There is a close relationship between the Poisson process and the uniform distribution. The following property can be proved. Under the condition that exactly r arrivals have occurred in the fixed time interval $(0, t)$, the r arrival epochs are statistically indistinguishable from r random points that are independently chosen in the interval $(0, t)$ according to the uniform distribution. We only verify the stated property for $r = 1$:

$$P(X_1 \leq x \mid N(t) = 1) = \frac{P(X_1 \leq x, N(t) = 1)}{P(N(t) = 1)} = \frac{P(N(x) = 1, N(t) = 1)}{P(N(t) = 1)}$$

$$= \frac{P(N(x) = 1, N(t) - N(x) = 0)}{P(N(t) = 1)} = \frac{\lambda x e^{-\lambda x} \times e^{-\lambda(t-x)}}{\lambda t e^{-\lambda t}} = \frac{x}{t} \text{ for } 0 \leq x \leq t,$$

where the fourth equality uses the independence of $N(x)$ and $N(t) - N(x)$. The relationship between the Poisson process and the uniform distribution explains why the times at which Poisson events occur show a tendency to cluster, see also Figure 4.4. It is inherent to randomly chosen points in an interval that these points are not evenly distributed over the interval, and so a tendency toward bunching is exactly how the Poisson process behaves. For example, this characteristic property of the Poisson process may be used to explain the surprisingly large number of shark attacks in Florida in the summer of

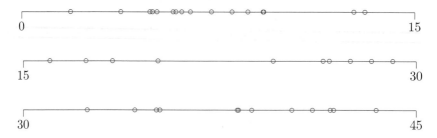

Figure 4.4. Events of a Poisson process with rate 1 in the time interval (0, 45).

1991. Unpredictable events such as shark attacks can be modeled by a Poisson process, with its bursty behavior. This means that there are periods with a much higher than average number of attacks, as well as periods with no attacks at all.

Problems

4.43 A reliability system has five operating components in parallel. The lifetimes of the components are independent random variables, each having an exponential distribution with expected value $\frac{1}{\lambda}$. The system fails as soon as only two components are working. What are the expected value and the standard deviation of the time until system failure? *Hint*: Verify first that $\min(X, Y)$ is exponentially distributed with parameter $\alpha + \beta$ when the random variables X and Y are independent and exponentially distributed with parameters α and β.

4.44 Limousines depart from the railway station to the airport from the early morning till late at night. The limousines leave from the railway station with independent interdeparture times that are exponentially distributed with an expected value of 20 minutes. Suppose you plan to arrive at the railway station at 3 o'clock in the afternoon. What are the expected value and the standard deviation of your waiting time at the railway station until a limousine leaves for the airport?

4.45 On Wednesday afternoon between 1 p.m. and 4:30 p.m., buses with tourists arrive in Gotham city to visit the castle in this picturesque town. The times between successive arrivals of buses are independent random variables, each having an exponential distribution with an expected value of 45 minutes. Each bus stays exactly 30 minutes on the parking lot of the castle. What is the probability mass function of the number of buses on the parking lot at 4 p.m.?

4.5 Exponential Distribution

4.46 On weeknight shifts between 6 p.m. and 10 p.m., 4.8 calls for medical emergencies arrive on average. It is reasonable to model the times between successive calls as independent random variables, each having the same exponential distribution. Let the random variable X measure the time from 6 p.m. until the first call occurs. What are the expected value and the median of X? What is the probability that the first call occurs between 6:20 p.m. and 6:45 p.m.? What is the probability of no calls between 7 p.m. and 7:20 p.m. and one or more calls between 7:20 p.m. and 7:45 p.m.?

4.47 You wish to cross a one-way traffic road on which cars drive at a constant speed and pass according to independent interarrival times having an exponential distribution with an expected value of $1/\lambda$ seconds. You can only cross the road when no car has come round the corner since c seconds. What is the probability distribution of the number of passing cars before you can cross the road when you arrive at an arbitrary moment? What property of the exponential distribution do you use?

4.48 The amount of time needed to wash a car at a car-washing station is exponentially distributed with an expected value of 15 minutes. You arrive at the car-washing station while it is occupied and one other car is waiting for a wash. The owner of this car informs you that the car in the washing station has already been there for 10 minutes. What is the probability that the car in the washing station will need no more than 5 minutes extra? What is the probability that you have to wait more than 20 minutes before your car can be washed?

4.49 A crucial component of a reliability system operates in a good state during an exponentially distributed time X with expected value $1/\mu$. After leaving the good state, the component enters a bad state. The system can still function properly in the bad state during a fixed time $a > 0$, but a failure of the system occurs after this time. The component is inspected every T time units, where $T > a$. It is replaced by a new one when the inspection reveals that the component is in the bad state or has failed. What is the probability that the replacement of a particular component is because of a system failure? What is the expected time between two replacements?

4.50 What is the probability that the closest integer to a random observation from the exponential density e^{-x} is odd? What is this probability given that the random observation is larger than a specific even integer r? Can you explain why both probabilities are the same?

4.51 In a video game with a time slot of fixed length T, signals occur according to a Poisson process with rate λ, where $T > \frac{1}{\lambda}$. In the time slot, you can push a button only once. You win if at least one signal occurs in the time slot and you push the button at the occurrence of the last signal. Your strategy is to let pass a fixed time s with $0 < s < T$ and push the button upon the first occurrence of a signal (if any) after time s. What is your probability of winning the game? What value of s maximizes this probability?

4.52 Suppose that events occur according to a Poisson process with rate λ. What is the probability distribution of the number of events that have occurred in the time interval $(0, a)$ given that n events have occurred in the time interval $(0, a + b)$?

4.53 A mini ferry carries cars across a river. Initially there are no cars waiting at the ferry. Cars arrive at the ferry according to a Poisson process with an average of 8 cars per hour. The ferry leaves as soon as there are two cars on the ferry or the first car to arrive has waited for 10 minutes, whichever occurs first. What is the probability that the ferry will leave with two cars? What is the expected value of the time until the ferry leaves?

4.54 The distances between major cracks on a particular highway can be modeled by independent random variables each having an exponential distribution with an expected value of 10 miles. What is the probability that there are no major cracks on a specific 15-mile stretch of the highway? What is the probability of two or more major cracks on that part of the highway?

4.55 A component in a repairable system has an expected time to failure of 250 hours. There are five spare parts available. The component and the spare parts each have the same exponentially distributed lifetime. What is the probability that the system will still be in operation after 1,000 hours? How many spare parts are needed to have a system reliability of 95% at 1,000 hours?

4.56 You go by bus to work each day. It takes 5 minutes to walk from your home to the bus stop. In order to get to work on time, you must take a bus at no later than 7:45 a.m. The interarrival times of the buses are independent random variables having an exponential distribution with an expected value of 10 minutes. What is the latest time you must leave home in order to be on time for work with a probability of at least 0.95?

4.57 In a traditional match between two soccer teams A and B, goals occur according to a Poisson process with a rate of $\frac{1}{30}$ per minute. The playing time of the match is 90 minutes. Each goal scored comes from team A with probability $\frac{12}{25}$ and from team B with probability $\frac{13}{25}$, independently of any other goal scored.

  (a) What is the probability of three or more goals during the match?

(b) What is the probability that exactly two goals will be scored in the first half of the match and exactly one goal in the second half?

(c) What are the probability of a draw and the probability of a win for team A?

4.58 Suppose that emergency response units are distributed throughout a large area according to a two-dimensional Poisson process. That is, the number of response units in any given bounded region has a Poisson distribution whose expected value is proportional to the area of the region, and the numbers of response units in disjoint regions are independent. An incident occurs at some point that is somewhere in the middle of the area. What is the probability of having at least one response unit within a distance r of the point at which the incident occurs?

4.6 Gamma, Weibull, and Beta Distributions

The gamma and Weibull distributions are important distributions, which have been found useful for describing random variables in inventory, reliability, and queueing applications, among others. The beta distribution is a general type of distribution and is a useful distribution in Bayesian inference.

4.6.1 Gamma Distribution

A positive continuous random variable X is said to have a *gamma distribution* with parameters $\alpha > 0$ and $\lambda > 0$ if X has the density function

$$f(x) = \lambda^\alpha \frac{x^{\alpha-1}}{\Gamma(\alpha)} e^{-\lambda x} \quad \text{for } x > 0$$

and $f(x) = 0$ otherwise, where the constant $\Gamma(\alpha)$ is the famous gamma integral

$$\Gamma(\alpha) = \int_0^\infty e^{-y} y^{\alpha-1} dy.$$

The gamma function $\Gamma(a) = \int_0^\infty e^{-y} y^{a-1} dy$ for $a > 0$ has the property

$$\Gamma(a+1) = a\Gamma(a) \quad \text{for } a > 0,$$

where $\Gamma(1) = 1$. This property is easily verified by partial integration and implies that $\Gamma(a) = (a-1)!$ if a is a positive integer. In the sequel, we will sometimes need the often-used value $\Gamma(\frac{1}{2}) = \sqrt{\pi}$.

The parameter α is a shape parameter and the parameter λ is a scale parameter of the gamma-distributed random variable X. The expected value and the variance of X are given by

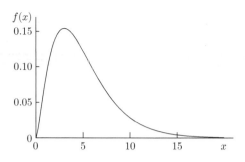

Figure 4.5. Gamma density ($\alpha = 2.5, \lambda = 0.5$).

$$E(X) = \frac{\alpha}{\lambda} \quad \text{and} \quad \text{var}(X) = \frac{\alpha}{\lambda^2}.$$

It is straightforward to derive these formulas, using the definition of the gamma function and the relation $\Gamma(a + 1) = a\Gamma(a)$ for $a > 0$. The details are left to the reader. Figure 4.5 displays the gamma density with $\alpha = 2.5$ and $\lambda = 0.5$. The graph in Figure 4.5 is representative of the shape of the gamma density if the shape parameter α is larger than 1; otherwise, the shape of the gamma density is similar to that of the exponential density in Figure 4.3. The gamma distribution is widely used in actuarial sciences for modeling the size of aggregate insurance claims, in climatological applications for representing variations in precipitation, in inventory applications for modeling demand sizes, and in queueing applications for modeling waiting times.

The gamma density with $\alpha = 1$ reduces to the exponential density. If α is given by an integer n, then the gamma density reduces to

$$\lambda^n \frac{x^{n-1}}{(n-1)!} e^{-\lambda x} \quad \text{for } x > 0.$$

This density is often called the *Erlang density* with parameters n and λ. A useful fact is that an Erlang-distributed random variable with parameters n and λ can be decomposed as the sum of n independent exponentially distributed random variables with parameter λ, see Section 4.5.

4.6.2 Weibull Density

A positive continuous random variable X is said to have a *Weibull distribution* with parameters $\alpha > 0$ and $\lambda > 0$ if X has the density function

$$f(x) = \alpha\lambda(\lambda x)^{\alpha-1} e^{-(\lambda x)^\alpha} \quad \text{for } x > 0$$

and $f(x) = 0$ otherwise. The parameter α is a shape parameter, and the parameter λ is a scale parameter. The Weibull density has a similar shape as the gamma density. The expected value and the variance of the random variable X are

$$E(X) = \frac{1}{\lambda}\Gamma\left(1 + \frac{1}{\alpha}\right) \quad \text{and} \quad \text{var}(X) = \frac{1}{\lambda^2}\left[\Gamma\left(1 + \frac{2}{\alpha}\right) - \left(\Gamma\left(1 + \frac{1}{\alpha}\right)\right)^2\right].$$

It is straightforward to derive these formulas, using the definition of the gamma function and a change of variable $u = (\lambda x)^\alpha$. The details are left to the reader. The cumulative distribution function $F(x) = P(X \le x)$ is

$$F(x) = 1 - e^{-(\lambda x)^\alpha} \quad \text{for } x \ge 0.$$

The Weibull distribution is the most commonly used distribution in reliability analysis. It is a useful probability model for fatigue strengths of materials and for lifetimes of devices with many components in which the first component to reach a critical stage determines the time to failure. Using advanced probability theory, it can be shown that under mild conditions the minimum of a large number of independent random strengths or lifetimes is approximately Weibull distributed (the weakest-link principle). This important result is anchored in the following property. Suppose that $X_1, \ldots, X_n$ are independent random variables, where any X_i has a Weibull distribution with shape parameter α and scale parameter λ_i. Then the random variable $T = \min(X_1, \ldots, X_n)$ has a Weibull distribution with shape parameter α and scale parameter $\lambda = \left(\sum_{i=1}^n \lambda_i^\alpha\right)^{1/\alpha}$. The proof is simple and goes as follows. Noting that $\min(a, b) > c$ if and only if $a > c$ and $b > c$ and using the independence of the random variables X_i, we get

$$P(T > x) = P(X_1 > x, \ldots, X_n > x) = P(X_1 > x) \cdots P(X_n > x)$$
$$= e^{-(\lambda_1 x)^\alpha} \cdots e^{-(\lambda_n x)^\alpha} = e^{-(\sum_{i=1}^n \lambda_i^\alpha) x^\alpha} = e^{-(\lambda x)^\alpha},$$

showing the desired result.

4.6.3 Beta Distribution

A positive continuous random variable X is said to have a *beta distribution* with parameters $\alpha > 0$ and $\beta > 0$ if X has the density function

$$f(x) = \frac{\Gamma(\alpha + \beta)}{\Gamma(\alpha)\Gamma(\beta)} x^{\alpha-1}(1 - x)^{\beta-1} \quad \text{for } 0 < x < 1$$

and $f(x) = 0$ otherwise. Both parameters α and β are shape parameters. The beta distribution is a flexible distribution, and the graph of the beta density

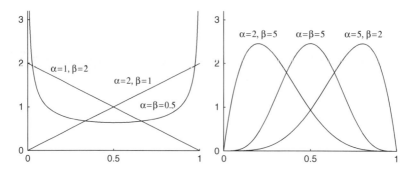

Figure 4.6. Several beta densities.

function can assume widely different shapes, depending on the values of α and β. An extreme case is the uniform distribution on $(0,1)$ corresponding to $\alpha = \beta = 1$. The graphs of several beta densities are given in Figure 4.6. The expected value and the variance of the random variable X are

$$E(X) = \frac{\alpha}{\alpha + \beta} \quad \text{and} \quad \text{var}(X) = \frac{\alpha\beta}{(\alpha + \beta)^2(\alpha + \beta + 1)}.$$

The beta density is often used to model the distribution of a random proportion, for example the proportion of defective items in a shipment. In Bayesian inference, the beta distribution is used as a description of uncertainty about the probability of success of an experiment. The beta distribution is also useful for the modeling of task duration in project planning. Also, the beta distribution is related to the gamma distribution. In Problem 5.47 of Chapter 5, you are asked to verify that $\frac{X}{X+Y}$ has a beta distribution with parameters α and β when X and Y are independent and gamma distributed with respective shape parameters α and β and scale parameter 1.

4.7 Normal Distribution

A continuous random variable X is said to have a *normal distribution* with parameters μ and $\sigma > 0$ if X has the density function

$$f(x) = \frac{1}{\sigma\sqrt{2\pi}} e^{-\frac{1}{2}(x-\mu)^2/\sigma^2} \quad \text{for } -\infty < x < \infty.$$

The parameter σ is a shape parameter, and the parameter μ is a scale parameter. The normal density with $\mu = 0$ and $\sigma = 1$ is called the *standard normal density* and is displayed in Figure 4.7. The normal density is symmetric around

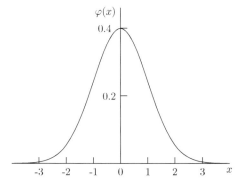

Figure 4.7. Standard normal density.

the point $x = \mu$. The normal distribution is also referred to frequently as the *Gaussian* distribution, named after Carl Friedrich Gauss (1777–1855), who discovered the normal curve in his analysis of errors of measurements made in astronomical observations and showed that the errors were fit well by a normal distribution. The expected value and the variance of the random variable X are given by

$$E(X) = \mu \text{ and } \text{var}(X) = \sigma^2.$$

Let us first verify that the normal density function indeed integrates to 1 over the interval $(-\infty, \infty)$. It suffices to verify this for $\mu = 0$ and $\sigma = 1$ (why?). To do so, we first note that $\frac{1}{\sqrt{2\pi}} \int_{-\infty}^{\infty} e^{-\frac{1}{2}x^2} dx = \frac{2}{\sqrt{2\pi}} \int_{0}^{\infty} e^{-\frac{1}{2}x^2} dx$. Thus we must show that the integral $\int_0^\infty e^{-\frac{1}{2}x^2} dx$ equals $\sqrt{\frac{1}{2}\pi}$. Using the change of variable $u = \frac{1}{2}x^2$, this integral can be written as the gamma integral $\frac{1}{\sqrt{2}} \int_0^\infty e^{-u} u^{-1/2} du = \frac{1}{\sqrt{2}} \Gamma(\frac{1}{2})$. Next, by $\Gamma(\frac{1}{2}) = \sqrt{\pi}$, the desired result follows. The celebrated formula $\int_0^\infty e^{-\frac{1}{2}x^2} dx = \sqrt{\frac{1}{2}\pi}$ can be proved without using the gamma integral. The reader is asked to do this in Problem 4.84. The formulas $E(X) = \mu$ and $\sigma^2(X) = \sigma^2$ will be verified in Rule 4.4 below.

The notation X is $N(\mu, \sigma^2)$ is often used as shorthand for X is a normally distributed random variable with parameters μ and σ. If $\mu = 0$ and $\sigma = 1$, the random variable X is said to have the *standard normal distribution*. The standard normal density function and the standard normal distribution function of an $N(0, 1)$ random variable have the special notation

$$\varphi(z) = \frac{1}{\sqrt{2\pi}} e^{-\frac{1}{2}z^2} \text{ and } \Phi(z) = \frac{1}{\sqrt{2\pi}} \int_{-\infty}^{z} e^{-\frac{1}{2}y^2} dy.$$

It is readily verified from the definition of $\Phi(z)$ that

$$\Phi(-z) = 1 - \Phi(z) \quad \text{for all } z.$$

No closed-form expression for the standard normal distribution function $\Phi(z)$ exists. The integral for $\Phi(z)$ looks terrifying, but the integral can be approximated with extreme precision by the quotient of two suitably chosen polynomials. For all practical purposes, the calculation of $\Phi(z)$ presents no difficulties at all and can be accomplished very quickly. The inverse function of $\Phi(z)$ is needed to find the percentiles of the standard normal distribution. The pth *percentile* of the standard normal distribution is denoted by z_p and is the unique solution of the equation[2]

$$\Phi(z_p) = p \quad \text{for } 0 < p < 1.$$

By the relation $\Phi(-z) = 1 - \Phi(z)$, we have $z_{1-p} = -z_p$ for all p. Important percentiles are $z_{0.95} = 1.6449$, $z_{0.975} = 1.9500$, and $z_{0.995} = 2.5758$.

The next rule enables us to get the formulas for the first two moments of an $N(\mu, \sigma^2)$-distributed random variable X and shows us how to calculate the cumulative distribution function of X.

Rule 4.4 *Let X be an $N(\mu, \sigma^2)$-distributed random variable. Define the random variable Z by*

$$Z = \frac{X - \mu}{\sigma}.$$

Then the random variable Z has the standard normal distribution with

$$E(Z) = 0 \quad \text{and} \quad \text{var}(Z) = 1.$$

Also, $E(X) = \mu$ and $\text{var}(X) = \sigma^2$. Moreover, for any x,

$$P(X \leq x) = \Phi\left(\frac{x - \mu}{\sigma}\right).$$

The proof is as follows. Since $\sigma > 0$, we have $P(Z \leq z) = P(X \leq \mu + \sigma z)$. Thus, using the change of variable $w = (x - \mu)/\sigma$,

$$P(Z \leq z) = \frac{1}{\sigma\sqrt{2\pi}} \int_{-\infty}^{\mu+\sigma z} e^{-\frac{1}{2}(x-\mu)^2/\sigma^2} \, dx = \frac{1}{\sqrt{2\pi}} \int_{-\infty}^{z} e^{-\frac{1}{2}w^2} \, dw = \Phi(z),$$

showing that Z has the standard normal distribution. The moments $E(Z^k)$ exist and are finite for all k, since $\int_{-\infty}^{\infty} |z^k| \varphi(z) \, dz < \infty$ for all k (verify!). In particular, $E(Z) = 0$, as follows by noting that the area under the graph of

---

[2] In general, the pth percentile of a continuous random variable X is given by any value x_p satisfying $P(X \leq x_p) = p$ for $0 < p < 1$. The 0.5th percentile is usually called the median.

$z\varphi(z)$ from $-\infty$ to 0 cancels the area under the graph from 0 to ∞. To get var(Z), note that $E(Z^2) = \frac{1}{\sqrt{2\pi}} \int_{-\infty}^{\infty} z^2 e^{-\frac{1}{2}z^2} dz$. By partial integration,

$$E(Z^2) = \frac{-1}{\sqrt{2\pi}} \int_0^{\infty} z\, de^{-\frac{1}{2}z^2} = \frac{-1}{\sqrt{2\pi}} z e^{-\frac{1}{2}z^2}\Big|_{-\infty}^{\infty} + \frac{1}{\sqrt{2\pi}} \int_{-\infty}^{\infty} e^{-\frac{1}{2}z^2} dz = 1,$$

using the fact that the standard normal density integrates to 1 over $(-\infty, \infty)$. This shows that $\text{var}(Z) = E(Z^2) - [E(Z)]^2 = 1$. Next, using Rule 4.2, we get the formulas $E(X) = \mu$ and $\text{var}(X) = \sigma^2$ for $X = \sigma Z + \mu$. The formula for $P(X \leq x)$ in terms of $\Phi(z)$ follows from

$$P(X \leq x) = P\left(\frac{X-\mu}{\sigma} \leq \frac{x-\mu}{\sigma}\right)$$

and the fact that $Z = (X - \mu)/\sigma$ has the standard normal distribution.

The normal distribution is the most important continuous distribution and has many applications. Although a normally distributed random variable can theoretically take on values in the interval $(-\infty, \infty)$, it still may provide a useful model for a variable that takes on only positive values, provided that the normal probability mass on the negative axis is negligible. Nearly all the mass of the $N(\mu, \sigma^2)$ density is within three standard deviations from the expected value. This follows from the useful relation

$$P(|X - \mu| > k\sigma) = 2[1 - \Phi(k)] \quad \text{for any } k > 0.$$

The reader is asked to verify this in Problem 4.65. The tail probability $P(|X - \mu| > k\sigma)$ has the values 0.3173, 0.0455, and 0.0027 for $k = 1, 2$, and 3. Thus, observations from a normal random variable will seldom be three or more standard deviations removed from the expected value.

The following properties contribute to the practical usefulness of the normal distribution model.

- If the random variable X has an $N(\mu, \sigma^2)$ distribution, then $aX + b$ is $N(a\mu + b, a^2\sigma^2)$ distributed for any constants a, b with $a \neq 0$.
- If X and Y are independent random variables that are $N(\mu_1, \sigma_1^2)$ and $N(\mu_2, \sigma_2^2)$ distributed, then $X + Y$ is $N(\mu_1 + \mu_2, \sigma_1^2 + \sigma_2^2)$ distributed.

The first property is easily verified by working out the probability distribution of $Y = aX + b$, see Problem 4.66. The second property requires more advanced tools and will be proved in Section 8.2.

Example 4.13 The annual rainfall in Amsterdam is normally distributed with an expected value of 799.5 mm and a standard deviation of 121.4 mm. What is the probability that the total rainfall in Amsterdam next year will be more than 1,000 mm?

Solution. Let the random variable X represent the annual rainfall in Amsterdam. The sought probability $P(X > 1,000)$ is evaluated as

$$P\left(\frac{X - 799.5}{121.4} > \frac{1,000 - 799.5}{121.4}\right) = 1 - \Phi(1.649) = 0.0496.$$

Example 4.14 In a periodic-review inventory system, the stock on hand of a certain item is raised to the level S at the beginning of each week. The demands for the item in the successive weeks are modeled by independent random variables, each having a normal distribution with an expected value of $\mu = 250$ units and a standard deviation of $\sigma = 45$ units. What is the required order-up level S so that the proportion of weeks in which a shortage occurs is no more than $\alpha = 0.05$?

Solution. The probability of a shortage in any given week is $P(X > S)$. Writing this probability as $P(\frac{X-\mu}{\sigma} > \frac{S-\mu}{\sigma})$, we have $P(X > S) = 1 - \Phi(\frac{S-\mu}{\sigma})$. The requirement that the proportion of weeks with a shortage must be no more than α boils down to solving S from $1 - \Phi(\frac{S-\mu}{\sigma}) = \alpha$, or, equivalently, $\Phi(\frac{S-\mu}{\sigma}) = 1 - \alpha$. In other words, $\frac{S-\mu}{\sigma}$ must be equal to the $(1-\alpha)$th percentile $z_{1-\alpha}$ of the standard normal distribution. This leads to

$$S = \mu + z_{1-\alpha}\,\sigma.$$

The value of the 95% percentile $z_{0.95}$ is 1.6449. Thus, for the numerical data $\mu = 250$, $\sigma = 45$, and $\alpha = 0.05$, the order-up level S should be 325 units.

4.7.1 Central Limit Theorem

The central limit theorem is intrinsically linked to the normal distribution. This theorem is queen among theorems in probability theory, and reads as follows:

Rule 4.5 (central limit theorem) *If $X_1, X_2, \ldots$ are independent random variables, each having the same probability distribution function with expected value μ and standard deviation σ, then*

$$\lim_{n \to \infty} P\left(\frac{X_1 + \cdots + X_n - n\mu}{\sigma\sqrt{n}} \leq x\right) = \Phi(x) \quad \text{for all } x.$$

An alternative formulation of the central limit theorem in terms of the sample mean is

$$\lim_{n \to \infty} P\left(\frac{(X_1 + \cdots + X_n)/n - \mu}{\sigma/\sqrt{n}} \leq x\right) = \Phi(x) \quad \text{for all } x.$$

A sketch of the proof of the central limit theorem will be given in Chapter 8. In words, the theorem states that the sum of n independent random variables, each having the same probability distribution with mean μ and standard deviation σ, has approximately a normal distribution with mean $n\mu$ and standard deviation $\sigma\sqrt{n}$ for n large enough. How large n should be depends strongly on the degree of asymmetry in the distribution of the X_k. An estimate for n can be given for a binomially distributed random variable X with parameters n and p. This random variable can be written as the sum of n independent random variables $X_1, \ldots, X_n$, where $X_k = 1$ with probability p and $X_k = 0$ with probability $1 - p$. As a rule of thumb, the approximation of the binomial distribution by the normal distribution can be used when $np(1 - p) \geq 25$. Also, the Poisson distribution, being a limiting distribution of the binomial, can be approximated by the normal distribution when the parameter λ of the Poisson distribution satisfies $\lambda \geq 25$.

The central limit theorem has an interesting history. The first version of this theorem was postulated by the French-born English mathematician Abraham de Moivre, who used the normal distribution to approximate the distribution of the number of heads resulting from many tosses of a fair coin. This discovery appeared in 1733 in his self-published pamphlet *"Approximato ad Summani Terminorum Binomi $(a + b)^n$ in Seriem Expansis"* and was first made public in 1738 in the second edition of his probability book *Doctrine of Chance*. De Moivre's finding was far ahead of its time, and was nearly forgotten until the famous French mathematician Pierre Simon Laplace rescued it from obscurity in his monumental work *Théorie Analytique des Probabilités*, which was published in 1812 and gave a decisive thrust to the further development of probability theory. Laplace expanded de Moivre's finding by approximating the binomial distribution with the normal distribution. But, as with de Moivre, Laplace's finding received little attention in his own time. It was not until the nineteenth century was at an end that the importance of the central limit theorem was discerned. In 1901, the Russian mathematician Aleksandr Lyapunov defined it in general terms and proved precisely how it worked mathematically. Nowadays, the central limit theorem is considered to be the unofficial sovereign of probability theory.

In the central limit theorem it is essential that the random variables X_k are independent, but it is not necessary for them to have the same distribution. Suppose that the X_k are independent random variables having the expected values μ_k and the standard deviations σ_k. Then, under a weak regularity condition, the normalized random variable

$$\frac{\sum_{k=1}^{n} X_k - \sum_{k=1}^{n} \mu_k}{\sqrt{\sum_{k=1}^{n} \sigma_k^2}}$$

also has the standard normal distribution as limiting distribution. This generalized version of the central limit theorem elucidates the reason that, in practice, so many random phenomena, such as the rate of return on a stock, the cholesterol level of an adult male, the duration of a pregnancy, etc., are approximately normally distributed. Each of these random quantities can be seen as the result of a large number of small independent random effects that add together. While the heights of human beings follow a normal distribution, weights do not. Heights are the result of the interaction of many factors outside one's control, while weights depend mainly on lifestyle. This example shows that the normal distribution should not be used mindlessly. In financial models it may be quite dangerous to assume blithely that data follow the normal distribution.

Next we give several examples to illustrate the central limit theorem.

Example 4.15 Your friend asserts to have rolled an average of 3.25 points per roll in 1,000 rolls of a fair die. Do you believe this?

Solution. Let the random variable X_i denote the outcome of the ith roll of the die. The random variables $X_1, \ldots, X_{1,000}$ are independent and uniformly distributed on the integers $1, \ldots, 6$, where $E(X_i) = 3.5$ and $\sigma(X_i) = 1.7078$. The average number of points per roll in 1,000 rolls is given by the sample mean $\frac{1}{1,000} \sum_{i=1}^{1,000} X_i$ and is approximately normally distributed with expected value 3.5 and standard deviation $1.7078/\sqrt{1,000} = 0.0540$. The value 3.25 lies $(3.5 - 3.25)/0.0540 = 4.629$ standard deviations below the expected value 3.5. The probability that a normal random variable takes on a value of more than 4.5 standard deviations below the expected value is extremely small. Thus, without doing any further calculations, you can say that the claim of your friend is highly implausible.

Example 4.16 Suppose that $X_1, \ldots, X_n$ are independent random variables that are uniformly distributed on $(0,1)$. Use the normal distribution to approximate the probability that the rounded sum $X_1 + \cdots + X_n$ equals the sum of the rounded X_i when the rounding is to the nearest integer.

Solution. Let the random variable D_i denote $D_i = X_i - \text{round}(X_i)$. Then, the sought probability is given by $P\left(-\frac{1}{2} \leq \sum_{i=1}^n D_i < \frac{1}{2}\right)$. This is easily seen by noting that the rounded sum is equal to the sum of the rounded X_i if and only if

$$\sum_{i=1}^n \text{round}(X_i) - \frac{1}{2} \leq \sum_{i=1}^n X_i < \sum_{i=1}^n \text{round}(X_i) + \frac{1}{2}.$$

4.7 Normal Distribution

The random variables $D_1, \ldots, D_n$ are independent and uniformly distributed on $(-\frac{1}{2}, \frac{1}{2})$ and thus have an expected value of 0 and a standard deviation of $\sqrt{1/12}$. Using the central limit theorem, it follows that

$$P\left(-\frac{1}{2} \le \sum_{i=1}^{n} D_i < \frac{1}{2}\right) \approx \Phi\left(\frac{\sqrt{3}}{\sqrt{n}}\right) - \Phi\left(-\frac{\sqrt{3}}{\sqrt{n}}\right)$$

for n large enough. The approximate values are 0.416, 0.271, 0.194, and 0.138 for $n = 10, 25, 50$, and 100. These approximations are excellent. The values obtained by one million simulation runs are 0.411, 0.269, 0.191, and 0.136.

Example 4.17 Consider again Example 3.20 in Section 3.8. How many bets must be placed by a player to have a probability of more than 99% of making a positive profit in the sales promotion of the lottery? What is the probability distribution of the net profit for a team of players who make 100,000 $5 bets?

Solution. In Example 3.20, we showed that the New York state lottery offered the player a favorable game with the sales promotion "Big Dipper Wednesday." The expected value of the profit per dollar bet is 19.47 cents. An answer to the question of how many bets are necessary for the probability of a positive profit being more than 99% can be given with the help of the central limit theorem. Let us assume that the player only takes betting actions consisting of 20 individual $5 bets on a game card. Such a betting action would cost $100. The probability that an individual bet results in k matching numbers is given by $p_k = \binom{4}{k}\binom{76}{20-k}/\binom{80}{20}$ for $k = 0, \ldots, 4$, see Example 3.20. The corresponding payoffs are $0, $0, $10, $50, and $550 for 0, 1, 2, 3, or 4 matching numbers during the sales promotion of the lottery. The expected value and the standard deviation of the net profit resulting from an individual $5 bet are then calculated as

$$\mu = 0.973615 \text{ dollars} \quad \text{and} \quad \sigma = 32.48248 \text{ dollars.}$$

The net profit resulting from n betting actions, each consisting of 20 individual $5 bets can be seen as the sum of $20n$ independent random variables each having expected value μ and standard deviation σ. For n sufficiently large, this sum is approximately normally distributed with expected value $20n\mu$ and standard deviation $\sigma\sqrt{20n}$. The smallest value of n for which the probability of having a positive net profit after n betting actions is at least 99% follows by solving x from the equation

$$1 - \Phi\left(\frac{-20x\mu}{\sigma\sqrt{20x}}\right) = 0.99.$$

This leads to $20x\mu/(\sigma\sqrt{20x}) = 2.32635$ and so $x = 301.19$. Thus, an estimated number of 302 betting actions are required to have a probability of at least 99% that the net profit is positive. Suppose that a team of players could make 1,250 betting actions on each of the four Wednesdays (recall that a new game is played every 4 or 5 minutes), giving a total of 5,000 betting actions. By the central limit theorem, the probability distribution of the net profit after 5,000 betting actions (100,000 individual bets) can be very well approximated by a normal distribution with an expected value of $100,000\mu = 97,361.50$ dollars and a standard deviation of $\sigma\sqrt{100,000} = 10,271.86$ dollars. The 0.025th and 0.975th percentiles of this normal distribution are calculated as 77,229 and 117,494 (the pth percentile of an $N(\mu, \sigma^2)$ distribution is given by $\mu + \sigma z_p$, where z_p is the pth percentile of the standard normal distribution). In other words, the net profit will be between $77,229 and $117,494 with a probability of about 95%. The story is that students who had taken a probability course in college applied their knowledge of probability to earn over $100,000 playing the lottery game with double payoffs.

Random-Walk Fluctuations and the Central Limit Theorem

In the coin-tossing experiment with a fair coin, it is typical to find that the actual difference in the number of heads and tails shows a tendency to grow. For instance, the chance of getting a split somewhere between 45 and 55 with 100 tosses is almost 73%. But for the difference of the number of heads and tails to be within a range of +5 to –5 after 10,000 tosses is much less likely, about 9%; and even quite unlikely, about 0.9%, after 1,000,000 tosses. The central limit theorem enables us to explain the growing fluctuations in the random walk describing the evolution of the number of heads tossed minus the number of tails tossed when a fair coin is repeatedly tossed. Suppose that a fair coin is tossed n times. Let the random variable X_i be equal to 1 if the ith toss results in heads, and 0 otherwise. Then, the random variable

$$E_n = \sum_{i=1}^{n} X_i - \sum_{i=1}^{n}(1 - X_i)$$

gives the difference between the number of heads and the number of tails in n tosses. Writing E_n as $E_n = 2\sum_{i=1}^{n} X_i - n$ and noting that $E(X_i) = \sigma(X_i) = \frac{1}{2}$, it follows that E_n is approximately $N(0, n)$ distributed for n large. Using the relation $P(|X - \mu| > c\sigma) = 2[1 - \Phi(c)]$ for any $N(\mu, \sigma^2)$-distributed random variable X, we get

$$P(|E_n| > c\sqrt{n}) \approx 2[1 - \Phi(c)]$$

4.7 Normal Distribution

for any constant $c > 0$ when n is sufficiently large. Thus, the range of the absolute difference between the number of heads and the number of tails has a tendency to grow proportionally with the square root of the number of tosses. This finding is otherwise not in conflict with the law of large numbers, which says that $\frac{1}{n} \times$(actual number of heads in n tosses minus actual number of tails in n tosses) goes to 0 when $n \to \infty$. A similar phenomenon as described in the coin-tossing experiment appears in lottery drawings. The difference between the number of times the most frequently drawn number comes up and the number of times the least frequently drawn number comes up in n drawings shows a tendency to increase proportionally with $\sqrt{n}$ as n increases. This phenomenon can be explained using the multivariate central limit theorem, which will be discussed in Chapter 6.

Problems

4.59 Gestation periods of humans are normally distributed with an expected value of 266 days and a standard deviation of 16 days. What is the percentage of births that are more than 20 days overdue?

4.60 The cholesterol level for an adult male of a specific racial group is normally distributed with an expected value of 5.2 mmol/L and a standard deviation of 0.65 mmol/L. Which cholesterol level is exceeded by 5% of the population?

4.61 The demand for some item follows a normal distribution. The inventory controller gives an estimate of 50 for the expected value of the demand and an estimate of 75 for the threshold level of demand that will only be exceeded with a probability of 5%. What is an estimate for the standard deviation of the demand?

4.62 The diameter of a 1 euro coin is normally distributed with an expected value of 23.25 mm and a standard deviation of 0.10 mm. A vending machine accepts only 1 euro coins with a diameter between 22.90 mm and 23.60 mm. What proportion of 1 euro coins will not be accepted by the vending machine?

4.63 The thickness of a protective coating on a conductor used in corrosive conditions is a random variable X that can be modeled by a normal distribution with an expected value of 25 microns and a standard deviation of 2.5 microns. What is the probability that the coating is less than 20 microns thick? It is decided to improve the precision of the process of attaching the coating. What value of the standard deviation should be aimed for so that the coating has a thickness of at least 20 microns with a probability of 99%?

4.64 The lengths of steel beams made in a steel mill follow a normal distribution with an expected value of 10 m and a standard deviation of 0.07 m. A customer requires beams that are no shorter than 9.85 m and no longer than 10.15 m. What proportion of the mill's output can be used to supply this customer?

4.65 Verify that $P(|X-\mu| > k\sigma) = 2[1-\Phi(k)]$ for any $k > 0$ if X is $N(\mu, \sigma^2)$ distributed.

4.66 Verify that $aX + b$ is $N(a\mu + b, a^2\sigma^2)$ distributed if X is $N(\mu, \sigma^2)$ distributed.

4.67 Somebody claims to have obtained 5,250 heads in 10,000 tosses of a fair coin. Do you believe this claim?

4.68 Let the random variable Z have the standard normal distribution.

(a) Verify that the probability density function of $|Z|$ is given by $\sqrt{2/\pi}\, e^{-\frac{1}{2}z^2}$ for $z > 0$ with $E(|Z|) = \sqrt{2/\pi}$ and $\sigma(|Z|) = \sqrt{1 - 2/\pi}$.

(b) Use the formula for the expected value in Problem 4.26 to verify that
$E[\max(Z - c, 0)] = \frac{1}{\sqrt{2\pi}} e^{-\frac{1}{2}c^2} - c[1 - \Phi(c)]$ for any $c \geq 0$.

4.69 Suppose that the random variables X and Y are independent and $N(\mu, \sigma^2)$ distributed. Prove that $E[\max(X, Y)] = \mu + \frac{\sigma}{\sqrt{\pi}}$ and $E[\min(X, Y)] = \mu - \frac{\sigma}{\sqrt{\pi}}$. *Hint:* Use the relations $\max(X, Y) + \min(X, Y) = X + Y$ and $\max(X, Y) - \min(X, Y) = |X - Y|$ together with the result (a) in Problem 4.68.

4.70 In "walking the line," a drunkard takes each time a unit step to the right or to the left with equal probabilities, independently of the previous steps. Let D_n be the distance of the drunkard from the starting point after n steps. Use the result of Problem 4.68 to verify that $E(D_n) \approx \sqrt{(2/\pi)n}$ and $\sigma(D_n) \approx \sqrt{(1 - 2/\pi)n}$ for large n. Also, verify $P(D_n \leq x) \approx \Phi\left(\frac{x}{\sqrt{n}}\right) - \Phi\left(\frac{-x}{\sqrt{n}}\right)$ for $x > 0$ when n is large.

4.71 Two instruments are used to measure the unknown length of a beam. If the true length of the beam is l, the measurement error made by the first instrument is normally distributed with mean 0 and standard deviation $0.006l$ and the measurement error made by the second instrument is normally distributed with mean 0 and standard deviation $0.004l$. The two measurement errors are independent of each other. What is the probability that the average value of the two measurements is within 0.5% of the actual length of the beam?

4.72 The lifetimes of two components in an electronic system are independent random variables X_1 and X_2, where X_i has a normal distribution with an expected value of μ_i time units and a standard deviation of σ_i time units.

What is the probability that the lifetimes of the two components expire within a time units from each other?

4.73 Use the central limit theorem to answer the following questions.

(a) What is, in Example 3.3, a good approximation for the probability that the profit of Joe and his brother will be $100 or more after 52 weeks?

(b) What is a good approximation for the probability that more than 80 rolls are required when you roll a fair die until the total sum exceeds 300?

4.74 Let $X_1, X_2, \ldots$ be a sequence of independent random variables each having the same distribution with expected value μ and standard deviation σ. Show that $P(|\frac{1}{n}(X_1 + \cdots + X_n) - \mu| > c) \approx 2\left[1 - \Phi\left(\frac{c\sqrt{n}}{\sigma}\right)\right]$ for n large.

4.75 Use the central limit theorem to verify the following results.

(a) The probability of getting r or more sixes in one throw of $6r$ fair dice tends to $\frac{1}{2}$ as $r \to \infty$.

(b) $e^{-n}\left(1 + \frac{n}{1!} + \frac{n^2}{2!} + \cdots + \frac{n^n}{n!}\right)$ tends to $\frac{1}{2}$ as $n \to \infty$.

4.76 An insurance company has 20,000 policyholders. The amount claimed yearly by a policyholder has an expected value of $150 and a standard deviation of $750. Give an approximation for the probability that the total amount claimed in the coming year will be larger than 3.3 million dollars.

4.77 In the 52 drawings of the 6/45 lottery in Orange Country last year, an even number was drawn 162 times and an odd number 150 times. Does this outcome cast doubts on the unbiased nature of the drawings?

4.78 The amounts of rainfall in Amsterdam during the months January to December are independent random variables with expected values of 62.1, 63.4, 58.9, 41.0, 48.3, 67.5, 65.8, 61.4, 82.1, 85.1, 89.0, and 74.9 mm and with standard deviations of 33.9, 27.8, 31.1, 24.1, 29.3, 33.8, 36.8, 32.1, 46.6, 42.4, 40.0, and 36.2 mm. What, approximately, is the probability that the total rainfall in Amsterdam next year will be larger than 1,000 mm?

4.79 The Nero Palace casino has a new, exciting gambling machine: the multiplying bandit. How does it work? The bandit has a lever or "arm" that the player may depress up to 10 times. After each pull, a H (heads) or a T (tails) appears, each with probability $\frac{1}{2}$. The game is over when heads appears for the first time, or when the player has pulled the arm 10 times. The player wins 2^k dollars if heads appears after k pulls ($1 \leq k \leq 10$), and wins $2^{11} = 2,048$ dollars if, after 10 pulls, heads has not appeared.

In order to play the game, the player must pay $15. Assume there are 5,000 games played each week. Use the central limit theorem to approximate the probability that the casino will lose money in a given week.

4.80 Suppose that a casino introduces a new game of chance. A player must stake one dollar on each play and gets back twice the stake with probability p and loses the stake with probability $1-p$, where $0 < p < \frac{1}{2}$. The casino owner wants to know the probability that the casino will have no profit after n plays for large n. Verify that this probability can be approximated by $\Phi(-c\sqrt{n})$ with $c = \frac{1}{2}(1-2p)/\sqrt{p(1-p)}$. Also, verify that the profit of the casino is steadily growing and is larger than $n(1-2p) - 2.326 \times 2\,[p(1-p)]^{\frac{1}{2}}\sqrt{n}$ after n plays with a probability of about 99%.

4.81 An insurance company will bring a new insurance product to the market. The company estimates that they can sell r policies, where r is quite large. It is assumed that the amounts claimed by the policyholders are independent random variables, each having the same distribution with mean μ and standard deviation σ. How should the company choose the premium for each policy so that, with a probability of about 99%, the profit of the company will be at least 10% of the total premium the company gets?

4.82 A space mission will take 150 days. A number of copies of a daily-use appliance must be taken along. The amount of time the appliance can be used is exponentially distributed with an expected value of two days. Use the central limit theorem to estimate how many copies of the appliance should be stocked so that the probability of a shortage during the mission is no more than 10^{-3}.

4.83 A new casino has just been opened. The casino owner makes the following promotional offer to induce gamblers to play at his casino. People who bet $10 on red get half their money back if they lose the bet, while they get the usual payout of $20 if they win the bet. This offer applies only to the first 2,500 bets. European roulette is played in the casino, and so a bet on red is lost with probability $\frac{19}{37}$ and is won with probability $\frac{18}{37}$. Use the central limit theorem to approximate the probability that the casino owner will lose more than 6,500 dollars on the promotional offer.

4.84 Prove that $I = \int_0^\infty e^{-\frac{1}{2}x^2}\,dx$ is equal to $\sqrt{\pi/2}$ by changing to polar coordinates $x = r\cos(\theta)$ and $y = r\sin(\theta)$ in $I^2 = \int_0^\infty \int_0^\infty e^{-\frac{1}{2}(x^2+y^2)}\,dx\,dy$. Use a change of variable to obtain from $\int_0^\infty e^{-\frac{1}{2}x^2}\,dx = \sqrt{\pi/2}$ that $\Gamma(\frac{1}{2}) = \sqrt{\pi}$.

4.8 Other Continuous Distributions

In this section we give several other continuous distributions. In particular, we discuss distributions that are important in statistics.

4.8.1 Lognormal Distribution

A positive continuous random variable X is said to have a *lognormal distribution* with parameters μ and $\sigma > 0$ if X has the density function

$$f(x) = \frac{1}{\sigma x \sqrt{2\pi}} e^{-\frac{1}{2}[\ln(x)-\mu]^2/\sigma^2} \quad \text{for } x > 0$$

and $f(x) = 0$ otherwise. A lognormally distributed random variable takes on only positive values. The graph of the lognormal density function with $\mu = 0$ and $\sigma = 1$ is displayed in Figure 4.8. It is not difficult to prove that the random variable X is lognormally distributed with parameters μ and σ if the random variable $\ln(X)$ is $N(\mu, \sigma^2)$ distributed, see Example 4.18 in Section 4.9. Hence, using the relation $P(X \leq x) = P(\ln(X) \leq \ln(x))$ for $x > 0$,

$$P(X \leq x) = \Phi\left(\frac{\ln(x) - \mu}{\sigma}\right) \quad \text{for } x > 0.$$

The expected value and the variance of the random variable X are

$$E(X) = e^{\mu + \frac{1}{2}\sigma^2} \quad \text{and} \quad \text{var}(X) = e^{2\mu + \sigma^2}\left(e^{\sigma^2} - 1\right).$$

A nice probabilistic derivation of the formulas for $E(X)$ and $\text{var}(X)$ is given in Remark 8.1 in Section 8.3. It is pointed out that the lognormal distribution has the property that X^a is lognormally distributed for any constant $a \neq 0$. Since $\ln(X^a) = a \ln(X)$ and $\ln(X)$ is normally distributed, this property follows from the fact that aV has a normal distribution when V is normally distributed.

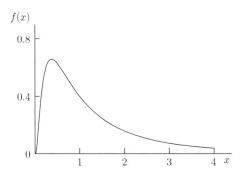

Figure 4.8. Lognormal density ($\mu = 0, \sigma = 1$).

The lognormal distribution provides a useful probability model for income distributions. The explanation is that its probability density function $f(x)$ is skewed to the left and tends very slowly to zero as x approaches infinity (assuming that $\sigma > 1$). In other words, most outcomes of this lognormal distribution will be relatively small, but very large outcomes occur occasionally. The lognormal distribution is also often used to model future stock prices after a longer period of time. Also, handling times of service requests at a call center and sizes of insurance claims often follow closely a lognormal distribution. In general, the lognormal distribution arises when the underlying random variable is the result of a large number of independent multiplicative effects. This can be explained with the help of the central limit theorem, using the fact that the logarithm of a product of terms is the sum of the logarithms of the terms.

Problems

4.85 Consider again Problem 4.83. Assume now that the casino owner makes an offer only to the most famous gambler in town. The casino owner lends \$1,000 to the gambler and proposes that he make 100 bets on red with this starting capital. The gambler is allowed to stake any amount of his bankroll at any bet. The gambler gets a quarter of the staked money back if he loses the bet, while he gets double the staked money back if he wins the bet. As reward, the gambler can keep to himself any amount in excess of a bankroll of \$1,000 when stopping after 100 bets. Suppose that the gambler decides to stake each time 5% of his current bankroll. Use the normal distribution to approximate the probability that the gambler takes home more than d dollars for $d = 0, 500, 1,000$, and $2,500$. *Hint*: Consider the logarithm of the size of the bankroll of the gambler. *Note*: This problem is an instance of the Kelly model, see Chapter 9.

4.86 A population of bacteria has the initial size s_0. In each generation, independently of each other, it is equally likely that the population increases by 25% or decreases by 20%. What is the approximate probability density of the size of the population after n generations with n large?

4.8.2 Chi-Square Distribution

A continuous random variable X is said to have a *chi-square distribution* with d *degrees of freedom* if it can be represented as

$$X = Z_1^2 + Z_2^2 + \cdots + Z_d^2,$$

4.8 Other Continuous Distributions

where $Z_1, Z_2, \ldots, Z_d$ are independent random variables, each having a standard normal distribution. The notation χ_d^2 distribution is often used for this distribution. The probability density function of X is given by

$$f(x) = \frac{(\frac{1}{2})^{\frac{1}{2}d}}{\Gamma(\frac{1}{2}d)} x^{\frac{1}{2}d-1} e^{-\frac{1}{2}x} \quad \text{for } x > 0$$

and $f(x) = 0$ otherwise, see Rule 8.7 in Section 8.3 for a proof. This density is a special case of the gamma density with shape parameter $\alpha = \frac{1}{2}d$ and scale parameter $\lambda = \frac{1}{2}$. Thus, the graph of the gamma density with $\alpha = 2.5$ and $\lambda = \frac{1}{2}$ in Figure 4.4 is also the graph of the chi-square density with $n = 5$. The expected value and the variance of the random variable X are

$$E(X) = d \quad \text{and} \quad \text{var}(X) = 2d,$$

as follows from the fact that the expected value and variance of a gamma density with shape parameter α and scale parameter λ are α/λ and α/λ^2.

The chi-square distribution plays an important role in statistics and is best known for its use in the so-called "chi-square tests." Also, the chi-square distribution arises in the analysis of random walks: if $V_1, \ldots, V_d$ are independent random variables that are $N(0, \sigma^2)$ distributed, then the random variable $W = \sqrt{V_1^2 + \cdots + V_d^2}$ has the density function

$$f_W(w) = \frac{(\frac{1}{2})^{\frac{1}{2}d-1}}{\sigma^d \Gamma(\frac{1}{2}d)} w^{d-1} e^{-\frac{1}{2}w^2/\sigma^2} \quad \text{for } w > 0.$$

The verification of this result is left as an exercise for the reader.

Problem

4.87 A shot is fired at a very large circular target. The horizontal and vertical coordinates of the point of impact are independent random variables each having a standard normal density. Here, the center of the target is taken as the origin. What is the density function of the distance from the center of the target to the point of impact? What are the expected value and the mode of the distance?

4.8.3 Student-t Distribution

A continuous random variable X is said to have a *Student-t distribution* with n degrees of freedom if it can be represented as

$$X = \frac{Z}{\sqrt{U/n}},$$

where Z has a standard normal distribution, U a chi-square distribution with n degrees of freedom, and the random variables Z and U are independent. In Problem 5.52 you are asked to show that the density function of X is

$$f(x) = c\left(1 + \frac{x^2}{n}\right)^{-\frac{1}{2}(n+1)} \quad \text{for } -\infty < x < \infty,$$

where $c = \frac{1}{\sqrt{\pi n}} \Gamma(\frac{1}{2}(n+1))/\Gamma(\frac{1}{2}n)$. In Figure 4.9, the Student-t density function is displayed for $n = 5$. The density function is very similar to that of the standard normal density, but it has a longer tail than the $N(0, 1)$ density. The Student-t distribution is named after William Gosset, who invented this distribution in 1908 and used the pen name "A. Student" in his publication. Gosset worked for the Guinness brewery in Dublin which, at that time, did not allow its employees to publish research papers. The expected value and the variance of X are undefined for $n = 1$, since $\int_{-\infty}^{\infty} |x|(1 + x^2)^{-1}\, dx = \infty$. For $n > 1$, the expected value and the variance of X are given by

$$E(X) = 0 \quad \text{and} \quad \text{var}(X) = \frac{n}{n-2},$$

where $\frac{n}{n-2}$ should be read as ∞ if $n = 2$. By the symmetry of the Student-t density around 0, we have $E(X) = 0$. The formula for $\text{var}(X)$ can be obtained as follows. By the independence of the random variables Z and U, we have that Z^2 and n/U are independent. Therefore, by the continuous analogue of Rule 3.8, $E(X^2) = E(Z^2)E(n/U)$, see also Rule 5.6(**b**) in Chapter 5. Obviously, $E(Z^2) = 1$. Using the results $\Gamma(\alpha + 1) = \alpha\Gamma(\alpha)$ and $\int_0^\infty \lambda^\alpha \frac{x^{\alpha-1}}{\Gamma(\alpha)} e^{-\lambda x}\, dx = 1$ for $\alpha > 0$, we get

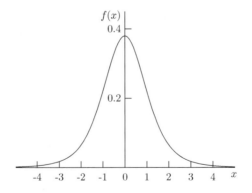

Figure 4.9. Student-t density for $n = 5$.

4.8 Other Continuous Distributions

$$E\left(\frac{n}{U}\right) = n \int_0^\infty \frac{1}{u}\left(\frac{1}{2}\right)^{n/2} \frac{u^{n/2-1}}{\Gamma(n/2)} e^{-\frac{1}{2}u}\, du = \frac{n}{n-2} \quad \text{for } n > 2,$$

which completes the derivation of the formula for var(X). The Student-t distribution is important in statistics, primarily when dealing with small samples from a *normal* population. In particular, this distribution is used in statistics for constructing an *exact* confidence interval in case the observations are generated from a normal distribution. This goes as follows. Suppose that $Y_1, \ldots, Y_n$ are independent samples from an $N(\mu, \sigma^2)$ distribution with unknown expected value μ. The construction of the confidence interval uses the sample mean $\overline{Y}(n)$ and the sample variance $\overline{S}^2(n)$, which are defined by

$$\overline{Y}(n) = \frac{1}{n}\sum_{k=1}^{n} Y_k \quad \text{and} \quad \overline{S}^2(n) = \frac{1}{n-1}\sum_{k=1}^{n}[Y_k - \overline{Y}(n)]^2.$$

It is stated without proof that the random variables $\overline{Y}(n)$ and $\overline{S}^2(n)$ are independent (the converse is also true: if $Y_1, \ldots, Y_n$ are independent random variables each having the same probability distribution function, then independence of $\overline{Y}(n)$ and $\overline{S}^2(n)$ implies that the Y_i are normally distributed). Moreover, it can be shown that $(\overline{Y}(n) - \mu)/(\sigma/\sqrt{n})$ has a standard normal distribution and $(n-1)\overline{S}^2(n)/\sigma^2$ has a chi-square distribution with $n-1$ degrees of freedom. Thus, the ratio

$$\frac{\overline{Y}(n) - \mu}{\sqrt{\overline{S}^2(n)/n}}$$

has a Student-t distribution with $n-1$ degrees of freedom. This important result holds for any value of n and enables us to give the following *exact* $100(1-\alpha)\%$ confidence interval:

$$P\left(\overline{Y}(n) - t_{n-1, 1-\frac{1}{2}\alpha}\sqrt{\overline{S}^2(n)/n} \le \mu \le \overline{Y}(n) + t_{n-1, 1-\frac{1}{2}\alpha}\sqrt{\overline{S}^2(n)/n}\right) = 1 - \alpha,$$

where $t_{n-1, 1-\frac{1}{2}\alpha}$ is the $(1 - \frac{1}{2}\alpha)$th percentile of the Student-t density function with $n-1$ degrees of freedom. That is, the area under the graph of this symmetric density function between the points $-t_{n-1, 1-\frac{1}{2}\alpha}$ and $t_{n-1, 1-\frac{1}{2}\alpha}$ equals $1 - \alpha$. This confidence interval for a sample from a normal population does not require a large n but can be used for any value of n. The statistic $(\overline{Y}(n) - \mu)/\sqrt{\overline{S}^2(n)/n}$ has the pleasant feature of being *robust*. This means that the statistic is not sensitive for small deviations from the normality assumption.

4.9 Inverse-Transformation Method and Simulation

A random-number generator, which generates random numbers between 0 and 1, is indispensable for simulating random observations from a random variable. Random-number generators are discussed in Appendix D. Consider the next example. Let R be a continuous random variable with probability density function $h(r) = re^{-\frac{1}{2}r^2}$ for $r > 0$ and $h(r) = 0$ otherwise. This is the Rayleigh density with parameter 1, a much used density in physics. How do we generate a random observation of R? To do this, we need the cumulative distribution function of the positive random variable R. Letting $H(r) = P(R \leq r)$, we have

$$H(r) = \int_0^r xe^{-\frac{1}{2}x^2}\,dx = -\int_0^r de^{-\frac{1}{2}x^2} = -e^{-\frac{1}{2}x^2}\Big|_0^r = 1 - e^{-\frac{1}{2}r^2}.$$

Suppose that u is a random number between 0 and 1, that is, u is a random observation from the uniform distribution on $(0, 1)$. Then the solution r to the equation

$$H(r) = u$$

is a random observation of R provided that this equation has a unique solution. Since $H(r)$ is strictly increasing on $(0, \infty)$ and is continuous, the equation $H(r) = u$ has a unique solution $r = r(u)$ for any u (this solution as a function of u is called the *inverse function* of $H(r)$). It is readily verified that the equation $H(r) = u$ has the explicit solution

$$r = \sqrt{-2\ln(1 - u)}.$$

It will be clear that the above approach is generally applicable to simulate a random observation of a continuous random variable when the cumulative distribution function of the random variable is strictly increasing and allows for an easily computable inverse function. For example, suppose that X is exponentially distributed with the cumulative distribution function $F(x) = 1 - e^{-\mu x}$ for $x > 0$. Then, a random observation from X is given by

$$x = -\frac{1}{\mu}\ln(1 - u),$$

where u is a random number from $(0, 1)$. This approach of generating random observations from a continuous distribution is known as the *inverse-transformation* method.[3]

---

[3] If X is a continuous random variable with a strictly increasing distribution function $F(x) = P(X \leq x)$ and U is uniformly distributed on $(0, 1)$, then the random variable $V = F^{-1}(U)$ has the same distribution as X, where F^{-1} is the inverse function of F. This result follows immediately from $P(V \leq x) = P(U \leq F(x)) = F(x)$.

4.9 Inverse-Transformation Method and Simulation

As a by-product of the discussion above, we find that the transformation $\sqrt{-2\ln(1-U)}$ applied to the uniform random variable U on $(0, 1)$ yields a random variable with probability density function $re^{-\frac{1}{2}r^2}$ on $(0, \infty)$. This result can be put in a more general framework. Suppose that X is a continuous random variable with probability density function $f(x)$. What is the probability density function of the random variable $Y = v(X)$ for a given function $v(x)$? A general formula can be given for the density function of $v(X)$ when the function $v(x)$ is either strictly increasing or strictly decreasing on the range of X. The function $v(x)$ then has a unique inverse function $a(y)$ (say). That is, for each attainable value y of $Y = v(X)$, the equation $v(x) = y$ has a unique solution $x = a(y)$. Note that $a(y)$ is strictly increasing (decreasing) if $v(x)$ is strictly increasing (decreasing). It is assumed that $a(y)$ is continuously differentiable.

Rule 4.6 *If the function $v(x)$ is strictly increasing or strictly decreasing, then the probability density of the random variable $Y = v(X)$ is given by*

$$f(a(y))|a'(y)|,$$

where $a(y)$ is the inverse function of $v(x)$.

The proof is simple and instructive. We first give the proof for the case that $v(x)$ is strictly increasing. Then, $v(x) \le v$ if and only if $x \le a(v)$. Thus,

$$P(Y \le y) = P(v(X) \le y) = P(X \le a(y)) = F(a(y)),$$

where $F(x)$ denotes the cumulative distribution function of X. Differentiating $P(Y \le y)$ leads to

$$\frac{d}{dy}P(Y \le y) = \frac{d}{da(y)}F(a(y))\frac{da(y)}{dy} = f(a(y))a'(y),$$

which gives the desired result, since $a'(y) > 0$ for a strictly increasing function $a(y)$. In the case of a strictly decreasing function $v(x)$, we have $v(x) \le v$ if and only if $x \ge a(v)$, and so

$$P(Y \le y) = P(v(X) \le y) = P(X \ge a(y)) = 1 - F(a(y)).$$

Differentiating $P(Y \le y)$ and noting that $a'(y) < 0$ yields the desired result.

Example 4.18 Let the random variable Y be defined by $Y = e^X$, where X is $N(\mu, \sigma^2)$ distributed. What is the density function of Y?

Solution. The inverse of the function $v(x) = e^x$ is $a(y) = \ln(y)$. The derivative of $a(y)$ is $1/y$. Next, by Rule 4.6, the density function of Y is

$$\frac{1}{\sigma\sqrt{2\pi}}e^{-\frac{1}{2}(\ln(y)-\mu)^2/\sigma^2}\frac{1}{y} \quad \text{for } y > 0.$$

That is, the random variable Y has a lognormal density with parameters μ and σ.

Example 4.18 was used to illustrate Rule 4.6. The density function of $Y = e^X$ can also be obtained directly by differentiation of the cumulative distribution function $P(Y \leq y) = P(X \leq \ln(y))$ for $y > 0$. In general, one best uses first principles to determine the probability density of a given function of a continuous random variable X. In Chapter 5 we will discuss an extremely useful extension of Rule 4.6 for two-dimensional vectors of jointly distributed random variables.

Example 4.19 Let the random variable X have the standard normal distribution. What is the density function of the random variable $Y = X^2$?

Solution. We use first principles to find the density function of $Y = X^2$. Note that Rule 4.6 cannot be used (explain!). The cumulative distribution function of Y is given by

$$P(Y \leq y) = P(-\sqrt{y} \leq X \leq \sqrt{y}) = P(X \leq \sqrt{y}) - P(X \leq -\sqrt{y})$$

$$= \int_{-\infty}^{\sqrt{y}} \frac{1}{\sqrt{2\pi}} e^{-\frac{1}{2}u^2} du - \int_{-\infty}^{-\sqrt{y}} \frac{1}{\sqrt{2\pi}} e^{-\frac{1}{2}u^2} du \quad \text{for } y > 0.$$

Differentiation of $P(Y \leq y)$ gives that Y has the density function

$$\frac{1}{\sqrt{2\pi}} y^{-\frac{1}{2}} e^{-\frac{1}{2}y} \quad \text{for } y > 0.$$

This is the gamma density with shape parameter $\frac{1}{2}$ and scale parameter $\frac{1}{2}$.

Other Methods to Simulate from Continuous Distributions

The discussion above touched upon the simulation of a random observation from a continuous probability distribution. The inverse-transformation method is a very useful method when the inversion of the cumulative distribution function requires little computational effort. For example, this is the case for the exponential distribution and the Weibull distribution, see Problem 4.92 below. However, for most probability distributions one has to resort to other methods in order to have low computing times. For the normal distribution there are several methods, such as Marsaglia's polar method, which is discussed in Problem 5.49. In the software tool Matlab, more sophisticated methods are used for simulating random observations from the normal distribution. Note that, by the result of Example 4.18, any simulation procedure for the normal distribution can also be used to simulate from the lognormal distribution.

4.9 Inverse-Transformation Method and Simulation

Tailor-made algorithms to simulate from specific probability distributions are often designed on the basis of the *acceptance–rejection method*. This powerful method is based on a simple geometrical idea. A probability density $f(x)$ from which it is difficult to simulate directly is bounded by $cr(x)$, where c is a constant and $r(x)$ is a probability density from which it is easy to simulate. Details of this method will be given in Section 7.2 of Chapter 7 and Section 10.5 of Chapter 10. The acceptance–rejection method can be used for both continuous and discrete distributions.

Problems

4.88 Let X be a continuous random variable with density function $f(x)$. Show that the random variable $Y = \frac{1}{X}$ has the density function $\frac{1}{y^2} f\left(\frac{1}{y}\right)$ for $y \neq 0$. Use this result to verify that the random variable $\frac{1}{X}$ has the same density as X when X has the two-sided density $f(x) = \frac{1}{\pi(1+x^2)}$ for $-\infty < x < \infty$ or the one-sided density $f(x) = \frac{2}{\pi(1+x^2)}$ for $x > 0$.

4.89 The speed of a molecule in a uniform gas at equilibrium is a random variable S whose probability density function is given by $f(s) = \sqrt{\frac{2}{\pi}} \frac{s^2}{c^3} e^{-\frac{1}{2}s^2/c^2}$ for $s > 0$, where $c > 0$ is a known constant. Determine the probability density of the kinetic energy $E = \frac{1}{2}mS^2$ of the molecule, where m is the mass of one gas molecule.

4.90 Let the random variable X have uniform density on $(-1, 1)$. What is the probability density of $Y = \ln(|X|^a)$ for any constant $a > 0$?

4.91 Let the variable Y be defined by $Y = 10^X$, where X is uniformly distributed on $(0, 1)$. Verify that Y has the density $g(y) = \frac{1}{y \ln(10)}$ for $1 < y < 10$ and $g(y) = 0$ otherwise. This is the so-called Benford density.

4.92 Verify that a random observation from the Weibull distribution with shape parameter α and scale parameter λ can be simulated by taking $X = \frac{1}{\lambda}[-\ln(1 - U)]^{1/\alpha}$, where U is a random number from the interval $(0, 1)$. Also, use the inverse-transformation method to simulate a random observation from the logistic distribution function $F(x) = e^x/(1+e^x)$ on $(-\infty, \infty)$.

4.93 Give a simple method to simulate from the gamma distribution with parameters α and λ when the shape parameter α is equal to the integer n. *Hint*: The sum of n independent random variables, each having an exponential density with parameter λ, has an Erlang density with parameters n and λ.

4.94 Let the probability density of the continuous random variable X be given by the triangular density with parameters a, b, and m, where $a < m \le b$. Verify that you can simulate a random observation x from the triangular density by generating a random number u from $(0, 1)$ and setting $x := a + (b-a)\sqrt{m_0 u}$ if $u \le m_0$ and $x := a + (b-a)[1 - \sqrt{(1-m_0)(1-u)}]$ if $u > m_0$, where $m_0 = (m-a)/(b-a)$. *Hint*: Consider the normalized variable $V = (X-a)/(b-a)$ and invert the cumulative distribution function of V.

4.95 How would you simulate a random observation from the positive random variable X whose density function is $p\lambda_1 e^{-\lambda_1 x} + (1-p)\lambda_2 e^{-\lambda_2 x}$ for $x > 0$, where $\lambda_1, \lambda_2 > 0$ and $0 < p < 1$? What about the random variable X whose density function is $p\lambda e^{-\lambda x}$ for $x > 0$ and $(1-p)\lambda e^{\lambda x}$ for $x < 0$?

4.10 Failure-Rate Function

The concept of a failure-rate function applies to a positive random variable and can best be explained by considering a random variable X that represents the lifetime or the time to failure of an item. It is assumed that the random variable X is continuous with cumulative distribution function $F(x) = P(X \le x)$ and probability density function $f(x)$. What is the probability that an item of age a will fail in the next Δa time units? For a continuity point a of $f(x)$ and Δa small, this probability is given by

$$P(X \le a + \Delta a \mid X > a) = \frac{P(a < X \le a + \Delta a)}{P(X > a)} \approx \frac{f(a)\Delta a}{1 - F(a)}.$$

This explains why the *failure-rate function* of X is defined as

$$r(x) = \frac{f(x)}{1 - F(x)}$$

for any x with $P(X > x) > 0$ and as $r(x) = 0$ otherwise. The term *hazard-rate function* is often used instead of failure-rate function. The function $r(x)$ is not a probability density, but $r(x)$ represents the conditional probability intensity that an item of age x will fail in the next moment. The function $1 - F(x)$ gives the probability of no failure before the item has reached age x and is often referred to as the *reliability function*. Another term for this function is the survival function. The cumulative distribution function $F(x)$ can be expressed in terms of the failure-rate function $r(x)$ as

$$F(x) = 1 - e^{-\int_0^x r(t)\,dt} \quad \text{for } x \ge 0.$$

4.10 Failure-Rate Function

To see this, note that $r(x)$ is the derivative of $-\ln(1 - F(x))$. Together, this and $F(0) = 0$ lead to $-\ln(1 - F(x)) = \int_0^x r(t)\,dt$ and so $1 - F(x) = e^{-\int_0^x r(t)\,dt}$. This shows that there is a one-to-one relationship between the failure-rate function and the cumulative distribution function.

To illustrate the concept of a failure-rate function, consider an exponentially distributed lifetime X with expected value $1/\mu$. Then $F(x) = 1 - e^{-\mu x}$ and $f(x) = \mu e^{-\mu x}$, and so

$$r(x) = \mu \quad \text{for all } x \geq 0.$$

That is, the exponential distribution has a constant failure rate, in agreement with the memoryless property discussed in Section 4.5. In other words, new is as good as used when an item has an exponentially distributed lifetime. This characteristic is fairly accurate for many kinds of electronic devices. More generally, if X has a Weibull distribution with shape parameter α and scale parameter λ, the failure-rate function is

$$r(x) = \alpha\lambda(\lambda x)^{\alpha - 1} \quad \text{for } x \geq 0,$$

as follows from the formulas for $F(x)$ and $f(x)$ in Section 4.6.2. The Weibull distribution has an increasing failure rate if $\alpha > 1$ and a decreasing failure rate if $0 < \alpha < 1$ (the Weibull distribution with $\alpha = 1$ reduces to the exponential distribution).

Most complex systems usually exhibit a failure rate that initially decreases to become nearly constant for a while, and then finally increases. This form of failure rate is known as the U-shaped failure rate or bathtub failure rate. An item with a bathtub failure rate has a fairly high failure rate when it is first put into operation. If the item survives the first period, then a nearly constant failure rate applies for some period. Finally, the failure rate begins to increase as wearout becomes a factor. More complicated probability distributions are required to model the bathtub-shaped failure-rate function.

Example 4.20 The time to failure of a pick-and-place machine is a random variable X with failure-rate function $r(x) = \frac{x}{1+x}$ for $x \geq 0$. What is the probability that a pick-and-place machine of age a will survive to age $a + s$?

Solution. Since $\int_0^x r(t)\,dt = x - \ln(1 + x)$ for $x > 0$, we have that $F(x) = P(X \leq x)$ is equal to $F(x) = 1 - e^{-[x - \ln(1+x)]}$. Noting that $e^{\ln(y)} = y$, the reliability function is

$$1 - F(x) = (1 + x)e^{-x} \quad \text{for } x \geq 0.$$

Thus, the sought conditional reliability $P(X > a + s \mid X > a)$ is given by

$$P(X > a + s \mid X > a) = \frac{1 - F(a + s)}{1 - F(a)} = \frac{1 + a + s}{1 + a} e^{-s}.$$

Mean Residual Life

While the failure-rate function at point t provides information about a very small interval after time t, the so-called *mean residual lifetime function* provides information about the whole interval after time t. The mean residual lifetime function $m(t)$ is defined as

$$m(t) = E(X - t \mid X > t)$$

for any t with $P(X > t) > 0$ and as $m(t) = 0$ otherwise. For fixed $t > 0$, the conditional density of $X - t$ given that $X > t$ is $f(x + t)/[1 - F(t)]$ for $x > 0$, as follows by differentiating $P(X - t > x \mid X > t) = [1 - F(x + t)]/[1 - F(t)]$ with respect to x. This leads to

$$m(t) = \frac{\int_0^\infty x f(x+t)\, dx}{1 - F(t)}.$$

Writing $\int_0^\infty x f(x+t)\, dx$ as $\int_0^\infty f(x+t)\, dx \int_0^x dv$ and interchanging the order of integration, it is a matter of simple algebra to show that $\int_0^\infty x f(x+t)\, dx$ is equal to $\int_t^\infty [1 - F(y)]\, dy$. Hence, for any t with $P(X > t) > 0$, the mean residual lifetime function $m(t)$ allows for the convenient representation

$$m(t) = \frac{\int_t^\infty [1 - F(y)]\, dy}{1 - F(t)}.$$

Example 4.21 The time to failure in years of the grasping function of a robot arm is a random variable X with failure-rate function $r(x) = \frac{1}{5}\sqrt{x}$ for $x \geq 0$. What is the reliability function? What is the mean residual life of the grasping function of a 3-year-old robot arm?

Solution. Let $F(x) = P(X \leq x)$. Since $\int_0^x r(t)\, dt = \frac{2}{15} x^{3/2}$ for $x \geq 0$, it follows that the reliability function is given by

$$1 - F(x) = e^{-2x^{1.5}/15} \quad \text{for } x \geq 0.$$

The mean residual life of the grasping function of a 3-year-old robot arm is equal to $m(3) = (1/e^{-2 \times 3^{1.5}/15}) \int_3^\infty e^{-2x^{1.5}/15}\, dx$. Using numerical integration, we obtain $m(3) = 2.273$ years.

Problems

4.96 The lifetime X of a vacuum tube satisfies $P(X \leq x) = x^\nu (2 - x)^\nu$ for $0 \leq x \leq 1$, where $0 < \nu < 1$. Verify that the failure-rate function $r(x)$ is $2\nu(1 - x)(2x - x^2)^{\nu - 1}/[1 - (2x - x^2)^\nu]$ for $0 < x < 1$ and is bathtub shaped.

4.97 Suppose that the positive random variables $X_1, \ldots, X_n$ are independent of each other and have respective failure-rate functions $r_1(x), \ldots, r_n(x)$. Prove that the random variable $V = \min(X_1, \ldots, X_n)$ has the failure-rate function $\sum_{k=1}^{n} r_k(x)$. In particular, conclude that V has an exponential distribution with parameter $\sum_{k=1}^{n} \lambda_k$ when each of the X_i is exponentially distributed with parameter λ_i.

4.98 Suppose that the positive random variable X has the failure-rate function $r(x) = (1+x)^{-1}$ for $x > 0$. Verify the remarkable result that the reliability function of X is also given by $(1+x)^{-1}$ for $x > 0$.

4.99 The lifetime X of a semiconductor has the failure-rate function $r(x) = \lambda \alpha x^{\alpha-1}/(1+x^\alpha)$ for $x \geq 0$, where the parameters α and λ are positive. What is the reliability function of the lifetime X? Verify that the failure-rate function first increases and then decreases when $\alpha > 1$.

4.100 Let X_1 and X_2 be positive continuous random variables with failure-rate functions $r_1(x)$ and $r_2(x)$. Suppose that $r_2(x) = 2r_1(x)$ for all x. Verify that $P(X_1 > s+t \mid X_1 > s) = \sqrt{P(X_2 > s+t \mid X_2 > s)}$ for all s, t.

4.101 The time to failure in hours of a particular device has a Weibull distribution with shape parameter $\alpha = 2.1$ and scale parameter $\lambda = \frac{1}{1,250}$. What is the mean residual life of the device given that it has survived to 1,000 hours?

4.11 Probability Distributions and Entropy[4]

An important concept in statistics is the principle of maximum entropy. This principle is often used to obtain prior probability distributions in Bayesian inference. For a probability mass function $p = (p_i, i \in S)$ on a countable set S, the entropy is defined by

$$h(p) = -\sum_{i \in S} p_i \log p_i.$$

For a continuous probability density function $p(x)$ on the interval I, the entropy is defined by

$$h(p) = -\int_I p(x) \log p(x) \, dx.$$

In the definition of entropy, $x \log x$ is understood to be zero whenever $x = 0$. The base of the logarithm is not important as long as the same base is used consistently. In statistics, the natural logarithm is preferred.

---
[4] This section can be skipped without loss of continuity.

The entropy of a probability distribution can be seen as a measure of the amount of ignorance in the distribution. The principle of maximum entropy states that the probability distribution which best represents the known information is the one with maximum entropy. This is expressed in the following result.

Rule 4.7 *The discrete uniform probability distribution maximizes the entropy among all discrete probability distributions $\{p_1, p_2, \ldots, p_n\}$ on the finite set $\{x_1, x_2, \ldots, x_n\}$, while the uniform density function maximizes the entropy among all continuous density functions $p(x)$ on the interval (a, b).*

This result can be proved by using the Lagrange multiplier method from calculus. We give a sketch of the proof. In the proof we use $e = 2.71828\ldots$ as the base of $\log x$. To maximize $-\sum_{i \geq 1} p_i \log p_i$ subject to the constraints $p_i \geq 0$ for all i and $\sum_{i=1}^{n} p_i = 1$, define the Lagrange function

$$F(p_1, \ldots, p_n, \lambda) = -\sum_{i=1}^{n} p_i \log p_i + \lambda \left(\sum_{i=1}^{n} p_i - 1 \right),$$

where λ is the Lagrange multiplier. Next put the partial derivatives $\partial F / \partial p_i$ and $\partial F / \partial \lambda$ equal to zero. This gives the equations

$$-1 - \log p_i + \lambda = 0 \quad \text{for } i = 1, 2, \ldots, n$$

together with $\sum_{i=1}^{n} p_i = 1$. The solution of these equations gives the solution of the original maximization problem. The first set of equations gives $p_i = e^{\lambda - 1}$ for all i. This is a positive constant value. The equation $\sum_{i=1}^{n} p_i = 1$ implies $e^{\lambda - 1} = \frac{1}{n}$ and so $p_i = \frac{1}{n}$ for all i. For the continuous case, the proof is more subtle. We must maximize $h(p) = -\int_a^b p(x) \log p(x) \, dx$ subject to $p(x) \geq 0$ for all x and $\int_a^b p(x) \, dx = 1$. To do so, form the Lagrange function $F(p, \lambda) = h(p) + \lambda (\int_a^b p(x) \, dx - 1)$. This function can be rewritten as $F(p, \lambda) = \int_a^b L(p(x), \lambda) \, dx - \lambda$, where $L(p, \lambda) = -p \log p + \lambda p$. Next, treat p as an indeterminate and put $\partial L / \partial p = 0$. This gives $-1 - \log p + \lambda = 0$, which leads to $p(x) = e^{\lambda - 1}$ for $x \in (a, b)$. Hence, $p(x)$ is a positive constant on (a, b). Since $\int_a^b p(x) \, dx = 1$, we get $p(x) = \frac{1}{b-a}$ for $x \in (a, b)$. We omit the details showing that the Lagrangian method indeed gives a maximum.

The principle of maximum entropy leads to the selection of a probability distribution that incorporates only what is actually known and introduces no unwarranted information. It is interesting to see which distribution is selected by the principle when the expected value and the standard deviation are given.

4.11 Probability Distributions and Entropy

Rule 4.8 *The exponential density with expected value μ has the maximum entropy among all continuous probability densities on $(0, \infty)$ with expected value μ. The normal density with expected value μ and standard deviation σ has the maximum entropy among all continuous probability densities on $(-\infty, \infty)$ with expected value μ and standard deviation σ.*

This rule can also be proved using the Lagrange multiplier method, but we omit the proof. To conclude this section, we give the following useful result for the class of discrete probability distributions satisfying a certain constraint on the state probabilities.

Rule 4.9 *Consider the class of probability mass functions $(p_1, p_2, \ldots, p_n)$ on the finite set $\{1, 2, \ldots, n\}$ satisfying the constraint $\sum_{i=1}^{n} p_i E_i = E$, where the E_i and E are given numbers such that $\min_i E_i < E < \max_i E_i$ and the E_j are not all equal. The maximum entropy distribution within this class is*

$$p_i^* = \frac{e^{-\beta E_i}}{\sum_{k=1}^{n} e^{-\beta E_k}} \quad \text{for } i = 1, \ldots, n,$$

where β is the unique solution of the equation $\sum_{k=1}^{n} e^{-\beta E_k}(E_k - E) = 0$.

In Problem 4.103 the reader is asked to prove Rule 4.9 using the Lagrange multiplier method.

As an illustration, suppose that a biased six-sided die is rolled 1,000 times. You have no information about the die, but you are told that the total score of the 1,000 rolls is 4,572. What is the best guess for the probability distribution of the numbers of dots on the faces of the die? There are infinitely many probability distributions $(p_1, \ldots, p_6)$ with $\sum_{i=1}^{6} i p_i = 4.572$. The principle of maximum entropy guides us to the best probability distribution that reflects our current knowledge. We apply Rule 4.9 with $E_i = i$ for $i = 1, \ldots, 6$ and $E = 4.572$. Using a numerical root-finding method, we find that the solution of the equation $\sum_{i=1}^{6} e^{-\beta i}(i - 4.572) = 0$ is $\beta = -0.402997$. The best guess for the maximum entropy distribution of the number of dots is given by $p_1^* = 0.0485$, $p_2^* = 0.0726$, $p_3^* = 0.1087$, $p_4^* = 0.1626$, $p_5^* = 0.2434$, and $p_6^* = 0.3641$.

Problems

4.102 Consider the class of all probability mass functions on $\{1, 2, \ldots\}$ that have a given number $\mu > 1$ as expected value. Verify that the maximum entropy distribution within this class is the geometric distribution $p_i = \left(1 - \frac{1}{\mu}\right)^{i-1} \frac{1}{\mu}$ for $i \geq 1$.

4.103 Give a sketch of the proof of Rule 4.9 using the Lagrange multiplier method.

4.104 A fast-food restaurant sells three different cheeseburgers. The prices are $4.50 for a regular cheeseburger, $6.25 for a double cheeseburger, and $7.50 for a big cheeseburger. You were asked to analyze the sales of the restaurant and found that, on average, customers are paying $5.75 for their cheeseburgers. However, you forgot to gather information about the probabilities of a customer ordering each of the three cheeseburgers. What is your best guess of these probabilities?

5
Jointly Distributed Random Variables

In experiments, one is often interested not only in individual random variables, but also in relationships between two or more random variables. For example, if the experiment is the testing of a new medicine, the researcher might be interested in the cholesterol level, blood pressure, and glucose level of a test person. Similarly, a political scientist investigating the behavior of voters might be interested in the income and level of education of a voter. There are many more examples in the physical sciences, medical sciences, and social sciences. In applications, one often wishes to make inferences about one random variable on the basis of observations of other random variables.

The purpose of this chapter is to familiarize you with the notations and the techniques relating to experiments whose outcomes are described by two or more real numbers. The discussion is restricted to the case of pairs of random variables. You will learn how to obtain the joint density function of two random variables and how to obtain the marginal densities from the joint density. In the field of statistics, one often needs the joint density of two new random variables that are functions of two given random variables. You will see that this can be nicely done by using the transformation rule from calculus. Important measures for two jointly distributed random variables are the covariance and the correlation coefficient. These statistics are discussed, together with the concept of regression to the mean.

5.1 Joint Probability Mass Function

If X and Y are two discrete random variables defined on the same sample space with probability measure P, then the mass function $p(x, y)$ defined by

$$p(x, y) = P(X = x, Y = y)$$

is called the *joint probability mass function* of X and Y. The quantity $P(X = x, Y = y)$ is the probability assigned by P to the intersection of the two sets $A = \{\omega : X(\omega) = x\}$ and $B = \{\omega : Y(\omega) = y\}$, with ω representing an element of the sample space. The so-called marginal probability mass functions of the random variables X and Y are defined by

$$p_X(x) = P(X = x) \quad \text{and} \quad p_Y(y) = P(Y = y).$$

The marginal probability mass functions can be obtained from the joint probability mass function by

$$p_X(x) = \sum_y P(X = x, Y = y) \quad \text{and} \quad p_Y(y) = \sum_x P(X = x, Y = y),$$

using the fact that the probability of the union of mutually exclusive events is the sum of the probabilities of the individual events.

Example 5.1 Two baseball teams A and B play a best-two-of-three series. The first team that wins two games is the overall winner. Each game is won by team A with probability p and by team B with probability $1-p$, independently of the other games. Let the random variable X be the number of games won by team A and Y be the number of games won by team B. What is the joint probability mass function of X and Y?

Solution. The random variables X and Y can be defined on the same sample space. Noting that a maximum of three games are played, each element of the sample space can be represented by $\omega = (\omega_1, \omega_2, \omega_3)$, where $\omega_i = a$ if the ith game is won by team A, $\omega_i = b$ if the ith game is won by team B, and $\omega_i = -$ if no ith game is played. The sample space has the elements $(a, a, -)$, $(b, b, -)$, (a, b, a), (b, a, a), (b, a, b), and (a, b, b). These elements get assigned the probabilities p^2, $(1-p)^2$, $p^2(1-p)$, $p^2(1-p)$, $p(1-p)^2$, and $p(1-p)^2$. We have $X(\omega) = i$ if i components of ω are equal to a and $Y(\omega) = j$ if j components of ω are equal to b. In Table 5.1, we give the joint probability mass function $p(x, y) = P(X = x, Y = y)$. For example, $P(X = 1, Y = 2) = P(\{(b, a, b), (a, b, b)\})$ and is thus $2p(1-p)^2$.

Table 5.1. *The mass function $p(x, y)$*

$x \setminus y$	0	1	2
0	0	0	$(1-p)^2$
1	0	0	$2p(1-p)^2$
2	p^2	$2p^2(1-p)$	0

Problems

5.1 You roll a pair of dice. What is the joint probability mass function of the low and high points rolled?

5.2 Let X be the number of hearts and Y be the number of diamonds in a bridge hand. What is the joint probability mass function of X and Y?

5.3 The joint probability mass function of the lifetimes X and Y of two connected components in a machine can be modeled by $p(x,y) = \frac{e^{-2}}{x!(y-x)!}$ for $x = 0, 1, \ldots$ and $y = x, x+1, \ldots$. What is the joint probability mass function of X and $Y - X$? What are the marginal distributions of X, $Y - X$, and Y?

5.4 In the final of the World Series baseball, two teams play a series consisting of at most seven games until one of the two teams has won four games. Two unevenly matched teams are pitted against each other and the probability that the weaker team will win any given game is equal to 0.45. Let X be the number of games won by the weaker team and Y be the number of games the final will take. What is the joint probability mass function of X and Y?

5.5 You choose three different numbers at random from the numbers $1, 2, \ldots, 10$. Let X be the smallest of these three numbers and Y be the largest. What is the joint probability mass function of X and Y? What are the marginal distributions of X and Y and what is the probability mass function of $Y - X$?

5.6 You repeatedly draw a number at random from the numbers $1, 2, \ldots, 10$. Let X be the number of drawings until the number 1 appears and Y be the number of drawings until the number 10 appears. What is the joint probability mass function of X and Y? What are the probability mass functions of $\min(X, Y)$ and $\max(X, Y)$?

5.7 You roll a fair die once. Let the random variable N be the number shown by the die. Two persons each toss N fair coins, independently of each other. Let X be the number of heads obtained by the first person and Y be the number of heads obtained by the second person. What is the joint probability mass function of X and Y? What is $P(X = Y)$? *Hint*: Use the identity $\sum_{k=0}^{n} \binom{n}{k}^2 = \binom{2n}{n}$.

5.8 You repeatedly toss two fair coins together until both coins show heads. Let X and Y denote the number of heads resulting from the tosses of the first and the second coin, respectively. What is the joint probability mass function of X and Y and what are the marginal distributions of X and Y? What is $P(X = Y)$? *Hint*: Use the identity $\sum_{k=0}^{\infty} \binom{m+k}{m} x^k = 1/(1-x)^{m+1}$ for $|x| < 1$.

5.2 Joint Probability Density Function

The next example provides a good starting point for a discussion of joint probability densities.

Example 5.2 A random point is picked inside a circular disk with radius r. Let the random variable X be the length of the line segment between the center of the disk and the randomly picked point, and the random variable Y (measured in radians) be the angle between this line segment and the horizontal axis. What is the joint distribution of X and Y?

Solution. The two continuous random variables X and Y are defined on a common probability space. The sample space consists of all points (x, y) in the two-dimensional plane such that $x^2 + y^2 \leq r^2$, where the point $(0, 0)$ represents the center of the disk. The probability $P(A)$ assigned to each well-defined subset A of the sample space is taken as the area of region A divided by πr^2. We can now determine the probability $P(X \leq a, Y \leq b)$ of the joint event that X takes on a value less than or equal to a and Y a value less than or equal to b. This event occurs only if the randomly picked point falls inside the disk segment with radius a and angle b. The area of this disk segment is $\frac{b}{2\pi}\pi a^2$. Dividing this by πr^2 gives

$$P(X \leq a, Y \leq b) = \frac{b}{2\pi} \frac{a^2}{r^2} \quad \text{for } 0 \leq a \leq r \text{ and } 0 \leq b \leq 2\pi.$$

We are now in a position to introduce the concept of joint density. Let X and Y be two random variables defined on the same sample space with probability measure P. The *joint cumulative probability distribution function* of X and Y is defined by $P(X \leq x, Y \leq y)$ for all x, y, where $P(X \leq x, Y \leq y)$ is shorthand for $P(\{\omega : X(\omega) \leq x \text{ and } Y(\omega) \leq y\})$ and the symbol ω represents an element of the sample space.

Definition 5.1 *The continuous random variables X and Y are said to have a joint probability density function $f(x, y)$ if the joint cumulative probability distribution function $P(X \leq a, Y \leq b)$ allows for the representation*

$$P(X \leq a, Y \leq b) = \int_{x=-\infty}^{a} \int_{y=-\infty}^{b} f(x, y)\, dx\, dy \quad \text{for } -\infty < a, b < \infty,$$

where the function $f(x, y)$ satisfies

$$f(x, y) \geq 0 \text{ for all } x, y \quad \text{and} \quad \int_{-\infty}^{\infty} \int_{-\infty}^{\infty} f(x, y)\, dx\, dy = 1.$$

5.2 Joint Probability Density Function

Just as in the one-dimensional case, $f(a, b)$ allows for the interpretation

$$f(a,b)\,\Delta a\,\Delta b \approx P\left(a - \frac{1}{2}\Delta a \leq X \leq a + \frac{1}{2}\Delta a,\, b - \frac{1}{2}\Delta b \leq Y \leq b + \frac{1}{2}\Delta b\right)$$

for small values of Δa and Δb when $f(x, y)$ is continuous at $(x, y) = (a, b)$. In words, the probability that (X, Y) falls into a small rectangle with sides of lengths Δa and Δb around the point (a, b) is approximately $f(a, b)\,\Delta a\,\Delta b$.

In general, the joint density function $f(x, y)$ is obtained by taking the second-order partial derivative of the joint cumulative distribution function $P(X \leq x, Y \leq y)$. Under mild regularity conditions, we have

$$f(x, y) = \frac{\partial}{\partial x}\left(\frac{\partial}{\partial y} P(X \leq x, Y \leq y)\right) = \frac{\partial}{\partial y}\left(\frac{\partial}{\partial x} P(X \leq x, Y \leq y)\right).$$

Thus it does not matter if we take the partial derivative with respect to x first or with respect to y first. In the specific case of Example 5.2, we find

$$f(x, y) = \begin{cases} \frac{1}{2\pi} \frac{2x}{r^2} & \text{for } 0 < x < r \text{ and } 0 < y < 2\pi, \\ 0 & \text{otherwise.} \end{cases}$$

In general, the joint probability density function is found by determining first the cumulative joint probability distribution function and taking next the second-order partial derivative. However, sometimes it is easier to find the joint probability density function by using its probabilistic interpretation. This is illustrated with the next example.

Example 5.3 The pointer of a spinner of radius r is spun three times. The three spins are performed independently of each other. With each spin, the pointer stops at an unpredictable point on the circle. The random variable L_i corresponds to the length of the arc from the top of the circle to the point where the pointer stops on the ith spin. The length of the arc is measured clockwise. Let $X = \min(L_1, L_2, L_3)$ and $Y = \max(L_1, L_2, L_3)$. What is the joint probability density function $f(x, y)$ of the two continuous random variables X and Y?

Solution. We can derive the joint probability density function $f(x, y)$ by using the interpretation that the probability $P(x < X \leq x + \Delta x, y < Y \leq y + \Delta y)$ is approximately equal to $f(x, y)\Delta x \Delta y$ provided that Δx and Δy are very small. The event $\{x < X \leq x + \Delta x, y < Y \leq y + \Delta y\}$ occurs only if one of the L_i takes on a value between x and $x + \Delta x$, one of the L_i a value between y and $y + \Delta y$, and the remaining L_i a value between x and y, where $0 < x < y$. For any i, the probability that the random variable L_i takes on a value between a and b equals $(b - a)/(2\pi r)$ for $0 < a < b < 2\pi r$. The values of L_1, L_2, and L_3 can be arranged in $3 \times 2 \times 1 = 6$ ways. Thus, by the independence of the spins,

$$P(x < X \leq x + \Delta x,\ y < Y \leq y + \Delta y) = 6\,\frac{\Delta x}{2\pi r} \times \frac{\Delta y}{2\pi r} \times \frac{(y-x)}{2\pi r}.$$

Hence, the joint probability density function of X and Y is given by

$$f(x, y) = \frac{6(y - x)}{(2\pi r)^3} \quad \text{for } 0 < x < y < 2\pi r$$

and $f(x, y) = 0$ otherwise.

The following important rule applies to any two random variables X and Y having a joint probability density function $f(x, y)$.

Rule 5.1 For any neat region C in the two-dimensional plane,

$$P\big((X, Y) \in C\big) = \iint_C f(x, y)\, dx\, dy.$$

This is a very useful result. A rigorous proof of Rule 5.1 is beyond the scope of this book. It is important to note that in calculating a double integral over a nonnegative integrand, it does not matter whether we integrate over x first or over y first. This is a basic fact from calculus. A double integral is usually computed as a repeated one-dimensional integral. Choosing the order of integration usually depends on the particular problem considered. In specific applications, care must be taken in determining the correct limits of integration when rewriting the double integral as a repeated one-dimensional integral. Sketching the region of integration may be helpful for that purpose. In many cases the region C is not a rectangle $\{a \leq x \leq b,\ c \leq y \leq d\}$, but of the general form

$$C = \{(x, y) : a \leq x \leq b,\ g_1(x) \leq y \leq g_2(x)\}.$$

Then you compute the double integral by holding x constant and integrating with respect to y as if this were a one-dimensional integral. This will lead to a function involving only x's, which can in turn be integrated. Thus,

$$\iint_C f(x, y)\, dx\, dy = \int_a^b \left[\int_{g_1(x)}^{g_2(x)} f(x, y)\, dy\right] dx.$$

Rule 5.1 can be used to determine the probability distribution function of any function $g(X, Y)$ of X and Y. This is demonstrated in the following three examples. In these examples, the solution approach is to write the sought probability in the form $P((X, Y) \in C)$ for an appropriately chosen domain of integration C and then apply Rule 5.1.

Example 5.4 An electronic system has two crucial components. The electronic system goes down if either of the two components fails. The joint

5.2 Joint Probability Density Function

density function of the lifetimes X and Y of the two components is given by $f(x, y) = \frac{1}{12}(1 + x + y)$ for $0 < x, y < 2$ and $f(x, y) = 0$ otherwise, where the lifetimes are measured in years. What is the probability that the electronic system will go down in its first year of operation?

Solution. The sought probability is $P(\min(X, Y) \leq 1)$. To find this probability, it is easier to compute the complementary probability $P(\min(X, Y) > 1)$. Note that $\min(X, Y) > 1$ if and only if both $X > 1$ and $Y > 1$. The probability $P(\min(X, Y) > 1)$ can be written as $P((X, Y) \in C)$, where the region C is given by $C = \{(x, y) : 1 < x < 2, 1 < y < 2\}$. Applying Rule 5.1, we find that $P(\min(X, Y) > 1)$ is equal to

$$\int_1^2 \int_1^2 \frac{1}{12}(1 + x + y) \, dx \, dy = \frac{1}{12} \int_1^2 dx \int_1^2 (1 + x + y) \, dy$$
$$= \frac{1}{12} \int_1^2 (1 + x + 1.5) \, dx = \frac{1}{3}.$$

Thus, the sought probability is $\frac{2}{3}$.

Example 5.5 Let the random variable X be a number chosen at random from $(0, 1)$ and Y be a number chosen at random from $(0, X)$. Take X and Y as the lengths of the sides of a rectangle. What is the probability density of the area of the rectangle?

Solution. Let $f(x, y)$ denote the joint probability density of the random variables X and Y. To find $f(x, y)$, fix x and y with $0 < y < x < 1$ and note that the probability $P(x < X \leq x + \Delta x, y < Y \leq y + \Delta y)$ can be evaluated as $P(x < X \leq x + \Delta x)P(y < Y \leq y + \Delta y \mid x < X \leq x + \Delta x)$ and so this probability is equal to $\Delta x \times \frac{\Delta y}{x}$ for Δx and Δy small. Thus, $f(x, y) = \frac{1}{x}$ for $0 < x < 1, 0 < y < x$ and $f(x, y) = 0$ otherwise. To find $P(XY \leq z)$, note that this probability can be written as $P((X, Y) \in C)$, where

$$C = \{(x, y) : 0 < x < 1, 0 < y < x \text{ and } xy \leq z\}.$$

To set up the double integral $\iint_C f(x, y) \, dx \, dy$ as a repeated one-dimensional integral, you first fix x and integrate y from 0 to $\min(x, z/x)$. Then, you integrate x from 0 to 1. This gives

$$P(XY \leq z) = \iint_C f(x, y) \, dx \, dy = \int_0^1 \left[\int_0^{\min(x, z/x)} \frac{1}{x} \, dy \right] dx$$
$$= \int_0^{\sqrt{z}} \left[\int_0^x \frac{1}{x} \, dy \right] dx + \int_{\sqrt{z}}^1 \left[\int_0^{z/x} \frac{1}{x} \, dy \right] dx.$$

Hence,

$$P(XY \le z) = \int_0^{\sqrt{z}} dx + \int_{\sqrt{z}}^1 \frac{z}{x^2} dx = \sqrt{z} - z + \sqrt{z}.$$

Differentiating $P(XY \le z)$ gives that the probability density $f(z)$ of the area of the rectangle is $f(z) = \frac{1}{\sqrt{z}} - 1$ for $0 < z < 1$ and $f(z) = 0$ otherwise.

Example 5.6 An insurance company is incurring losses under collision insurance and losses under liability insurance. These two losses can be modeled by continuous random variables X and Y with the joint probability density function $f(x, y) = \frac{3}{5}(5 - 3x - 2y)$ for $x, y > 0$, $x + y < 1$ and $f(x, y) = 0$ otherwise. The losses are expressed in units of millions of dollars. What is the density function of the total losses? What is the probability that the losses under collision insurance are larger than the losses under liability insurance?

Solution. The probability $P(X + Y \le z)$ can be written as $P((X, Y) \in C)$, where the region $C = \{(x, y) : x, y \ge 0, x + y < z\}$ for fixed z. Applying Rule 5.1, we find

$$P(X + Y \le z) = \iint_C f(x, y) \, dx \, dy = \int_0^z \left[\int_0^{z-x} \frac{3}{5}(5 - 3x - 2y) \, dy \right] dx$$

$$= \frac{3}{5} \int_0^z [(5 - 3x)(z - x) - (z - x)^2] \, dx \quad \text{for } 0 \le z \le 1.$$

This leads to $P(X + Y \le z) = 1.5z^2 - 0.5z^3$ for $0 \le z \le 1$. Therefore, the density function $f(z)$ of $X + Y$ is given by $f(z) = 3z - 1.5z^2$ for $0 < z < 1$ and $f(z) = 0$ otherwise. The probability $P(X > Y)$ can be calculated as

$$\int_0^1 \left[\int_0^{\min(x, 1-x)} \frac{3}{5}(5 - 3x - 2y) \, dy \right] dx$$

$$= \frac{3}{5} \int_0^{0.5} \left[x(5 - 3x) - x^2 \right] dx + \frac{3}{5} \int_{0.5}^1 \left[(1 - x)(5 - 3x) - (1 - x)^2 \right] dx.$$

This gives $P(X > Y) = 0.425 + 0.2 = 0.625$.

Uniform Distribution Over a Region

The following result is very useful. Suppose that a random point (X, Y) is picked inside a bounded region R in the two-dimensional plane. Then the joint density function of X and Y is the uniform density function

$$f(x, y) = \frac{1}{\text{area of region } R} \quad \text{for } (x, y) \in R.$$

The proof is simple. For any subset $C \subseteq R$,

$$P((X, Y) \in C) = \frac{\text{area of } C}{\text{area of } R},$$

being the mathematical definition of the selection of a random point inside the region R. Integral calculus tells us that the area of $C = \iint_C dx\, dy$. Thus, for any subset $C \subseteq R$,

$$P((X, Y) \in C) = \iint_C \frac{1}{\text{area of } R}\, dx\, dy,$$

showing that the uniform density function $f(x, y) = 1/(\text{area of region } R)$ for $(x, y) \in R$ is the density function of the random point (X, Y). The extension of this result to higher dimensions is obvious. As an illustration, we give the next example.

Example 5.7 Suppose that (X, Y, Z) is a random point inside the unit cube $\Omega = \{(x, y, z) : 0 \le x, y, z \le 1\}$. What is $P(X^2 > YZ)$?

Solution. The unit cube has volume 1 and so the random point (X, Y, Z) has the joint density function $f(x, y, z) = 1$ for $(x, y, z) \in \Omega$ and $f(x, y, z) = 0$ otherwise. Let $C = \{(x, y, z) \in \Omega : x^2 > yz\}$. Then

$$P(X^2 > YZ) = \iiint_C f(x, y, z)\, dx\, dy\, dz.$$

This leads to

$$P(X^2 > YZ) = \int_0^1 dz \int_0^1 dy \int_{\sqrt{yz}}^1 dx = \int_0^1 dz \int_0^1 (1 - \sqrt{yz})\, dy$$

$$= \int_0^1 \left(1 - \frac{2}{3}\sqrt{z}\right) dz = \frac{5}{9}.$$

Problems

5.9 Let the joint probability density function of the random variables X and Y be given by $f(x, y) = ce^{-2x}$ for $0 < y \le x < \infty$ and $f(x, y) = 0$ otherwise. Determine the constant c. What is the density function of $Y - X$?

5.10 Let the joint probability density function of the random variables X and Y be given by $f(x, y) = cxe^{-2x(1+y)}$ for $x, y > 0$ and $f(x, y) = 0$ otherwise. Determine the constant c. What is the density function of XY?

5.11 Let the joint probability density function of the random variables X and Y be given by $f(x, y) = c\sqrt{x+y}$ for $0 < x, y < 1$ and $f(x, y) = 0$ otherwise. Determine the constant c. What is the density function of $X + Y$?

5.12 The joint probability density function of the random variables X and Y is given by $f(x, y) = x + y$ for $0 \leq x, y \leq 1$ and $f(x, y) = 0$ otherwise. Consider the circle centered at the origin and passing through the point (X, Y). What is the probability that the circumference of the circle is larger than π?

5.13 Let the random variable X be the smallest and Y be the largest of two points that are randomly chosen in the interval $(0, a)$ for $a > 0$. What is the joint probability density function of X and Y?

5.14 Suppose that (X, Y, Z) is a random point inside the unit cube $\{(x, y, z) : 0 \leq x, y, z \leq 1\}$. What is $P(X + Y < Z)$? Use a symmetry argument to find $P(\max(X, Y) > Z)$.

5.15 There are two alternative routes for a ship passage. The sailing times for the two routes are random variables X and Y that have the joint probability density function $f(x, y) = \frac{1}{10}e^{-\frac{1}{2}(y+3-x)}$ for $5 < x < 10$, $y > x - 3$ and $f(x, y) = 0$ otherwise. What is $P(X < Y)$?

5.16 The joint probability density function of the random variables X and Y is given by $f(x, y) = \frac{1}{2}(x+y)e^{-(x+y)}$ for $x, y > 0$ and $f(x, y) = 0$ otherwise. What is the density function of the random variable $X + Y$?

5.17 Let (X, Y) be a random point inside the unit square $\{(x, y) : 0 \leq x, y \leq 1\}$. What is $P(\max(X, Y) > a \min(X, Y))$ for $a > 1$?

5.18 An electronic device contains two circuits. The second circuit is a backup for the first one and is switched on only when the first circuit has failed. The electronic device goes down when the second circuit fails. The joint density function of the operating times X and Y of the first and the second circuits is given by $f(x, y) = 24/(x + y)^4$ for $x, y > 1$ and $f(x, y) = 0$ otherwise. What is the density function of the time $X + Y$ until the device goes down?

5.19 The lifetimes X and Y of two components in a machine have the joint density function $f(x, y) = \frac{1}{4}(2y + 2 - x)$ for $0 < x < 2$, $0 < y < 1$ and $f(x, y) = 0$ otherwise. The machine goes down when both components have failed. What is the probability density of the time until the machine goes down?

5.20 A stick is broken into three pieces at two randomly chosen points. What is the joint density of the smallest breakpoint and the largest breakpoint? What is the probability that no piece is longer than half the length of the stick?

5.21 What is the probability that the equation $Ax^2 + Bx + 1 = 0$ has two real roots when the random variables A and B have the joint density function

$f(a, b) = a + b$ for $0 < a, b < 1$? What is the probability that the equation $Ax^2 + Bx + C = 0$ has two real roots when the random variables A, B, and C have the joint density function $f(a, b, c) = \frac{2}{3}(a + b + c)$ for $0 < a, b, c < 1$?

5.3 Marginal Probability Densities

If the two random variables X and Y have a joint probability density function $f(x, y)$, then each of the random variables X and Y has a probability density itself. Using the fact that $\lim_{n \to \infty} P(A_n) = P(\lim_{n \to \infty} A_n)$ for any increasing sequence of events A_n, it follows that

$$P(X \leq a) = \lim_{b \to \infty} P(X \leq a, Y \leq b) = \int_{-\infty}^{a} \left[\int_{-\infty}^{\infty} f(x, y) \, dy \right] dx.$$

This representation shows that X has probability density function

$$f_X(x) = \int_{-\infty}^{\infty} f(x, y) \, dy \quad \text{for } -\infty < x < \infty.$$

In the same way, the random variable Y has probability density function

$$f_Y(y) = \int_{-\infty}^{\infty} f(x, y) \, dx \quad \text{for } -\infty < y < \infty.$$

The probability density functions $f_X(x)$ and $f_Y(y)$ are called the *marginal probability density functions* of X and Y. The following interpretation can be given to the marginal density $f_X(x)$ at the point $x = a$ when a is a continuity point of $f_X(x)$. For Δa small, $f_X(a) \Delta a$ gives approximately the probability that (X, Y) falls into a vertical strip in the two-dimensional plane with width Δa and around the vertical line $x = a$. A similar interpretation applies to $f_Y(b)$ for any continuity point b of $f_Y(y)$.

Example 5.6 (continued) What are the marginal densities of the losses X under collision insurance and the losses Y under liability insurance?

Solution. For any $0 < x < 1$, the joint density function $f(x, y)$ is zero for y outside the interval $(0, 1 - x)$. Thus, in order to get $f_X(x)$, we integrate $f(x, y)$ over y from 0 to $1 - x$. Thus, by $f_X(x) = \int_0^{1-x} f(x, y) \, dy$, we get

$$f_X(x) = \int_0^{1-x} \frac{3}{5}(5 - 3x - 2y) \, dy = \frac{3}{5}(1 - x)(4 - 2x) \quad \text{for } 0 < x < 1.$$

In the same way, we get from $f_Y(y) = \int_0^{1-y} f(x, y) \, dx$ that

$$f_Y(y) = \int_0^{1-y} \frac{3}{5}(5 - 3x - 2y) \, dx = \frac{3}{5}(1 - y)(3.5 - 0.5y) \quad \text{for } 0 < y < 1.$$

Example 5.8 A random point (X, Y) is chosen inside the unit circle. What is the marginal density of X?

Solution. Denote by $C = \{(x,y) : x^2 + y^2 \leq 1\}$ the unit circle. The joint probability density function $f(x,y)$ of X and Y is $f(x,y) = 1/(\text{area of } C)$ for $(x,y) \in C$ and $f(x,y) = 0$ otherwise. Hence,

$$f(x,y) = \frac{1}{\pi} \quad \text{for } (x,y) \in C.$$

To obtain $f_X(x)$ for $-1 < x < 1$, note that $f(x,y)$ is equal to zero for those y satisfying $y^2 > 1 - x^2$. Thus, for fixed x, we integrate y from $-\sqrt{1-x^2}$ to $\sqrt{1-x^2}$. This gives

$$f_X(x) = \int_{-\sqrt{1-x^2}}^{\sqrt{1-x^2}} \frac{1}{\pi} \, dy = \frac{2}{\pi}\sqrt{1-x^2} \quad \text{for } -1 < x < 1.$$

Can you explain why the marginal density of X is not the uniform density on $(-1, 1)$? *Hint*: Interpret $P(x < X \leq x + \Delta x)$ as the area of a vertical strip in the unit circle.

Problems

5.22 The joint density function of the random variables X and Y is given by $f(x,y) = 4xe^{-2x(1+y)}$ for $x, y > 0$. What are the marginal densities of X and Y?

5.23 Let the joint density function $f(x,y)$ of the random variables X and Y be equal to $4e^{-2x}$ for $0 < y \leq x < \infty$ and 0 otherwise. What are the marginal densities of X and Y?

5.24 Let the random variable X be the portion of a flood insurance claim for flooding damage to a house and Y the portion of the claim for flooding damage to the rest of the property. The joint density of X and Y is $f(x,y) = 3 - 2x - y$ for $0 < x, y < 1$ and $x + y < 1$. What are the marginal densities of X and Y?

5.25 A random point (X, Y) is chosen inside the equilateral triangle having $(0,0)$, $(1,0)$, and $(\frac{1}{2}, \frac{1}{2}\sqrt{3})$ as corner points. Determine the marginal densities of X and Y. Before determining the function $f_X(x)$, can you explain why $f_X(x)$ must be largest at $x = \frac{1}{2}$?

5.26 The joint density of the random variables X and Y satisfies $f(x,y) = \frac{1}{x}$ for (x,y) inside the triangle with vertices $(0,0)$, $(1,0)$, and $(1,1)$ and $f(x,y) = 0$ otherwise. What are the marginal densities of X and Y?

5.3 Marginal Probability Densities

5.27 An unreliable electronic system has two components hooked up in parallel. The lifetimes X and Y of the two components have the joint density $f(x, y) = e^{-y}$ for $0 < x \leq y < \infty$. The system goes down when both components have failed. What is the joint density of X and $Y - X$? What are the marginal densities of X and $Y - X$? What is the density function of the time until the system goes down?

5.3.1 Independence of Jointly Distributed Random Variables

A general condition for the independence of the jointly distributed random variables X and Y is stated in Definition 3.2. In terms of the marginal densities, the continuous analogue of Rule 3.7 for the discrete case is:

Rule 5.2 *The jointly distributed random variables X and Y are independent if and only if*

$$f(x, y) = f_X(x)f_Y(y) \quad \text{for all } x, y.$$

This rule follows directly from Definition 3.2 and the basic relation $f(x, y) = \frac{\partial^2}{\partial x \partial y} P(X \leq x, Y \leq y)$.

We next prove the convolution formula for independent continuous random variables.

Rule 5.3 (convolution formula) *Let X and Y be two independent continuous random variables having the marginal density functions $f_X(x)$ and $f_Y(y)$. Then the sum $X + Y$ has the probability density function*

$$f_{X+Y}(z) = \int_{-\infty}^{\infty} f_X(z - y) f_Y(y) \, dy \quad \text{for } -\infty < z < \infty.$$

In particular, $f_{X+Y}(z) = \int_0^z f_X(z - y) f_Y(y) \, dy$ for $z > 0$ when X and Y are nonnegative.

The proof goes as follows. The probability $P(X + Y \leq z)$ can be written as $P((X, Y) \in C)$, where the region $C = \{(x, y) : -\infty < x, y < \infty, x + y \leq z\}$. By the independence of X and Y, the joint density function of X and Y is $f(x, y) = f_X(x) f_Y(y)$. Applying Rule 5.1, we find

$$P(X + Y \leq z) = \iint_C f(x, y) \, dx \, dy = \int_{-\infty}^{\infty} \left[\int_{-\infty}^{z-y} f_X(x) \, dx \right] f_Y(y) \, dy$$

$$= \int_{-\infty}^{\infty} P(X \leq z - y) f_Y(y) \, dy \quad \text{for } -\infty < z < \infty.$$

Then, by differentiating $P(X + Y \leq z)$ with respect to z, we get the convolution formula for $f_{X+Y}(z)$. Differentiation under the integral sign is justified by a general result from real analysis. If X and Y are nonnegative, then $P(X \le z - y) = 0$ for $y > z$ and $f_Y(y) = 0$ for $y < 0$, and the expression for $f_{X+Y}(z)$ reduces to $\int_0^z f_X(z - y) f_Y(y)\, dy$ for $z > 0$.

Example 5.2 (continued) Let the random variable X be the length of the line segment between a randomly picked point inside a circular disk with radius r and the center of the disk, and let Y be the angle between this line segment and the horizontal axis. Are the random variables X and Y independent?

Solution. To answer the question, we need the marginal densities of X and Y. The joint density of X and Y was determined as $f(x, y) = \frac{x}{\pi r^2}$ for $0 < x < r$, $0 < y < 2\pi$ and $f(x, y) = 0$ otherwise. Then it follows from $f_X(x) = \int_0^{2\pi} \frac{x}{\pi r^2}\, dy$ that

$$f_X(x) = \frac{2x}{r^2} \quad \text{for } 0 < x < r$$

and $f_X(x) = 0$ otherwise. Using $f_Y(y) = \int_0^r \frac{x}{\pi r^2}\, dx$, we have

$$f_Y(y) = \frac{1}{2\pi} \quad \text{for } 0 < y < 2\pi$$

and $f_Y(y) = 0$ otherwise. We can now conclude that $f(x, y) = f_X(x) f_Y(y)$ for all x, y, and so the distance X and the angle Y are independent. This is a somewhat surprising result, because there is dependence between the components of the randomly picked point inside the circle.

Example 5.9 Let X and Y be two independent random variables, each having the standard normal distribution. What is the probability distribution function of $X^2 + Y^2$? How could you use this result to prove that the variance of an $N(\mu, \sigma^2)$-distributed random variable is σ^2?

Solution. Since X and Y are independent random variables with densities $f_X(x) = \frac{1}{\sqrt{2\pi}} e^{-\frac{1}{2}x^2}$ and $f_Y(y) = \frac{1}{\sqrt{2\pi}} e^{-\frac{1}{2}y^2}$, the joint density of X and Y is

$$f(x, y) = \frac{1}{2\pi} e^{-\frac{1}{2}(x^2 + y^2)} \quad \text{for } -\infty < x, y < \infty.$$

For any given $z > 0$,

$$P(X^2 + Y^2 \le z) = \iint_C f(x, y)\, dx\, dy = \iint_C \frac{1}{2\pi} e^{-\frac{1}{2}(x^2 + y^2)}\, dx\, dy,$$

where $C = \{(x, y) : x^2 + y^2 \le z\}$ is a disk of radius $\sqrt{z}$. In the Cartesian (rectangular) coordinates x and y, the disk C is described by the inequalities $-\sqrt{z} \le x \le \sqrt{z}$ and $-\sqrt{z - x^2} \le y \le \sqrt{z - x^2}$. This gives an unpleasant integral to compute. However, a disk of radius $\sqrt{z}$ can be defined in terms of

5.3 Marginal Probability Densities

polar coordinates r and θ with $0 \leq \theta \leq 2\pi$ and $0 \leq r \leq \sqrt{z}$, where r (the distance) and θ (the angle) are defined by $x = r\cos(\theta)$ and $y = r\sin(\theta)$. When changing to polar coordinates, the integral gets constant limits of integration and becomes easier to compute. The term $x^2 + y^2$ is converted into $r^2 \cos^2(\theta) + r^2 \sin^2(\theta) = r^2$, using the identity $\cos^2(\theta) + \sin^2(\theta) = 1$. Some care is required with $dx\,dy$ when changing to polar coordinates. An infinitesimally small area element $\Delta x\,\Delta y$ in Cartesian coordinates is not converted into $\Delta r\,\Delta\theta$ but becomes $r\,\Delta r\,\Delta\theta$ in polar coordinates. Hence, by changing to polar coordinates,

$$P(X^2 + Y^2 \leq z) = \iint_C \frac{1}{2\pi} e^{-\frac{1}{2}(x^2+y^2)}\,dx\,dy = \int_0^{2\pi} \int_0^{\sqrt{z}} \frac{1}{2\pi} e^{-\frac{1}{2}r^2}\,r\,dr\,d\theta$$

$$= \int_0^{\sqrt{z}} e^{-\frac{1}{2}r^2}\,r\,dr\, \frac{1}{2\pi} \int_0^{2\pi} d\theta = \int_0^{\sqrt{z}} e^{-\frac{1}{2}r^2}\,r\,dr.$$

By the change of variable $u = r^2$, the integral $\int_0^{\sqrt{z}} e^{-\frac{1}{2}r^2}\,r\,dr$ can be computed as $\int_0^z \frac{1}{2} e^{-\frac{1}{2}u}\,du = 1 - e^{-\frac{1}{2}z}$. Hence,

$$P(X^2 + Y^2 \leq z) = 1 - e^{-\frac{1}{2}z} \quad \text{for } z > 0.$$

That is, $X^2 + Y^2$ has an exponential distribution with expected value 2. Thus, $E(X^2) = E(Y^2) = 1$ for the $N(0, 1)$-distributed random variables X and Y. This implies that $E[(\frac{V-\mu}{\sigma})^2] = 1$ for an $N(\mu, \sigma^2)$-distributed random variable V. This in turn gives $E[(V - \mu)^2] = \sigma^2$, proving that an $N(\mu, \sigma^2)$-distributed random variable has variance σ^2.

The result that $X^2 + Y^2$ has an exponential distribution with parameter $\lambda = \frac{1}{2}$ if X and Y are independent $N(0, 1)$ random variables is useful for simulation purposes. It can be used to simulate a random observation from a chi-squared random variable. Recall that a chi-squared random variable with n degrees of freedom is distributed as $X_1^2 + X_2^2 + \cdots + X_n^2$ with $X_1, X_2, \ldots, X_n$ independent $N(0, 1)$ random variables, see Section 4.8.2. Let u be a random number from $(0, 1)$, then $-2\ln(u)$ is a random observation from the exponential distribution with mean 2, see Section 4.9. Hence, taking $n = 2k$, it follows that $-2[\ln(u_1) + \cdots + \ln(u_k)] = -2\ln(u_1 \times \cdots \times u_k)$ is a random observation from the chi-squared distribution with $2k$ degrees of freedom when $u_1, \ldots, u_k$ are independent random numbers from $(0, 1)$. For the case of $2k + 1$ degrees of freedom, one additional random observation from the standard normal distribution is needed.

To conclude this section, we give a very important result for the minimum of independent random variables that are exponentially distributed.

Rule 5.4 Suppose that X and Y are independent random variables, where X is exponentially distributed with expected value $\frac{1}{\alpha}$ and Y is exponentially distributed with expected value $\frac{1}{\beta}$. Then,

$$P(\min(X,Y) \le z) = 1 - e^{-(\alpha+\beta)z} \quad \text{for } z \ge 0,$$

that is, $\min(X,Y)$ is exponentially distributed with expected value $\frac{1}{\alpha+\beta}$. Also,

$$P(X < Y) = \frac{\alpha}{\alpha+\beta}.$$

The first relation is easiest proved by using Definition 3.2. Since $\min(a,b) > c$ only if $a > c$ and $b > c$, we have $P(\min(X,Y) > z) = P(X > z, Y > z)$. Also, $P(X > z, Y > z) = P(X > z)P(Y > z)$, by the independence of X and Y. Thus, $P(\min(X,Y) > z) = P(X > z)P(Y > z) = e^{-\alpha z}e^{-\beta z} = e^{-(\alpha+\beta)z}$. To prove the second relation, let $C = \{(x,y) : 0 < x < y < \infty\}$ and note that

$$P(X < Y) = \iint_C f(x,y)\,dx\,dy = \int_0^\infty \left[\int_x^\infty f_X(x)f_Y(y)\,dy\right] dx.$$

This gives $P(X < Y) = \int_0^\infty \alpha e^{-\alpha x}\,dx \int_x^\infty \beta e^{-\beta y}\,dy = \int_0^\infty \alpha e^{-\alpha x} e^{-\beta x}\,dx$. Hence, by $\int_0^\infty (\alpha+\beta)e^{-(\alpha+\beta)x}\,dx = 1$, we have $P(X < Y) = \alpha/(\alpha+\beta)$.

An immediate consequence of Rule 5.4 is that the minimum of n independent random variables $X_1, \ldots, X_n$ is exponentially distributed with expected value $1/\sum_{i=1}^n \lambda_i$ when each X_i is exponentially distributed with expected value $1/\lambda_i$. Further, the probability that a particular X_r will be the smallest among the X_i is $\lambda_r / \sum_{i=1}^n \lambda_i$.

Problems

5.28 Let X and Y be two independent random variables, each having uniform distribution on $(0,1)$. Take X and Y as the lengths of the sides of a rectangle. What are the probability density and the expected value of the area of the rectangle?

5.29 An internet router is able to simultaneously route packets on two lines to their destination. The packet delays on each line are independent random variables, each having an exponential density with expected value $1/\lambda$. What is the probability density of the difference of the packet delays on the two lines?

5.30 The random variables X and Y are independent and uniformly distributed on $(0,1)$. Verify that both $\frac{1}{2}(X+Y)$ and $|X-Y|$ have a triangular density on (0, 1) with respective parameters ($a = 0, b = 1, m = 0.5$) and ($a = 0, b = 1, m = 1$).

5.31 Suppose that the random variables X and Y are independent and uniformly distributed on (0, 1). Let the random variable F be the fractional part of $X + Y$. Prove that F is uniformly distributed on (0, 1).

5.32 Two ships will arrive in a particular harbor that can handle only one ship at a time. Independently of each other, one of the ships will arrive at a random moment in the next 24 hours and the other ship will arrive at a random moment in the next 36 hours. Each ship requires a handling time of seven hours. What is the probability that the ship arriving second will have to wait?

5.33 Suppose that X and Y are independent and uniformly distributed on (0, 1). What is the probability density of $Z = X/Y$? What is the probability that the first significant (nonzero) digit of Z equals 1? What about the digits $2, \ldots, 9$?

5.34 Suppose that X and Y are independent random variables, each having an exponential distribution with expected value $1/\lambda$. Let $Z = X/Y$. Can you explain why the density function of Z does not depend on λ. Verify that Z has the so-called one-sided Cauchy density $\frac{2}{\pi(1+z^2)}$ for $z > 0$.

5.35 Suppose that X and Y are independent random variables, each having an exponential distribution with expected value $1/\lambda$. Verify that $\max(X, Y)$ has the same distribution as $X + \frac{1}{2}Y$.

5.36 Suppose that $X_1, \ldots, X_n$ are independent random variables, each having uniform distribution on the interval (0, 1).
   (a) Use induction to prove that $P(X_1 + \cdots + X_n \leq s) = \frac{s^n}{n!}$ for $0 \leq s \leq 1$.
   (b) Let N be the smallest n such that $X_1 + \cdots + X_n > 1$. Verify that $E(N) = e$.

5.37 A dealer draws successively random numbers from (0, 1) until the sum exceeds a prespecified value a with $0 < a < 1$. The dealer has to beat the sum a without exceeding 1. What is the probability that the dealer wins the game?

5.38 Let X_1, X_2, and X_3 be independent random variables that are uniformly distributed on (0, 1). What is the probability $P(X_1 > X_2 + X_3)$? What is the probability that the largest of the three random variables is greater than the sum of the other two?

5.39 Suppose that X and Y are independent random variables that are exponentially distributed with parameter λ. Let $V = \min(X, Y)$ and $W = \max(X, Y)$. What is the joint density function of V and W? What is $P(W - V > z)$?

5.3.2 Substitution Rule

The expected value of a given function of jointly distributed random variables X and Y can be calculated by the two-dimensional substitution rule. In the continuous case, we have

Rule 5.5 *If the random variables X and Y have a joint probability density function $f(x,y)$, then*

$$E[g(X,Y)] = \int_{-\infty}^{\infty} \int_{-\infty}^{\infty} g(x,y) f(x,y) \, dx \, dy$$

for any function $g(x,y)$ provided that the integral is absolutely convergent or has a nonnegative integrand.

The proof of Rule 5.5 is technical and will be omitted.

Rule 5.6 *Let X and Y be two random variables defined on the same probability space. It is assumed that $E(X)$ and $E(Y)$ exist and are finite. Then,*

(a) *For any constants a, b,*

$$E(aX + bY) = aE(X) + bE(Y).$$

(b) *For independent random variables X and Y,*

$$E(XY) = E(X)E(Y).$$

These results are valid for any type of random variables. We give a proof for the case that X and Y are continuous and have a joint density function $f(x,y)$. By Rule 5.5, $E(aX+bY) = \int_{-\infty}^{\infty}\int_{-\infty}^{\infty}(ax+by)f(x,y)\,dx\,dy$ and so

$$E(aX+bY) = \int_{-\infty}^{\infty}\int_{-\infty}^{\infty} axf(x,y)\,dx\,dy + \int_{-\infty}^{\infty}\int_{-\infty}^{\infty} byf(x,y)\,dx\,dy$$

$$= \int_{-\infty}^{\infty} ax\,dx \int_{-\infty}^{\infty} f(x,y)\,dy + \int_{-\infty}^{\infty} by\,dy \int_{-\infty}^{\infty} f(x,y)\,dx.$$

Thus, $E(aX+bY) = a\int_{-\infty}^{\infty} xf_X(x)\,dx + b\int_{-\infty}^{\infty} yf_Y(y)\,dy = aE(X)+bE(Y)$. The manipulations with the integrals are justified by the assumption that both $\int_{-\infty}^{\infty}|x|f_X(x)\,dx < \infty$ and $\int_{-\infty}^{\infty}|y|f_Y(y)\,dy < \infty$. To prove **(b)**, we first verify that $E(|XY|)$ is finite. By the independence of X and Y, $f(x,y) = f_X(x)f_Y(y)$ for all x,y, see Rule 5.2. It then follows that

$$E(|XY|) = \int_{-\infty}^{\infty}\int_{-\infty}^{\infty} |xy|f_X(x)f_Y(y)\,dx\,dy$$

$$= \int_{-\infty}^{\infty} |x|f_X(x)\,dx \int_{-\infty}^{\infty} |y|f_Y(y)\,dy < \infty.$$

Since $E(|XY|) < \infty$, the manipulations with the integrals can be repeated with $|xy|$ replaced by xy. This leads to $E(XY) = E(X)E(Y)$.

To illustrate the substitution rule, consider the following two questions. What is the expected value of the distance between two points that are chosen at random in the interval $(0, 1)$? What is the expected value of the distance of a point chosen at random inside the unit square to any vertex of the unit square? To answer these questions, let X and Y be two independent random variables that are uniformly distributed on $(0, 1)$. The answer to the first question is $E(|X - Y|)$ and the answer to the second question is $E(\sqrt{X^2 + Y^2})$. The joint density function of X and Y is given by $f(x, y) = 1$ for all $0 < x, y < 1$, by the independence of X and Y. Using the substitution rule, it follows that $E(|X - Y|) = \int_0^1 \left[\int_0^1 |x - y| \, dy \right] dx$ and so

$$E(|X - Y|) = \int_0^1 dx \left[\int_0^x (x - y) \, dy + \int_x^1 (y - x) \, dy \right]$$

$$= \int_0^1 \left[\frac{1}{2} x^2 + \frac{1}{2} - \frac{1}{2} x^2 - x(1 - x) \right] dx = \frac{1}{3}.$$

Also, by the substitution rule, $E(\sqrt{X^2 + Y^2}) = \int_0^1 \int_0^1 \sqrt{x^2 + y^2} \, dx \, dy$. This double integral is numerically computed as 0.7652.

As another illustration of Rule 5.5, consider Example 5.2 again. In this example, a point is picked at random inside a circular disk with radius r and the point $(0, 0)$ as center. What is the expected value of the rectangular distance from the randomly picked point to the center of the disk? This rectangular distance is given by $|X\cos(Y)| + |X\sin(Y)|$ (the rectangular distance from point (a, b) to $(0, 0)$ is defined by $|a| + |b|$). For the function $g(x, y) = |x\cos(y)| + |x\sin(y)|$, we find

$$E[g(X, Y)] = \int_0^r \int_0^{2\pi} (x|\cos(y)| + x|\sin(y)|) \frac{x}{\pi r^2} \, dx \, dy$$

$$= \frac{1}{\pi r^2} \int_0^{2\pi} |\cos(y)| \, dy \int_0^r x^2 \, dx + \frac{1}{\pi r^2} \int_0^{2\pi} |\sin(y)| \, dy \int_0^r x^2 \, dx.$$

Hence, $E[g(X, Y)] = \frac{r^3}{3\pi r^2} \left[\int_0^{2\pi} |\cos(y)| \, dy + \int_0^{2\pi} |\sin(y)| \, dy \right] = \frac{8r}{3\pi}$.

The same ideas hold in the discrete case. If the random variables X and Y have the joint probability mass function $p(x, y) = P(X = x, Y = y)$, then

$$E[g(X, Y)] = \sum_x \sum_y g(x, y) p(x, y).$$

Problems

5.40 The lengths X and Y of the sides of a rectangle have the joint density function $f(x, y) = x+y$ for $0 < x, y < 1$ and $f(x, y) = 0$ otherwise. What is the expected value of the area of the rectangle?

5.41 An unreliable electronic system has two components hooked up in parallel. The lifetimes X and Y of the two components are independent of each other and each have an exponential density with expected value 1. The system goes down when both components have failed. The system is inspected every T time units. At inspection, any failed unit is replaced. What is the expected amount of time the system is down between two inspections?

5.42 Consider again Problem 5.19. What is the expected value of the time until the system goes down? What is the expected value of the time between the failures of the two components?

5.43 Consider again Problem 5.12. What is the expected value of the area of the circle?

5.44 Consider again Problem 5.3. What is the expected value of the random variables $X + Y$ and XY?

5.4 Transformation of Random Variables

In statistical applications, one sometimes needs the joint density of two random variables V and W that are defined as functions of two other random variables X and Y having a joint density $f(x, y)$. Suppose that the random variables V and W are defined by $V = g(X, Y)$ and $W = h(X, Y)$ for given functions g and h. What is the joint probability density function of V and W? An answer to this question will be given under the assumption that the transformation is one-to-one. That is, it is assumed that the equations $v = g(x, y)$ and $w = h(x, y)$ can be solved uniquely to yield functions $x = a(v, w)$ and $y = b(v, w)$. It is also assumed that the partial derivatives of the functions $a(v, w)$ and $b(v, w)$ with respect to v and w are continuous in (v, w). Then the following transformation rule holds:

Rule 5.7 (transformation rule) *The joint probability density function of V and W is given by*

$$f(a(v, w), b(v, w))|J(v, w)|,$$

where the Jacobian $J(v, w)$ is given by the determinant

$$\begin{vmatrix} \frac{\partial a(v,w)}{\partial v} & \frac{\partial a(v,w)}{\partial w} \\ \frac{\partial b(v,w)}{\partial v} & \frac{\partial b(v,w)}{\partial w} \end{vmatrix} = \frac{\partial a(v, w)}{\partial v}\frac{\partial b(v, w)}{\partial w} - \frac{\partial a(v, w)}{\partial w}\frac{\partial b(v, w)}{\partial v}.$$

5.4 Transformation of Random Variables

The proof of this rule is omitted. This transformation rule looks intimidating, but is easy to use in many applications. Using Rule 5.7, we can give an elegant proof of the following important result for the normal distribution.

Rule 5.8 *Let X and Y be independent $N(0, 1)$ random variables. Then,*

(a) *The random variables $X + Y$ and $X - Y$ are independent and $N(0, 2)$ distributed.*
(b) *The random variable $\frac{X}{Y}$ has the density function $\frac{1}{\pi(1+z^2)}$ for $-\infty < z < \infty$. This density is called the two-sided Cauchy density.*[1]

To prove **(a)**, use the transformation $V = X + Y$ and $W = X - Y$. The inverse functions $a(v, w)$ and $b(v, w)$ are $x = \frac{v+w}{2}$ and $y = \frac{v-w}{2}$. Hence,

$$J(v, w) = \begin{vmatrix} \frac{1}{2} & \frac{1}{2} \\ \frac{1}{2} & -\frac{1}{2} \end{vmatrix} = -\frac{1}{2}.$$

Since X and Y are independent $N(0, 1)$ random variables, it follows from Rule 5.2 that their joint density function is

$$f_{X,Y}(x, y) = \frac{1}{\sqrt{2\pi}} e^{-\frac{1}{2}x^2} \times \frac{1}{\sqrt{2\pi}} e^{-\frac{1}{2}y^2} \quad \text{for } -\infty < x, y < \infty.$$

Applying Rule 5.7, we get that $\frac{1}{\sqrt{2\pi}} e^{-\frac{1}{2}(\frac{v+w}{2})^2} \frac{1}{\sqrt{2\pi}} e^{-\frac{1}{2}(\frac{v-w}{2})^2} \times \frac{1}{2}$ is the joint density function of V and W, which gives

$$f_{V,W}(v, w) = \frac{1}{\sqrt{2}\sqrt{2\pi}} e^{-\frac{1}{2}v^2/2} \times \frac{1}{\sqrt{2}\sqrt{2\pi}} e^{-\frac{1}{2}w^2/2} \quad \text{for } -\infty < v, w < \infty.$$

This implies that the marginal density functions $f_V(v)$ and $f_W(w)$ of V and W are $N(0, 2)$ densities. Also, $f_{V,W}(v, w) = f_V(v)f_W(w)$ for all v, w. Thus, by Rule 5.2, $V = X + Y$ and $W = X - Y$ are independent.

To prove **(b)**, use the transformation $V = Y$ and $W = X/Y$. The inverse functions $a(v, w)$ and $b(v, w)$ are given by $x = vw$ and $y = v$. Hence,

$$J(v, w) = \begin{vmatrix} w & v \\ 1 & 0 \end{vmatrix} = -v.$$

Applying Rule 5.7, we obtain that the joint density function of V and W is

$$f_{V,W}(v, w) = \frac{1}{\sqrt{2\pi}} e^{-\frac{1}{2}(vw)^2} \times \frac{1}{\sqrt{2\pi}} e^{-\frac{1}{2}v^2} |v| \quad \text{for } -\infty < v, w < \infty.$$

---

[1] The two-sided Cauchy density is similar in shape to the standard normal density, but its tails are much fatter. The expected value and the standard deviation of the Cauchy density are undefined. The Cauchy distribution has applications in physics, among others.

The density of $W = X/Y$ is $f_W(w) = \int_{-\infty}^{\infty} f_{V,W}(v, w)\, dv$. This gives

$$f_W(w) = \frac{1}{2\pi} \int_{-\infty}^{\infty} |v| e^{-\frac{1}{2}(1+w^2)v^2}\, dv = \frac{2}{2\pi} \int_0^{\infty} v e^{-\frac{1}{2}(1+w^2)v^2}\, dv.$$

Using the change of variable $u = \frac{1}{2}v^2$, we have that $\int_0^{\infty} v e^{-\frac{1}{2}(1+w^2)v^2}\, dv$ is equal to $\int_0^{\infty} e^{-(1+w^2)u}\, du = \frac{1}{1+w^2}$. This shows that X/Y has the two-sided Cauchy density.

Problems

5.45 The random variables X and Y are independent and exponentially distributed with parameter μ. Let $V = X + Y$ and $W = \frac{X}{X+Y}$. What is the joint density function of V and W? Prove that V and W are independent.

5.46 The random variables X and Y are independent and uniformly distributed on $(0, 1)$. Let $V = X+Y$ and $W = X/Y$. What is the joint density function of V and W? Are V and W independent?

5.47 The random variables V and W are defined by $V = Z_1^2 + Z_2^2$ and $W = Z_1^2 - Z_2^2$, where Z_1 and Z_2 are independent random variables each having the standard normal distribution. What is the joint density function of V and W? Are V and W independent?

5.48 The continuous random variables X and Y have the joint density function $f(x, y) = \frac{1}{\pi} x e^{-\frac{1}{2}x(1+y^2)}$ for $x, y > 0$ and $f(x, y) = 0$ otherwise. Show that the random variables $Y\sqrt{X}$ and X are independent.

5.49 Let (X, Y) be a point chosen at random inside the unit circle. Define V and W by $V = X\sqrt{-2\ln(Q)/Q}$ and $W = Y\sqrt{-2\ln(Q)/Q}$, where $Q = X^2 + Y^2$. Verify that the random variables V and W are independent and $N(0, 1)$ distributed. This result is the basis of Marsaglia's polar method for generating random observations from the normal distribution. *Hint*: Use the relations $V^2 + W^2 = -2\ln(X^2 + Y^2)$ and $\frac{W}{V} = \frac{Y}{X}$ to find the joint density of (V, W).

5.50 A point (V, W) inside the unit circle is constructed as follows. A number R is chosen at random between 0 and 1 and an angle Θ is randomly chosen between 0 and 2π. Then, a point (V, W) is chosen at random on the circumference of the circle with radius R through $V = R\cos(\Theta)$ and $W = R\sin(\Theta)$. Use the transformation formula to find the joint density of V and W. What are the marginal densities of V and W? Can you intuitively explain why the point (V, W) is not uniformly distributed over the unit circle? *Hint*: $\int_0^x \frac{dt}{\sqrt{1+t^2}} = \ln(x + \sqrt{1 + x^2})$ for $x > 0$.

5.51 Suppose that the independent random variables X and Y are gamma distributed with scale parameter 1, where X has shape parameter α and Y has shape parameter β. Verify that $\frac{X}{X+Y}$ has a beta distribution with parameters α and β. *Hint*: Use the transformation $V = \frac{X}{X+Y}$ and $W = X + Y$.

5.52 Let Z and Y be independent random variables, where Z is $N(0, 1)$ distributed and Y is χ_ν^2 distributed. Use the transformation $V = Y$ and $W = Z/\sqrt{Y/\nu}$ to prove that the density of $W = Z/\sqrt{Y/\nu}$ is the Student-t density $c\left(1 + \frac{w^2}{\nu}\right)^{-(\nu+1)/2}$ with $c = \frac{1}{\sqrt{\pi\nu}}\Gamma\left(\frac{1}{2}(\nu+1)\right)/\Gamma\left(\frac{1}{2}\nu\right)$.

5.4.1 Order Statistics[2]

Suppose that $X_1, X_2, \ldots, X_n$ is a sequence of independent continuous random variables, each having probability density $f(x)$. In statistics it is often useful to consider the "ordered" sequence of random variables. Denote by $X_{(1)}$ and $X_{(n)}$ the random variables

$$X_{(1)} = \min(X_1, X_2, \ldots, X_n) \text{ and } X_{(n)} = \max(X_1, X_2, \ldots, X_n).$$

More generally, let $X_{(j)}$ represent the jth smallest of $X_1, X_2, \ldots, X_n$. This is an example of a transformation that is not one-to-one. The ordered random variables $X_{(j)}$ are called the *order statistics* of $X_1, X_2, \ldots, X_n$.

Rule 5.9 *The joint density function of the order statistics $Y_1 = X_{(1)}, Y_2 = X_{(2)}, \ldots, Y_n = X_{(n)}$ is*

$$g(y_1, y_2, \ldots, y_n) = n! f(y_1) f(y_2) \cdots f(y_n) \quad \text{for } y_1 < y_2 < \ldots < y_n$$

and $g(y_1, y_2, \ldots, y_n) = 0$ otherwise.

Using first principles, the proof goes as follows. Fix $y_1 < \ldots < y_n$, where the y_i are continuity points of $f(y)$. The order statistics $X_{(1)}, \ldots, X_{(n)}$ will take on the values $y_1, \ldots, y_n$ if and only if for some permutation $(i_1, \ldots, i_n)$ of $(1, \ldots, n)$ we have $X_{i_1} = y_1, \ldots, X_{i_n} = y_n$. By the independence of the X_k, we have for any permutation $(i_1, \ldots, i_n)$ that the probability of the event $\{y_1 - \frac{1}{2}dy_1 < X_{i_1} < y_1 + \frac{1}{2}dy_1, \ldots, y_n - \frac{1}{2}dy_n < X_{i_n} < y_n + \frac{1}{2}dy_n\}$ is approximately equal to $f(y_1)dy_1 \times \cdots \times f(y_n)dy_n$ for $dy_1, \ldots, dy_n$ close to zero. This implies that

---

[2] This section can be skipped without loss of continuity.

$$P\left(y_1 - \tfrac{1}{2}dy_1 < X_{(1)} < y_1 + \tfrac{1}{2}dy_1, \ldots, y_n - \tfrac{1}{2}dy_n < X_{(n)} < y_n + \tfrac{1}{2}dy_n\right)$$
$$\approx n! f(y_1) \cdots f(y_n) dy_1 \cdots dy_n,$$

which yields the desired result.

The marginal density of the order statistic $Y_j = X_{(j)}$ is

$$f_{(j)}(y) = \frac{n!}{(j-1)!(n-j)!} f(y) [F(y)]^{j-1} [1 - F(y)]^{n-j}.$$

This result can be obtained by integrating the joint density, but it is simpler to use a direct approach. The event $\{y - \tfrac{1}{2}dy < Y_j < y + \tfrac{1}{2}dy\}$ requires that one of the random variables $X_1, \ldots, X_n$ falls into the infinitesimal interval $(y - \tfrac{1}{2}dy, y + \tfrac{1}{2}dy)$, $j-1$ of them fall to the left of y, and $n-j$ of them fall to the right of y. The number of possible partitions of the X_k in these three groups is $\binom{n}{1}\binom{n-1}{j-1} = j\binom{n}{j}$. Thus, $P\left(y - \tfrac{1}{2}dy < Y_j < y + \tfrac{1}{2}dy\right) \approx \frac{n!}{(j-1)!(n-j)!} f(y) dy [F(y)]^{j-1} [1 - F(y)]^{n-j}$, yielding the desired result.

Rule 5.10 *Suppose that the independent random variables $X_1, X_2, \ldots, X_n$ are uniformly distributed on $(0, a)$. Then the joint density of the order statistics $Y_1 = X_{(1)}, Y_2 = X_{(2)}, \ldots, Y_n = X_{(n)}$ is $\frac{n!}{a^n}$ for $0 < y_1 < y_2 < \ldots < y_n < a$. Also, $E(Y_j) = j\frac{a}{n+1}$ for $j = 1, \ldots, n$.*

In Rule 5.10, $E(Y_j) = j\frac{a}{n+1}$ is easily verified by noting that the density of Y_j is the beta density $f_{(j)}(y) = \frac{n!}{(j-1)!(n-j)!} \frac{1}{a} \left(\frac{y}{a}\right)^{j-1} \left(1 - \frac{y}{a}\right)^{n-j}$ for $0 < y < a$.

There is a noteworthy relation between the Poisson process and the order statistics for the uniform distribution. Under the condition that a Poisson process has exactly n occurrences of events in a given time interval $(0, a)$, the joint distribution of the n event times is the same as the joint distribution of the order statistics of n independent random variables that are uniformly distributed on $(0, a)$. This result is very useful in the analysis of queueing and inventory models. Its proof is rather technical and will be omitted. For the case of $n = 1$, the result was proved in Remark 4.1 of Section 4.5.

Problems

5.53 Let $U_{(1)}$ the smallest and $U_{(n)}$ be the largest of n independent random variables $U_1, \ldots, U_n$ that are uniformly distributed on $(0, 1)$. What is the joint density function of $U_{(1)}$ and $U_{(n)}$?

5.54 Consider again Problem 5.53. Use the transformation method to verify that the density function of the range $R_n = U_{(n)} - U_{(1)}$ is given by $n(n-1) x^{n-2}(1-x)$ for $0 < x < 1$.

5.5 Covariance and Correlation Coefficient

Let the random variables X and Y be defined on the same probability space. A basic rule in probability is that the expected value of the sum $X + Y$ equals the sum of the expected values of X and Y. Does a similar rule hold for the variance of the sum $X + Y$? To answer this question, we first introduce the concept of covariance.

Definition 5.2 *The covariance $cov(X, Y)$ of two random variables X and Y is defined by*

$$cov(X, Y) = E\big[(X - E(X))(Y - E(Y))\big]$$

whenever the expectations exist.

Note that $cov(X, X) = var(X)$. It is also immediate from the definition that

$$cov(aX, bY) = ab \, cov(X, Y)$$

for any constants a and b. The formula for $cov(X, Y)$ can be written in the equivalent form

$$cov(X, Y) = E(XY) - E(X)E(Y),$$

by writing $(X - E(X))(Y - E(Y))$ as $XY - XE(Y) - YE(X) + E(X)E(Y)$ and using the linearity property of the expectation operator. Using the fact that $E(XY) = E(X)E(Y)$ for independent random variables (see Rule 5.6), the alternative formula for $cov(X, Y)$ has as direct consequence:

Rule 5.11 *If X and Y are independent random variables, then*

$$cov(X, Y) = 0.$$

However, the converse of this result is not always true. As a counterexample, let X be uniformly distributed on $(-1, 1)$ and let $Y = X^2$. Then, by $E(X) = 0$ and $E(X^3) = 0$, we have $cov(X, Y) = E(X^3) - E(X)E(X^2) = 0$, but X and Y are not independent.

We are now in a position to state the next rule.

Rule 5.12 *For any two random variables X and Y,*

$$var(X + Y) = var(X) + 2cov(X, Y) + var(Y).$$

If the random variables X and Y are independent, then
$$\text{var}(X+Y) = \text{var}(X) + \text{var}(Y).$$
The proof of this rule is a matter of some algebra. Put for abbreviation $\mu_X = E(X)$ and $\mu_Y = E(Y)$. Using the definition of variance and the linearity of the expectation operator, we find
$$\text{var}(X+Y) = E[(X+Y-\mu_X-\mu_Y)^2]$$
$$= E[(X-\mu_X)^2] + 2E[(X-\mu_X)(Y-\mu_Y)] + E[(Y-\mu_Y)^2],$$
showing the first part of Rule 5.12. The second part follows from Rule 5.11.

Rule 5.12 can be extended to the case of finitely many random variables:
$$\text{var}(a_1 X_1 + \cdots + a_n X_n) = \sum_{i=1}^{n} a_i^2 \text{var}(X_i) + 2 \sum_{i=1}^{n-1} \sum_{j=i+1}^{n} a_i a_j \text{cov}(X_i, X_j)$$
for any constants $a_1, \ldots, a_n$. You are asked to verify this in Problem 5.72.

The units of $\text{cov}(X, Y)$ are not the same as the units of $E(X)$ and $E(Y)$. Therefore, it is often more convenient to use the *correlation coefficient* of X and Y. This statistic is defined as
$$\rho(X, Y) = \frac{\text{cov}(X, Y)}{\sqrt{\text{var}(X)}\sqrt{\text{var}(Y)}},$$
provided that $\text{var}(X) > 0$ and $\text{var}(Y) > 0$. The correlation coefficient is a dimensionless quantity with the property that
$$-1 \leq \rho(X, Y) \leq 1.$$
You are asked to verify this property in Problem 5.71. The random variables X and Y are said to be *uncorrelated* if $\rho(X, Y) = 0$. It follows from Rule 5.11 that independent random variables X and Y are always uncorrelated. However, the converse of this result is not always true, as was demonstrated below Rule 5.11. Nevertheless, $\rho(X, Y)$ is often used as a measure of the dependence of X and Y. In Problem 5.69, you are asked to verify that $|\rho(X, Y)| = 1$ if and only if $Y = aX + b$ for some constants a and b with $a \neq 0$, where $\rho(X, Y) = 1$ if $a > 0$ and $\rho(X, Y) = -1$ if $a < 0$.

Problems

5.55 Consider again Problem 5.3. What is the correlation coefficient of X and Y?

5.56 The continuous random variables X and Y have the joint density $f(x, y) = 4y^2$ for $0 < x < y < 1$ and $f(x, y) = 0$ otherwise. What is the correlation coefficient of X and Y? Can you intuitively explain why this correlation coefficient is positive?

5.57 You wish to invest in two funds, A and B, both having the same expected return. The returns of the funds are negatively correlated with correlation coefficient ρ_{AB}. The standard deviations of the returns on funds A and B are given by σ_A and σ_B. Suppose a fraction f of your money is invested in fund A and a fraction $1 - f$ in fund B. What value of f minimizes the standard deviation of the portfolio's return?

5.58 Let X, Y, and Z be independent random variables each having expected value 0 and standard deviation 1. Define $V = X + Z$ and $W = Y + Z$. What is $\rho(V, W)$?

5.59 You want to invest in two stocks A and B. The rates of return on these stocks in the coming year depend on the development of the economy. The economic prospects for the coming year consist of three equally likely case scenarios: a strong economy, a normal economy, and a weak economy. If the economy is strong, the rate of return on stock A will be equal to 34% and the rate of return on stock B will be equal to -20%. If the economy is normal, the rate of return on stock A will be equal to 9.5% and the rate of return on stock B will be equal to 4.5%. If the economy is weak, stocks A and B will have rates of return of -15% and 29%.

(a) Is the correlation coefficient of the rates of return on the stocks A and B positive or negative?

(b) How can you divide the investment amount between two stocks if you desire a portfolio with a minimum variance? What are the expected value and standard deviation of the rate of return on this portfolio?

5.60 The continuous random variables X and Y have joint density function $f(x, y) = 6(x - y)$ for $0 < y < x < 1$ and $f(x, y) = 0$ otherwise. What is the correlation coefficient of X and Y?

5.61 What is the correlation coefficient of the Cartesian coordinates of a random point (X, Y) inside the unit circle $C = \{(x, y) : x^2 + y^2 \leq 1\}$?

5.62 What is the correlation coefficient of the Cartesian coordinates of a random point (X, Y) inside the region $D = \{(x, y) : |x| + |y| \leq 1\}$?

5.63 Suppose that (X, Y) is a random point inside the unit square $\{(x, y) : 0 \leq x, y \leq 1\}$. What is the correlation coefficient of the random variables $V = \min(X, Y)$ and $W = \max(X, Y)$?

5.64 You roll a pair of dice. What is the correlation coefficient of the high and low points rolled?

5.65 Let X be a randomly chosen integer from $1, \ldots, 100$ and Y be a randomly chosen integer from $1, \ldots, X$. What is the correlation coefficient $\rho(X, Y)$?

5.66 Let X be a randomly chosen number from the interval $(0, 1)$ and Y a randomly chosen number from the interval $(0, X)$. What is the correlation coefficient $\rho(X, Y)$?

5.67 Pete and Bill repeatedly play a particular game. Each performance of the game results in a win for Pete with probability p, in a win for Bill with probability q, and in a tie with probability $r = 1-p-q$. Let X be the number of duels until the first win of Pete and Y be the number of duels until the first win of Bill. Verify that $\text{cov}(X, Y) = -1/(1-r)$. *Hint*: Use the identities $\sum_{k=1}^{\infty} k a^{k-1} = \frac{1}{(1-a)^2}$ and $\sum_{k=1}^{\infty} k^2 a^{k-1} = \frac{1+a}{(1-a)^3}$ for $|a| < 1$.

5.68 You generate three random numbers from $(0, 1)$. Let X be the smallest of these three numbers and Y be the largest. What is the correlation coefficient $\rho(X, Y)$?

5.69 Let the random variables X and Y be defined on the same probability space. Put for abbreviation $\sigma_1 = \sigma(X)$, $\sigma_2 = \sigma(Y)$, and $\rho = \rho(X, Y)$. Prove that $|\rho| = 1$ if and only if $Y = aX + b$ for some constants a and b with $a \neq 0$, where $\rho = 1$ if $a > 0$ and $\rho = -1$ if $a < 0$. *Hint*: To prove the "only if" part, evaluate the variance of $V = \frac{Y}{\sigma_2} - \rho \frac{X}{\sigma_1}$.

5.70 Let the random variables X and Y be defined on the same probability space. Verify that $\rho(aX + b, cY + d) = \rho(X, Y)$ for any constants a, b, c, and d with $ac > 0$.

5.71 Let X and Y be defined on the same probability space.
  (a) Prove the Cauchy–Schwarz inequality $[E(XY)]^2 \leq E(X^2)E(Y^2)$. *Hint*: Evaluate $E[(X - tY)^2]$ to conclude that $E(X^2) - 2tE(XY) + t^2 E(Y^2)$ is nonnegative for all t and minimize this function of t.
  (b) Verify that $-1 \leq \rho(X, Y) \leq 1$. *Hint*: Apply the Cauchy–Schwarz inequality with X replaced by $X - E(X)$ and Y replaced by $Y - E(Y)$.
  (c) Use the Cauchy–Schwarz inequality to prove the inequality $P(X = 0) \leq \text{var}(X)/E(X^2)$ when X is a nonnegative random variable with $E(X) > 0$. *Hint*: Define the random variable Y to be 1 if $X > 0$ and 0 if $X = 0$.

5.72 Let the random variables $X_1, \ldots, X_n$ have a common probability space.
  (a) Show $\text{var}\left(\sum_{i=1}^n a_i X_i\right) = \sum_{i=1}^n a_i^2 \text{var}(X_i) + 2 \sum_{i=1}^{n-1} \sum_{j=i+1}^n a_i a_j \text{cov}(X_i, X_j)$ for any constants $a_1, \ldots, a_n$.
  (b) Suppose that $\sigma^2(X_i) = \sigma^2$ for all i and $\text{cov}(X_i, X_j) = c$ for all $i \neq j$. Verify that the variance of $\overline{X}_n = (1/n) \sum_{k=1}^n X_k$ is given by

$\frac{\sigma^2}{n} + (1 - \frac{1}{n})c$ (in investment theory, the term σ^2/n is referred to as the nonsystematic risk and the term $(1 - 1/n)c$ is referred to as the systematic risk; the nonsystematic risk can be significantly reduced by diversifying to a large number of stocks).

(c) Let the random variables $Y_1, \ldots, Y_m$ be defined on the same probability space as $X_1, \ldots, X_n$. Verify that, for any constants $a_1, \ldots, a_n$ and $b_1, \ldots, b_m$,

$$\text{cov}\left(\sum_{i=1}^n a_i X_i, \sum_{j=1}^m b_j Y_j\right) = \sum_{i=1}^n \sum_{j=1}^m a_i b_j \text{cov}(X_i, Y_j).$$

(d) Suppose that $X_1, \ldots, X_n$ are independent random variables each having the same variance σ^2. Let $\overline{X}_n = \frac{1}{n}\sum_{k=1}^n X_k$. Verify that $\text{cov}(\overline{X}_n, X_i - \overline{X}_n) = 0$.

(e) Suppose that the random variables X_1 and X_2 have the same variance. Verify that the covariance of the random variables $X_1 - X_2$ and $X_1 + X_2$ is zero.

5.73 Let $\mathbf{C} = (\sigma_{ij})$ be the covariance matrix of a random vector $(X_1, \ldots, X_n)$, where $\sigma_{ii} = \text{var}(X_i)$ and $\sigma_{ij} = \text{cov}(X_i, X_j)$ for $i \neq j$. Verify that the matrix $\mathbf{C}$ is symmetric and positive semi-definite.

5.74 Let X and Y be two independent $N(0, 1)$ random variables. Define V and W as $V = X + Y$ and $W = Y - aX$ for some constant $0 < a < 1$. Verify that the correlation coefficients between X and V and between V and W are both positive, but the correlation coefficient of X and W is negative, showing that correlation need not be transitive.

5.75 Suppose that the random variables X and Y are independent and $N(0, 1)$ distributed. Verify that the correlation coefficient of $\max(X, Y)$ and $\min(X, Y)$ is equal to $\frac{1}{\pi - 1}$. *Hint*: Use the result of Problem 4.69.

5.76 Suppose that X and Y are independent random variables each having the same exponential distribution with expected value $1/\lambda$. What is the correlation coefficient of $\max(X, Y)$ and $\min(X, Y)$?

5.5.1 Linear Predictor and Regression

Suppose that X and Y are two dependent random variables. In statistical applications, it is often the case that we can observe the random variable X but we want to know the dependent random variable Y. A basic question in statistics is: what is the best *linear predictor* of Y with respect to X? That is, for which linear function $y = \alpha + \beta x$ is $E[(Y - \alpha - \beta X)^2]$ minimal? The answer is

$$y = \mu_Y + \rho_{XY}\frac{\sigma_Y}{\sigma_X}(x - \mu_X),$$

where $\mu_X = E(X), \mu_Y = E(Y), \sigma_X = \sqrt{\text{var}(X)}, \sigma_Y = \sqrt{\text{var}(Y)}$, and $\rho_{XY} = \rho(X, Y)$. The derivation is based on the linearity of the expectation operator. Writing $Y - \alpha - \beta X = Y - \mu_Y - \beta(x - \mu_X) + (\mu_Y - \alpha - \beta\mu_X)$, we get

$$E[(Y - \alpha - \beta X)^2] = E[(Y - \mu_Y - \beta(X - \mu_X))^2] + (\mu_Y - \alpha - \beta\mu_X)^2 + 2(\mu_Y - \alpha - \beta\mu_X)E[Y - \mu_Y - \beta(X - \mu_X)].$$

Thus, $E[(Y-\alpha-\beta X)^2] = \sigma_Y^2 + \beta^2\sigma_X^2 - 2\beta\text{cov}(X,Y) + (\mu_Y-\alpha-\beta\mu_X)^2$. In order to minimize this quadratic function in α and β, we put the partial derivatives with respect to α and β equal to zero. Noting that $\text{cov}(X, Y) = \rho_{XY}\sigma_X\sigma_Y$, we find after some algebra that

$$\beta = \frac{\rho_{XY}\sigma_Y}{\sigma_X} \quad \text{and} \quad \alpha = \mu_Y - \frac{\rho_{XY}\sigma_Y}{\sigma_X}\mu_X.$$

For these values of α and β, we have

$$E[(Y - \alpha - \beta X)^2] = \sigma_Y^2(1 - \rho_{XY}^2).$$

This minimum is sometimes called the residual variance of Y.

The phenomenon of *regression to the mean* can be explained with the help of the best linear predictor. Think of X as the height of a 25-year-old father and think of Y as the height his newborn son will have at the age of 25 years. It is reasonable to assume that $\mu_X = \mu_Y = \mu$, $\sigma_X = \sigma_Y = \sigma$, and that $0 < \rho < 1$ with $\rho = \rho(X, Y)$. The best linear predictor $\hat{Y}$ of Y then satisfies

$$\hat{Y} = \mu + \rho(X - \mu).$$

If the height of the father scores above the mean, then the best linear prediction is that the height of the son will score closer to the mean. Very tall fathers tend to have somewhat shorter sons and very short fathers somewhat taller ones! Regression to the mean shows up in a wide variety of places: it helps to explain why great movies often have disappointing sequels, and disastrous presidents often have better successors.

Problems

5.77 John and Pete are going to throw the discus in their very last turn at a champion game. The distances thrown by John and Pete are independent random variables D_1 and D_2 that are $N(\mu_1, \sigma_1^2)$ and $N(\mu_2, \sigma_2^2)$ distributed. What is the best linear prediction of the distance thrown by John given that the difference between the distances of the throws of John and Pete is d?

6
Multivariate Normal Distribution

Do the one-dimensional normal distribution and the one-dimensional central limit theorem allow for a generalization to dimension two or higher? The answer is yes. Just as the one-dimensional normal density is completely determined by its expected value and variance, the bivariate normal density is completely specified by the expected values and the variances of its marginal densities and by its correlation coefficient. The bivariate normal distribution appears in many applied probability problems. This probability distribution can be extended to the multivariate normal distribution in higher dimensions. Many natural phenomena can be modeled by the multivariate normal distribution. The multidimensional central limit theorem shows that this distribution arises when you take the sum of a large number of independent random vectors. To specify the multivariate normal distribution, all you have to do is compute a vector of expected values and a matrix of covariances. A nice feature of the multivariate normal distribution is its mathematical tractability. The tractability is based on the result that a multivariate normal distribution can be decomposed into univariate normal distributions.

The purpose of this chapter is to give a first introduction to the multivariate normal distribution. Several practical applications will be discussed, including the drunkard's walk in higher dimensions and the chi-square test.

6.1 Bivariate Normal Distribution

A random vector (X, Y) is said to have a *standard bivariate normal distribution* with parameter ρ if it has the joint probability density function

$$f(x, y) = \frac{1}{2\pi\sqrt{1-\rho^2}} e^{-\frac{1}{2}(x^2 - 2\rho xy + y^2)/(1-\rho^2)} \quad \text{for } -\infty < x, y < \infty,$$

where ρ is a constant with $-1 < \rho < 1$. Before showing that ρ can be interpreted as the correlation coefficient of X and Y, we derive the marginal densities of X and Y. Therefore, we first decompose $f(x, y)$ as

$$f(x, y) = \frac{1}{\sqrt{2\pi}} e^{-\frac{1}{2}x^2} \frac{1}{\sqrt{1-\rho^2}\sqrt{2\pi}} e^{-\frac{1}{2}(y-\rho x)^2/(1-\rho^2)}.$$

Next observe that, for *fixed* x,

$$g(y) = \frac{1}{\sqrt{1-\rho^2}\sqrt{2\pi}} e^{-\frac{1}{2}(y-\rho x)^2/(1-\rho^2)}$$

is an $N(\rho x, 1-\rho^2)$ density. This implies that $\int_{-\infty}^{\infty} g(y)\,dy = 1$ and so

$$f_X(x) = \int_{-\infty}^{\infty} f(x, y)\,dy = \frac{1}{\sqrt{2\pi}} e^{-\frac{1}{2}x^2} \quad \text{for } -\infty < x, y < \infty.$$

In other words, the marginal density $f_X(x)$ of X is the standard normal density. Also, for reasons of symmetry, the marginal density $f_Y(y)$ of Y is the standard normal density. Next we prove that ρ is the correlation coefficient $\rho(X, Y)$ of X and Y. Since $\text{var}(X) = \text{var}(Y) = 1$, it suffices to verify that $\text{cov}(X, Y) = \rho$. To do so, we use again the above decomposition of the bivariate normal density $f(x, y)$. By $E(X) = E(Y) = 0$, we have $\text{cov}(X, Y) = E(XY)$ and so $\text{cov}(X, Y) = \int_{-\infty}^{\infty} \int_{-\infty}^{\infty} xy f(x, y)\,dx\,dy$. Using the shorthand notation $\tau = \sqrt{1-\rho^2}$, it now follows that

$$\text{cov}(X, Y) = \int_{x=-\infty}^{\infty} x \frac{1}{\sqrt{2\pi}} e^{-\frac{1}{2}x^2} dx \int_{y=-\infty}^{\infty} y \frac{1}{\tau\sqrt{2\pi}} e^{-\frac{1}{2}(y-\rho x)^2/\tau^2} dy$$

$$= \int_{-\infty}^{\infty} \rho x^2 \frac{1}{\sqrt{2\pi}} e^{-\frac{1}{2}x^2} dx = \rho,$$

where the second equality uses the fact that the expected value of an $N(\rho x, \tau^2)$ random variable is ρx and the third equality uses the fact that $E(Z^2) = \sigma^2(Z) = 1$ for a standard normal random variable Z.

A random vector (X, Y) is said to have a *bivariate normal distribution* with parameters $(\mu_1, \mu_2, \sigma_1^2, \sigma_2^2, \rho)$ if the standardized random vector

$$\left(\frac{X - \mu_1}{\sigma_1}, \frac{Y - \mu_2}{\sigma_2} \right)$$

has the standard bivariate normal distribution with parameter ρ. In this case the joint density $f(x, y)$ of the random variables X and Y is given by

$$f(x, y) = \frac{1}{2\pi \sigma_1 \sigma_2 \sqrt{1-\rho^2}} e^{-\frac{1}{2}\left[\left(\frac{x-\mu_1}{\sigma_1}\right)^2 - 2\rho\left(\frac{x-\mu_1}{\sigma_1}\right)\left(\frac{y-\mu_2}{\sigma_2}\right) + \left(\frac{y-\mu_2}{\sigma_2}\right)^2\right]/(1-\rho^2)}.$$

6.1 Bivariate Normal Distribution

While the formula for the joint density function of the bivariate normal distribution looks intimidating, we usually do not use this formula itself. Instead, we use the structural properties of the bivariate normal distribution.

Rule 6.1 *Suppose that the random vector (X, Y) has a bivariate normal distribution with parameters $(\mu_1, \mu_2, \sigma_1^2, \sigma_2^2, \rho)$. Then,*

(a) *The marginal densities $f_X(x)$ and $f_Y(y)$ of X and Y are the $N(\mu_1, \sigma_1^2)$ density and the $N(\mu_2, \sigma_2^2)$ density.*
(b) *The correlation coefficient of X and Y is given by $\rho(X, Y) = \rho$.*

The result (a) follows directly from the fact that $(X-\mu_1)/\sigma_1$ and $(Y-\mu_2)/\sigma_2$ are $N(0, 1)$ distributed, as was verified above. Also, it was shown above that the covariance of $(X - \mu_1)/\sigma_1$ and $(Y - \mu_2)/\sigma_2$ equals ρ. To get result (b), note that $\text{cov}(aX + b, cY + d) = ac\,\text{cov}(X, Y)$ and so

$$\rho = \text{cov}\left(\frac{X - \mu_1}{\sigma_1}, \frac{Y - \mu_2}{\sigma_2}\right) = \frac{1}{\sigma_1\sigma_2}\text{cov}(X, Y) = \rho(X, Y).$$

In general, uncorrelatedness is a necessary but not sufficient condition for independence of two random variables, see Rule 5.11. However, for a bivariate normal distribution, uncorrelatedness is a necessary and sufficient condition for independence:

Rule 6.2 *Bivariate normal random variables X and Y are independent if and only if they are uncorrelated.*

This important result is a direct consequence of Rule 5.2, since the above representation of the bivariate normal density $f(x, y)$ reveals that $f(x, y) = f_X(x)f_Y(y)$ if and only if $\rho = 0$.

As already pointed out, the bivariate normal distribution has the important property that its marginal distributions are one-dimensional normal distributions. The following characterization of the bivariate normal distribution can be given.

Rule 6.3 *Random variables X and Y have a bivariate normal distribution if and only if $aX + bY$ is normally distributed for any constants a and b.*[7]

The "only if" part of this result can be proved by elementary means. The reader is asked to do this in Problem 6.1. The proof of the "if" part is more advanced

---

[7] To be precise, this result requires the following convention: if X is normally distributed and $Y = aX + b$ for constants a and b, then (X, Y) is said to have a bivariate normal distribution. This is a singular bivariate distribution: the probability mass of the two-dimensional vector (X, Y) is concentrated on the one-dimensional line $y = ax + b$. Also, a random variable X with $P(X = \mu) = 1$ for a constant μ is said to have a degenerate $N(\mu, 0)$ distribution with its mass concentrated at a single point.

and requires the technique of moment-generating functions, see Problem 8.23 in Chapter 8. To conclude that (X, Y) has a bivariate normal distribution it is not sufficient that X and Y are normally distributed, but normality of $aX + bY$ should be required for all constants a and b not both equal to zero. A counterexample is as follows. Let the random variable Y be equal to X with probability 0.5 and equal to $-X$ with probability 0.5, where X has a standard normal distribution. Then the random variable Y also has a standard normal distribution. It is readily verified that $\text{cov}(X, Y) = 0$. This would imply that X and Y are independent if (X, Y) has a bivariate normal distribution. However, X and Y are obviously dependent, showing that (X, Y) does not have a bivariate normal distribution.

Problems

6.1 Prove that $aX + bY$ is normally distributed for any constants a and b if (X, Y) has a bivariate normal distribution. How do you calculate $P(X > Y)$?

6.2 The rates of return on two stocks A and B have a bivariate normal distribution with parameters $\mu_1 = 0.08$, $\mu_2 = 0.12$, $\sigma_1 = 0.05$, $\sigma_2 = 0.15$, and $\rho = -0.50$. What is the probability that the average of the returns on stocks A and B will be larger than 0.11?

6.3 Suppose that the probability density function $f(x, y)$ of the random variables X and Y is given by the standard bivariate normal density with parameter ρ. Verify that the probability density function $f_Z(z)$ of the ratio $Z = X/Y$ is given by the generalized Cauchy density

$$\int_{-\infty}^{\infty} |y| f(zy, y) \, dy = \frac{(1/\pi)\sqrt{1-\rho^2}}{(z-\rho)^2 + 1 - \rho^2} \quad \text{for } -\infty < z < \infty.$$

This result generalizes the result in part **(b)** of Rule 5.8.

6.4 Let the random variables X and Y have a bivariate normal distribution with parameters $(\mu_1, \mu_2, \sigma_1^2, \sigma_2^2, \rho)$, and let $\Phi(x)$ denote the standard normal distribution function. Use the decomposition of the standard bivariate normal density to verify that $P(X \leq a, Y \leq b)$ can be calculated as

$$\frac{1}{\sqrt{2\pi}} \int_{-\infty}^{(a-\mu_1)/\sigma_1} \Phi\left(\frac{(b-\mu_2)/\sigma_2 - \rho x}{\sqrt{1-\rho^2}}\right) e^{-\frac{1}{2}x^2} dx.$$

Next verify that $P(X \leq \mu_1, Y \leq \mu_2) = \frac{1}{4} + \frac{1}{2\pi}\text{arctg}(a)$ with $a = \rho/\sqrt{1-\rho^2}$.

6.5 Let (X, Y) have a bivariate normal distribution. Define the random variables V and W by $V = a_1 X + b_1 Y + c_1$ and $W = a_2 X + b_2 Y + c_2$, where a_i, b_i, and c_i are constants for $i = 1, 2$. Argue that (V, W) has a bivariate normal distribution.

6.6 Suppose that (X, Y) has a standard bivariate normal distribution with parameter ρ. Specify the joint probability density of X and $X + Y$.

6.7 Suppose that (X, Y) has a standard bivariate normal distribution with parameter ρ, where $-1 < \rho < 1$. Let $V = (Y - \rho X)/\sqrt{1 - \rho^2}$. Verify that X and V are independent $N(0, 1)$ random variables.

6.8 Suppose that the random variables Z_1 and Z_2 are independent and $N(0, 1)$ distributed. Define the random variables X_1 and X_2 by $X_1 = \mu_1 + \sigma_1 Z_1$ and $X_2 = \mu_2 + \sigma_2 \rho Z_1 + \sigma_2 \sqrt{1 - \rho^2} Z_2$, where $\mu_1, \mu_2, \sigma_1, \sigma_2$, and ρ are constants with $\sigma_1 > 0$, $\sigma_2 > 0$, and $-1 < \rho < 1$. Prove that (X_1, X_2) has a bivariate normal distribution with parameters $(\mu_1, \mu_2, \sigma_1^2, \sigma_2^2, \rho)$.

6.9 Let (X, Y) have a bivariate normal distribution with $\sigma^2(X) = \sigma^2(Y)$. Prove that the random variables $X + Y$ and $X - Y$ are independent and normally distributed. This result generalizes the result in part (**a**) of Rule 5.8.

6.1.1 The Drunkard's Walk

The drunkard's walk is one of the most useful probability models in the physical sciences. Let us formulate this model in terms of a particle moving on the two-dimensional plane. The particle starts at the origin $(0, 0)$. At each step the particle travels a unit distance in a randomly chosen direction between 0 and 2π, independently of the other steps. What is the joint density function of the (x, y)-coordinates of the position of the particle after n steps?

Let the random variable Θ denote the direction taken by the particle at any step. At each step the x-coordinate of the position of the particle changes by an amount that is distributed as $\cos(\Theta)$ and the y-coordinate by an amount that is distributed as $\sin(\Theta)$. The continuous random variable Θ has a uniform distribution on $(0, 2\pi)$. Let X_k and Y_k be the changes of the x-coordinate and the y-coordinate of the position of the particle in the kth step. Then the position of the particle after n steps can be represented by the random vector (S_{n1}, S_{n2}), where

$$S_{n1} = X_1 + \cdots + X_n \quad \text{and} \quad S_{n2} = Y_1 + \cdots + Y_n.$$

For each n, the random vectors $(X_1, Y_1), \ldots, (X_n, Y_n)$ are independent and have the same distribution. The reader will not be surprised to learn that the two-dimensional version of the central limit theorem applies to the random vector $(S_{n1}, S_{n2}) = (X_1 + \cdots + X_n, Y_1 + \cdots + Y_n)$. In general form, the two-dimensional version of the central limit theorem reads as

$$\lim_{n\to\infty} P\left(\frac{S_{n1} - n\mu_1}{\sigma_1\sqrt{n}} \le x, \frac{S_{n2} - n\mu_2}{\sigma_2\sqrt{n}} \le y\right)$$

$$= \frac{1}{2\pi\sqrt{(1-\rho^2)}} \int_{-\infty}^{x} \int_{-\infty}^{y} e^{-\frac{1}{2}(v^2 - 2\rho vw + w^2)/(1-\rho^2)} \, dv \, dw,$$

where $\mu_1 = E(X_i)$, $\mu_2 = E(Y_i)$, $\sigma_1^2 = \sigma^2(X_i)$, $\sigma_2^2 = \sigma^2(Y_i)$, and $\rho = \rho(X, Y)$. In the particular case of the drunkard's walk, we have

$$\mu_1 = \mu_2 = 0, \ \sigma_1 = \sigma_2 = \frac{1}{\sqrt{2}} \ \text{and} \ \rho = 0.$$

The derivation of this result is simple and instructive. The random variable Θ has the uniform density function $f(\theta) = \frac{1}{2\pi}$ for $0 < \theta < 2\pi$. Applying the substitution rule gives

$$\mu_1 = E[\cos(\Theta)] = \int_0^{2\pi} \cos(\theta) f(\theta) \, d\theta = \frac{1}{2\pi} \int_0^{2\pi} \cos(\theta) \, d\theta = 0.$$

In the same way, $\mu_2 = 0$. Using the substitution rule again,

$$\sigma_1^2 = E[\cos^2(\Theta)] = \int_0^{2\pi} \cos^2(\theta) f(\theta) \, d\theta = \frac{1}{2\pi} \int_0^{2\pi} \cos^2(\theta) \, d\theta.$$

In the same way, $\sigma_2^2 = \frac{1}{2\pi} \int_0^{2\pi} \sin^2(\theta) \, d\theta$. Invoking the celebrated formula $\cos^2(\theta) + \sin^2(\theta) = 1$ from trigonometry, we obtain $\sigma_1^2 + \sigma_2^2 = 1$. Thus, for reasons of symmetry, $\sigma_1^2 = \sigma_2^2 = \frac{1}{2}$. Finally,

$$\text{cov}(X_1, Y_1) = E[(\cos(\Theta) - 0)(\sin(\Theta) - 0)] = \frac{1}{2\pi} \int_0^{2\pi} \cos(\theta) \sin(\theta) \, d\theta.$$

This integral is equal to zero, since $\cos(x + \frac{\pi}{2}) \sin(x + \frac{\pi}{2}) = -\cos(x) \sin(x)$ for each of the ranges $0 \le x \le \frac{\pi}{2}$ and $\pi \le x \le \frac{3\pi}{2}$. This verifies that $\rho = 0$.

Next we can formulate two interesting results that will be proved below using the two-dimensional central limit theorem. The first result states that

$$P(S_{n1} \le x, S_{n2} \le y) \approx \frac{1}{\pi n} \int_{-\infty}^{x} \int_{-\infty}^{y} e^{-(t^2 + u^2)/n} \, dt \, du$$

for n large. In other words, the position of the particle after n steps has approximately the bivariate normal density function

$$\phi_n(x, y) = \frac{1}{\pi n} e^{-(x^2 + y^2)/n}$$

when n is large. That is, the probability of finding the particle after n steps in a small rectangle with sides Δa and Δb around the point (a, b) is approximately

6.1 Bivariate Normal Distribution

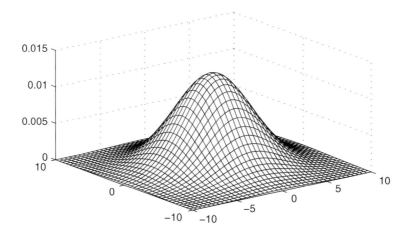

Figure 6.1. The density of the particle's position after 25 steps.

equal to $\phi_n(a,b)\Delta a \Delta b$ for n large. In Figure 6.1, we display the bivariate normal density function $\phi_n(x,y)$ for $n = 25$. The correlation coefficient of the bivariate normal density $\phi_n(x,y)$ is zero. Therefore, in accordance with our intuition, the coordinates of the position of the particle after many steps are practically independent of each other. Moreover, by the decomposition

$$\phi_n(x,y) = \frac{1}{\sqrt{n/2}\sqrt{2\pi}} e^{-\frac{1}{2}x^2/\frac{1}{2}n} \times \frac{1}{\sqrt{n/2}\sqrt{2\pi}} e^{-\frac{1}{2}y^2/\frac{1}{2}n},$$

each of the coordinates of the position of the particle after n steps is approximately $N(0, \frac{1}{2}n)$ distributed for n large. The second result states that

$$E(D_n) \approx \frac{1}{2}\sqrt{\pi n}$$

for n large, where the random variable D_n is defined by

D_n — the distance from the origin to the position of the particle after n steps.

The proof of these results goes as follows. Rewrite $P(S_{n1} \leq x, S_{n2} \leq y)$ as

$$P\left(\frac{S_{n1} - n\mu_1}{\sigma_1\sqrt{n}} \leq \frac{x - n\mu_1}{\sigma_1\sqrt{n}}, \frac{S_{n2} - n\mu_2}{\sigma_2\sqrt{n}} \leq \frac{y - n\mu_2}{\sigma_2\sqrt{n}}\right).$$

Substituting the values of μ_1, μ_2, σ_1, σ_2, and ρ, it follows next from the two-dimensional central limit theorem that

$$P(S_{n1} \leq x, S_{n2} \leq y) \approx \frac{1}{2\pi} \int_{-\infty}^{x/\sqrt{n/2}} \int_{-\infty}^{y/\sqrt{n/2}} e^{-\frac{1}{2}(v^2+w^2)} \, dv \, dw$$

for n large. By the change of variables $t = v\sqrt{n/2}$ and $u = w\sqrt{n/2}$, the first result is obtained. To find the approximation for $E(D_n)$, note that

$$D_n = \sqrt{S_{n1}^2 + S_{n2}^2}.$$

Using the substitution rule, we have

$$E(D_n) \approx \frac{1}{\pi n} \int_{-\infty}^{\infty} \int_{-\infty}^{\infty} \sqrt{x^2 + y^2} \, e^{-(x^2+y^2)/n} \, dx \, dy$$

for n large. To evaluate this integral, we convert Cartesian coordinates to polar coordinates $x = r\cos(\theta)$ and $y = r\sin(\theta)$ with $dx\,dy = r\,dr\,d\theta$ and use the identity $\cos^2(\theta) + \sin^2(\theta) = 1$. Thus, the integral is evaluated as

$$\int_0^{\infty} \int_0^{2\pi} \sqrt{r^2\cos^2(\theta) + r^2\sin^2(\theta)} \, e^{-[r^2\cos^2(\theta)+r^2\sin^2(\theta)]/n} \, r \, dr \, d\theta$$

$$= \int_0^{\infty} \int_0^{2\pi} r^2 e^{-r^2/n} \, dr \, d\theta = 2\pi \int_0^{\infty} r^2 e^{-r^2/n} \, dr.$$

Obviously,

$$\int_0^{\infty} r^2 e^{-r^2/n} \, dr = -\frac{n}{2} \int_0^{\infty} r \, de^{-r^2/n} = -\frac{n}{2} r e^{-r^2/n} \Big|_0^{\infty} + \frac{n}{2} \int_0^{\infty} e^{-r^2/n} \, dr$$

$$= \frac{1}{2}\frac{n}{2}\sqrt{n/2}\sqrt{2\pi} \int_{-\infty}^{\infty} \frac{1}{\sqrt{n/2}\sqrt{2\pi}} e^{-\frac{1}{2}r^2/(\frac{1}{2}n)} \, dr = \frac{1}{4}n\sqrt{n\pi},$$

using the fact that the $N(0, \frac{1}{2}n)$ density integrates to 1 over $(-\infty, \infty)$. Putting the pieces together, we get

$$E(D_n) \approx \frac{1}{2}\sqrt{\pi n}$$

for large n (say, $n \geq 10$). Since $P(D_n \leq u) \approx \iint_C \frac{1}{\pi n} e^{-(x^2+y^2)/n} \, dx \, dy$ with domain $C = \{(x, y) : \sqrt{x^2 + y^2} \leq u\}$, a slight modification of the above analysis shows that $P(D_n \leq u) \approx \frac{2}{n} \int_0^u r e^{-r^2/n} \, dr$. Hence, for n large,

$$P(D_n \leq u) \approx 1 - e^{-u^2/n} \quad \text{for } u > 0.$$

The approximate probability density of D_n is $\frac{2u}{n} e^{-u^2/n}$ for $u > 0$. This is the Rayleigh density.

6.1.2 Drunkard's Walk in Dimension Three or Higher

This section summarizes some results for the drunkard's walk in higher dimensions. When the drunkard's walk occurs in three-dimensional space,

6.1 Bivariate Normal Distribution

it can be shown that the joint probability density function of the (x, y, z)-coordinates of the position of the particle after n steps is approximately given by the trivariate normal probability density function

$$\frac{1}{(2\pi n/3)^{3/2}} e^{-\frac{3}{2}(x^2+y^2+z^2)/n}$$

for n large. Hence, for n large, the coordinates of the particle after n steps are practically independent of each other and are each approximately $N(0, \frac{1}{3}n)$ distributed. This result extends to the drunkard's walk in dimension d. Then, for n large, each of the coordinates of the particle after n steps has an approximate $N(0, \frac{1}{d}n)$ distribution, while the probability distribution of the distance D_n between the origin and the position of the particle after n steps satisfies

$$P(D_n \leq u) \approx \int_0^{u/\sqrt{n}} \frac{d^{\frac{1}{2}d}}{2^{\frac{1}{2}d-1}\Gamma(\frac{1}{2}d)} e^{-\frac{1}{2}dr^2} r^{d-1} \, dr \quad \text{for } u > 0,$$

where $\Gamma(a)$ is the gamma function. The probability distribution of D_n is related to the chi-squared distribution with d degrees of freedom, see also Section 4.8.2. It is a matter of some algebra to derive from the approximate distribution of D_n that

$$E(D_n) \approx \frac{\alpha_d}{d^{\frac{1}{2}} 2^{\frac{1}{2}d-1}\Gamma(\frac{1}{2}d)} \sqrt{n},$$

where $\alpha_m = \int_0^\infty x^m e^{-\frac{1}{2}x^2} \, dx$. Using partial integration, it is not difficult to verify that the α_m can be computed recursively from

$$\alpha_m = (m-1)\alpha_{m-2} \quad \text{for } m = 2, 3, \ldots$$

with $\alpha_0 = \sqrt{\frac{1}{2}\pi}$ and $\alpha_1 = 1$. In particular, using the fact that $\Gamma(\frac{3}{2}) = \frac{1}{2}\sqrt{\pi}$, we find for the three-dimensional space that $E(D_n) \approx \sqrt{\frac{8n}{3\pi}}$ for n large.

Problems

6.10 Two particles carry out a drunkard's walk on the two-dimensional plane, independently of each other. Both particles start at the origin $(0, 0)$. One particle performs n steps and the other m steps. Can you give an intuitive explanation of why the expected distance between the final positions of the two particles is equal to $\frac{1}{2}\sqrt{\pi}\sqrt{n+m}$?

6.11 The rectangular distance from the origin to the position of the particle after n steps is defined by $R_n = |S_{n1}| + |S_{n2}|$. What are the expected value and the probability density of the random variable R_n for n large?

6.12 Use the model of the drunkard's walk in dimension three to estimate the average travel time of a photon from the sun's core to its surface. The sun's radius measures 70,000 km. A photon travels with the speed of light (300,000 km/s) and has a countless number of collisions on its way to the sun's surface. Under the assumption that the distance traveled by a photon between two collisions is 10^{-1} mm, verify that the average number of collisions that a photon undergoes before reaching the sun's surface is about 5.773×10^{23}. Next, conclude that the average travel time of a photon from the sun's core to its surface can be estimated as 6,000 years.

6.2 Multivariate Normal Distribution

The multivariate normal distribution is a very useful probability model to describe dependencies between two or more random variables. In finance, the multivariate normal distribution is frequently used to model the joint distribution of the returns in a portfolio of assets.

First we give a general definition of the multivariate normal distribution.

Definition 6.1 *A d-dimensional random vector $(S_1, S_2, \ldots, S_d)$ is said to be multivariate normal distributed if, for any d-tuple of real numbers $\alpha_1, \ldots, \alpha_d$, the one-dimensional random variable*

$$\alpha_1 S_1 + \alpha_2 S_2 + \cdots + \alpha_d S_d$$

has a (univariate) normal distribution.

Recall the convention that a degenerate random variable X with $P(X = \mu) = 1$ for a constant μ is considered as an $N(\mu, 0)$-distributed random variable. Definition 6.1 implies that each of the individual random variables $S_1, \ldots, S_d$ is normally distributed. Let us define the vector $\mu = (\mu_i)$, $i = 1, \ldots, d$ of expected values and the matrix $\mathbf{C} = (\sigma_{ij})$, $i, j = 1, \ldots, d$ of covariances by

$$\mu_i = E(X_i) \quad \text{and} \quad \sigma_{ij} = \text{cov}(X_i, X_j).$$

Note that $\sigma_{ii} = \text{var}(X_i)$. The multivariate normal distribution is called *nonsingular* if the determinant of the covariance matrix $\mathbf{C}$ is nonzero; otherwise, the distribution is called singular. By a basic result from linear algebra, a singular covariance matrix **C** means that the probability mass of the multivariate normal distribution is concentrated on a subspace with dimension lower than d. In applications, the covariance matrix of the multivariate normal distribution is often singular. In the example of the drunkard's walk on the two-dimensional plane, however, the approximate multivariate normal distribution of the position of the particle after n steps has the nonsingular covariance matrix

$$\begin{pmatrix} \frac{1}{2}n & 0 \\ 0 & \frac{1}{2}n \end{pmatrix}.$$

A very useful result for practical applications is the fact that the multivariate normal distribution is *uniquely determined* by the vector of expected values and the covariance matrix. Note that the covariance matric C is symmetric and positive semi-definite, see Problem 5.73.

Further study of the multivariate normal distribution requires matrix analysis and advanced methods in probability theory, such as the theory of the so-called characteristic functions. Linear algebra is indispensable for multivariate analysis in probability and statistics. The following important result for the multivariate normal distribution is stated without proof: the random variables $S_1, \ldots, S_d$ can be expressed as linear combinations of independent standard normal random variables. That is,

$$S_i = \mu_i + \sum_{j=1}^{d} a_{ij} Z_j \quad \text{for } i = 1, \ldots, d,$$

where $Z_1, \ldots, Z_d$ are independent random variables each having the standard normal distribution (the same $Z_1, \ldots, Z_d$ apply to each of the S_i). The matrix $\mathbf{A} = (a_{ij})$ satisfies $\mathbf{C} = \mathbf{A}\mathbf{A}^T$, with $\mathbf{A}^T$ denoting the transpose of the matrix **A** (you may verify this result directly by writing out $\text{cov}(S_i, S_j)$ from the decomposition formula for the S_i). Moreover, using the fact that the covariance matrix **C** is symmetric and positive semi-definite, it follows from a basic diagonalization result in linear algebra that the matrix **A** can be computed from

$$\mathbf{A} = \mathbf{U}\mathbf{D}^{1/2},$$

where the matrix $\mathbf{D}^{1/2}$ is a diagonal matrix with the square roots of the eigenvalues of the covariance matrix **C** on its diagonal (these eigenvalues are real and nonnegative). The orthogonal matrix **U** has the normalized eigenvectors of the matrix **C** as column vectors (Cholesky decomposition is a convenient method to compute the matrix **A** when **C** is nonsingular). The decomposition result for the vector $(S_1, \ldots, S_d)$ is particularly useful when simulating random observations from the multivariate normal distribution. Fast codes are widely available to simulate from the one-dimensional standard normal distribution.

Remark 6.1 The result that the S_i are distributed as $\mu_i + \sum_{j=1}^{d} a_{ij}Z_j$ has a useful corollary. By taking the inproduct of the vector $(S_1 - \mu_1, \ldots, S_d - \mu_d)$ with itself, it is a matter of basic linear algebra to prove that $\sum_{j=1}^{d}(S_j - \mu_j)^2$ is distributed as $\sum_{j=1}^{d} \lambda_j Z_j^2$. Moreover, if the eigenvalues λ_k of the covariance matrix $\mathbf{C}$ are 0 or 1, then the random variable $\sum_{j=1}^{d} \lambda_j Z_j^2$ has a chi-squared distribution. These rather technical matters are not proved but are intended to give you better insight into the chi-square test, which will be discussed in Section 6.4.

Remark 6.2 If the covariance matrix $\mathbf{C}$ of the multivariate normal distribution is nonsingular, it is possible to give an explicit expression for the corresponding multivariate probability density function. To do so, let us define the matrix $\mathbf{Q} = (q_{ij})$ by

$$q_{ij} = \frac{\sigma_{ij}}{\sigma_i \sigma_j} \quad \text{for } i,j = 1, \ldots, d,$$

where σ_ℓ is shorthand for $\sqrt{\sigma_{\ell\ell}}$. Denote by γ_{ij} the (i,j)th element of the inverse matrix $\mathbf{Q}^{-1}$ and let the polynomial $Q(x_1, \ldots, x_d)$ denote

$$Q(x_1, \ldots, x_d) = \sum_{i=1}^{d} \sum_{j=1}^{d} \gamma_{ij} x_i x_j.$$

Then the standardized vector $\left(\frac{S_1 - \mu_1}{\sigma_1}, \ldots, \frac{S_d - \mu_d}{\sigma_d}\right)$ can be shown to have the standard multivariate normal probability density function

$$\frac{1}{(2\pi)^{d/2}\sqrt{\det(\mathbf{Q})}} e^{-\frac{1}{2}Q(x_1, \ldots, x_d)}.$$

This multidimensional density function reduces to the standard bivariate normal probability density function from Section 6.1 when $d = 2$.

6.3 Multidimensional Central Limit Theorem

The central limit theorem is the queen of all theorems in probability theory. The one-dimensional version is discussed extensively in Chapter 4. The analysis of the drunkard's walk on the two-dimensional plane used the two-dimensional version. The multidimensional version of the central limit theorem is as follows. Suppose that

6.3 Multidimensional Central Limit Theorem

$$\mathbf{X}_1 = (X_{11}, \ldots, X_{1d}), \mathbf{X}_2 = (X_{21}, \ldots, X_{2d}), \ldots, \mathbf{X}_n = (X_{n1}, \ldots, X_{nd})$$

are independent random vectors of dimension d. The random vector $\mathbf{X}_k$ has the one-dimensional random variable X_{kj} as its jth component. The random vectors $\mathbf{X}_1, \ldots, \mathbf{X}_n$ are said to be independent if

$$P(\mathbf{X}_1 \in A_1, \ldots, \mathbf{X}_n \in A_n) = P(\mathbf{X}_1 \in A_1) \cdots P(\mathbf{X}_n \in A_n)$$

for any subsets $A_1, \ldots, A_n$ of the d-dimensional Euclidean space. Note that, for fixed k, the random variables $X_{k1}, \ldots, X_{kd}$ need not be independent. Also assume that $\mathbf{X}_1, \ldots, \mathbf{X}_n$ have the same individual distributions, that is, $P(\mathbf{X}_1 \in A) = \ldots = P(\mathbf{X}_n \in A)$ for any subset A of the d-dimensional space. Under this assumption, let

$$\mu_j^{(0)} = E(X_{1j}) \quad \text{and} \quad \sigma_{ij}^{(0)} = \text{cov}(X_{1i}, X_{1j})$$

for $i, j = 1, \ldots, d$, assuming that the expectations exist. For $j = 1, \ldots, d$, we now define the random variable S_{nj} by

$$S_{nj} = X_{1j} + X_{2j} + \cdots + X_{nj}.$$

Multidimensional central limit theorem *For n large, the random vector $\mathbf{S}_n = (S_{n1}, S_{n2}, \ldots, S_{nd})$ has approximately a multivariate normal distribution. The vector μ of expected values and the covariance matrix $\mathbf{C}$ of the multivariate normal distribution are given by*

$$\mu = \left(n\mu_1^{(0)}, \ldots, n\mu_d^{(0)}\right) \quad \text{and} \quad \mathbf{C} = \left(n\sigma_{ij}^{(0)}\right),$$

assuming that each of the independent random vectors $\mathbf{X}_k$ has the same probability distribution.

The proof of this result is beyond the scope of this book. In the next section we discuss two applications of the multidimensional central limit theorem. In the first application, we will use the fact that the assumption of the same probability distribution for each of the random vectors $\mathbf{X}_k$ may be weakened in the multidimensional central limit theorem. The assumption of independence, however, is crucial.

Problems

6.13 Let $\mathbf{X} = (X_1, \ldots, X_n)$ and $\mathbf{Y} = (Y_1, \ldots, Y_m)$ be independent random vectors each having a multivariate normal distribution. Prove that the random vector $(\mathbf{X}, \mathbf{Y})$ also has a multivariate normal distribution.

6.14 The annual rates of return on the three stocks A, B, and C have a trivariate normal distribution. The rate of return on stock A has expected value 7.5% and standard deviation 7%, the rate of return on stock B has expected value 10% and standard deviation 12%, and the rate of return on stock C has expected value 20% and standard deviation 18%. The correlation coefficient of the rates of return on stocks A and B is 0.7, the correlation coefficient is -0.5 for the stocks A and C, and the correlation coefficient is -0.3 for the stocks B and C. An investor has $100,000 in cash. Any cash that is not invested in the three stocks will be put in a riskless asset that offers an annual interest rate of 5%.

  (a) Suppose that the investor puts $20,000 in stock A, $20,000 in stock B, $40,000 in stock C, and $20,000 in the riskless asset. What are the expected value and the standard deviation of the portfolio's value next year?
  (b) Can you find a portfolio whose risk is smaller than the risk of the portfolio from (a) but whose expected return is not less than that of the portfolio from (a)?
  (c) For the investment plan from (a), find the probability that the portfolio's value next year will be less than $112,500 and the probability that the portfolio's value next year will be more than $125,000.

6.3.1 Predicting Election Results

A nice application of the multivariate normal distribution is the prediction of election results based on probability statements made by polled voters. Think of a polling method in which voters are not asked to choose a favorite party, but instead are asked to indicate how likely they are to vote for each party. Suppose that there are three parties A, B, and C. Let us assume that a representative group of n voters is polled. A probability distribution (p_{iA}, p_{iB}, p_{iC}) with $p_{iA} + p_{iB} + p_{iC} = 1$ describes the voting behavior of respondent i for $i = 1, \ldots, n$. That is, p_{iP} is the probability that respondent i will vote for party P on election day. Let the random variable S_{nA} be the number of respondents of the n interviewed voters who actually vote for party A on election day. The random variables S_{nB} and S_{nC} are defined in a similar manner. The vector $\mathbf{S}_n = (S_{nA}, S_{nB}, S_{nC})$ can be written as the sum of n random vectors $\mathbf{X}_1 = (X_{1A}, X_{1B}, X_{1C}), \ldots, \mathbf{X}_n = (X_{nA}, X_{nB}, X_{nC})$, where the random variable X_{iP} is defined by

$$X_{iP} = \begin{cases} 1 & \text{if respondent } i \text{ votes for party } P, \\ 0 & \text{otherwise.} \end{cases}$$

The random vector $\mathbf{X}_i = (X_{iA}, X_{iB}, X_{iC})$ describes the voting behavior of respondent i. The simplifying assumption is made that the random vectors $\mathbf{X}_1, \ldots, \mathbf{X}_n$ are independent. These random vectors do not have the same individual distributions. However, under the crucial assumption of independence, the multidimensional central limit theorem can be shown to remain valid and thus the random vector (S_{nA}, S_{nB}, S_{nC}) has approximately a multivariate normal distribution for n large. This multivariate normal distribution has

$$\mu = \left(\sum_{i=1}^{n} p_{iA}, \sum_{i=1}^{n} p_{iB}, \sum_{i=1}^{n} p_{iC} \right)$$

as vector of expected values and

$$\mathbf{C} = \begin{pmatrix} \sum_{i=1}^{n} p_{iA}(1-p_{iA}) & -\sum_{i=1}^{n} p_{iA}p_{iB} & -\sum_{i=1}^{n} p_{iA}p_{iC} \\ -\sum_{i=1}^{n} p_{iA}p_{iB} & \sum_{i=1}^{n} p_{iB}(1-p_{iB}) & -\sum_{i=1}^{n} p_{iB}p_{iC} \\ -\sum_{i=1}^{n} p_{iA}p_{iC} & -\sum_{i=1}^{n} p_{iB}p_{iC} & \sum_{i=1}^{n} p_{iC}(1-p_{iC}) \end{pmatrix}$$

as covariance matrix (this matrix is singular, since for each row the sum of the elements is zero). The vector μ of expected values is obvious, but a few words of explanation are in order for the covariance matrix $\mathbf{C}$. By the independence of $X_{1A}, \ldots, X_{nA}$,

$$\sigma^2(S_{nA}) = \sigma^2 \left(\sum_{i=1}^{n} X_{iA} \right) = \sum_{i=1}^{n} \sigma^2(X_{iA}) = \sum_{i=1}^{n} p_{iA}(1-p_{iA}).$$

Since

$$\text{cov}(S_{nA}, S_{nB}) = E\left[\left(\sum_{i=1}^{n} X_{iA} \right) \left(\sum_{j=1}^{n} X_{jB} \right) \right] - E\left(\sum_{i=1}^{n} X_{iA} \right) E\left(\sum_{j=1}^{n} X_{jB} \right),$$

it follows from the independence of X_{iA} and X_{jB} for $j \neq i$ and the fact that X_{iA} and X_{iB} cannot both be positive that

$$\text{cov}(S_{nA}, S_{nB}) = \sum_{i=1}^{n} \sum_{j \neq i} E(X_{iA}X_{jB}) - \left(\sum_{i=1}^{n} p_{iA} \right) \left(\sum_{j=1}^{n} p_{jB} \right)$$

$$= \sum_{i=1}^{n} \sum_{j \neq i} p_{iA}p_{jB} - \sum_{i=1}^{n} \sum_{j=1}^{n} p_{iA}p_{jB} = -\sum_{i=1}^{n} p_{iA}p_{iB}.$$

Similarly, the other terms in matrix $\mathbf{C}$ are explained.

It is standard fare in statistics to simulate random observations from the multivariate normal distribution. This means that computer simulation provides a fast and convenient tool to estimate probabilities of interest, such

Table 6.1. *Voting probabilities*

No. of voters	(p_{iA}, p_{iB}, p_{iC})
230	(0.20, 0.80, 0)
140	(0.65, 0.35, 0)
60	(0.70, 0.30, 0)
120	(0.45, 0.55, 0)
70	(0.90, 0.10, 0)
40	(0.95, 0, 0.05)
130	(0.60, 0.35, 0.05)
210	(0.20, 0.55, 0.25)

as the probability that party A will receive the most votes or the probability that the two parties A and C will receive more than half of the votes.

As a numerical illustration, suppose that a representative group of $n = 1{,}000$ voters is polled. The probabilities assigned by each of the 1,000 voters to parties A, B, and C are summarized in Table 6.1: the vote of each of 230 persons will go to parties A, B, and C with probabilities 0.20, 0.80, and 0, the vote of each of 140 persons will go to parties A, B, and C with probabilities 0.65, 0.35, and 0, and so on. Each person votes independently of the other persons. Let the random variable S_A be defined as

$$S_A = \text{the number of votes for party } A \text{ when the 1,000 voters}$$
$$\text{actually vote on election day.}$$

Similarly, the random variables S_B and S_C are defined. How do we calculate probabilities such as the probability that party A will become the largest party and the probability that parties A and C together will get the majority of the votes? These probabilities are given by $P(S_A > S_B, S_A > S_C)$ and $P(S_A + S_C > S_B)$. Simulating from the trivariate normal approximation for the random vector (S_A, S_B, S_C) provides a simple and fast method to get approximate values for these probabilities. The random vector (S_A, S_B, S_C) has approximately a trivariate normal distribution. Using the data from Table 6.1, the vector of expected values and the covariance matrix of this trivariate normal distribution are calculated as

$$\mu = (454, 485, 61) \quad \text{and} \quad \mathbf{C} = \begin{pmatrix} 183.95 & -167.65 & -16.30 \\ -167.65 & 198.80 & -31.15 \\ -16.30 & -31.15 & 47.45 \end{pmatrix}.$$

In order to simulate random observations from this trivariate normal distribution, the eigenvalues $\lambda_1, \lambda_2, \lambda_3$ and the corresponding normalized eigenvectors

$\mathbf{u}_1, \mathbf{u}_2, \mathbf{u}_3$ of matrix $\mathbf{C}$ must first be calculated. Using standard software, we find the eigenvalues $\lambda_1 = 70.6016$, $\lambda_2 = 359.5984$, and $\lambda_3 = 0$ with the corresponding eigenvectors

$\mathbf{u}_1 = (0.4393, 0.3763, -0.8157)$, $\mathbf{u}_2 = (-0.6882, 0.7246, -0.0364)$, and
$\mathbf{u}_3 = (0.5774, 0.5774, 0.5774)$.

The diagonal matrix $\mathbf{D}^{1/2}$ has $\sqrt{\lambda_1}, \sqrt{\lambda_2}$, and $\sqrt{\lambda_3}$ on its diagonal and the orthogonal matrix $\mathbf{U}$ has $\mathbf{u}_1, \mathbf{u}_2$, and $\mathbf{u}_3$ as column vectors. The matrix product $\mathbf{UD}^{1/2}$ gives the desired decomposition matrix

$$\mathbf{A} = \begin{pmatrix} 3.6916 & -13.0508 & 0 \\ 3.1622 & 13.7405 & 0 \\ -6.8538 & -0.6897 & 0 \end{pmatrix}.$$

Thus, the random vector (S_A, S_B, S_C) can be represented approximately as

$$S_A = 454 + 3.6916 Z_1 - 13.0508 Z_2,$$
$$S_B = 485 + 3.1622 Z_1 + 13.7405 Z_2,$$
$$S_C = 61 - 6.8538 Z_1 - 0.6897 Z_2,$$

where Z_1 and Z_2 are independent random variables each having the standard normal distribution. Note that the condition $S_A + S_B + S_C = 1,000$ is preserved in this decomposition. Using the decomposition of the multivariate normal distribution, it is standard fare to simulate random observations from the trivariate normal approximation to (S_A, S_B, S_C). A simulation study with 100,000 random observations leads to the following estimates:

P(party A becomes the largest party) $= 0.123 \, (\pm 0.002)$,
P(party B becomes the largest party) $= 0.877 \, (\pm 0.002)$,
P(parties A and C get the majority of the votes) $= 0.855 \, (\pm 0.002)$,

where the numbers in parentheses indicate the 95% confidence intervals. How accurate is the model underlying these predictions? They are based on an approximately multivariate normal distribution. To find out how well this approximative model works, we simulate directly from the data in Table 6.1. In each simulation run, 230 random draws are done from the probability mass function $(0.2, 0.8, 0)$, 140 random draws from $(0.65, 0.35, 0)$, and so on. Performing 100,000 simulation runs, we obtain the estimates $0.120 \, (\pm 0.002)$, $0.872 \, (\pm 0.002)$, and $0.851 \, (\pm 0.002)$ for the three probabilities above. The approximative values of the multivariate normal model are very close to the simulated values. This justifies the use of this model, which is computationally much less demanding than the simulation approach.

In the above analysis we have silently assumed that each respondent is going to actually vote. If this is not the case, then the analysis needs only a minor modification. Now person i is asked to specify the probability p_i that he/she will actually vote and the conditional probabilities $\overline{p}_{iA}$, $\overline{p}_{iB}$, and $\overline{p}_{iC}$ that the vote will go to party A, party B, or party C, respectively, when the person is going to vote. Also, introduce an imaginary party D for the non-voters. Then repeat the above analysis for the case of four parties A, B, C, and D, where the voting probabilities of the voters are given by $p_{iA} = p_i \times \overline{p}_{iA}$, $p_{iB} = p_i \times \overline{p}_{iB}$, $p_{iC} = p_i \times \overline{p}_{iC}$, and $p_{iD} = 1 - p_i$.

6.3.2 The r/s Lottery

In the 6/45 lottery, six balls are drawn out of a drum with 45 balls numbered from 1 to 45. More generally, in the r/s lottery, r balls are drawn from a drum with s balls. For the r/s lottery, define the random variable S_{nj} by

S_{nj} = the number of times ball number j is drawn in n drawings

for $j = 1, \ldots, s$. Letting

$$X_{kj} = \begin{cases} 1 & \text{if ball number } j \text{ is drawn at the } k\text{th drawing,} \\ 0 & \text{otherwise,} \end{cases}$$

we can represent S_{nj} in the form

$$S_{nj} = X_{1j} + X_{2j} + \cdots + X_{nj}.$$

Thus, by the multidimensional central limit theorem, the random vector $\mathbf{S}_n = (S_{n1}, \ldots, S_{ns})$ has approximately a multivariate normal distribution. The quantities $\mu_j^{(0)} = E(X_{1j})$ and $\sigma_{ij}^{(0)} = \mathrm{cov}(X_{1i}, X_{1j})$ are given by

$$\mu_j^{(0)} = \frac{r}{s}, \quad \sigma_{jj}^{(0)} = \frac{r}{s}\left(1 - \frac{r}{s}\right) \quad \text{and} \quad \sigma_{ij}^{(0)} = -\frac{r(s-r)}{s^2(s-1)} \quad \text{for } i \neq j.$$

It is left to the reader to verify this, using the relations $P(X_{1j} = 1) = \frac{r}{s}$ for all j and $P(X_{1i} = 1, X_{1j} = 1) = \frac{r}{s} \times \frac{r-1}{s-1}$ for all i, j with $i \neq j$. The covariance matrix $\mathbf{C} = \left(n\sigma_{ij}^{(0)}\right)$ is singular. The reason is that the sum of the elements of each row is zero. The matrix $\mathbf{C}$ has rank $s - 1$.

An interesting random walk for the lottery is the process that describes how the random variable $\max_{1 \leq j \leq s} S_{nj} - \min_{1 \leq j \leq s} S_{nj}$ behaves as a function of n. This random variable gives the difference between the number of occurrences of the most-drawn ball number and that of the least-drawn ball number in the first n drawings. Simulation experiments reveal that the sample paths of the random walk exhibit the tendency to increase proportionally to $\sqrt{n}$ as n gets larger. The central limit theorem is at work here. In particular, it can be proved that a constant c exists such that

$$E\left[\max_{1\leq j\leq s} S_{nj} - \min_{1\leq j\leq s} S_{nj}\right] \approx c\sqrt{n}$$

for n large. Using computer simulation, we find for c the estimate 1.52 in the 6/45 lottery and the estimate 1.48 in the 6/49 lottery.

6.4 Chi-Square Test

The chi-square (χ^2) test is tailored to measure how well an assumed distribution fits given data, when the data are the result of independent repetitions of an experiment with a finite number of possible outcomes. Let's consider an experiment with d possible outcomes $j = 1, \ldots, d$, where the outcome j occurs with probability p_j for $j = 1, \ldots, d$. It is assumed that the probabilities p_j are not estimated from the data but are known. Suppose that n independent repetitions of the experiment are done. Define the random variable X_{kj} by

$$X_{kj} = \begin{cases} 1 & \text{if the outcome of the } k\text{th experiment is } j, \\ 0 & \text{otherwise.} \end{cases}$$

Then the random vectors $\mathbf{X}_1 = (X_{11}, \ldots, X_{1d}), \ldots, \mathbf{X}_n = (X_{n1}, \ldots, X_{nd})$ are independent and have the same probability distribution. Let the random variable N_j represent the number of times that the outcome j will appear in the n repetitions of the experiment. That is,

$$N_j = X_{1j} + \cdots + X_{nj} \quad \text{for } j = 1, \ldots, d.$$

A convenient measure of the distance between the random variables N_j and their expected values np_j is the weighted sum of squares

$$\sum_{j=1}^{d} w_j (N_j - np_j)^2$$

for appropriately chosen weights $w_1, \ldots, w_d$. How do we choose the constants w_j? Naturally, we want to make the distribution of the weighted sum of squares as simple as possible. This is achieved by choosing $w_j = (np_j)^{-1}$. For large n, the test statistic

$$D = \sum_{j=1}^{d} \frac{(N_j - np_j)^2}{np_j}$$

has approximately a chi-square distribution with $d-1$ degrees of freedom (one degree of freedom is lost because of the linear relationship $\sum_{j=1}^{d} N_j - n$). We briefly outline the proof of this very useful result that goes back to Karl Pearson (1857–1936), one of the founders of modern statistics. Using the multidimensional central limit theorem, it can be shown that, for large n, the random vector

$$\left(\frac{N_1 - np_1}{\sqrt{np_1}}, \ldots, \frac{N_d - np_d}{\sqrt{np_d}} \right)$$

has approximately a multivariate normal distribution with the zero vector as its vector of expected values and the matrix $\mathbf{C} = \mathbf{I} - \sqrt{\mathbf{p}}\sqrt{\mathbf{p}}^T$ as its covariance matrix, where $\mathbf{I}$ is the identity matrix and the column vector $\sqrt{\mathbf{p}}$ has $\sqrt{p_j}$ as its jth component. Using the fact that $\sum_{j=1}^{d} p_j = 1$, the reader familiar with linear algebra may easily verify that one of the eigenvalues of the matrix $\mathbf{C}$ is equal to zero, and all other $d-1$ eigenvalues are equal to 1. Thus, by appealing to the result stated in Remark 6.1, the random variable

$$\sum_{j=1}^{d} \left(\frac{N_j - np_j}{\sqrt{np_j}} \right)^2$$

is approximately distributed as the sum of the squares of $d-1$ independent $N(0, 1)$ random variables and thus has an approximate chi-square distribution with $d-1$ degrees of freedom (χ^2_{d-1} distribution).

To get an idea of how well the chi-squared approximation performs, consider the following situation. Somebody claims to have rolled a fair die 1,200 times and found that the outcomes 1, 2, 3, 4, 5, and 6 occurred 196, 202, 199, 198, 202, and 203 times. Since the frequencies of the outcomes are very close to the expected values, the suspicion is that the data are fabricated. Assuming a fair die, the test statistic D takes on the value $\frac{1}{200}(4^2 + 2^2 + 1^2 + 2^2 + 2^2 + 3^2) = 0.19$. We immediately notice that the value 0.19 lies far below the expected value 5 of the χ^2_5 distribution. To test whether the data are fabricated, we determine the probability that the test statistic $D = \sum_{j=1}^{6}(N_j - 200)^2/200$ is less than or equal to 0.19, where N_j is the number of occurrences of outcome j in 1,200 rolls of a fair die. The chi-squared approximation to this probability is $P(\chi^2_5 \leq 0.19) = 0.00078$. This approximation is very close to the simulated value $P(D \leq 0.19) = 0.00083$ obtained from four million simulation runs of 1,200 rolls of a fair die. This very small probability indicates that the data are indeed most likely fabricated. The finding that $P(\chi^2_5 \leq 0.19)$ is an excellent approximation to the exact value of $P(D \leq 0.19)$ confirms the widely used rule of thumb that the chi-squared approximation can be applied when $np_j \geq 5$ for all j.

6.4 Chi-Square Test

In the above discussion, we assumed that the probabilities $p_1, \ldots, p_d$ of the possible outcomes $1, \ldots, d$ are known beforehand. In applications, however, you sometimes have to estimate one or more parameters in order to get the probabilities $p_1, \ldots, p_d$, as is the case in Problems 6.17–6.20. Then you must lower the number of degrees of freedom of the chi-square test statistic by one for each parameter estimated from the data.

Problems

6.15 The observed frequencies of birth months for the children born in England and Wales in the year 2010 are 60,179, 54,551, 59,965, 57,196, 59,444, 59,459, 62,166, 60,598, 62,986, 64,542, 60,745, and 61,334 for the months January to December. Use the chi-square test to make clear that the assumption of equally likely birth dates is not satisfied in reality.

6.16 In a classical experiment, Gregor Mendel observed the shape and color of peas that resulted from certain crossbreedings. A sample of 556 peas was studied, with the result that 315 produced round yellow, 108 produced round green, 101 produced wrinkled yellow, and 32 produced wrinkled green. According to Mendelian theory, the frequencies should be in the ratio 9 : 3 : 3 : 1. What do you conclude from a chi-square test?

6.17 Vacancies in the US Supreme Court over the 78-year period 1933–2010 have the following history: 48 years with 0 vacancies, 23 years with 1 vacancy, 7 years with 2 vacancies, and 0 years with $\geq$ 3 vacancies. Use a chi-square test to investigate how closely the observed frequencies conform to Poisson frequencies.

6.18 In Von Bortkiewicz's classical study on the distribution of 196 soldiers kicked to death by horses among 14 Prussian cavalry corps over the 20 years 1875 to 1894, the data are as follows. In 144 corps-years no deaths occurred, 91 corps-years had one death, 32 corps-years had two deaths, 11 corps-years had three deaths, and 2 corps-years had four deaths. Use a chi-square test to investigate how closely the observed frequencies conform to the expected Poisson frequencies. Assume that the probability of death is constant for all years and corps.

6.19 A total of 64 matches were played during the 2010 soccer World Cup in South Africa. The number of goals per match was distributed as follows. There were 7 matches with zero goals, 17 matches with one goal, 13 matches with two goals, 14 matches with three goals, 7 matches with four goals, 5 matches with five goals, and 1 match with seven goals. Use the chi-square test to find whether the distribution of the number of goals per match can be described by a Poisson distribution.

6.20 In a famous physics experiment performed by Rutherford, Chadwick, and Ellis in 1920, the number of α-particles emitted by a piece of radioactive material was counted during 2,608 time intervals of 7.5 seconds each. There were 57 intervals with 0 particles, 203 intervals with 1 particle, 383 intervals with 2 particles, 525 intervals with 3 particles, 532 intervals with 4 particles, 408 intervals with 5 particles, 273 intervals with 6 particles, 139 intervals with 7 particles, 45 intervals with 8 particles, 27 intervals with 9 particles, 10 intervals with 10 particles, 4 intervals with 11 particles, 0 intervals with 12 particles, 1 interval with 13 particles, and 1 interval with 14 particles. Use a chi-square test to investigate how closely the observed frequencies conform to Poisson frequencies.

6.21 In the Dutch National Lottery, six "main numbers" and one "bonus number" are drawn from the numbers 1 to 45 and in addition one color is drawn from six distinct colors. For each ticket you are asked to mark six distinct numbers and one color. You win (a) the jackpot (first prize) by matching the six main numbers and the color, (b) the second prize by matching the six main numbers, but not the color, (c) the third prize by matching five main numbers, the color, and the bonus number, (d) the fourth prize by matching five main numbers and the bonus number but not the color, (e) the fifth prize by matching five main numbers and the color, but not the bonus number, and (f) the sixth prize by matching only five main numbers. A total of 98,364,597 tickets filled in during a half-year period resulted in 2 winners of the jackpot, 6 winners of the second prize, 9 winners of the third prize, 35 winners of the fourth prize, 411 winners of the fifth prize, and 2,374 winners of the sixth prize. Use a chi-square test to find out whether or not the tickets were randomly filled in.

7
Conditioning by Random Variables

In Chapter 2, conditional probabilities were introduced by conditioning upon the occurrence of an event B of nonzero probability. In applications, this event B is often of the form $Y = b$ for a discrete random variable Y. However, when the random variable Y is continuous, the condition $Y = b$ has probability zero for any number b. In this chapter we will develop techniques for handling a condition provided by the observed value of a continuous random variable. You will see that the conditional probability density function of X given $Y = b$ for continuous random variables is analogous to the conditional probability mass function of X given $Y = b$ for discrete random variables. The conditional distribution of X given $Y = b$ enables us to define the natural concept of conditional expectation of X given $Y = b$. This concept allows for an intuitive understanding and is of utmost importance. In statistical applications, it is often more convenient to work with conditional expectations instead of the correlation coefficient when measuring the strength of the relationship between two dependent random variables. In applied probability problems, the computation of the expected value of a random variable X is often greatly simplified by conditioning on an appropriately chosen random variable Y. Learning the value of Y provides additional information about the random variable X and for that reason the computation of the conditional expectation of X given $Y = b$ is often simple. Much attention will be paid to the law of conditional probability and the law of conditional expectation. These laws are extremely useful when solving applied probability problems. Among other things, they will be used in the solving of stochastic optimization problems. In the final section we explain Bayesian inference for continuous models and give several statistical applications.

7.1 Conditional Distributions

Suppose that the random variables X and Y are defined on the same sample space Ω with probability measure P. A basic question for dependent random variables X and Y is: if the observed value of Y is y, what distribution now describes the distribution of X? We first answer this question for the discrete case. Conditioning on a discrete random variable is nothing other than conditioning on an event having nonzero probability. The analysis for the continuous case involves some technical subtleties, because the probability that a continuous random variable will take on a particular value is always zero. In this section, definitions will be given and intuition will be developed.

7.1.1 Conditional Probability Mass Function

Let X and Y be two discrete random variables with joint probability mass function $p(x, y) = P(X = x, Y = y)$. The *conditional probability mass function* of X given that $Y = y$ is defined by

$$P(X = x \mid Y = y) = \frac{P(X = x, Y = y)}{P(Y = y)}$$

for any fixed y with $P(Y = y) > 0$. This definition is just $P(A \mid B) = \frac{P(AB)}{P(B)}$ written in terms of random variables, where $A = \{\omega : X(\omega) = x\}$ and $B = \{\omega : Y(\omega) = y\}$ with ω denoting an element of the sample space. For any fixed y, we have that $\sum_x P(X = x, Y = y) = P(Y = y)$ and so the function $P(X = x \mid Y = y)$ satisfies

$$P(X = x \mid Y = y) \geq 0 \text{ for all } x \text{ and } \sum_x P(X = x \mid Y = y) = 1.$$

This shows that $P(X = x \mid Y = y)$ as a function of x is indeed a probability mass function. The notation $p_X(x \mid y)$ is often used for $P(X = x \mid Y = y)$.

Example 7.1 A carnival tent offers the following game. Passers-by can pick balls from a bowl for payment. The bowl contains 50 balls and each ball has inside a piece of paper. Five of the pieces of paper entitle the player to a main prize and 15 of the pieces of paper entitle them to a consolation prize. Suppose that a carnival-goer picks three balls at random from the bowl. Let the random variable X be the number of main prizes won and Y be the number of consolation prizes won. What are the conditional probability mass functions of X and Y?

Solution. The joint probability mass function of X and Y is given by

$$P(X = x, Y = y) = \frac{\binom{5}{x}\binom{15}{y}\binom{30}{3-x-y}}{\binom{50}{3}} \quad \text{for } 0 \leq x \leq 3 \text{ and } 0 \leq y \leq 3 - x.$$

The marginal mass functions $P(X = x) = \sum_{y=0}^{3-x} P(X = x, Y = y)$ and $P(Y = y) = \sum_{x=0}^{3-y} P(X = x, Y = y)$ are given by

$$P(X = x) = \frac{\binom{5}{x}\binom{45}{3-x}}{\binom{50}{3}} \quad \text{and} \quad P(Y = y) = \frac{\binom{15}{y}\binom{35}{3-y}}{\binom{50}{3}}$$

for $0 \leq x \leq 3$ and $0 \leq y \leq 3$. Therefore, for any fixed y, the conditional probability of having won x main prizes given that y consolation prizes have been won is equal to

$$P(X = x \mid Y = y) = \frac{\binom{5}{x}\binom{30}{3-x-y}}{\binom{35}{3-y}} \quad \text{for } x = 0, \ldots, 3 - y.$$

Also, for any fixed x, the conditional probability of having won y consolation prizes given that x main prizes have been won is equal to

$$P(Y = y \mid X = x) = \frac{\binom{15}{y}\binom{30}{3-x-y}}{\binom{45}{3-x}} \quad \text{for } y = 0, \ldots, 3 - x.$$

Problems

7.1 Consider again Example 5.1. What is the conditional probability mass function of the number of games won by team A given that team B is the overall winner?

7.2 Let $P(X = x, Y = y) = \binom{y}{x} \left(\frac{1}{6}\right)^x \left(\frac{1}{3}\right)^{y-x}$ for $x = 0, 1, \ldots, y$ and $y = 1, 2, \ldots$ be the joint probability mass function of the random variables X and Y. What is the conditional mass function of X?

7.3 A fair die is rolled repeatedly. Let the random variable X be the number of rolls until the face value 1 appears and Y be the number of rolls until the face value 6 appears. What are the conditional mass functions $P(X = x \mid Y = 2)$ and $P(X = x \mid Y = 20)$?

7.4 Three different numbers are chosen at random from the numbers $1, 2, \ldots, 10$. Let X be the smallest of these three numbers and Y the largest. What are the conditional mass functions of X and Y?

7.5 A fair die is rolled until a six appears. Each time the die is rolled, a fair coin is tossed. Let X be the number of rolls of the die and Y the number of heads resulting from the tosses of the coin. What is the conditional mass function of X? *Hint*: Use the identity $\sum_{n=r}^{\infty} \binom{n}{r} a^n = a^r/(1-a)^{r+1}$ for $0 < a < 1$ to find $P(Y = j)$ when $j \neq 0$.

7.6 Two dice are rolled. Let the random variable X be the smallest of the two outcomes and let Y be the largest of the two outcomes. What are the conditional mass functions of X and Y?

7.7 You simultaneously roll 24 dice. Next you roll only those dice that showed the face value six in the first roll. Let the random variable X be the number of sixes in the first roll and Y the number of sixes in the second roll. What is the conditional mass function of X?

7.8 Let X be the number of hearts and Y the number of diamonds in a bridge hand. What are the conditional mass functions of X and Y?

7.1.2 Conditional Probability Density Function

What is the continuous analogue of the conditional probability mass function when X and Y are continuous random variables with a joint probability density function $f(x, y)$? In this situation, we have the complication that $P(Y = y) = 0$ for each real number y. Nevertheless, this situation also allows for a natural definition of the concept of conditional distribution. Toward this end, we need the probabilistic interpretations of the joint density function $f(x, y)$ and the marginal densities $f_X(x)$ and $f_Y(y)$ of the random variables X and Y. For small values of $\Delta a > 0$ and $\Delta b > 0$,

$$P\left(a - \tfrac{1}{2}\Delta a \leq X \leq a + \tfrac{1}{2}\Delta a \,\Big|\, b - \tfrac{1}{2}\Delta b \leq Y \leq b + \tfrac{1}{2}\Delta b\right)$$

$$= \frac{P\left(a - \tfrac{1}{2}\Delta a \leq X \leq a + \tfrac{1}{2}\Delta a, b - \tfrac{1}{2}\Delta b \leq Y \leq b + \tfrac{1}{2}\Delta b\right)}{P\left(b - \tfrac{1}{2}\Delta b \leq Y \leq b + \tfrac{1}{2}\Delta b\right)}$$

$$\approx \frac{f(a, b) \Delta a \Delta b}{f_Y(b) \Delta b} = \frac{f(a, b)}{f_Y(b)} \Delta a$$

provided that (a, b) is a continuity point of $f(x, y)$ and $f_Y(b) > 0$. This leads to the following definition.

Definition 7.1 *If X and Y are continuous random variables with joint probability density function $f(x, y)$ and $f_Y(y)$ is the marginal density function of Y, then the conditional probability density function of X given that $Y = y$ is defined by*

$$f_X(x \mid y) = \frac{f(x, y)}{f_Y(y)} \quad \text{for } -\infty < x < \infty$$

for any fixed y with $f_Y(y) > 0$.

7.1 Conditional Distributions

Since $f_Y(y) = \int_{-\infty}^{\infty} f(x, y)\, dx$, we have for any fixed y that

$$f_X(x \mid y) \geq 0 \text{ for all } x \quad \text{and} \quad \int_{-\infty}^{\infty} f_X(x \mid y)\, dx = 1.$$

This shows that $f_X(x \mid y)$ as a function of x is indeed a probability density function. Similarly, the conditional probability density function of the random variable Y given that $X = x$ is defined by $f_Y(y \mid x) = \frac{f(x,y)}{f_X(x)}$ for any fixed x with $f_X(x) > 0$.

A probabilistic interpretation can be given for $f_X(a \mid b)$: given that the observed value of Y is b, the probability of the random variable X taking on a value in a small interval of length Δa around the point a is approximately equal to $f_X(a \mid b)\Delta a$ if a is a continuity point of $f_X(x \mid b)$.

The concept of a conditional probability distribution function is defined as follows. For any fixed y with $f_Y(y) > 0$, the conditional probability that the random variable X takes on a value less than or equal to x given that $Y = y$ is denoted by $P(X \leq x \mid Y = y)$ and is defined as

$$P(X \leq x \mid Y = y) = \int_{-\infty}^{x} f_X(u \mid y)\, du \quad \text{for all } x.$$

Example 7.2 An insurance company is incurring losses under collision insurance and under liability insurance. The losses can be modeled by the random variables X and Y with the joint probability density function $f(x, y) = \frac{3}{5}(5 - 3x - 2y)$ for $x, y > 0$, $x + y < 1$ and $f(x, y) = 0$ otherwise. The losses are expressed in units of millions of dollars. What is the conditional density of Y given that $X = x$? What is the probability that the losses under liability insurance are more than 0.5 million dollars given that the losses under collision insurance are 0.25 million dollars?

Solution. To get $f_Y(y \mid x)$, we need the marginal density of X. The density $f_X(x)$ is given by $\int_0^{1-x} \frac{3}{5}(5 - 3x - 2y)\, dy = \frac{3}{5}(1 - x)(4 - 2x)$ for $0 < x < 1$. Hence, for any $0 < x < 1$,

$$f_Y(y \mid x) = \frac{5 - 3x - 2y}{(1 - x)(4 - 2x)} \quad \text{for } 0 < y < 1 - x$$

and $f_Y(y \mid x) = 0$ otherwise. The sought probability is given by

$$\int_{0.5}^{0.75} f_Y(y \mid 0.25)\, dy = \int_{0.5}^{0.75} \frac{5 - 0.75 - 2y}{0.75 \times 3.5}\, dy = 0.2857.$$

Example 7.3 An electronic device contains two circuits. The second circuit is a backup for the first one and is switched on only when the first circuit has failed. The electronic device goes down when the second circuit fails. The joint density function of the operating times X and Y of the first and the second circuits is given by $f(x,y) = 24/(x+y)^4$ for $x, y > 1$ and $f(x,y) = 0$ otherwise. What is the probability that the second circuit will work at more than t time units given that the first circuit has failed after s time units?

Solution. To answer the posed question, we need the marginal density of the lifetime X of the first circuit. This density is given by

$$f_X(x) = \int_1^\infty \frac{24}{(x+y)^4}\,dy = \frac{8}{(x+1)^3} \quad \text{for } x > 1$$

and $f_X(x) = 0$ otherwise. Hence, the conditional density of the lifetime of the second circuit given that the first circuit has failed after s time units is

$$f_Y(y \mid s) = \frac{f(s,y)}{f_X(s)} = \frac{3(s+1)^3}{(s+y)^4} \quad \text{for } y > 1$$

and $f_Y(y \mid s) = 0$ otherwise. The probability that the second circuit will work at more than t time units given that the first circuit has failed after s time units is

$$P(Y > t \mid X = s) = \int_t^\infty f_Y(y \mid s)\,dy = \frac{(s+1)^3}{(s+t)^3} \quad \text{for } t \geq 1.$$

Example 7.4 Suppose that the random variables X and Y have a bivariate normal distribution with parameters $(\mu_1, \mu_2, \sigma_1^2, \sigma_2^2, \rho)$. What are the conditional probability densities of X and Y?

Solution. The joint density function $f(x,y)$ is specified in Section 6.1. In this section, we also found that the marginal density $f_Y(y)$ of Y is the $N(\mu_1, \sigma_1^2)$ density. Substituting the expressions for these densities into the definition of $f_X(x \mid y)$, it follows that the conditional probability density $f_X(x \mid y)$ of X given that $Y = y$ is the

$$N\left(\mu_1 + \rho\frac{\sigma_1}{\sigma_2}(y - \mu_2),\, \sigma_1^2(1 - \rho^2)\right)$$

density (verify!). Similarly, the conditional probability density $f_Y(y \mid x)$ of Y given that $X = x$ is the

$$N\left(\mu_2 + \rho\frac{\sigma_2}{\sigma_1}(x - \mu_1),\, \sigma_2^2(1 - \rho^2)\right)$$

density. Note that the expected values of the conditional densities are linear functions of the conditioning variable, and the conditional variances are constants.

Example 7.5 A very tasty looking toadstool growing in the forest is nevertheless so poisonous that it is fatal to squirrels that consume more than half of it. Squirrel 1, however, does partake of it, and later on squirrel 2 does the

7.1 Conditional Distributions

same. What is the probability that both squirrels survive? Assume that the first squirrel consumes a uniformly distributed amount of the toadstool, and the second squirrel a uniformly distributed amount of the remaining part of the toadstool.

Solution. Let the random variable X be the proportion of the toadstool consumed by squirrel 1 and Y the proportion of the toadstool consumed by squirrel 2. To answer the question, we need the joint density $f(x, y)$ of X and Y. By the definition $f_Y(y \mid x) = f(x, y)/f_X(x)$, we have

$$f(x, y) = f_X(x) f_Y(y \mid x).$$

Since $f_X(x) = 1$ for $0 < x < 1$ and $f_Y(y \mid x) = \frac{1}{1-x}$ for $0 < y < 1 - x$, we get

$$f(x, y) = \frac{1}{1 - x} \quad \text{for } 0 < x < 1 \text{ and } 0 < y < 1 - x.$$

The probability of both squirrels surviving can now be evaluated as

$$P\left(X \le \frac{1}{2}, Y \le \frac{1}{2}\right) = \int_0^{\frac{1}{2}} \int_0^{\frac{1}{2}} f(x, y)\, dx\, dy = \int_0^{\frac{1}{2}} \frac{dx}{1 - x} \int_0^{\frac{1}{2}} dy$$

$$= \frac{1}{2} \int_{\frac{1}{2}}^1 \frac{du}{u} = \frac{1}{2} \ln(2) = 0.3466.$$

Simulating from a Joint Density

The relation $f_X(x \mid y) = \frac{f(x,y)}{f_Y(y)}$ can be written in the more insightful form

$$f(x, y) = f_Y(y) f_X(x \mid y),$$

in analogy with $P(AB) = P(B)P(A \mid B)$. This representation of $f(x, y)$ may be helpful in simulating a random observation from the joint probability distribution of X and Y. First, a random observation for Y is generated from the density function $f_Y(y)$. If the value of this observation is y, then a random observation for X is generated from the conditional density function $f_X(x \mid y)$. To illustrate this, suppose that the joint density function of X and Y is given by $f(x, y) = e^{-x}$ for $0 < y < x$ and $f(x, y) = 0$ otherwise. Then, by $f_Y(y) = \int_y^\infty f(x, y)\, dx$, the marginal density function of Y is the exponential density function $f_Y(y) = e^{-y}$ for $y > 0$. By $f_X(x \mid y) = f(x, y)/f_Y(y)$, the conditional density function $f_X(x \mid y) = e^{-(x-y)}$ for $x > y$. This is an exponential density function shifted to the point y. Hence, by generating two random observations v_1 and v_2 from the exponential density e^{-v}, a random observation (x, y) from the density $f(x, y)$ is given by $x = v_1 + v_2$ and $y = v_1$. A random observation from the exponential density can be obtained by the inverse-transformation method discussed in Section 4.9.

Problems

7.9 Let X and Y be two continuous random variables with the joint probability density $f(x, y) = xe^{-x(y+1)}$ for $x, y > 0$ and $f(x, y) = 0$ otherwise. What are the conditional density functions $f_X(x \mid y)$ and $f_Y(y \mid x)$? What is $P(Y > 1 \mid X = 1)$?

7.10 Let X and Y be two continuous random variables with the joint probability density $f(x, y) = x - y + 1$ for $0 < x, y < 1$ and $f(x, y) = 0$ otherwise. What are the conditional densities $f_Y(y \mid x)$ and $f_X(x \mid y)$? What are $P(X > 0.5 \mid Y = 0.25)$ and $P(Y > 0.5 \mid X = 0.25)$?

7.11 Let X and Y be two continuous random variables with the joint probability density $f(x, y) = \frac{1}{x}$ for $0 < y \leq x < 1$ and $f(x, y) = 0$ otherwise. What are the conditional density functions $f_X(x \mid y)$ and $f_Y(y \mid x)$?

7.12 Consider again Problem 5.27. What are the conditional density functions $f_X(x \mid y)$ and $f_Y(y \mid x)$?

7.13 A point X is randomly chosen in $(0, 1)$ and then a point Y is randomly chosen in $(1 - X, 1)$. What are the probabilities $P(X + Y > 1.5)$ and $P(Y > 0.5)$?

7.14 The candidates A and B participate in an election. The random variables X and Y denote the proportion of votes cast for the candidates A and B. The joint density function of X and Y is $f(x, y) = 3(x + y)$ for $x, y > 0$ and $x + y < 1$ and $f(x, y) = 0$ otherwise. What is the conditional density function $f_X(x \mid y)$?

7.15 Suppose that X and Y are continuous random variables such that the density function of X is $f_X(x) = 2x$ for $0 < x < 1$ and the conditional distribution of Y given that $X = x$ is the uniform distribution on $(0, x)$. Show that the conditional density of X given that $Y = y$ is the uniform density on $(y, 1)$.

7.16 Let X and Y be two continuous random variables with joint probability density $f(x, y) = \frac{2y}{x^2}$ for $0 < y \leq x < 1$ and $f(x, y) = 0$ otherwise. What are the conditional density functions $f_X(x \mid y)$ and $f_Y(y \mid x)$? How do you simulate a random observation from $f(x, y)$?

7.17 Let the discrete random variable X and the continuous random variable Y be defined on the same sample space, where the probability mass function $p_X(x) = P(X = x)$ and the conditional density function $f_Y(y \mid x) = \frac{d}{dy} P(Y \leq y \mid X = x)$ are given. Assume that $f_Y(y \mid x)$ is a continuous function of y for any x. Explain why

$$p_X(x \mid y) = \frac{p_X(x) f_Y(y \mid x)}{\sum_u p_X(u) f_Y(y \mid u)}$$

is a natural definition for the conditional probability mass function of X given the observed value $Y = y$.

7.18 A receiver gets as input a random signal that is represented by a discrete random variable X, where X takes on the value $+1$ with probability p and the value -1 with probability $1 - p$. The output Y is a continuous random variable which is equal to the input X plus random noise, where the random noise has an $N(0, \sigma^2)$ distribution. You can only observe the output. What is the conditional probability of $X = 1$ given the observed value $Y = y$?

7.19 An insurance policy is written to cover a loss X, where X has the density function $f(x) = 1.5x(2-x)$ for $0 < x < 2$. The time in hours to process a claim of size x is uniformly distributed on $(x, 2x)$. What is the probability that a randomly chosen claim on this policy is processed in no more than one hour?

7.20 Let X and Y be two independent random variables each having an exponential density with expected value $1/\lambda$. Show that the conditional density of X given that $X + Y = u$ is the uniform density on $(0, u)$.

7.2 Law of Conditional Probability for Random Variables

For discrete random variables X and Y, the unconditional probability $P(X = a)$ can be calculated from

$$P(X = a) = \sum_b P(X = a \mid Y = b)P(Y = b).$$

This *law of conditional probability* is a special case of Rule 2.2 in Chapter 2. In the situation of continuous random variables X and Y, the continuous analogue of the law of conditional probability is:

Rule 7.1 *For continuous random variables X and Y,*

$$P(X \leq a) = \int_{-\infty}^{\infty} P(X \leq a \mid Y = y) f_Y(y) \, dy.$$

To prove this rule, integrate both sides of $f(x, y) = f_Y(y) f_X(x \mid y)$ with respect to y in order to obtain the basic relation

$$f_X(x) = \int_{-\infty}^{\infty} f_X(x \mid y) f_Y(y) \, dy.$$

Hence, by $P(X \leq a) = \int_{-\infty}^{a} f_X(x) \, dx$, we can write $P(X \leq a)$ as

$$\int_{-\infty}^{a} dx \int_{-\infty}^{\infty} f_X(x \mid y) f_Y(y) \, dy = \int_{-\infty}^{\infty} f_Y(y) \, dy \int_{-\infty}^{a} f_X(x \mid y) \, dx$$
$$= \int_{-\infty}^{\infty} P(X \leq a \mid Y = y) f_Y(y) \, dy,$$

which verifies Rule 7.1. The interchange of the order of integration is justified by the nonnegativity of the integrand.

The importance of the continuous analogue of the law of conditional probability can hardly be overestimated. In applications, the conditional probability $P(X \leq a \mid Y = y)$ is often calculated without explicitly using the joint distribution of X and Y, but through a direct physical interpretation of the conditional probability in the context of the concrete application. To illustrate this, let's return to Example 7.5 and calculate the probability that squirrel 2 will survive. This probability can be obtained as

$$P(Y \leq 0.5) = \int_0^1 P(Y \leq 0.5 \mid X = x) f_X(x) \, dx$$
$$= \int_0^{0.5} \frac{0.5}{1-x} \, dx + \int_{0.5}^1 1 \, dx = 0.5 \ln(2) + 0.5 = 0.8466.$$

In the next example, the law of conditional probability is used for the situation of a discrete random variable X and a continuous random variable Y. A precise definition of $P(X \leq a \mid Y = y)$ for this situation requires some technical machinery and will not be given. However, in the context of the concrete problem, it is immediately obvious what is meant by the conditional probability.

Example 7.6 Every morning at exactly the same time, Mr. Johnson rides the metro to work. Each time he waits for the metro at the same place in the metro station. If the metro arrives at the station, then the exact spot where it comes to a stop is a surprise. From experience, Mr. Johnson knows that the distance between him and the nearest metro door once the metro has stopped is uniformly distributed between 0 and 2 meters. Mr. Johnson is able to find a place to sit with probability $1 - \sqrt{0.5y}$ if the nearest door is y meters from where he is standing. On any given morning, what is the probability that Mr. Johnson will succeed in finding a place to sit down?

Solution. The probability is not $1 - \sqrt{0.5 \times 1} = 0.293$ as some people believe (they substitute the expected value of the distance to the nearest door for y into the formula $1 - \sqrt{0.5y}$). The correct value is obtained as follows. Define the

7.2 Law of Conditional Probability for Random Variables

random variable X as equal to 1 when Mr. Johnson finds a seat and 0 otherwise. Let the random variable Y be the distance from Mr. Johnson's waiting place to the nearest metro door. We have

$$P(X = 1 \mid Y = y) = 1 - \sqrt{0.5y}.$$

The random variable Y has the probability density function $f_Y(y) = \frac{1}{2}$ for $0 < y < 2$. Hence the unconditional probability that Mr. Johnson will succeed in finding a place to sit down on any given morning is equal to

$$P(X = 1) = \int_0^2 \left(1 - \sqrt{0.5y}\right) \frac{1}{2}\, dy = \frac{1}{3}.$$

The next example deals with the continuous version of the game of chance discussed in Example 2.11.

Example 7.7 Each of two players A and B in turn draws a random number once or twice. The random numbers are independent samples from the uniform distribution on $(0, 1)$. For each player, the decision whether to go for a second draw depends on the result of the first draw. The object of the game is to have the highest total score, from one or two draws, without going over 1. A player gets a final score of zero if the player draws two numbers and the sum of these numbers exceeds 1. Player A takes the first draw of one or two random numbers and then waits for the opponent's score. Player B has the advantage of knowing the score of player A. What strategy maximizes the probability of player A winning? What is the value of this probability?

Solution. It is obvious how player B acts. Player B stops after the first draw if the score of this draw is larger than the final score of player A and continues otherwise. In analyzing the problem, it is natural to condition on the outcome of the first draw of player A. Denote by the random variable U the number player A gets at the first draw. Define the conditional probability

$S(a) =$ the winning probability of player A if player A
stops after the first draw given that $U = a$.

Also, define $C(a)$ as the winning probability of player A if player A continues for a second draw given that $U = a$. It will be shown that there is a number a^* such that $C(a) > S(a)$ for all $0 \le a < a^*$, $C(a^*) = S(a^*)$ and $C(a) < S(a)$ for all $a^* < a \le 1$. Then the optimal strategy for player A is to stop after the first draw if this draw gives a number larger than a^* and to continue otherwise. By the law of conditional probability, the overall winning probability of player A under this strategy is given by

$$P_A = \int_0^{a^*} C(a) f_U(a)\, da + \int_{a^*}^1 S(a) f_U(a)\, da,$$

where $f_U(a) = 1$ for all $0 < u < 1$. It remains to find $S(a)$, $C(a)$, and a^*.

To find $S(a)$, we reason as follows. Under the condition that player A has stopped with a score of a, player B can win in two possible ways: (i) the first draw of player B results in a number $y > a$: (ii) the first draw of player B results in a number $y \le a$ and his second draw gives a number between $a - y$ and $1 - y$. Letting the uniform random variable Y be the number player B gets at the first draw and using again the law of conditional probability, we have

$$1 - S(a) = \int_a^1 1 \times f_Y(y)\, dy + \int_0^a [1 - y - (a - y)] f_Y(y)\, dy$$
$$= 1 - a + (1 - a)a = 1 - a^2,$$

showing that $S(a) = a^2$. To obtain $C(a)$, let the uniform random variable V be the number player A gets at the second draw. If player A has a total score of $a + v$ after the second draw with $a + v \le 1$, then player A will win with probability $(a + v)^2$, in view of the result $S(x) = x^2$ (the probability of player A winning depends on his final score and not on the number of draws made by the player). Hence,

$$C(a) = \int_0^{1-a} (a + v)^2 f_V(v)\, dv + \int_{1-a}^1 0 \times f_V(v)\, dv$$
$$= \int_a^1 w^2\, dw = \frac{1}{3}(1 - a^3).$$

The function $S(a) - C(a)$ is negative for $a = 0$ and positive for $a = 1$. Also, this continuous function is strictly increasing on $(0, 1)$ (the derivative is positive). Therefore, the optimal stopping level for the first player is the unique solution to the equation

$$a^2 = \frac{1}{3}(1 - a^3)$$

on $(0, 1)$. The numerical solution of this equation is $a^* = 0.53209$. The winning probability of player A can next be calculated as $P_A = 0.4538$.

A challenging problem is to analyze the three-player game. The reader is asked to do this in Problem 7.35 and to verify that the optimal stopping level for the first player is the solution $a^* = 0.64865$ to the equation $a^4 = \frac{1}{5}(1 - a^5)$. The optimal strategy for the second player in the three-player game is to stop after the first draw only if the score of this draw exceeds both the final score of the first player and the stopping level 0.53209 of the first player in the two-player game. It is noted that the solution of the continuous game provides an excellent approximation to the optimal stopping levels for the players in a television game show that involves spinning a wheel with the money values $1, 2, \ldots, R$ on it. In this game show each of the players in turn spins the wheel once or twice and the goal is to have the highest total score without going over R.

Conditional Probability and Rejection Sampling

Conditional probability can be exploited to simulate from "difficult" probability densities for which the inverse-transformation method cannot be used.

Rule 7.2 *Let $f(x)$ be a given probability density function. Suppose that the continuous random variable Y has a probability density $g(y)$ such that, for some constant $c > 0$, $f(y) \leq cg(y)$ for all y. If the random variable U is independent of Y and uniformly distributed on $(0, 1)$, then*

$$P\left(Y \leq x \mid U \leq \frac{f(Y)}{cg(Y)}\right) = \int_{-\infty}^{x} f(y)dy \quad \text{for all } x.$$

In words, under the condition that $U \leq \frac{f(Y)}{cg(Y)}$, the conditional density function of Y is equal to the density function $f(x)$.

The proof goes as follows. By conditioning on Y and using the fact that $P(U \leq u) = u$ for $0 < u < 1$, it follows that

$$P\left(U \leq \frac{f(Y)}{cg(Y)}\right) = \int_{-\infty}^{+\infty} \frac{f(y)}{cg(y)} g(y) dy = \frac{1}{c}.$$

Also, by conditioning on Y,

$$P\left(Y \leq x, U \leq \frac{f(Y)}{cg(Y)}\right) = \int_{-\infty}^{x} \frac{f(y)}{cg(y)} g(y) dy = \frac{1}{c} \int_{-\infty}^{x} f(y) dy.$$

The conditional probability $P\left(Y \leq x \mid U \leq \frac{f(Y)}{cg(Y)}\right)$ is the ratio of $\frac{1}{c} \int_{-\infty}^{x} f(y) dy$ and $\frac{1}{c}$, which verifies the desired result.

Rule 7.2 underlies the so-called acceptance–rejection method for simulating from a "difficult" density $f(x)$ via an "easy" density $g(y)$. This method proceeds as follows:

Step 1. Generate a random observation y from the density $g(y)$ and generate a random number u from $(0, 1)$.

Step 2. If $u \leq \frac{f(y)}{cg(y)}$, then accept y as a sample from $f(x)$; otherwise, return to step 1.

Intuitively, the acceptance–rejection method generates a random point $(Y, U \times cg(Y))$ under the graph of $cg(y)$ and accepts the point only when it also falls under the graph of $f(y)$, as is the case when $U \times cg(Y) \leq f(Y)$. In the proof of Rule 7.2 we found that $P\left(U \leq \frac{f(Y)}{cg(Y)}\right) = \frac{1}{c}$, which shows that the number of iterations of the acceptance–rejection method is geometrically distributed with an expected value of c. The constant c satisfies $c \geq 1$, as follows by integrating both sides of $f(x) \leq cg(x)$ over x and noting that the densities $f(x)$ and $g(x)$ integrate to 1. To illustrate the acceptance–rejection method, consider the "difficult" density function $f(x) = \sqrt{2/\pi}\, e^{-\frac{1}{2}x^2}$ for $x > 0$ (being the density of $|Z|$ with Z the standard normal random variable). Take as "easy" density the exponential density $g(x) = e^{-x}$ for $x > 0$. Then $f(x)/g(x) \leq c$ for all $x > 0$ with $c = \sqrt{2e/\pi}$ (verify that $\max_{x>0} f(x)/g(x)$ is $\sqrt{2e/\pi} \approx 1.32$). A random observation from $g(x) = e^{-x}$ is obtained by using the inverse-transformation method discussed in Section 4.9.

Problems

7.21 Let $X_1, X_2, \ldots$ be independent random variables that are uniformly distributed on $(0, 1)$. The random variable N is defined as the smallest $n \geq 2$ for which $X_n > X_1$. What is the probability mass function of N?

7.22 You draw at random a number p from the interval $(0, 1)$. Then you toss n times a biased coin giving heads with probability p. Let X be the number of heads that will appear. Verify that X has the discrete uniform distribution on $0, 1, \ldots, n$.

7.23 The length of time required to unload a ship has an $N(\mu, \sigma^2)$ distribution. The crane to unload the ship has just been overhauled and the time for which it will operate until the next breakdown has an exponential distribution with an expected value of $1/\lambda$. What is the probability of no breakdown during the unloading of the ship?

7.24 Every morning there is a limousine service from Gotham city to the airport. The car has capacity for 10 passengers. The number of reservations for a trip follows a discrete uniform distribution between 5 and 10. The probability that a person with a reservation shows up for the trip is 0.80. Passengers without reservations are also accepted, as long as the car is not full. The number of walk-up passengers for a trip is independent of the number of reservations and has a discrete probability distribution with $p_0 = 0.25$, $p_1 = 0.45$, $p_2 = 0.20$, and $p_3 = 0.10$. For a given trip, let the random variable V denote the number of passengers with reservations who show up and let W denote the number of walk-up passengers who are accepted. What are the probability mass functions of V and W?

7.25 You first roll a die once. You next roll the die as many times as the outcome of the first roll. Let the random variable X be the total number of sixes in the rolls of the die. What is the probability mass function of X?

7.26 A random observation X is simulated from the gamma density with shape parameter r and scale parameter $(1-p)/p$, where r is a positive integer and $0 < p < 1$. Then a random observation N is simulated from the Poisson distribution whose expected value is given by the simulated value from the gamma density. Verify that $N+r$ is a random observation from the negative binomial distribution with parameters r and p.

7.27 An opaque bowl contains B balls, where B is given. Each ball is red or white. The number of red balls in the bowl is unknown, but has a binomial distribution with parameters B and p. You randomly select r balls out of the urn without replacing any. Use the law of conditional probability to obtain the probability distribution of the number of red balls among the selected balls. Does the result surprise you? Can you give a direct probabilistic argument for the result obtained?

7.28 The independent random variables X_1 and X_2 are $N(\mu_1, \sigma_1^2)$ and $N(\mu_2, \sigma_2^2)$ distributed. Let the random variable V be distributed as X_1 with given probability p and as X_2 with probability $1-p$. What is the probability density of V? Is this probability density the same as the probability density of the random variable $W = pX_1 + (1-p)X_2$?

7.29 Use twice the law of conditional probability to calculate the probability that the quadratic equation $Ax^2 + Bx + C = 0$ will have two real roots when A, B, and C are independent samples from the uniform distribution on $(0, 1)$.

7.30 You repeatedly generate a random number from $(0, 1)$ until you obtain a number less than $a = \frac{1}{2}$. The first number is picked at random from $(0, 1)$ and each subsequent number is picked at random from the interval between 0 and the previously chosen number. What is the probability that you have to pick exactly two numbers and what is the probability that you have to pick exactly three numbers?

7.31 A particle makes jumps either upwards or downwards at times $n = 1, 2, \ldots$, where the jumps are independent of each other and the position of the particle changes each time according to the standard normal distribution. The particle starts at the zero level. What is the expected number of crossings of the zero level during the first n jumps? *Hint*: Let I_k be 1 if the zero level is crossed at the $(k+1)$th jump and 0 otherwise.

7.32 You leave work at random times between 5.45 p.m. and 6.00 p.m. to take the bus home. Bus numbers 1 and 3 bring you home. You take the first bus that arrives. Bus number 1 arrives exactly every 15 minutes starting from the hour, and the interarrival times of buses number 3 are independent random variables having an exponential distribution with an expected value of 15 minutes. What is the probability that you take bus number 1 home on any given day? Use the law of conditional probability and the memoryless property of the exponential distribution to verify that this probability equals $1 - e^{-1}$. Can you give an intuitive explanation why the probability is larger than $\frac{1}{2}$?

7.33 Consider again the stochastic game in Example 7.7. It is now assumed that each player must act without knowing what the other player has done. What strategy should player A follow to ensure that his probability of winning is at least 50%, regardless of what player B will do? Assume that the player whose total score is closest to 1 is the winner when the scores of both players exceed 1. *Hint*: Define $P(a, b)$ as the win probability of player A when players A and B use threshold values a and b for stopping after the first draw.

7.34 Two players A and B each receive a random number between 0 and 1. Each player gets a once-only opportunity to discard his number and receive a new random number between 0 and 1. This choice is made without knowing the other player's number or whether the other player chose to replace his number. The player ending up with the higher number wins. The strategy used by player A is to choose another random number only if the first number received is below the threshold value a. Show that by taking $a = \frac{1}{2}(\sqrt{5} - 1)$, player A ensures himself of a win probability of at least 50%, regardless of what player B will do. *Hint*: Condition on the initial numbers received by the players to find the win probability $P(a, b)$ of player A when the players A and B use threshold values a and b.

7.35 (difficult) Consider the three-player variant of Example 7.7. What is the optimal strategy for the first player A and what is the overall win probability of player A? Next use the probability distribution function of the final score of player A to find the overall win probabilities of the players B and C. *Hint*: Given that player A gets score a at the first draw, let $S(a)$ be the overall win probability of player A if player A stops after the first draw and let $C(a)$ be the overall win probability if player A continues. Verify first that $S(a) = a^2 \times a^2$ for $a \geq a_2$, $S(a) = \left(a^2 + \frac{1}{2}a_2^2 - \frac{1}{2}a^2\right) \times a^2$ for $0 < a < a_2$, $C(a) = \frac{1}{5}(1 - a^5)$ for $a \geq a_2$, and $C(a) = \frac{1}{10}(a_2^5 - a^5) + \frac{1}{6}(a_2^5 - a_2^2 a^3) + \frac{1}{5}(1 - a_2^5)$ for $0 < a < a_2$, where a_2 is the optimal stopping level for the first player in the two-player game.

7.3 Law of Conditional Expectation

In the case that the random variables X and Y have a discrete distribution, the *conditional expectation* of X given that $Y = y$ is defined by

7.3 Law of Conditional Expectation

$$E(X \mid Y = y) = \sum_x x\, P(X = x \mid Y = y)$$

for each y with $P(Y = y) > 0$ under the assumption that $E(X)$ is finite. The assumption $E(|X|) < \infty$ implies that the series $\sum_x x P(X = x \mid Y = y)$ is absolutely convergent for each y with $P(Y = y) > 0$. This fact is easily verified. The formula $E(|X|) = \sum_x |x| P(X = x)$ can be rewritten as

$$\sum_x |x| \sum_y P(X = x, Y = y) = \sum_x |x| \sum_y P(X = x \mid Y = y) P(Y = y)$$

$$= \sum_y \left[\sum_x |x| P(X = x \mid Y = y) \right] P(Y = y).$$

The interchange of the order of summation is justified by the fact that the terms in the summation are nonnegative. Since $E(|X|) < \infty$, the series between brackets is finite for each y with $P(Y = y) > 0$, as was to be verified.

In the case that X and Y are continuously distributed with joint probability density function $f(x, y)$, the *conditional expectation* of X given that $Y = y$ is defined by

$$E(X \mid Y = y) = \int_{-\infty}^{\infty} x f_X(x \mid y)\, dx$$

for each y with $f_Y(y) > 0$, provided that $E(X)$ is finite. In the same way as above, it can be verified that this assumption implies that the integral defining $E(X \mid Y = y)$ converges absolutely for the "relevant" values of y.

A law of utmost importance is the *law of conditional expectation*. In the discrete case the law reads as

$$E(X) = \sum_y E(X \mid Y = y)\, P(Y = y),$$

while for the continuous case the law reads as

$$E(X) = \int_{-\infty}^{\infty} E(X \mid Y = y) f_Y(y)\, dy.$$

In both cases it is assumed that $E(X)$ is finite. In words, the law of conditional expectation says that the unconditional expected value of X may be obtained by first conditioning on an appropriate random variable Y to get the conditional expected value of X given that $Y = y$ and then taking the expectation of this quantity with respect to Y. We sketch the proof for the continuous case. Substituting $f_X(x) = \int_{-\infty}^{\infty} f(x, y)\, dy$ into $E(X) = \int_{-\infty}^{\infty} x f_X(x)\, dx$ and using the relation $f(x, y) = f_Y(y) f_X(x \mid y)$, we get

$$E(X) = \int_{-\infty}^{\infty} x f_X(x) dx = \int_{-\infty}^{\infty} x \, dx \int_{-\infty}^{\infty} f_Y(y) f_X(x \mid y) \, dy$$
$$= \int_{-\infty}^{\infty} f_Y(y) \, dy \int_{-\infty}^{\infty} x f_X(x \mid y) \, dx = \int_{-\infty}^{\infty} E(X \mid Y = y) f_Y(y) \, dy.$$

The interchange of the order of integration is justified by the finiteness of $\int_{-\infty}^{\infty} |x| f_X(x \mid y) \, dx$ for all relevant values of y.

Example 7.8 The relationship between household expenditure and net income of households in Fantasia is given by the joint density function

$$f(x, y) = \begin{cases} c(x-10)(y-10) & \text{for } 10 < x < y < 30, \\ 0 & \text{otherwise,} \end{cases}$$

where the normalizing constant $c = \frac{1}{20,000}$. What is the expected value of the household expenditure of a randomly chosen household given that the income of the household is y?

Solution. To answer the question, let the random variables X and Y represent the household expenditure and the net income of a randomly selected household. How do we find $E(X \mid Y = y)$? We first determine the marginal density of the conditioning variable Y:

$$f_Y(y) = \int_{-\infty}^{\infty} f(x, y) \, dx = c \int_{10}^{y} (x-10)(y-10) \, dx$$
$$= \frac{1}{2} c (y-10)^3 \quad \text{for } 10 < y < 30.$$

Using the relation $f_X(x \mid y) = \frac{f(x,y)}{f_Y(y)}$, we next obtain, for fixed y, the conditional density function

$$f_X(x \mid y) = \begin{cases} \frac{2(x-10)}{(y-10)^2} & \text{for } 10 < x < y, \\ 0 & \text{otherwise.} \end{cases}$$

This gives the desired result

$$E(X \mid Y = y) = \int_{10}^{y} x \frac{2(x-10)}{(y-10)^2} \, dx = 10 + \frac{2}{3}(y-10) \quad \text{for } 10 < y < 30.$$

The law of conditional expectation is nicely illustrated with the next example. This example is closely related to Example 3.5.

Example 7.9 Eleven closed boxes are put in random order in front of you. One of these boxes contains a devil's penny and the other 10 boxes contain money. You know which dollar amounts $a_1, \ldots, a_{10}$ are in the 10 boxes. You may mark as many boxes as you wish. The marked boxes are opened simultaneously. You win the money from these boxes if the box with the devil's penny is not

7.3 Law of Conditional Expectation

among the opened boxes; otherwise, you win nothing. How many boxes should you mark to maximize your expected gain? What is the expected value of the gain of the game?

Solution. Suppose you mark m boxes. Let the random variable G be the gain of the game. The random variable G can be represented as

$$G = X_1 + \cdots + X_m,$$

where X_k is the gain obtained from the ith of the m marked boxes. The random variables $X_1, \ldots, X_m$ are interchangeable and thus identically distributed. To find $E(G) = \sum_{k=1}^{m} E(X_k)$, we condition upon the random variable Y to be defined as 1 if the box with the devil's penny is not among the m marked boxes and 0 otherwise. By the law of conditional expectation, $E(X_k) = E(X_k \mid Y = 0)P(Y = 0) + E(X_k \mid Y = 1)P(Y = 1)$. Since $E(X_k \mid Y = 0) = 0$ and $P(Y = 1) = 1 - \frac{m}{11}$ (why?), we get

$$E(X_k) = E(X_k \mid Y = 1)\left(1 - \frac{m}{11}\right).$$

The conditional distribution of the random variable X_k given that $Y = 1$ is the discrete uniform distribution on $a_1, \ldots, a_{10}$. Hence, for all k,

$$E(X_k \mid Y = 1) = \frac{1}{10}\sum_{i=1}^{10} a_i \quad \text{and} \quad E(X_k) = \left(1 - \frac{m}{11}\right)\frac{1}{10}\sum_{i=1}^{10} a_i.$$

Inserting the expression for $E(X_k)$ into $E(G) = \sum_{k=1}^{m} E(X_k)$ gives

$$E(G) = m\left(1 - \frac{m}{11}\right)\frac{1}{10}\sum_{i=1}^{10} a_i.$$

The function $x(1 - \frac{x}{11})$ of the continuous variable x is maximal for $x = 5.5$. The maximal value of $E(G)$ as a function of m is attained for both $m = 5$ and $m = 6$. Hence, the optimal decision is to mark five or six boxes in order to maximize the expected gain of the game. The maximal value of the expected gain is $\frac{3}{11}\sum_{i=1}^{10} a_i$ dollars. The standard deviation of the gain differs for $m = 5$ and $m = 6$. Proceeding along the same lines as for $E(G)$ and using the interchangeability of the random variables X_i, it is readily derived that

$$E(G^2) = \left(1 - \frac{m}{11}\right)\left[\frac{m}{10}\sum_{i=1}^{10} a_i^2 + \frac{m(m-1)}{10 \times 9}\sum_{i=1}^{10} a_i \sum_{j=1, j \neq i}^{10} a_j\right].$$

The derivation uses the result $E(X_k X_l \mid Y = 1) = \frac{1}{10 \times 9}\sum_{i=1}^{10} a_i \sum_{j=1, j \neq i}^{10} a_j$ for $k \neq l$ and the relation $E(G^2) = \sum_{k=1}^{m} E(X_k^2) + 2\sum_{k=1}^{m-1}\sum_{l=k+1}^{m} E(X_k X_l)$.

The special case of $a_i = i$ for all i is an interesting case and describes the game that is played with the 11 cards ace, two, three, ..., nine, ten, and a joker, where the joker is in the role of the devil's penny and any other card counts for its face value (ace = 1). Then the maximal expected gain is $15. The standard deviation of the gain is $14.14 if you pick five cards and $16.73 if you pick six cards. The standard deviation of the gain is about the same as the expected value, showing that the actual gain in the game is quite erratic.

In Section 7.4 we give more examples showing the practical usefulness of the law of conditional expectation.

Problems

7.36 Consider again Example 7.1. What is the expected value of the number of consolation prizes won, given that no main prize has been won?

7.37 Consider again Example 7.3. What is the expected value of the lifetime of the second circuit, given that the first circuit has failed after s time units?

7.38 You choose three different numbers at random from the numbers $1, 2, \ldots, 100$. Let X be the smallest of these three numbers and Y the largest. What are the conditional expected values $E(X \mid Y = y)$ and $E(Y \mid X = x)$?

7.39 You generate three random numbers from $(0, 1)$. Let X be the smallest of these three numbers and Y the largest. What are the conditional expected values $E(X \mid Y = y)$ and $E(Y \mid X = x)$?

7.40 Verify that $E(X \mid Y = y) = E(X)$ for all y if X and Y are independent. Also, verify that $\text{cov}(X, Y) = 0$ if $E(X \mid Y = y)$ does not depend on y.

7.41 Let (X, Y) have a bivariate normal density with parameters $(\mu_1, \mu_2, \sigma_1^2, \sigma_2^2, \rho)$, where $\sigma_1^2 = \sigma_2^2$. What is $E(X \mid X + Y = v)$? *Hint*: Use the result of Problem 6.9.

7.42 Let the joint density function $f(x, y)$ of the random variables X and Y be equal to 1 for $-y < x < y$, $0 < y < 1$ and 0 otherwise. What are the conditional expected values $E(X \mid Y = y)$ and $E(Y \mid X = x)$?

7.43 Suppose that n Bernoulli trials each having success probability p are performed. What is the conditional expected value of the number of trials until the first success given that $r \geq 1$ successes have occurred?

7.44 In the game of Fast Pig, you may roll as many dice as you wish in any round. The roll of the dice contributes nothing to your score if one or more of the dice come up with a 1; otherwise, your score is increased by the total number of points rolled. How many dice should be rolled in order to maximize the expected number of points gained in a round? What are the expected value and the standard deviation of the number of points gained?

7.45 The percentage of zinc content and iron content in ore from a certain location has the joint density $f(x, y) = \frac{1}{75}(5x + y - 30)$ for $2 < x < 3$, $20 < y < 30$ and $f(x, y) = 0$ otherwise. What is the expected value of the zinc content in a sample of ore given that the iron content is y?

7.46 Someone purchases a liability insurance policy. The probability that a claim will be made on the policy is 0.1. In case of a claim, the size of the claim has an exponential distribution with an expected value of $1,000,000. The maximum insurance policy payout is $2,000,000. What is the expected value of the insurance payout?

7.47 Let X and Y have the joint density function $f(x, y)$.

(a) Explain why the natural definition for the conditional expected value of X given that $a < Y < b$ is

$$E(X \mid a < Y < b) = \frac{1}{P(a < Y < b)} \int_{-\infty}^{\infty} x \, dx \int_{a}^{b} f(x, w) \, dw.$$

Also, explain why the natural definition for the conditional expected value of X given that $X > Y$ is

$$E(X \mid X > Y) = \frac{1}{P(X > Y)} \int_{-\infty}^{\infty} x \, dx \int_{-\infty}^{x} f(x, w) \, dw.$$

(b) Verify that $E(X \mid a < Y < b) = \frac{\rho}{\sqrt{2\pi}} (e^{-\frac{1}{2}a^2} - e^{-\frac{1}{2}b^2}) / [\Phi(b) - \Phi(a)]$ and $E(X \mid X > Y) = \sqrt{(1-\rho)/\pi}$ when (X, Y) has a standard bivariate normal density with correlation coefficient ρ.

7.48 Let X be a continuous random variable and A be any event with $P(A) > 0$. The conditional expected value $E(X \mid A)$ is defined as $\int_{-\infty}^{\infty} x f(x \mid A) \, dx$, where $f(x \mid A)$ is the derivative of the conditional distribution function $P(X \leq x \mid A)$. What are $E(U_1 \mid U_1 > U_2)$ and $E(U_2 \mid U_1 > U_2)$ when the independent random variables U_1 and U_2 are uniformly distributed on $(0, 1)$?

7.3.1 The Regression Curve and Conditional Expectation

For two dependent random variables X and Y, let

$$m(x) = E(Y \mid X = x).$$

The curve of the function $y = m(x)$ is called the *regression curve* of Y on X. It is a better measure for the dependence between X and Y than the correlation coefficient (recall that dependence does not necessarily imply a nonzero correlation coefficient).[1] In statistical applications it is often the case that we can observe the random variable X but we want to know the dependent random variable Y. The function value $y = m(x)$ can be used as a prediction of the value of the random variable Y, given the observation x of the random variable X. The function $m(x) = E(Y \mid X = x)$ is an optimal prediction function in the sense that this function minimizes

$$E\big[(Y - g(X))^2\big]$$

over all functions $g(x)$. We only sketch the proof of this result. For any random variable U, the minimum of $E[(U-c)^2]$ over all constants c is achieved for the constant $c = E(U)$. This follows by differentiating $E[(U - c)^2] = E(U^2) - 2cE(U) + c^2$ with respect to c. Using the law of conditional expectation, $E[(Y - g(X))^2]$ can be expressed as

$$E\big[(Y - g(X))^2\big] = \int_{-\infty}^{\infty} E\big[(Y - g(X))^2 \big| X = x\big] f_X(x)\, dx.$$

For every x, the inner side of the integral is minimized by $g(x) = E(Y|X = x)$, yielding that $m(X)$ is the minimum mean-squared-error predictor of Y from X. By the law of conditional expectation, the statistic $m(X)$ has the same expected value as Y. The predictor $m(X)$ has the nice feature that $\text{var}[m(X)] \le \text{var}(Y)$. We omit the technical proof of this result. An intuitive explanation of the result is that the conditional distribution of Y given the value of X involves more information than the distribution of Y alone.

For the case that X and Y have a bivariate normal distribution with parameters $(\mu_1, \mu_2, \sigma_1^2, \sigma_2^2, \rho)$, it follows from the result in Example 7.4 that

$$m(x) = \mu_2 + \rho \frac{\sigma_2}{\sigma_1}(x - \mu_1).$$

In this case the optimal prediction function $m(x)$ coincides with the best linear prediction function discussed in Section 5.5.1. The best linear prediction function uses only the expected values, variances, and correlation coefficient of the random variables X and Y.

Example 7.10 In a classical study on the heights of fathers and their grown sons, Sir Francis Galton (1822–1911) measured the heights of 1,078 fathers and sons. Galton observed that tall fathers tend to have tall sons but the sons

---

[1] It is not generally true that $E(Y \mid X = x) = E(Y)$ for all x and $E(X \mid Y = y) = E(X)$ for all y are sufficient conditions for the independence of X and Y. This is readily verified from the example in which (X, Y) has the probability mass function $p(x, y) = \frac{1}{8}$ for $(x, y)=(1,1)$, $(1,-1)$, $(-1,1)$, $(-1,-1)$ and $p(x, y) = \frac{1}{2}$ for $(x, y)=(0,0)$.

are not, on average, as tall as the fathers. Also, short fathers have short sons who, however, are not as short on average as their fathers. Galton called this phenomenon *regression to the mean*. To explain this phenomenon, let the random variable X represent the fathers' heights in inches and Y represent the sons' heights in inches. Galton modeled the distribution of (X, Y) by a bivariate normal distribution with parameters

$$E(X) = 67.7, \ E(Y) = 68.7, \ \sigma(X) = 2.7, \ \sigma(Y) = 2.7, \ \rho(X, Y) = 0.5.$$

Using the above results, the prediction for Y given that $X = x$ is given by

$$E(Y \mid X = x) = 68.7 + \frac{0.5 \times 2.7}{2.7}(x - 67.7) = 34.85 + 0.5x.$$

The line $y = 34.85 + 0.5x$ is midway between the line $y = x + 1$ and the line $y = 68.7$. For example, if a father is very tall and has a height of 73.1 inches, his son is not predicted to be 74.1 inches but only 71.4 inches tall.

7.4 Conditional Expectation as a Computational Tool

In applied probability problems, the law of conditional expectation is a very useful result to calculate unconditional expectations. Beginning students often have difficulties in choosing the conditioning variable when they do a mathematical analysis. However, in a simulation program this "difficult" step would offer no difficulties at all and would be naturally done. Our advice to students is: if the first step in the analysis looks difficult to you, think of what you would do in a simulation program of the problem.

Example 7.11 A reliability system has two identical units, where one unit is in full operation and the other unit is in cold standby. The operating time of an active unit has an exponential density with expected value $1/\mu$. Upon failure of the operating unit, the standby unit is put into operation provided a standby unit is available. The time to replace a failed unit by a new one is fixed and is equal to $\tau > 0$. A system failure occurs if no standby unit is available at the moment the operating unit fails. What is the expected value of the time until the first system failure?

Solution. Let the random variable X denote the time until the first system failure. Also, let L_1 be the operating time of the first unit put into operation and L_2 be the operating time of the other unit. As one would do in a simulation program, we condition on the random variable L_2 to find $E(X)$. By the law of conditional expectation,

$$E(X) = E(X \mid L_2 \leq \tau)P(L_2 \leq \tau) + E(X \mid L_2 > \tau)P(L_2 > \tau).$$

Since L_1 and L_2 are independent, $E(X \mid L_2 \leq \tau) = \frac{1}{\mu} + E(L_2 \mid L_2 \leq \tau)$. Hence,

$$E(X \mid L_2 \leq \tau) = \frac{1}{\mu} + \frac{1}{P(L_2 \leq \tau)} \int_0^\tau x\mu e^{-\mu x}\, dx$$

$$= \frac{1}{\mu} + \frac{1}{\mu(1 - e^{-\mu\tau})}\left(1 - e^{-\mu\tau} - \mu\tau e^{-\mu\tau}\right).$$

If $L_2 > \tau$, then the continuation of the process beyond the moment the failed unit has been replaced is a probabilistic replica of the process beginning at time zero. This fact uses the memoryless property of the exponential distribution (explain!). Hence, letting C be distributed as X,

$$E(X \mid L_2 > \tau) = E(L_1 + \tau + C) = \frac{1}{\mu} + \tau + E(X).$$

Hence, by $E(X) = E(X \mid L_2 \leq \tau)(1 - e^{-\mu\tau}) + E(X \mid L_2 > \tau)e^{-\mu\tau}$, we get

$$E(X) = \frac{1}{\mu} + e^{-\mu\tau}[\tau + E(X)] + \frac{1}{\mu}\left(1 - e^{-\mu\tau} - \mu\tau e^{-\mu\tau}\right).$$

Next we obtain the desired result

$$E(X) = \frac{2 - e^{-\mu\tau}}{\mu(1 - e^{-\mu\tau})}.$$

The next example underlines that thinking recursively can be very rewarding for the calculation of expected values.

Example 7.12 In any drawing of the 6/45 lottery, six different numbers are chosen at random from the numbers 1 to 45. What is the expected value of the number of drawings until each of the numbers 1 to 45 has been drawn?

Solution. Define the random variable X_i as the remaining number of drawings that are needed to obtain each of the numbers 1 to 45 when i numbers are still missing. To find $E(X_{45})$, we use a recurrence relation. The conditional distribution of X_i given that the next drawing yields k missing numbers is the same as the unconditional distribution of $1 + X_{i-k}$. Hence, by the law of conditional expectation,

$$E(X_i) = \sum_{k=0}^{\min(i,6)} [1 + E(X_{i-k})] \frac{\binom{i}{k}\binom{45-i}{6-k}}{\binom{45}{6}} \quad \text{for } i = 1, 2, \ldots, 45,$$

where $E(X_0) = 0$. Using this recursion equation, we compute $E(X_i)$ for $i = 1, 2, \ldots, 45$. This results in $E(X_{45}) = 31.497$.

7.4 Conditional Expectation as a Computational Tool   285

Problems

7.49 A farming operation is located in a remote area that is more or less unreachable in the winter. As early as September, the farmer must order fuel oil for the coming winter. The amount of fuel oil he needs each winter is random, and depends on the severity of the winter weather to come. The winter will be normal with probability 2/3 and very cold with probability 1/3. The number of gallons of oil the farmer needs to get through the winter is $N(\mu_1, \sigma_1^2)$ distributed in a normal winter and $N(\mu_2, \sigma_2^2)$ distributed in a very cold winter. The farmer decides in September to stock up Q gallons of oil for the coming winter. What is the probability that he will run out of oil in the coming winter? What is the expected value of the number of gallons the farmer will come up short for the coming winter? What is the expected value of the number of gallons he will have left over at the end of the winter?

7.50 Nobel airlines has a direct flight from Amsterdam to Palermo. This particular flight uses an aircraft with $N = 150$ seats. The number of people who seek to reserve seats for a given flight has a Poisson distribution with expected value $\lambda = 170$. The airline management has decided to book up to $Q = 165$ passengers in order to protect themselves against no-shows. The probability of a booked passenger not showing up is $q = 0.07$. The booked passengers act independently of each other. What is the expected value of the number of people who show up for a given flight? What is the expected value of the number of people who show up but cannot be seated due to overbooking?

7.51 A buffer contains a geometrically distributed number of messages waiting to be transmitted over a communication channel, where the parameter p of the geometric distribution is known. Your message is one of the waiting messages. The messages are transmitted one by one in a random order. Let the random variable X be the number of messages that are transmitted before your message. What are the expected value and the standard deviation of X?

7.52 The transmission time of a message requires a geometrically distributed number of time slots, where the geometric distribution has parameter a with $0 < a < 1$. In each time slot one new message arrives with probability p and no message arrives with probability $1 - p$. What are the expected value and the standard deviation of the number of newly arriving messages during the transmission time of a message?

7.53 A fair coin is tossed no more than n times, where n is fixed in advance. You stop the coin-toss experiment when the proportion of heads exceeds $\frac{1}{2}$ or n tosses have been done, whichever occurs first. Use the law of conditional expectation to calculate, for $n = 5, 10, 25$, and 50, the expected value of the proportion of heads at the moment the coin-toss experiment is stopped. *Hint*: Define the random variable $X_k(i)$ as the proportion of heads upon stopping, given that k tosses are still possible and heads have turned up i times so far. Set up a recursion for $E[X_k(i)]$.

7.54 Consider again Problem 7.32. Use the law of conditional expectation to verify that the expected value of your waiting time until the next bus arrival is $\frac{15}{e}$.

7.55 You spin a game-board spinner in a round box whose circumference is marked with a scale from 0 to 1. When the spinner comes to rest, it points to a random number between 0 and 1. After your first spin, you have to decide whether to spin the spinner for a second time. Your payoff is $1,000 times the total score of your spins, as long as this score does not exceed 1; otherwise, your payoff is zero. What strategy maximizes the expected value of your payoff? What is the expected value of your payoff under the optimal strategy?

7.56 The following game is offered in a particular carnival tent. The carnival master has a red and a blue beaker, each containing 10 balls numbered $1, \ldots, 10$. He shakes the beakers thoroughly and picks at random one ball from each beaker. Then he tells you the value r of the ball picked from the red beaker and asks you to guess whether the unknown value b of the ball picked from the blue beaker is larger than r or smaller than r. If you guess correctly, you get b dollars. If $r = b$, you get $\frac{1}{2}b$ dollars. If you are wrong about which is larger, you get nothing. You have to pay $4.50 for the privilege of playing the game. Is this a fair game?

7.57 Use a recursion scheme to calculate the expected value of the reward obtained under the stopping rules in Problems 3.24 and 3.25, where the stopping level has the value 20 in Problem 3.24 and the value 6 in Problem 3.25.

7.58 Consider the following dice game in which you repeatedly roll two dice. The dollar reward for each roll is the dice total, provided that the total is not the same as in the previous roll; otherwise, the game is over and you lose the total reward gathered so far. You stop rolling the dice when you have gathered a total reward of $35 or more. Use a recursion scheme to find the expected value of the total reward under this stopping rule.

7.59 Consider the gambler's ruin problem given in Example 2.9. Verify that the expected duration of the game is $\frac{a}{q-p} - \frac{a+b}{q-p} \frac{1-(q/p)^a}{1-(q/p)^{a+b}}$ if $p \neq q$ and ab if $p = q$. Use this result to calculate the expected value of the number of bets in the following roulette game. Your initial bankroll is $50 and you stake each time $25 on red in European roulette. The game is stopped

7.4 Conditional Expectation as a Computational Tool 287

when you have gone broke or reached a bankroll of $250. In European roulette you get back twice your stake with probability $\frac{18}{37}$ and lose your stake with probability $\frac{19}{37}$.

7.60 Suppose c cars start in a random order along an infinitely long one-lane highway. They are all going at different but constant speeds and cannot pass each other. If a faster car ends up behind a slower car, it must slow down to the speed of the slower car. Eventually, the cars will clump up in traffic jams. Use a recursion to find the expected number of clumps of cars, where a clump may consist of a single car. *Hint*: Condition on the position of the slowest car.

7.4.1 Conditional Probability and Stochastic Optimization[2]

The law of conditional expectation is a key ingredient of the recursive approach of stochastic dynamic optimization. This approach is a computational tool for optimization problems in which a sequence of interrelated decisions must be made in order to maximize reward or minimize cost. As an example, consider the game of rolling a fair die at most five times. You may stop whenever you want, and receive as a reward the number shown on the die at the time you stop. What is the stopping rule that maximizes your expected payoff in this optimal stopping game? To answer this question, the idea is to consider a sequence of nested problems having planning horizons of increasing length. For the one-roll problem in which only one roll is permitted, the solution is trivial. You stop after the first roll and your expected payoff is $1 \times \frac{1}{6} + 2 \times \frac{1}{6} + \cdots + 6 \times \frac{1}{6} = 3.5$. In the two-roll problem, you stop after the first roll if the outcome of this roll is larger than the expected value 3.5 of the amount you get if you do not stop but continue with what is a one-roll game. Hence, in the two-roll problem, you stop if the first roll gives a 4, 5, or 6; otherwise, you continue. The expected payoff in the two-roll game is $\frac{1}{6} \times 4 + \frac{1}{6} \times 5 + \frac{1}{6} \times 6 + \frac{3}{6} \times 3.5 = 4.25$. Next consider the three-roll problem. If the first roll in the three-roll problem gives an outcome larger than 4.25, then you stop; otherwise, you do not stop and continue with what is a two-roll game. Hence, the expected payoff in the three-roll problem is $\frac{1}{6} \times 5 + \frac{1}{6} \times 6 + \frac{4}{6} \times 4.25 = 4.67$. Knowing this expected payoff, we can solve the four-roll problem. In the four-roll problem you stop after the first roll if this roll gives a 5 or 6; otherwise, you continue. The expected payoff in the four-roll problem is $\frac{1}{6} \times 5 + \frac{1}{6} \times 6 + \frac{4}{6} \times 4.6667 = 4.944$. Finally, we find the optimal strategy for the original five-roll problem. In this problem you stop after the

---

[2] This specialized section can be skipped at first reading.

first roll if this roll gives a 5 or 6; otherwise, you continue. The maximal expected payoff in the original problem is $\frac{1}{6} \times 5 + \frac{1}{6} \times 6 + \frac{4}{6} \times 4.944 = 5.129$.

The above method of backward induction is usually called the method of stochastic dynamic programming. It decomposes the original problem into a series of nested problems having planning horizons of increasing length. Each nested problem is simple to solve, and the solutions of the nested problems are linked by a recursion. The above argument can be formalized as follows. For any k with $1 \leq k \leq 5$, define the function $f_k(i)$, $i = 0, 1, \ldots, 6$ as the maximal expected payoff that can be achieved when k rolls are still allowed and the outcome of the last roll is i. This function is called the *value function*. It enables us to compute the desired maximal expected payoff $f_5(0)$ and the optimal strategy for achieving this expected payoff in the five-roll problem. This is done by applying the recursion

$$f_k(i) = \max\left\{i, \frac{1}{6}\sum_{j=1}^{6} f_{k-1}(j)\right\} \quad \text{for } i = 0, 1, \ldots, 6,$$

where k runs from 1 to 5 and the recursion is initialized with $f_0(j) = j$. The reasoning behind this recursion is as follows. Suppose the last roll showed the outcome i and $k-1$ rolls are still allowed. If you stop, your reward is i. If you continue and act optimally from the next step on, the expected reward is $\frac{1}{6}\sum_{j=1}^{6} f_{k-1}(j)$, by the law of conditional expectation. Taking the maximum of i and $\frac{1}{6}\sum_{j=1}^{6} f_{k-1}(j)$ gives $f_k(i)$.

The method of backward induction is very versatile and does not require that the outcomes of the successive experiment are independent of each other. As an example, take the following game. You take cards, one at a time, from a thoroughly shuffled deck of 26 red and 26 black cards. You may stop whenever you want and your payoff is the number of red cards drawn minus the number of black cards drawn. What is the maximal expected value of the payoff? The approach is again to decompose the original problem into a sequence of smaller nested problems. Define the value function $E(r,b)$ as the maximal expected payoff you can still achieve if r red and b black cards are left in the deck. If you stop in the state with r red and b black cards left in the deck, your reward is $26 - r - (26 - b) = b - r$. If you continue in this state and act optimally from the next step on, your expected reward is $\frac{r}{r+b}E(r-1,b) + \frac{b}{r+b}E(r,b-1)$, by the law of conditional expectation. Thus, we can establish the recursion scheme

$$E(r,b) = \max\left\{b - r, \frac{r}{r+b}E(r-1,b) + \frac{b}{r+b}E(r,b-1)\right\}.$$

The maximal expected payoff $E(26, 26)$ is obtained by "backward" calculations, starting with $E(r, 0) = 0$ and $E(0, b) = b$. It has the value

$E(26, 26) = 2.6245$. The optimal decisions in the various states can be summarized through threshold values β_k: stop if the number of red cards drawn minus the number of black cards drawn is β_k or more after the kth draw; otherwise, continue. The numerical values of the β_k are $\beta_1 = 2$, $\beta_2 = 3$, $\beta_3 = 4$, $\beta_4 = 5$, $\beta_5 = 6$, $\beta_6 = 5$, $\beta_7 = 6$, $\beta_8 = 7$, $\beta_9 = 6$, $\beta_{2m} = 5$, and $\beta_{2m+1} = 4$ for $5 \leq m \leq 11$, $\beta_{2m} = 3$ and $\beta_{2m+1} = 4$ for $12 \leq m \leq 16$, $\beta_{2m} = 3$ and $\beta_{2m+1} = 2$ for $17 \leq m \leq 21$, $\beta_{44} = 1$, $\beta_{45} = 2$, $\beta_{46} = 1$, $\beta_{47} = 2$, $\beta_{48} = 1$, $\beta_{49} = 0$, $\beta_{50} = 1$, and $\beta_{51} = 0$. In the problems below, we give several other problems that can be solved by the method of backward induction.

Problems

7.61 Consider again Problem 7.53, but assume now that you want to find the expected payoff of an optimal stopping rule. What is the maximal expected payoff for $n = 25, 50, 100$, and $1,000$?

7.62 Consider a game of rolling two fair dice at most six times. You may stop whenever you want and your payoff is the score of the two dice upon stopping. Use the method of backward induction to find the maximal expected payoff.

7.63 Suppose that somebody chooses at random n different numbers from the numbers 1 to N, where n and N are given with $n < N$. Each of the chosen numbers is written on a slip of paper and the n slips of paper are put into a hat. This happens all unseen to you. Next you are asked to take the slips of paper out of the hat, one by one. You stop picking numbers when you come to a number that you guess to be the largest of the numbers in the hat. You cannot go back to a previously picked number. What is the optimality equation when you want to maximize the probability of picking the largest number in the hat? Solve the optimality equation for $n = 10$ and $N = 100$.

7.64 Consider the following modification of Problem 7.63. The n numbers on the slips of paper need not be different and are randomly chosen from the numbers 1 to N, independently of each other. Give the optimality equation when the goal is to maximize the probability of picking the largest number in the hat and solve this equation for $n = 10$ and $N = 100$.

7.65 A game machine can be used to drop balls into bins. The balls are dropped one at a time and any ball will land at random into one of b bins. You can stop dropping balls whenever you wish. At the end of the game you win \$1 for every bin with exactly one ball and you lose \$0.50 for every bin with two or more balls. Empty bins do not count.

(a) Give the optimality equation for the computation of the optimal stopping rule and the maximal expected value of your net winnings.

(b) Take $b = 25$ and calculate the maximal expected value of your net winnings. Compare the expected value of your net winnings under the optimal stopping rule with the expected value of your net winnings under the one-stage-look-ahead rule from Problem 3.26.

7.66 Consider the following dice game in which you repeatedly roll two dice. The dollar reward for each roll is the dice total, provided that this total is different from the total in the previous roll; otherwise, the game is over and you lose the total reward gathered so far. You can stop rolling the dice whenever you wish. You have decided beforehand to stop anyway when your total reward is more than M, where M is a given (large) number. Formulate the optimality equation for the computation of an optimal stopping rule when the goal is to maximize the expected value of the total reward.

7.4.2 Conditioning and the Poissonization Trick[3]

In this section we discuss a beautiful trick that goes back to the famous probabilist A. N. Kolmogorov. This trick, which is almost like magic, allows you to answer all kinds of tough-looking questions. The key to the Poissonization trick is the so-called splitting property of the Poisson process. Recall from Section 4.5.1 that the Poisson process is a stochastic process in which events occur one at a time and the interoccurrence times of the events are independent random variables having a common exponential distribution. Rule 4.3 explains the name of the Poisson process: the number of events occurring in any given time interval has a Poisson distribution. Let $1/\lambda$ be the expected time between the occurrences of two consecutive events. In other words, the events occur at rate λ. Suppose that, at the occurrence of an event, this event is classified as a type-i event with probability p_i for $i = 1, 2, \ldots, n$ with $\sum_{i=1}^{n} p_i = 1$, independently of the other events. This classification results in n subprocesses of events, each subprocess having events of a specific classification. The reader is asked to take for granted that these subprocesses are *independent* Poisson processes with respective rates $\lambda p_1, \lambda p_2, \ldots, \lambda p_n$. The independence property of the subprocesses is remarkable and is crucial in the Poissonization trick. An intuitive explanation of the independence property is provided by the splitting result of Rule 3.13, see also the discussion of the Poisson process in Section 11.1.1

---

[3] This specialized section can be skipped without loss of continuity.

7.4 Conditional Expectation as a Computational Tool

The Poissonization trick has the feature of creating independence between random variables in the context of an imaginary Poisson process. This will be explained in the following two examples.

Example 7.13 In the casino game of craps, two dice are rolled repeatedly. What is the probability that a total of 2, 3, 4, 5, 6, 8, 9, 10, 11, and 12 will be rolled before rolling a 7?

Solution. The problem is transformed into a problem in which the rolls of the two dice occur at times generated by a Poisson process with rate $\lambda = 1$. The sought probability in the original model where the rolls occur at fixed times is the same as in the model where the rolls occur at times generated by a Poisson process. Each roll results in a total of j with probability p_j, where $p_j = \frac{j-1}{36}$ for $1 \le j \le 7$ and $p_j = p_{14-j}$ for $8 \le j \le 12$. Then, for any fixed j, the rolls resulting in a total of j occur according to a Poisson process with rate $\lambda p_j = p_j$. Moreover, and very importantly, these subprocesses are independent of each other. The analysis now proceeds as follows. Fix t and suppose that a roll with a total of 7 occurs for the first time at time t. For any $j \ne 7$, the probability that at least one roll with a total of j will occur in $(0, t)$ is $1 - e^{-p_j t}$. Hence, by the independence of the separate Poisson processes, the product

$$(1 - e^{-p_2 t}) \cdots (1 - e^{-p_6 t})(1 - e^{-p_8 t}) \cdots (1 - e^{-p_{12} t})$$

represents the probability of a total of 2, 3, 4, 5, 6, 8, 9, 10, 11, and 12 showing up in $(0, t)$. The probability density of the time until a total of 7 will occur for the first time is equal to $p_7 e^{-p_7 t}$. Hence, by the law of conditional probability, the sought probability is given by

$$\int_0^\infty (1 - e^{-p_2 t}) \cdots (1 - e^{-p_6 t})(1 - e^{-p_8 t}) \cdots (1 - e^{-p_{12} t}) p_7 e^{-p_7 t} \, dt.$$

Numerical integration gives the value 0.005258 for the probability that a total of 2, 3, 4, 5, 6, 8, 9, 10, 11, and 12 will be rolled before rolling a 7. In some casinos the payout for this craps side bet is \$175 for a \$1 bet. This corresponds to a house edge of 7.99%. Another craps side bet is the following. You win if some total is rolled twice or more before a 7. A win pays 2 for 1. You are asked to verify that your win probability is

$$1 - \int_0^\infty (e^{-p_2 t} + p_2 t e^{-p_2 t}) \cdots (e^{-p_6 t} + p_6 t e^{-p_6 t})$$
$$\times (e^{-p_8 t} + p_8 t e^{-p_2 t}) \cdots (e^{-p_{12} t} + p_{12} t e^{-p_{12} t}) p_7 e^{-p_7 t} \, dt = 0.47107.$$

This corresponds to a house edge of 5.79%.

The Poissonization trick may also be used to calculate expected values. Before showing this, we state a useful result for expected values. For any nonnegative random variable T,

$$E(T) = \int_0^\infty P(T > t)\,dt.$$

The reader was asked to prove this result in Problem 4.26. A probabilistic proof is as follows. For any $t > 0$, let the indicator variable $I_t = 1$ if $T > t$ and $I_t = 0$ otherwise. Then $T = \int_0^\infty I_t\,dt$ and so $E(T) = E(\int_0^\infty I_t\,dt)$. Since I_t is nonnegative, it is allowed to interchange the order of expectation and integration. Noting that $E(I_t) = 1 \times P(I_t = 1) + 0 \times P(I_t = 0)$ and $P(I_t = 1) = P(T > t)$, the result for $E(T)$ follows next.

In the next example, we apply the Poissonization trick to an instance of the coupon-collector problem with unequal probabilities.

Example 7.14 Two dice are rolled repeatedly. What is the expected value of the number of rolls until each of the dice totals $2, 3, \ldots, 12$ has appeared?

Solution. As in Example 7.13, imagine that the rolls of the two dice occur at epochs generated by a Poisson process with rate $\lambda = 1$. In the transformed model, let N be the number of rolls until each possible dice total has appeared. To find $E(N)$, define T as the first epoch at which each possible dice total has occurred. Then $T = X_1 + X_2 + \cdots + X_N$, where the X_i are the interoccurrence times of the Poisson process generating the rolls. The X_i are independent random variables having an exponential density with expected value $1/\lambda = 1$. We have $E(T) = E(N)E(X_1)$, as can be seen by using Wald's equation from Problem 3.61. Since $E(X_1) = 1$, we get

$$E(N) = E(T).$$

To find $E(T)$, note that $T = \max_{2 \le j \le 12} T_j$, where T_j is the first epoch at which the Poisson process generates a roll with a dice total of j. The rolls resulting in a dice total of j occur according to a Poisson process with rate $\lambda p_j = p_j$, where $p_j = \frac{j-1}{36}$ for $1 \le j \le 7$ and $p_j = p_{14-j}$ for $8 \le j \le 12$. Moreover, these Poisson subprocesses are independent of each other and so the random variables T_j are independent of each other. By $P(T \le t) = P(T_j \le t)$ for $2 \le j \le 12$, we get $P(T \le t) = P(T_2 \le t) \cdots P(T_{12} \le t)$, using the independence of the T_j. Since $P(T_j \le t) = 1 - P(T_j > t)$ and $P(T_j > t)$ is nothing other than the probability $e^{-p_j t}$ that the Poisson subprocess with rate p_j has no event in $(0, t)$, we get

$$P(T \le t) = (1 - e^{-p_2 t}) \times (1 - e^{-p_3 t}) \times \cdots \times (1 - e^{-p_{12} t}) \quad \text{for } t > 0.$$

7.4 Conditional Expectation as a Computational Tool

By the relations $E(T) = \int_0^\infty P(T > t)\,dt$ and $E(T) = E(N)$, we obtain

$$E(N) = \int_0^\infty \left[1 - (1 - e^{-p_2 t}) \times (1 - e^{-p_3 t}) \times \cdots \times (1 - e^{-p_{12} t})\right] dt,$$

yielding the desired expression for the expected value of the number of rolls until each of the possible dice totals has appeared. Using numerical integration, we find the numerical value 61.217 for $E(N)$.

Problems

7.67 Suppose that balls are placed sequentially into one of b bins, where the bin for each ball is selected at random. Use the Poissonization trick to find the probability that none of the bins contains more than m balls at the moment each bin has received one or more balls. *Hint*: Let A_i be the event that the ith bin is the last bin to receive its first ball and none of the other bins contains more than m balls when this happens.

7.68 A beer company brings a new beer with the brand name Babarras to the market and prints one of the letters of this brand name underneath each bottle cap. Each of the letters A, B, R, and S must be collected a certain number of times in order to get a nice beer glass. The quotas for the letters A, B, R, and S are 3, 2, 2, and 1. Underneath the cap of each bottle the letter A appears with probability 0.15, the letter B with probability 0.10, the letter R with probability 0.40, and the letter S with probability 0.35, independently of the other bottles. What is the expected number of bottles that must be purchased in order to form the word Babarras?

7.69 In the game of solitaire Knock 'm Down, you receive 12 tokens and a card with 11 sections numbered $2, \ldots, 12$. You place the tokens on the sections according to $(a_2, \ldots, a_{12}) = (0, 0, 1, 1, 2, 3, 2, 2, 1, 0, 0)$, where a_k is the number of tokens allocated to section k. Then two dice are rolled. If the dice total is k, you remove one token from section k when that section is not empty. The two dice are rolled repeatedly until you have removed all tokens. Use the Poissonization trick to find the expected value of the number of rolls needed to remove all tokens.

7.70 In the coupon-collector problem, you receive a coupon each time you purchase a certain product, where the coupon is equally likely to be any one of n types. Use the Poissonization trick to find an explicit expression for the expected number of purchases needed to collect two complete sets of coupons. Verify by numerical integration that this expression has the value 24.134 when $n = 6$ (fair die).

7.71 A fair die is rolled repeatedly. What is the probability that each of the even numbers will show up before an odd number appears?

7.72 A bag contains 5 red balls, 10 green balls, and 20 blue balls. The balls are drawn out of the bag at random, one by one and without replacement. What is the probability of having drawn at least one green ball and at least one blue ball before drawing the first red ball. *Hint*: The sought probability does not change when it is assumed that the balls are drawn with replacement rather than without replacement.

7.73 Consider the following horse-race dice game with six players. Each player has its own horse. The horses are labeled $1, 2, \ldots, 6$. Each horse has its own racing line consisting of seven panels numbered from 0 to 6. The finish line is in panel 6 and the first horse reaching this panel wins. Each time a fair die is rolled. After each roll the horse whose label is equal to the rolled number is moved one panel ahead. Suppose that the starting position of the horse with label k is s_k, where $s_1 = 0$, $s_2 = 1$, $s_3 = 2$, $s_4 = 2$, $s_5 = 1$, and $s_6 = 0$. Use the Poissonization trick to find the win probabilities of the horses. What is the expected duration of the game when the starting position of each horse is panel 0?

7.74 Three players A, B, and C play a series of games until one of the players has won five points. Each game counts for one point and is won by player A with probability $\frac{4}{9}$, by player B with probability $\frac{3}{9}$, and by player C with probability $\frac{2}{9}$. Since the players are not equally strong, player A starts with 0 points, player B with 1 point, and player C with 2 points. Use the Poissonization trick to find the win probabilities of the players and the expected duration of the game.

7.75 A deck of cards consists of 20 different cards, divided into four suits of five cards each. Each time one card is randomly drawn from the deck. The drawn card is put back into the deck and the process repeats itself. What is the expected number of picks until each card of some of the suits has been obtained?

7.76 Suppose that balls are placed sequentially into one of b bins, where the bin for each ball is selected at random. Verify that the expected number of balls needed until each bin contains at least m balls is given by $b \int_0^\infty \left[1 - \left(1 - \sum_{k=0}^{m-1} e^{-u} \frac{u^k}{k!} \right)^b \right] du$.

7.5 Bayesian Inference – Continuous Case

Bayesian statistics takes a fundamentally different viewpoint than classical statistics. Where the frequentist view of probability and of statistical inference is based on the idea of a random experiment that can be repeated very often, the Bayesian view of probability and of statistical inference is based on personal assessments of probability and on data of a few performances of a random experiment. In practice, many statisticians use ideas from both methods. Unlike the frequentist approach, the Bayesian approach to statistical inference treats population parameters not as fixed, unknown constants but as random variables – subject to change as additional data arise. Probability distributions have to be assigned to these parameters by the investigator before observing data. These distributions are called prior distributions and are inherently subjective. They represent the investigator's uncertainty about the true value of the unknown parameters. Assigning probabilities by degree of belief is consistent with the idea of a fixed but unknown value of the parameter. Hence, in the Bayesian approach, you first specify a prior distribution for the population parameters and you revise your beliefs about the population parameters by learning from data that were collected. The so-obtained posterior probability distribution reflects your new beliefs about the population parameters. Bayes' rule is the recipe for computing the posterior distribution based on the prior distribution and the observed data. How this rule works was discussed in Section 2.5 for the case of a discrete prior distribution $\{p_0(\theta_i), i = 1, \ldots, n\}$ for the population parameter Θ. Letting the likelihood $L(x \mid \theta)$ denote the conditional probability of obtaining the data x given the value θ of the parameter Θ, the prior distribution is updated according to Bayes' rule:

$$p(\theta_i \mid x) = \frac{L(x \mid \theta_i)p_0(\theta_i)}{\sum_{k=1}^{n} L(x \mid \theta_k)p_0(\theta_k)} \quad \text{for } i = 1, \ldots, n.$$

This is how it all works for discrete problems. In Section 2.5, we have given several examples of the Bayesian approach for the discrete model and we discussed differences between Bayesian and frequentist inference. In the second half of the twentieth century there was much controversy between the two schools of thought, but nowadays many statisticians use a combination of Bayesian and frequentist reasoning.

Bayes' Rule for the Continuous Case

What is the formulation of Bayes' rule for the continuous case? For that purpose we go back to results in Section 7.1.2. Let X and Y be two continuous random variables with joint density $f(x, y)$ and marginal densities $f_X(x)$ and $f_Y(y)$. Denote by $f_X(x \mid y)$ the conditional density of X given $Y = y$ and by $f_Y(y \mid x)$ the conditional density of Y given $X = x$. Then,

$$f_Y(y \mid x) = \frac{f_X(x \mid y)f_Y(y)}{\int f_X(x \mid y')f_Y(y')\,dy'} \quad \text{for all } y.$$

To explain this continuous version of Bayes' rule, we use the relations

$$f(x,y) = f_Y(y \mid x) f_X(x) \text{ and } f(x,y) = f_X(x \mid y) f_Y(y).$$

These relations imply that $f_Y(y \mid x) = f_X(x \mid y) f_Y(y)/f_X(x)$. Noting that $f_X(x) = \int f_X(x \mid y') f_Y(y') \, dy'$, the Bayes' formula follows.

In continuous Bayesian problems, the hypotheses are represented by a continuous parameter Θ having a prior density $f_0(\theta)$. Let the likelihood function $L(x \mid \theta)$ represent the conditional density of the data x given θ. Then, given data x, the prior density of Θ is updated to the posterior density

$$f(\theta \mid x) = \frac{L(x \mid \theta) f_0(\theta)}{\int_\Theta L(x \mid \theta') f_0(\theta') \, d\theta'} \quad \text{for all } \theta.$$

The denominator of this formula is not a function of θ, but it is a constant for fixed x. In the literature, the Bayes' formula is often written as

$$f(\theta \mid x) \propto L(x \mid \theta) f_0(\theta) \quad \text{for all } \theta,$$

where the symbol $\propto$ stands for proportionality and the proportionality constant is such that the conditional density $f(\theta \mid x)$ integrates to one for fixed x. In the case that the elements of Θ are high-dimensional vectors, it is usually not computationally feasible to find the proportionality constant by numerical integration. Also, in the high-dimensional case, it will typically be very difficult to obtain the marginal density of any component of Θ by numerical integration. One then has to resort to Markov chain Monte Carlo simulation to find the posterior density. This method has increased enormously the applicability of Bayesian analysis and will be discussed in Chapter 10.

Bayes' Rule for the Mixed Case

In mixed Bayesian problems the update of the prior density $f_0(\theta)$ is based on the discrete outcome E of a random experiment rather than on an outcome of a continuous variable. The adaptation of Bayes' formula is obvious. Letting the likelihood $L(E \mid \theta)$ denote the conditional probability of the outcome E given θ, the prior density is updated according to

$$f(\theta \mid E) \propto L(E \mid \theta) f_0(\theta) \quad \text{for all } \theta,$$

where the proportionality constant is such that $f(\theta \mid E)$ integrates to 1. The motivation of this definition of the posterior density $f(\theta \mid E)$ proceeds along similar lines as the motivation of Definition 7.1.

To illustrate, suppose that the prior density of the success probability in a Bernoulli experiment is given by a beta(α, β) density and that n independent repetitions of the Bernoulli experiment have led to s successes and $r =$

7.5 Bayesian Inference – Continuous Case

$n - s$ failures. Then the above relation gives that the posterior density is proportional to $\theta^{\alpha-1}(1-\theta)^{\beta-1}\theta^{s}(1-\theta)^{r} = \theta^{\alpha+s-1}(1-\theta)^{\beta+r-1}$ and thus has a beta$(\alpha+s, \beta+r)$ density. If the prior density and the posterior density have the same distributional form, then the prior and posterior are called conjugate distributions, and the prior is called a *conjugate prior* for the likelihood function.

In the next example we give an application of a mixed Bayesian problem. This example is a continuous version of Problem 2.85 and illustrates that Bayesian analysis is particularly useful for less repeatable experiments.

Example 7.15 You wonder who is the better player of the tennis players Alassi and Bicker. You have a continuous prior density $f_0(\theta)$ for the probability θ that Alassi is the better player. Then you learn about a tournament at which a best-of-five series of matches is played between Alassi and Bicker over a number of days. In such an encounter, the first player to win three matches is the overall winner. It turns out that Alassi wins the best-of-five contest. How should you update your prior density $f_0(\theta)$?

Solution. The prior density $f_0(\theta)$ expresses your uncertainty about the probability θ of Alassi being the better player. A typical approach for overcoming concerns about the subjective nature of prior densities is to consider several choices for the prior density. We will consider the following two choices for the prior density $f_0(\theta)$:

Case A. $f_0(\theta)$ is the uniform density on (0.4, 0.6), that is,

$$f_0(\theta) = 5 \quad \text{for } 0.4 < \theta < 0.6.$$

Case B. $f_0(\theta)$ is the triangular density on (0.4, 0.6) with $\theta = 0.5$ as its mode, that is,

$$f_0(\theta) = \begin{cases} 100(\theta - 0.4) & \text{for } 0.4 < \theta \leq 0.5 \\ 100(0.6 - \theta) & \text{for } 0.5 < \theta < 0.6. \end{cases}$$

How should you adjust the prior when you have learned about the outcome E that Alassi has won the best-of-five contest? For that we need the conditional probability $L(E \mid \theta)$ that Alassi is the first player to win three matches in the contest when the probability of Alassi winning any given match has the fixed value θ. Assuming independence between the matches, it is easily verified that

$$L(E \mid \theta) = \theta^3 + \binom{3}{2}\theta^2(1-\theta)\theta + \binom{4}{2}\theta^2(1-\theta)^2\theta.$$

On the basis of the information that Alassi has won the best-of-five contest, the prior density $f_0(\theta)$ is updated to $f(\theta \mid E)$. The posterior density $f(\theta \mid E)$ is proportional to $L(E \mid \theta)f_0(\theta)$. In this example the proportionality constant is easily determined by integration. The proportionality constant has the value 2 for both cases A and B. The posterior density is now completely specified. We can now answer the question of how the prior probability of Alassi being the better player changes when we learn that Alassi has won the best-of-five contest. The prior probability that Alassi is the better player is equal to 0.5 for both cases A and B. The posterior probability of Alassi being the better player is given by

$$\int_{0.5}^{0.6} f(\theta \mid E)\,d\theta = 2 \int_{0.5}^{0.6} L(E \mid \theta) f_0(\theta)\,d\theta.$$

This posterior probability has the value 0.5925 for case A and the value 0.5620 for case B.

Suppose you are not only told that Alassi won the best-of-five contest, but also informed that Alassi won this contest by winning the first three matches. Then the formula for $L(E \mid \theta)$ becomes $L(E \mid \theta) = \theta^3$ and the calculations yield that the posterior probability of Alassi being the better player has the value 0.6452 for case A and the value 0.5984 for case B.

In physics, a fundamental problem is a probabilistic assessment of the true value of a physical quantity from experimental data. A question such as "what is the probability that the true value of the mass of the top quark is between two specified bounds?" is a natural question for physicists. The next example shows how useful Bayesian analysis can be for the probabilistic assessment of the true value of a physical quantity. Also, in this example we encounter the concept of a Bayesian credible interval. Such an interval is a probability interval for the true value of an unknown quantity.

Example 7.16 You want to know more precisely the true weight of a particular steel beam. The uniform density on (71, 75) is assumed as the prior density of the weight of the beam, where the weight is measured in kilograms. Your steel beam is weighed several times on a scale. The scale is known to give results that are normally distributed without bias but with a standard deviation of $\sqrt{2}$ kg. The 10 measurements of the weighing process are 72.883, 71.145, 73.677, 73.907, 71.879, 75.184, 75.182, 73.447, 73.963, and 73.747 kg. How should you update the prior density of the weight of your steel beam?

Solution. The prior density $f_0(\theta)$ of the weight of the beam is given by $f_0(\theta) = 0.25$ for $71 < \theta < 75$. For ease of notation, denote the 10 measurements of the weight by $x_1, \ldots, x_{10}$ and let $\mathbf{x} = (x_1, \ldots, x_{10})$. The likelihood density function $L(\mathbf{x} \mid \theta)$ is given by the product $h(x_1 \mid \theta) \times \cdots \times h(x_{10} \mid \theta)$, where $h(x \mid \theta) = \frac{1}{2\sqrt{\pi}} e^{-\frac{1}{2}(x-\theta)^2/2}$. This leads to

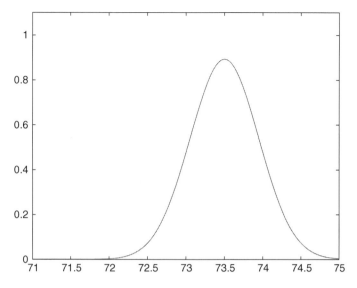

Figure 7.1. The posterior density of the weight of the beam.

$$L(\mathbf{x} \mid \theta) = (2\sqrt{\pi})^{-10} e^{-\frac{1}{2} \sum_{i=1}^{10} (x_i - \theta)^2 / 2} \quad \text{for all } \theta.$$

The posterior density $f(\theta \mid \mathbf{x})$ is proportional to $L(\mathbf{x} \mid \theta) f_0(\theta)$. Using numerical integration, we find the value 4.407×10^7 for the proportionality constant. This completes the specification of the posterior density $f(\theta \mid \mathbf{x})$ of the weight of the beam. The graph of $f(\theta \mid \mathbf{x})$ is displayed in Figure 7.1. Setting the derivative of $f(\theta \mid \mathbf{x})$ equal to zero yields that the posterior density reaches its maximum at $\theta = (1/10) \sum_{i=1}^{10} x_i = 73.501$. The mode of the posterior density is called the Bayesian maximum a posteriori estimate. Next we determine an interval which contains 95% (say) of the area under the curve of the posterior density. A *Bayesian 95% credible interval* for θ is any interval (θ_1, θ_2) for which

$$\int_{\theta_1}^{\theta_2} f(\theta \mid \mathbf{x}) \, d\theta = 0.95.$$

Among all such intervals, one might choose the one determined by the 0.025 percentile and the 0.975 percentile of $f(\theta \mid \mathbf{x})$. In other words, for $p = 0.025$ and $p = 0.975$, one numerically solves for α in the equation

$$\int_{71}^{\alpha} f(\theta \mid \mathbf{x}) \, d\theta = p.$$

This gives the 0.025 percentile 72.625 and the 0.975 percentile 74.375, and so $(72.625, 74.375)$ is a Bayesian 95% credible interval for the weight of the beam. The Bayesian credible interval tells you that you believe the true weight of the beam lies between 72.625 and 74.375 with probability 0.95. This is an interpretation which is in accordance with how most people would interpret a confidence interval. A frequentist confidence interval cannot be interpreted in this way, but refers to the statistical properties of the way the interval is constructed, see Appendix D.

Problems

7.77 Consider again Example 7.16, but assume now that the prior density of the true weight of the beam is an $N(\mu_0, \sigma_0^2)$ density. Prove that the posterior density of the weight is also a normal density. What is a Bayesian 95% credible interval for the true weight of the beam and what value of θ maximizes the posterior density when $\mu_0 = 73$ kg and $\sigma_0 = 0.7$ kg? *Hint*: Verify that the likelihood $L(\mathbf{x} \mid \theta)$ is proportional to $e^{-\frac{1}{2}n(\theta-\bar{x})^2/\sigma^2}$ for all θ, where $n = 10$, $\sigma = \sqrt{2}$, and $\bar{x} = (1/n)\sum_{i=1}^{n} x_i$.

7.78 A job applicant will have an IQ test. The prior density of his IQ is the $N(100, 125)$ density. If the true value of the IQ is x, then the test will result in a score that is $N(x, 56.25)$ distributed. The test gives a score of 123 points. What is the posterior density $f(\theta \mid \text{data})$ of the IQ of the job applicant? What is a Bayesian 95% credible interval for θ?

7.79 An astronomer will perform an experiment to measure the distance to a star. The error in the measurement of the true value of the distance is given by an $N(0, \sigma^2)$ density with $\sigma = 20$ light years. The astronomer's prior belief about the true value of the distance is described by a normal distribution with an expected value of 150 light years and a standard deviation of 25 light years. The measurement gives a distance of 140 light years. Verify that the posterior density is a normal density. What is the Bayesian maximum a posteriori probability estimate for the distance of the star and what is a Bayesian 95% credible interval for the distance?

7.80 A pollster conducts a poll attempting to predict whether the Liberal Party or the National Party will win the coming election in a country with two parties. On the basis of a previous poll, the pollster's prior belief about the proportion of Liberal voters is a beta(a, b) density with $a = 474$ and $b = 501$. In the new poll, 527 voters are for the Liberal Party and 573 voters are for the National Party. What is a Bayesian 95% credible interval for the proportion of Liberal voters? What is the posterior probability that the Liberal Party will win the election, assuming that the poll is representative of the population at the time of the election?

7.81 The lifetime of a light bulb has an exponential density. The parameter of this density is a fixed but unknown constant. Your prior belief about the true parameter value is described by a gamma density with shape parameter α and scale parameter λ. You decide to observe m light bulbs over the time interval $(0, T)$. It appears that r of the m light bulbs have failed at the respective times $t_1, \ldots, t_r$, while the other $m - r$ light bulbs are still functioning at time T. What is the posterior density of the true parameter value of the lifetime density?

8
Generating Functions

Generating functions were introduced by the Swiss genius Leonhard Euler (1707–1783) in the eighteenth century to facilitate calculations in counting problems. However, this important concept is also extremely useful in applied probability, as was first demonstrated by the work of Abraham de Moivre (1667–1754), who discovered the technique of generating functions independently of Euler. In modern probability theory, generating functions are an indispensable tool in combination with methods from numerical analysis.

The purpose of this chapter is to give the basic properties of generating functions and to show the utility of this concept. First, the generating function is defined for a discrete random variable on nonnegative integers. Next, we consider the more general moment-generating function, which is defined for any random variable. The generating function and the moment-generating function provide a powerful tool for both theoretical and computational purposes. In particular, this tool can be used to prove the central limit theorem. A sketch of the proof will be given in Section 8.4.

8.1 Generating Functions

We first introduce the concept of a generating function for a discrete random variable X whose possible values belong to the set of nonnegative integers.

Definition 8.1 *If X is a nonnegative, integer-valued random variable, then the generating function of X is defined by*

$$G_X(z) = \sum_{k=0}^{\infty} z^k P(X = k) \quad \text{for } |z| \leq 1.$$

The power series $G_X(z)$ is absolutely convergent for any $|z| \leq 1$ (why?). For any z, we can interpret $G_X(z)$ as

$$G_X(z) = E(z^X),$$

as follows by applying the substitution rule. The probability mass function of X is uniquely determined by the generating function of X. To see this, use the fact that the derivative of an infinite series is obtained by differentiating the series term by term. Hence,

$$\frac{d^r}{dz^r}G_X(z) = \sum_{k=r}^{\infty} k(k-1)\cdots(k-r+1)z^{k-r}P(X=k) \quad \text{for } r = 1, 2, \ldots.$$

In particular, by taking $z = 0$,

$$P(X = r) = \frac{1}{r!}\frac{d^r}{dz^r}G_X(z)\bigg|_{z=0} \quad \text{for } r = 1, 2, \ldots.$$

This proves that the generating function of X *uniquely* determines the probability mass function of X. This basic result explains the importance of the generating function. In many applications, it is relatively easy to obtain the generating function of a random variable X, even when the probability mass function is not explicitly given. Several examples will be given in the next section. Once we know the generating function of a random variable X, it is a simple matter to obtain the factorial moments of the random variable X. The rth factorial moment of the random variable X is defined by $E[X(X-1)\cdots(X-r+1)]$ for $r = 1, 2, \ldots$. In particular, the first factorial moment of X is the expected value of X. The variance of X is determined by the first and second factorial moments of X. Putting $z = 1$ in the above expression for the rth derivative of $G_X(z)$, we obtain

$$E[X(X-1)\cdots(X-r+1)] = \frac{d^r}{dz^r}G_X(z)\bigg|_{z=1} \quad \text{for } r = 1, 2, \ldots.$$

In particular, we find the following result.

Rule 8.1 Let $G_X(z)$ be the generating function of a nonnegative, integer-valued random variable X. Then,

$$E(X) = G'_X(1) \text{ and } E\left(X^2\right) = G''_X(1) + G'_X(1).$$

Example 8.1 Suppose that the random variable X has a Poisson distribution with expected value λ. Verify that the generating function of X is given by

$$G_X(z) = e^{-\lambda(1-z)} \quad \text{for } |z| \leq 1.$$

What are the expected value and the standard deviation of X?

Solution. Applying the definition of a generating function and using the series expansion $e^x = \sum_{n=0}^{\infty} x^n/n!$, we find

$$\sum_{k=0}^{\infty} z^k e^{-\lambda} \frac{\lambda^k}{k!} = e^{-\lambda} \sum_{k=0}^{\infty} \frac{(\lambda z)^k}{k!} = e^{-\lambda} e^{\lambda z},$$

which verifies that $G_X(z) = e^{-\lambda(1-z)}$. Differentiating $G_X(z)$ gives $G'_X(1) = \lambda$ and $G''_X(1) = \lambda^2$. Thus, $E(X) = \lambda$ and $E(X^2) = \lambda^2 + \lambda$. This implies that both the expected value and the variance of a Poisson-distributed random variable with parameter λ are given by λ, in agreement with earlier results in Example 3.12.

8.1.1 Convolution Rule

The importance of the concept of a generating function comes up especially when calculating the probability mass function of a sum of independent random variables that are nonnegative and integer valued.

Rule 8.2 *Let X and Y be two nonnegative, integer-valued random variables. If the random variables X and Y are independent, then*

$$G_{X+Y}(z) = G_X(z) G_Y(z) \quad \text{for } |z| \leq 1.$$

Rule 8.2 is known as the *convolution rule* for generating functions and can be extended directly to the case of a finite sum of independent random variables. The proof is simple. If X and Y are independent, then the random variables $U = z^X$ and $V = z^Y$ are independent for any fixed z, see Rule 3.6. Also, by Rule 3.8, $E(UV) = E(U)E(V)$ for independent U and V. Thus,

$$E(z^{X+Y}) = E(z^X z^Y) = E(z^X) E(z^Y),$$

proving that $G_{X+Y}(z) = G_X(z) G_Y(z)$. The converse of the statement in Rule 8.2 is, in general, not true. The random variables X and Y are not necessarily independent if $G_{X+Y}(z) = G_X(z) G_Y(z)$. It is left to the reader to verify that a counterexample is provided by the random vector (X, Y) that takes on the values $(1, 1)$, $(2, 2)$, and $(3, 3)$ each with probability $\frac{1}{9}$ and the values $(1, 2)$, $(2, 3)$, and $(3, 1)$ each with probability $\frac{2}{9}$.

Example 8.2 Suppose that X and Y are independent random variables that are Poisson distributed with respective parameters λ and μ. What is the probability mass function of $X + Y$?

Solution. Since X and Y are independent, it follows from Rule 8.2 that $G_{X+Y}(z) = G_X(z) G_Y(z)$. Using the result from Example 8.1, we then get

$$G_{X+Y}(z) = e^{-\lambda(1-z)}e^{-\mu(1-z)} = e^{-(\lambda+\mu)(1-z)} \quad \text{for } |z| \leq 1.$$

Since a Poisson-distributed random variable with parameter $\lambda + \mu$ has the generating function $e^{-(\lambda+\mu)(1-z)}$ and the generating function $G_{X+Y}(z)$ uniquely determines the probability mass function of $X + Y$, it follows that $X + Y$ has a Poisson distribution with parameter $\lambda + \mu$.

A nice application of the convolution rule for generating functions is the calculation of the probability mass function of the total score of a roll of d dice. For $d = 2$, this probability mass function is readily calculated by hand using the sample space of the equally likely outcomes of a roll of two dice. This manual approach can also be taken for a roll of $d = 3$ dice, but the calculations become quite tedious. However, for larger values of d, it is much easier to get the probability mass function by the generating function approach. We illustrate this for $d = 3$. Let X_1, X_2, and X_3 denote the three face values resulting from a roll of the three dice. Then each X_i has the generating function $\frac{1}{6}z + \frac{1}{6}z^2 + \frac{1}{6}z^3 + \frac{1}{6}z^4 + \frac{1}{6}z^5 + \frac{1}{6}z^6$. By the convolution rule, the total score $S = X_1 + X_2 + X_3$ has the generating function $G_S(z) = (\frac{1}{6}z + \frac{1}{6}z^2 + \frac{1}{6}z^3 + \frac{1}{6}z^4 + \frac{1}{6}z^5 + \frac{1}{6}z^6)^3$. Using standard mathematical software to expand powers of polynomials, we find that

$$G_S(z) = \frac{1}{216}z^3 + \frac{1}{72}z^4 + \frac{1}{36}z^5 + \frac{5}{108}z^6 + \frac{5}{72}z^7 + \frac{7}{72}z^8$$
$$+ \frac{25}{216}z^9 + \frac{1}{8}z^{10} + \frac{1}{8}z^{11} + \frac{25}{216}z^{12} + \frac{7}{72}z^{13}$$
$$+ \frac{5}{72}z^{14} + \frac{5}{108}z^{15} + \frac{1}{36}z^{16} + \frac{1}{72}z^{17} + \frac{1}{216}z^{18}.$$

The probability mass function of the score can be read off directly from the expansion. This example shows just how useful generating functions can be to solve counting problems. Could you have explained beforehand without doing any calculations why the sum k must have the same probability of occurring as the sum $21 - k$ for $k = 3, \ldots, 10$?

In the next section we discuss a generally applicable method for numerically inverting generating functions for which an explicit expression can be given.

Problems

8.1 Suppose that the random variable X has a binomial distribution with parameters n and p and that the random variable Y has a negative binomial distribution with parameters r and p.

(a) What are the generating functions of X and Y?

(b) Use the result of (a) to get the expected values and the variances of X and Y.

8.2 Let X be a nonnegative, integer-valued random variable. You do not know the probability mass function of X, but you do have an explicit expression for the generating function $G_X(z)$. Show that X will take on an even value with probability $\frac{1}{2}[G_X(-1) + 1]$.

8.3 You have a symmetric die with the numerals $1, 2, \ldots, 5$ on the six faces, where two faces have the numeral 1. You roll this die n times. Use the result of Problem 8.2 to find the probability that the total score will be an even number of points.

8.4 You successively take n independent samples from a continuously distributed random variable. A sample is a record if it exceeds each of the earlier samples (the first sample is by definition a record). Let the random variable I_k be equal to 1 if the kth sample is a record and 0 otherwise.

(a) Prove that the random variables $I_1, \ldots, I_n$ are independent. *Hint*: Verify that $P(I_k = 1) = \frac{1}{k}$ for all k and $P(I_{i_r} = 1, \ldots, I_{i_1} = 1) = \frac{1}{i_r} \times \cdots \times \frac{1}{i_1}$ for all $1 \leq i_1 < \cdots < i_r \leq n$ and $1 \leq r \leq n$.

(b) What is the generating function of the total number of records? Use computer algebra for the product of polynomials to calculate the probability mass function of the number of records when $n = 10$.

8.5 Suppose that the nonnegative, integer-valued random variables X and Y are independent and have the same probability distribution. Verify that X and Y are Poisson distributed when the sum $X + Y$ is Poisson distributed.

8.6 Suppose the number of claims an insurance company will receive is a random variable N having a Poisson distribution with expected value λ. The claim sizes are independent random variables with a common probability mass function a_k for $k = 1, 2, \ldots$. Let the total claim size S be defined by $S = \sum_{i=1}^{N} X_i$, where $X_1, X_2, \ldots$ represent the individual claim sizes. Prove that the generating function of the random sum S is given by $e^{-\lambda[1-A(z)]}$, where $A(z)$ is the generating function of the individual claim sizes. Also, verify that $E(S) = E(N)E(X_1)$ and $\text{var}(S) = E(N)\text{var}(X_1) + \text{var}(N)[E(X_1)]^2$ with $E(N) = \text{var}(N) = \lambda$. *Note*: The probability distribution of the random sum S is called the *compound Poisson distribution*.

8.7 For any $t \geq 0$, let the random variable $N(t)$ be the number of occurrences of some event up to time t, where $N(0) = 0$. Suppose that

(i) For any $s, t \geq 0$, the random variable $N(t + s) - N(s)$ is independent of $N(s)$ and its probability mass function depends only on t.

(ii) There is a constant $\lambda > 0$ such that $P(N(t+h) - N(t) = 1) = \lambda h + o(h)$ and $P(N(t+h) - N(t) \geq 2) = o(h)$ as $h \to 0$ for any $t \geq 0$, where $o(h)$ is the standard notation for an unspecified function with $o(h)/h \to 0$ as $h \to 0$.

Prove that $N(t)$ is Poisson distributed with expected value λt for any $t > 0$.
Hint: Let $g_z(t) = E(z^{N(t)})$ and evaluate $g_z(t+h) - g_z(t)$ for h small.

8.8 You roll a single die. Then you simultaneously roll as many dice as the face value shown up on the roll of the die. Use the generating-function approach to find the probability that the sum of face values shown up on the simultaneous roll is less than or equal to 12.

8.1.2 Inversion of the Generating Function

In many applications, it is possible to derive an explicit expression for the generating function of a random variable X whose probability mass function is not readily available and has a complicated form. Is this explicit expression for the generating function of practical use, apart from calculating the moments of X? The answer is yes! If an explicit expression for the generating function of the random variable X is available, then the numerical values of the (unknown) probability mass function of X can be calculated by appealing to the discrete fast Fourier transform method from numerical analysis (this algorithm functions in the seemingly mystical realm of complex numbers, which world nonetheless is of great real-world significance). An explanation of how this extremely powerful method works is beyond the scope of this book. However, it is useful to know that this method exists. In practice, it is often used to calculate convolutions of discrete probability distributions.

An example of a complicated probability mass function whose generating function allows for a simple explicit expression is provided by the coupon-collector problem, discussed in earlier chapters. This problem is as follows. Each time you buy a certain product (chewing gum, for example) you receive a coupon (a baseball card, for example), which is equally likely to be any one of n types. Let the random variable X be the number of purchases required to get a complete set of coupons. The random variable X can be written as the sum of n independent random variables $X_1, \ldots, X_n$, where X_i is the number of purchased products needed to go from $i-1$ different coupons to i different coupons. The random variable X_i is geometrically distributed with parameter $a_i = \frac{n-i+1}{n}$ for all i and has generating function $a_i z/(1 - z + a_i z)$. Then the convolution Rule 8.2 gives

$$G_X(z) = \frac{a_1 a_2 \cdots a_n z^n}{(1 - z + a_1 z)(1 - z + a_2 z) \cdots (1 - z + a_n z)}.$$

The coupon-collector problem with $n = 365$ distinct coupons enables us to calculate how many persons are needed to have a group of persons in which all 365 possible birthdays (excluding February 29) are represented with a probability of at least 50% when each birthday is equally likely. Using the discrete fast Fourier transform method, we can calculate that the group should consist of 2,287 randomly picked persons. Then the probability of having each birthday represented is 0.5004.

The next example shows that the probability distribution of the waiting time for success runs in Bernoulli trials can also be studied through the generating-function approach (other approaches to analyze runs are discussed in Chapters 2 and 10). In this example and the subsequent example, we use the method of *first-step analysis*. A frequently helpful trick is to condition on what happens in the first step or the first few steps of the experiment.

Example 8.3 For a sequence of independent Bernoulli trials with success probability p, let the random variable X be the number of trials it would take to have a run of r successes in a row. What are the expected value and the variance of X?

Solution. The key to the solution is the following observation. The random variable X equals r if the first r trials result in a success. The conditional distribution of X is the same as the unconditional distribution of $k + X$ if the first $k - 1$ trials result in a success and are followed by a nonsuccessful trial, where $1 \le k \le r$. Thus, by the law of conditional expectation, the generating function $G_X(z) = E(z^X)$ is given by

$$z^r p^r + \sum_{k=1}^{r} E(z^{k+X}) p^{k-1}(1-p) = z^r p^r + E(z^X) \sum_{k=1}^{r} z^k p^{k-1}(1-p).$$

Since $\sum_{k=1}^{r} a^{k-1} = (1 - a^r)/(1 - a)$ for $|a| < 1$, it follows that

$$G_X(z) = \frac{p^r z^r (1 - pz)}{1 - z + (1-p) p^r z^{r+1}}.$$

The probability mass function of X can be obtained by numerical inversion of this explicit expression. Another method to calculate this probability distribution is the recursive method discussed in Example 2.10. It is interesting to note that the probability $P(X \le n)$ can also be interpreted as the probability of seeing a run of r successes in a row somewhere during n trials of the Bernoulli experiment.

Using the formulas $E(X) = G'_X(1)$ and $E(X^2) = G''_X(1) + G'_X(1)$, it is a matter of some algebra to conclude that

$$E(X) = \frac{1 - p^r}{(1 - p)p^r} \quad \text{and} \quad \text{var}(X) = \frac{1 - p^{2r+1} - (1 - p)p^r(2r + 1)}{(1 - p)^2 p^{2r}}.$$

In particular, for $p = 0.5$,

$$E(X) = 2^{r+1} - 2 \quad \text{and} \quad \text{var}(X) = 2^{2r+2}[1 - (2r + 1)2^{-(r+1)} - 2^{-(2r+1)}].$$

As a sanity check, the formulas for $E(X)$ and $\text{var}(X)$ reduce to $\frac{1}{p}$ and $\frac{1-p}{p^2}$ when $r = 1$, in agreement with the expected value and the variance of the geometric distribution.

Example 8.4 In the Chow–Robbins game you repeatedly toss a fair coin. You can stop whenever you want. Your payoff is the proportion of heads obtained at the time you stop. There is no limit on the number of tosses. What is the expected gain under the simple stopping rule that prescribes stopping as soon as the number of heads exceeds the number of tails?

Solution. Let p_n be the probability that on the nth toss the number of heads exceeds the number of tails for the first time. Noting that $p_n = 0$ for n even, we have that the expected gain under the stopping rule is

$$\sum_{k=1}^{\infty} \frac{k}{2k - 1} p_{2k-1}.$$

The p_n can be obtained by the generating-function approach. Let the random variable X be the number of tosses until the number of heads exceeds the number of tails for the first time. To find $P(z) = \sum_{n=0}^{\infty} p_n z^n$, we use again the method of first-step analysis. By conditioning on the outcome of the first toss, we have that $X = 1$ with probability 0.5 and X is distributed as $1 + X_1 + X_2$ with probability 0.5, where X_1 and X_2 are independent and have the same distribution as X. This gives

$$E(z^X) = 0.5z + 0.5E(z^{1+X_1+X_2})$$

and so $P(z) = \frac{1}{2}z + \frac{1}{2}zP^2(z)$. The solution of this quadratic equation is

$$P(z) = \frac{1 - \sqrt{1 - z^2}}{z} \quad \text{for } |z| \leq 1$$

(the minus sign must be taken in view of $P(0) = 0$). This generating function can be inverted analytically by using the Taylor-series expansion

$$(1 + x)^a = \sum_{k=0}^{\infty} \binom{a}{k} x^k \quad \text{for } -1 < x < 1$$

for any non-integer number a. The generalized binomial coefficient $\binom{a}{k}$ is defined as 1 for $k = 0$ and as $a(a-1)\cdots(a-k+1)/k!$ for $k = 1, 2, \ldots$. In particular,

$$\binom{0.5}{k} = (-1)^{k-1} \left(\frac{1}{2}\right)^{2k-1} \frac{(2k-2)!}{k!\,(k-1)!} \quad \text{for } k = 1, 2, \ldots.$$

Using the series expansion of $(1+x)^a$ with $a = 0.5$ and $x = -z^2$, we find that $P(z)$ can be evaluated as $\frac{1}{z}\left(1 - \sum_{k=0}^{\infty} \binom{0.5}{k}(-1)^k z^{2k}\right)$. This leads to

$$P(z) = \sum_{k=1}^{\infty} \left(\frac{1}{2}\right)^{2k-1} \frac{(2k-2)!}{k!\,(k-1)!} z^{2k-1}.$$

This gives the explicit expression $p_{2k-1} = (\frac{1}{2})^{2k-1}\frac{(2k-2)!}{k!(k-1)!}$ for $k \geq 1$ and so the expected gain $\sum_{k=1}^{\infty}\frac{k}{2k-1}p_{2k-1}$ is

$$\sum_{k=1}^{\infty}\frac{k}{2k-1}\left(\frac{1}{2}\right)^{2k-1}\frac{(2k-2)!}{k!\,(k-1)!} = \frac{1}{2}\sum_{k=1}^{\infty}\frac{1}{2k-1}\left(\frac{1}{2}\right)^{2k-2}\binom{2k-2}{k-1}.$$

Using the Taylor-series expansion $\arcsin(x) = \sum_{k=1}^{\infty}\frac{1}{2k-1}(\frac{1}{2})^{2k-2}\binom{2k-2}{k-1}x^{2k-1}$ for $|x| \leq 1$, we get that the expected gain is equal to $\frac{1}{2}\arcsin(1) = \frac{\pi}{4}$. It is remarkable that the expected gain of the simple stopping rule is quite close to the expected gain of the optimal stopping rule in the Chow–Robbins game. The optimal stopping rule is rather complex and is characterized by integers $\beta_1, \beta_2, \ldots$ such that you stop after the nth toss when the number of heads minus the number of tails is larger than or equal to β_n. These threshold values are difficult to compute and the maximum expected gain is only known to be between the bounds $0.7929530\ldots$ and $0.7929556\ldots$.

Problems

8.9 You participate in a game that consists of a series of independent plays. Any play results in a win with probability p, in a loss with probability q, and in a draw with probability r, where $p + q + r = 1$. One point is added to your score each time a play is won; otherwise, your score remains unchanged. The game ends as soon as you lose a play. Let the random variable X denote your total score when the game is ended. Use the generating-function approach to find an explicit expression for the probability mass function of X.

8.10 Independently of each other, you generate integers at random from $0, 1, \ldots, 9$ until a zero is obtained. Use the method of first-step analysis to obtain the generating function of the sum of the generated integers.

8.11 You toss a fair coin repeatedly. For fixed $r > 1$, let the random variable X be the number of tosses you need to obtain either r heads in a row or r tails in a row. What is the generating function of X? Give explicit expressions for the expected value and the variance of X. *Hint*: Define X_1 as the number of tosses until either $r - 1$ heads in a row appear or r tails in a row and X_2 as the number of tosses until either $r - 1$ tails in a row appear or r heads in a row.

8.12 Consider again Example 8.4. Let X be the number of tosses of a fair coin until the number of heads exceeds the number of tails for the first time. Use the generating function of X to find $E(X)$. Calculate the tail probability $P(X > n)$ for $n = 5, 10, 25, 100, 1{,}000, 5{,}000$, and $10{,}000$. Conclude that the probability mass function of X has a very long tail.

8.2 Branching Processes and Generating Functions

The family name is inherited by sons only. Take a father who has one or more sons. In turn, each of his sons will have a random number of sons, each son of the second generation will have a random number of sons, and so forth. What is the probability that the family name will ultimately die out? The process describing the survival of family names is an example of a so-called branching process. Branching processes arise naturally in many situations. In physics, the model of branching processes can be used to study neutron chain reactions. A chance collision of a nucleus with a neutron yields a random number of new neutrons. Each of these secondary neutrons may hit some other nuclei, producing more additional neutrons, and so forth. In genetics, the model can be used to estimate the probability of long-term survival of genes that are subject to mutation. All of these examples possess the following structure. There is a population of individuals able to produce offspring of the same kind. Each individual will, by the end of its lifetime, have produced j new offspring with probability p_j for $j = 0, 1, \ldots$. All offspring behave independently. The number of individuals initially present, denoted by X_0, is called the size of the 0th generation. All offspring of the 0th generation constitute the first generation, and their number is denoted by X_1. In general, let X_n denote the size of the nth generation. We are interested in the probability that the population will eventually die out. To avoid uninteresting cases, it is assumed that $0 < p_0 < 1$. In order to find the *extinction probability*, it is no restriction to assume that $X_0 = 1$ (why?). Define the probability u_n by

$$u_n = P(X_n = 0).$$

Obviously, $u_0 = 0$ and $u_1 = p_0$. Noting that $X_n = 0$ implies $X_{n+1} = 0$, it follows that $u_{n+1} \geq u_n$ for all n. Since u_n is an increasing sequence of numbers, $\lim_{n \to \infty} u_n$ exists. Denote this limit by u_∞. The probability u_∞ is the sought extinction probability. This requires some explanation. The probability that extinction will ever occur is defined as $P(X_n = 0$ for some $n \geq 1)$. However, $\lim_{n \to \infty} P(X_n = 0) = P(X_n = 0$ for some $n \geq 1)$, using the fact that $\lim_{n \to \infty} P(A_n) = P(\bigcup_{n=1}^\infty A_n)$ for any increasing sequence of events A_n. The probability u_∞ can be computed by using the generating function $P(z) = \sum_{j=0}^\infty p_j z^j$ of the offspring distribution p_j. To do so, we first argue that

$$u_n = \sum_{k=0}^\infty (u_{n-1})^k p_k \quad \text{for } n = 2, 3, \ldots.$$

This relation can be explained using the law of conditional probability. Fix $n \geq 2$. Now, condition on $X_1 = k$ and use the fact that the k subpopulations generated by the distinct offspring of the original parent behave independently and follow the same distributional law. The probability that any particular one of those subpopulations will die out within $n - 1$ generations is u_{n-1} by definition. Thus, the probability that all k subpopulations will die out within $n - 1$ generations is equal to $P(X_n = 0 \mid X_1 = k) = (u_{n-1})^k$ for $k \geq 1$. This relation is also true for $k = 0$, since $X_1 = 0$ implies that $X_n = 0$ for all $n \geq 2$. The equation for u_n follows next by using the fact that

$$P(X_n = 0) = \sum_{k=0}^\infty P(X_n = 0 \mid X_1 = k) p_k,$$

by the law of conditional probability.

Using the definition of the generating function $P(z) = \sum_{k=0}^\infty p_k z^k$, the recursion equation for u_n can be rewritten as

$$u_n = P(u_{n-1}) \quad \text{for } n = 2, 3, \ldots.$$

Next, by letting $n \to \infty$ on both sides of this equation and using a continuity argument, it can be shown that the sought probability u_∞ satisfies the equation

$$u = P(u).$$

This equation may have more than one solution. However, it can be shown that u_∞ is the smallest positive root of the equation $u = P(u)$. It may happen that $u_\infty = 1$, that is, the population is sure to die out ultimately. The case of $u_\infty = 1$ can only happen if the expected value of the offspring distribution p_j is less than or equal to 1. The proof of this fact is omitted. As an illustration, consider the numerical example with $p_0 = 0.25$, $p_1 = 0.25$, and $p_2 = 0.5$. The equation $u = P(u)$ then becomes the quadratic equation $u = \frac{1}{4} + \frac{1}{4}u + \frac{1}{2}u^2$. This equation has roots $u = 1$ and $u = \frac{1}{2}$. The smallest root gives the extinction probability $u_\infty = \frac{1}{2}$.

Problems

8.13 In a branching process with one ancestor, the number of offspring of each individual has the shifted geometric distribution $\{p_k = p(1-p)^k, k = 0, 1, \ldots\}$ with parameter $p \in (0, 1)$. What is the probability of extinction as a function of p?

8.14 Every adult male in a certain society is married. 20% of the married couples have no children. The other 80% have two or three children with respective probabilities $\frac{1}{3}$ and $\frac{2}{3}$, each child being equally likely to be a boy or a girl. What is the probability that the male line of a father with one son will eventually die out?

8.15 A population of bacteria begins with a single individual. In each generation, each individual dies with probability $\frac{1}{3}$ or splits in two with probability $\frac{2}{3}$. What is the probability that the population will die out by generation 3 and what is the probability that the population will die out eventually? What are these probabilities if the initial population consists of two individuals?

8.3 Moment-Generating Functions

How do we generalize the concept of a generating function when the random variable is not integer valued and nonnegative? The idea is to work with $E(e^{tX})$ instead of $E(z^X)$. Since e^{tX} is a nonnegative random variable, $E(e^{tX})$ is defined for any value of t. However, it may happen that $E(e^{tX}) = \infty$ for some values of t. For any nonnegative random variable X, we have that $E(e^{tX}) < \infty$ for any $t \leq 0$ (why?), but $E(e^{tX})$ need not be finite when $t > 0$. To illustrate this, consider the nonnegative random variable X having the one-sided Cauchy density function $f(x) = (2/\pi)/(1 + x^2)$ for $x > 0$. Then $E(e^{tX}) = \int_0^\infty e^{tx} f(x)\, dx = \infty$ for any $t > 0$, since $e^{tx} \geq 1 + tx$ and $\int_0^\infty \frac{x}{1+x^2}\, dx = \infty$. In the case that the random variable X can take on both positive and negative values, it may happen that $E(e^{tX}) = \infty$ for all $t \neq 0$. An example is provided by the random variable X having the two-sided Cauchy density function $f(x) = (1/\pi)/(1+x^2)$ for $-\infty < x < \infty$. Fortunately, most random variables X of practical interest have the property that $E(e^{tX}) < \infty$ for all t in a neighborhood of 0.

Definition 8.2 *A random variable X is said to have a moment-generating function if $E(e^{tX}) < \infty$ for all t in an interval of the form $-\delta < t < \delta$ for some $\delta > 0$. For those t with $E(e^{tX}) < \infty$, the moment-generating function of X is defined and denoted by*

$$M_X(t) = E(e^{tX}).$$

If the random variable X has a probability density function $f(x)$, then

$$M_X(t) = \int_{-\infty}^{\infty} e^{tx} f(x) dx.$$

As an example, consider the case of an exponentially distributed random variable X. The density function $f(x)$ of X is equal to $\lambda e^{-\lambda x}$ for $x > 0$ and 0 otherwise. Then $M_X(t) = \lambda \int_0^{\infty} e^{-(\lambda - t)x} dx$. This integral is finite only if $\lambda - t > 0$. Thus, $M_X(t)$ is defined only for $t < \lambda$ and is then given by $M_X(t) = \lambda/(\lambda - t)$.

The explanation of the name "moment-generating function of a random variable X" is as follows. Assuming that $E(e^{tX})$ is finite in an interval $(-\delta, \delta)$ for some $\delta > 0$, we have

$$M_X(t) = 1 + tE(X) + t^2 \frac{E(X^2)}{2!} + t^3 \frac{E(X^3)}{3!} + \cdots$$

for $-\delta < t < \delta$. This result follows by using the expansion $E(e^{tX}) = E\left(\sum_{n=0}^{\infty} t^n \frac{X^n}{n!}\right)$ and interchanging the order of expectation and summation, the latter being justified by the finiteness of $E(e^{tX})$ in a neighborhood of $t = 0$. The series expansion of $M_X(t)$ shows that the moments $E(X^r)$ are finite for $r = 1, 2, \ldots$. Differentiating the series expansion of $M_X(t)$ and next taking $t = 0$, we obtain

$$E(X^r) = \frac{d^r}{dt^r} M_X(t)\Big|_{t=0} \quad \text{for } r = 1, 2, \ldots.$$

In particular, we find the following result.

Rule 8.3 *Let X be a random variable such that $M_X(t) = E(e^{tX})$ is finite in some neighborhood of $t = 0$. Then,*

$$E(X) = M_X'(0) \text{ and } E(X^2) = M_X''(0).$$

8.3 Moment-Generating Functions

A moment-generating function not only determines the moments of a random variable X, but also determines uniquely the probability distribution of X. The following uniqueness theorem holds for the moment-generating function.

Rule 8.4 *If the moment-generating functions $M_X(t)$ and $M_Y(t)$ of the random variables X and Y exist and $M_X(t) = M_Y(t)$ for all t satisfying $-\delta < t < \delta$ for some $\delta > 0$, then the random variables X and Y have the same probability distribution.*

The proof of this rule is beyond the scope of this book. Also, we have the following very useful rule.

Rule 8.5 *Let X and Y be two random variables with moment-generating functions $M_X(t)$ and $M_Y(t)$. If the random variables X and Y are independent, then*

$$M_{X+Y}(t) = M_X(t)M_Y(t)$$

for all t in a neighborhood of $t = 0$.

The proof is easy. If X and Y are independent, then the random variables e^{tX} and e^{tY} are independent for any fixed t, see Rule 3.6. Since $E(UV) = E(U)E(V)$ for independent random variables U and V, it follows that

$$E[e^{t(X+Y)}] = E(e^{tX}e^{tY}) = E(e^{tX})E(e^{tY}).$$

Example 8.5 Suppose that the random variable X has an $N(\mu, \sigma^2)$ distribution. Verify that

$$M_X(t) = e^{\mu t + \frac{1}{2}\sigma^2 t^2} \quad \text{for } -\infty < t < \infty$$

and use this result to show that $E(X) = \mu$ and $\sigma^2(X) = \sigma^2$.

Solution. Let $Z = (X - \mu)/\sigma$. Then Z has the $N(0, 1)$ distribution and its moment-generating function $M_Z(t) = E(e^{tZ})$ is given by

$$\frac{1}{\sqrt{2\pi}} \int_{-\infty}^{\infty} e^{tz} e^{\frac{1}{2}z^2} \, dz = e^{\frac{1}{2}t^2} \frac{1}{\sqrt{2\pi}} \int_{-\infty}^{\infty} e^{-\frac{1}{2}(z-t)^2} \, dz = e^{\frac{1}{2}t^2},$$

where the last equality uses the fact that for fixed t, the function $\frac{1}{\sqrt{2\pi}} e^{-\frac{1}{2}(z-t)^2}$ is the probability density function of an $N(t, 1)$ distribution. This implies that the integral of this function over the interval $(-\infty, \infty)$ equals 1. The expression for $M_X(t)$ follows next from

$$E(e^{tX}) = E(e^{t(\mu+\sigma Z)}) = e^{t\mu} E(e^{t\sigma Z}) = e^{t\mu} e^{\frac{1}{2}\sigma^2 t^2}.$$

The first and second derivatives of $M_X(t)$ at the point $t = 0$ are given by $M_X'(0) = \mu$ and $M_X''(0) = \mu^2 + \sigma^2$, showing that the expected value and the variance of an $N(\mu, \sigma^2)$-distributed random variable are equal to μ and σ^2.

Remark 8.3 The moment-generating function $M_X(t)$ of the normal distribution also enables us to derive the expected value and the variance of the lognormal distribution. If X is $N(\mu, \sigma^2)$ distributed, then $Y = e^X$ has a lognormal density with parameters μ and σ. Taking $t = 1$ in the moment-generating function $M_X(t) = E(e^{tX}) = e^{\mu t + \frac{1}{2}\sigma^2 t^2}$, we obtain $E(Y)$. Also, by $e^{2X} = Y^2$, we obtain $E(Y^2)$ by putting $t = 2$ in $M_X(t) = E(e^{tX})$.

Using the result of Example 8.5, we easily verify that a linear combination of independent normal random variables has, again, a normal distribution.

Rule 8.6 *Suppose that the random variables $X_1, \ldots, X_n$ are independent and normally distributed, where X_i has an $N(\mu_i, \sigma_i^2)$ distribution. Then, for any constants $a_1, \ldots, a_n$, the random variable $U = a_1 X_1 + \cdots + a_n X_n$ has an $N(\mu, \sigma^2)$ distribution with*

$$\mu = a_1 \mu_1 + \cdots + a_n \mu_n \quad \text{and} \quad \sigma^2 = a_1^2 \sigma_1^2 + \cdots + a_n^2 \sigma_n^2.$$

It suffices to prove this result for $n = 2$. The general result follows next, by induction. Using Rule 8.5 and the result from Example 8.5, we find

$$E\left[e^{t(a_1 X_1 + a_2 X_2)}\right] = E(e^{ta_1 X_1}) E(e^{ta_2 X_2}) = e^{\mu_1 a_1 t + \frac{1}{2}\sigma_1^2 (a_1 t)^2} e^{\mu_2 a_2 t + \frac{1}{2}\sigma_2^2 (a_2 t)^2}$$
$$= e^{(a_1 \mu_1 + a_2 \mu_2) t + \frac{1}{2}(a_1^2 \sigma_1^2 + a_2^2 \sigma_2^2) t^2},$$

proving the desired result with an appeal to the uniqueness Rule 8.4.

Rule 8.6 shows that the class of normal distributions is closed. A similar result holds for the gamma distribution, see Problem 8.19.

Example 8.6 Suppose that the random variable X has a gamma distribution with shape parameter α and scale parameter λ. Verify that

$$M_X(t) = \left(\frac{\lambda}{\lambda - t}\right)^\alpha \quad \text{for } t < \lambda.$$

Solution. Fix t with $t < \lambda$ and note that

$$M_X(t) = \int_0^\infty e^{tx} \frac{\lambda^\alpha}{\Gamma(\alpha)} x^{\alpha-1} e^{-\lambda x} \, dx = \int_0^\infty \frac{\lambda^\alpha}{\Gamma(\alpha)} x^{\alpha-1} e^{-(\lambda-t)x} \, dx$$
$$= \left(\frac{\lambda}{\lambda - t}\right)^\alpha \int_0^\infty \frac{(\lambda - t)^\alpha}{\Gamma(\alpha)} x^{\alpha-1} e^{-(\lambda-t)x} \, dx.$$

Using the fact that $(\lambda - t)^\alpha x^{\alpha-1} e^{-(\lambda-t)x} / \Gamma(\alpha)$ is a gamma density for any fixed t with $t < \lambda$ and thus integrates to 1, the desired result follows.

Remark 8.4 A special case of the gamma distribution arises by taking a positive integer n for the shape parameter α. This special case is called the Erlang distribution with shape parameter n and scale parameter λ. A very useful result in queueing and reliability analysis is the fact that a random variable having the Erlang distribution with shape parameter n and scale parameter λ can be decomposed as the sum of n independent exponential random variables, each having the same parameter λ. This result can be seen as follows. As shown before, an exponential random variable with parameter λ has the moment-generating function $\lambda/(\lambda-t)$ for $t < \lambda$. Thus, by Rule 8.5, the sum of n independent exponential random variables each having parameter λ has the moment-generating function $[\lambda/(\lambda-t)]^n$ for $t < \lambda$. It now follows from Example 8.6 and the uniqueness Rule 8.4 that this sum is gamma distributed with shape parameter n and scale parameter λ.

Rule 8.7 Let $Z_1, \ldots, Z_n$ be independent random variables, each having a standard normal distribution. Define the chi-squared random variable U by $U = Z_1^2 + \cdots + Z_n^2$. Then the random variable U has a gamma density with shape parameter $\frac{1}{2}n$ and scale parameter $\frac{1}{2}$.

The proof goes as follows. Letting Z be an $N(0, 1)$ random variable, the integral representation $\frac{1}{\sqrt{2\pi}} \int_{-\infty}^{\infty} e^{tx^2} e^{-\frac{1}{2}x^2} dx$ of $E(e^{tZ^2})$ can be written as

$$\frac{1}{\sqrt{2\pi}} \int_{-\infty}^{\infty} e^{-\frac{1}{2}(1-2t)x^2} dx = \frac{1}{\sqrt{1-2t}} \frac{1}{(1/\sqrt{1-2t})\sqrt{2\pi}} \int_{-\infty}^{\infty} e^{-\frac{1}{2}x^2/(1/\sqrt{1-2t})^2} dx$$

and so $E(e^{tZ^2}) = \frac{1}{\sqrt{1-2t}}$ for $t < \frac{1}{2}$. Then, by applying Rule 8.5,

$$M_U(t) = \frac{1}{\sqrt{1-2t}} \cdots \frac{1}{\sqrt{1-2t}} = \frac{1}{(1-2t)^{n/2}} \quad \text{for } t < \frac{1}{2}.$$

Comparing this expression with the moment-generating function of the gamma density in Example 8.6 and using the uniqueness Rule 8.4, it follows that the chi-squared-distributed random variable U has a gamma density with shape parameter $\frac{1}{2}n$ and scale parameter $\frac{1}{2}$.

Problems

8.16 Suppose that the random variable X is uniformly distributed on the interval (a, b). What is the moment-generating function of X?

8.17 Suppose that the random variable X has the moment-generating function $E(e^{tX}) = \frac{1}{2}\frac{e^t-1}{t} + \frac{1}{2}$ for $-\infty < t < \infty$. Argue that X is a mixed random variable and specify the probability distribution of X.

8.18 Determine the moment-generating function of the random variable X having as density function the so-called *Laplace density* function $f(x) = \frac{1}{2}ae^{-a|x|}$ for $-\infty < x < \infty$, where a is a positive constant. Use the moment-generating function of X to find $E(X)$ and var(X).

8.19 Let $X_1, \ldots, X_n$ be independent random variables, each having a gamma density with the same scale parameter λ. Let α_i be the shape parameter of the density of X_i. Verify that $X_1 + \cdots + X_n$ has a gamma density with shape parameter $\alpha_1 + \cdots + \alpha_n$ and scale parameter λ.

8.20 Suppose that the random variables X and Y are independent and have the same probability distribution. Use the moment-generating function to prove that X and Y are normally distributed if $X + Y$ has a normal distribution.

8.21 Suppose that the random variable X has the so-called *logistic density* function $f(x) = e^x/(1 + e^x)^2$ for $-\infty < x < \infty$. Use the moment-generating function to find the expected value and variance of X.

8.22 Suppose that the moment-generating function $M_X(t)$ of the continuous random variable X has the property $M_X(t) = e^t M_X(-t)$ for all t. What is $E(X)$? Do you think that this property uniquely determines the density of X? *Hint*: Relate $1 - X$ to X.

8.23 The moment-generating function of two jointly distributed random variables X and Y is defined by $M_{X,Y}(s,t) = E(e^{sX+tY})$, provided that this integral is finite for all (s,t) in a neighborhood of $(0,0)$. A basic result is that $M_{X,Y}(s,t)$ uniquely determines the joint distribution of X and Y.[1]

  (a) Show that $M_{X,Y}(s,t) = e^{\frac{1}{2}(s^2+2\rho st+t^2)}$ if (X,Y) has a standard bivariate normal distribution with correlation coefficient ρ.

  (b) Suppose that the jointly distributed random variables X and Y have the property that $aX + bY$ is normally distributed for any constants a and b. Prove that (X,Y) has a bivariate normal density.

8.4 Central Limit Theorem Revisited

We cannot withhold from the reader at least a glimpse of the steps involved in the proof of the central limit theorem, which plays such a prominent

---

[1] Using this uniqueness result, it can be shown that X and Y are independent if and only if $M_{X,Y}(s,t) = M_X(s)M_Y(t)$ for all s,t. We omit the proof.

8.4 Central Limit Theorem Revisited

role in probability theory. The mathematical formulation of the central limit theorem is as follows. Suppose that $X_1, X_2, \ldots$ are independent and identically distributed random variables with expected value μ and standard deviation σ. Then

$$\lim_{n \to \infty} P\left(\frac{X_1 + \cdots + X_n - n\mu}{\sigma \sqrt{n}} \leq x\right) = \frac{1}{\sqrt{2\pi}} \int_{-\infty}^{x} e^{-\frac{1}{2}y^2} dy \quad \text{for all } x.$$

We make this result plausible under the assumption that the moment-generating function of the X_i exists and is finite for all t in some neighborhood of $t = 0$.[2] To do so, consider the standardized variables

$$Y_i = \frac{X_i - \mu}{\sigma} \quad \text{for } i = 1, 2, \ldots.$$

Then $E(Y_i) = 0$ and $\text{var}(Y_i) = 1$. Letting

$$Z_n = \frac{Y_1 + \cdots + Y_n}{\sqrt{n}},$$

we have $Z_n = (X_1 + \cdots + X_n - n\mu)/(\sigma \sqrt{n})$. Denoting by $M_{Z_n}(t) = E(e^{tZ_n})$ the moment-generating function of Z_n, it will be proved in a moment that

$$\lim_{n \to \infty} M_{Z_n}(t) = e^{\frac{1}{2}t^2}$$

for all t in a neighborhood of $t = 0$. In other words,

$$\lim_{n \to \infty} M_{Z_n}(t) = E(e^{tZ})$$

when Z is a standard normal random variable. This result implies

$$\lim_{n \to \infty} P(Z_n \leq x) = P(Z \leq x) \quad \text{for all } x,$$

by a deep continuity theorem for moment-generating functions. This theorem linking the convergence of moment-generating functions to the convergence of cumulative distribution functions must be taken for granted by the reader.

To verify that the moment-generating function of Z_n converges to the moment-generating function of the standard normal random variable, let $M_Y(t)$ be the moment-generating function of the Y_i. Using the assumption that $Y_1, \ldots, Y_n$ are independent and identically distributed, it follows that

$$E(e^{tZ_n}) = E(e^{t(Y_1 + \cdots + Y_n)/\sqrt{n}}) = E(e^{(t/\sqrt{n})Y_1}) \cdots E(e^{(t/\sqrt{n})Y_n})$$

---

[2] In order to prove the central limit theorem without this assumption, one must use the so-called characteristic function rather than the moment-generating function. The characteristic function of a random variable X is defined by $E(e^{itX})$ for any real t, where i is the complex number with $i^2 = -1$. This complex-valued function is well-defined for each t.

and so
$$M_{Z_n}(t) = \left[M_Y(t/\sqrt{n})\right]^n \quad \text{for } n = 1, 2, \ldots.$$

Since $M_Y(t) = 1 + tE(Y_1) + \frac{t^2}{2!}E(Y_1^2) + \frac{t^3}{3!}E(Y_1^3) + \cdots$ in some neighborhood of $t = 0$, and using the fact that $E(Y_1) = 0$ and $\text{var}(Y_1) = 1$ so that $E(Y_1^2) = 1$, it follows that

$$M_Y(t) = 1 + \frac{1}{2}t^2 + \epsilon(t)$$

in a neighborhood of $t = 0$, where $\epsilon(t)$ tends to zero faster than t^2 as $t \to 0$. That is, $\lim_{t \to 0} \frac{\epsilon(t)}{t^2} = 0$. Now fix t and let $\epsilon_n = \epsilon(t/\sqrt{n})$. Then

$$M_{Z_n}(t) = \left(1 + \frac{1}{2}\frac{t^2}{n} + \frac{n\epsilon_n}{n}\right)^n \quad \text{for } n = 1, 2, \ldots.$$

Since $\lim_{u \to 0} \epsilon(u)/u^2 = 0$, we have that $\lim_{n \to \infty} n\epsilon_n = 0$. Using the fact that $\lim_{n \to \infty} (1 + \frac{a}{n})^n = e^a$ for any constant a, it is now a matter of standard manipulation in analysis to obtain the desired result $\lim_{n \to \infty} M_{Z_n}(t) = e^{\frac{1}{2}t^2}$.

9
Additional Topics in Probability

This chapter discusses a number of additional topics in probability. Useful bounds and inequalities are presented, with the emphasis on Chernoff bounds. Also, a proof of the strong law of large numbers is given. The strong law of large numbers is used to establish the renewal–reward theorem for stochastic processes having the property that the process probabilistically restarts itself at certain points in time. The renewal–reward theorem has many applications in applied probability. You will learn how to apply this beautiful theorem to a variety of interesting probability problems. Before we present the renewal–reward theorem, you will encounter the famous Kelly betting system, which is an exciting application of the strong law of large numbers.

9.1 Bounds and Inequalities

In probability applications it is sometimes important to have a bound on the probability that a sum of independent random variables takes on a value that is far from its expected value. The most useful bounds are the so-called Chernoff bounds. These bounds are typically sharper than bounds such as Markov's inequality and Chebyshev's inequality. We first state Markov's inequality, which is the basis for many other inequalities. This inequality was proved by the Russian mathematician A. A. Markov in 1889.

Rule 9.1 (Markov's inequality) *Let Y be a nonnegative random variable. Then,*

$$P(Y \geq a) \leq \frac{E(Y)}{a} \quad \text{for any constant } a > 0.$$

The proof is surprisingly simple. For fixed $a > 0$, let the indicator variable I be equal to 1 if $Y \geq a$ and 0 otherwise. Then $E(I) = P(Y \geq a)$. Moreover, by the nonnegativity of Y, we have that $Y \geq aI$. Markov's inequality now follows from

$$E(Y) \geq aE(I) = aP(Y \geq a).$$

A corollary of Markov's inequality is Chebyshev's inequality. This inequality is named after the Russian mathematician P. L. Chebyshev, who proved the inequality in 1867. Another proof was later given by A. A. Markov, who was a student of P. L. Chebyshev.

Rule 9.2 (Chebyshev's inequality) *Let X be a random variable with expected value μ and standard deviation σ. Then the two-sided Chebyshev's inequality states that*

$$P(|X - \mu| \geq c) \leq \frac{\sigma^2}{c^2} \quad \text{for any constant } c > 0.$$

The one-sided Chebyshev's inequality states that, for any constant $c > 0$,

$$P(X \geq \mu + c) \leq \frac{\sigma^2}{\sigma^2 + c^2} \quad \text{and} \quad P(X \leq \mu - c) \leq \frac{\sigma^2}{\sigma^2 + c^2}.$$

The two-sided Chebyshev's inequality follows directly from Markov's inequality by taking $Y = (X - \mu)^2$ and $a = c^2$:

$$P(|X - \mu| \geq c) = P\big((X - \mu)^2 \geq c^2\big) \leq \frac{E(X - \mu)^2}{c^2} = \frac{\sigma^2}{c^2}.$$

To obtain the one-sided Chebyshev's inequality, we first assume that $\mu = 0$ and use the following trick. Write $P(X \geq c)$ as $P(X + b \geq c + b)$ for the parameter $b > 0$. Noting that $c + b > 0$, it follows that $X + b \geq c + b$ implies $(X + b)^2 \geq (c + b)^2$ and so $P(X + b \geq c + b) \leq P\big((X + b)^2 \geq (c + b)^2\big)$. Thus,

$$P(X \geq c) \leq P\big((X + b)^2 \geq (c + b)^2\big)$$

for all $b > 0$. Using Markov's inequality, we next get that

$$P(X \geq c) \leq \frac{E\big((X + b)^2\big)}{(c + b)^2} = \frac{\sigma^2 + b^2}{(c + b)^2} \quad \text{for all } b > 0,$$

where the last equality uses the assumption $\mu = 0$. Since $(\sigma^2 + b^2)/(c + b)^2$ as a function of b is minimal for $b = \frac{\sigma^2}{c}$ (verify!), we get after some algebra

$$P(X \geq c) \leq \frac{\sigma^2}{\sigma^2 + c^2}.$$

Next consider the case that $\mu \neq 0$. Both random variables $X - \mu$ and $\mu - X$ have expected value 0 and standard deviation σ. Then the latter inequality implies

9.1 Bounds and Inequalities

that both $P(X - \mu \geq c)$ and $P(\mu - X \geq c)$ are less than or equal to $\frac{\sigma^2}{\sigma^2+c^2}$, which proves the one-sided Chebyshev's inequality.

Example 9.1 A webstore receives on average 7,500 orders per week with a standard deviation of 150 orders. What is a lower bound for the probability that the webstore will receive between 7,000 and 8,000 orders next week? What is an upper bound for the probability that 8,000 or more orders will be received next week?

Solution. Let the random variable X be the number of orders the webstore will receive next week. Then $E(X) = 7{,}500$ and $\sigma^2(X) = 22{,}500$. The probability $P(7{,}000 < X < 8{,}000)$ can be written as $P(-500 < X - 7{,}500 < 500) = P(|X - 7{,}500| < 500)$. Using the two-sided Chebyshev's inequality, it then follows that $P(7{,}000 < X < 8{,}000)$ is given by

$$1 - P(|X - 7{,}500| \geq 500) \geq 1 - \frac{22{,}500}{500^2} = 0.91.$$

The probability $P(X \geq 8{,}000)$ can be written as $P(X \geq 7{,}500 + 500)$. The one-sided Chebyshev's inequality then gives

$$P(X \geq 8{,}000) \leq \frac{22{,}500}{22{,}500 + 500^2} = 0.0826.$$

Remark 9.1 An immediate consequence of Chebyshev's inequality is the following bound. Suppose that $X_1, \ldots, X_n$ are independent random variables, each having the same distribution with expected value μ and standard deviation σ. Then, for any $\epsilon > 0$,

$$P\left(\left|\frac{1}{n}\sum_{k=1}^{n} X_k - \mu\right| \geq \epsilon\right) \leq \frac{\sigma^2}{n\epsilon^2}.$$

To see this, write $P(|\frac{1}{n}\sum_{k=1}^{n} X_k - \mu| \geq \epsilon)$ as $P(|\sum_{k=1}^{n} X_k - n\mu| \geq n\epsilon)$ and note that $E(X_1 + \cdots + X_n) = n\mu$ and $\sigma^2(X_1 + \cdots + X_n) = n\sigma^2$. Next the bound follows by applying Chebyshev's inequality.[1] An immediate consequence of the bound is the so-called *weak law of large numbers*:

$$\lim_{n \to \infty} P\left(\left|\frac{1}{n}\sum_{k=1}^{n} X_k - \mu\right| \geq \epsilon\right) = 0 \text{ for any } \epsilon > 0.$$

A closer examination of the proof shows that the weak law of large numbers remains valid when the assumption that the independent random variables

---

[1] The central limit theorem improves this bound: $P(|\frac{1}{n}\sum_{k=1}^{n} X_k - \mu| \geq \epsilon) \approx 2[1 - \Phi(\epsilon\sqrt{n}/\sigma)]$ for large n, see Problem 4.74.

$X_1, X_2, \ldots$ have the same distribution is weakened to the assumption that $E(X_i) = \mu$ for all i and $\sigma^2(X_i)$ is bounded in i.

Chernoff Bounds

These bounds are named after the American mathematician and statistician H. Chernoff. The basic form of the Chernoff bound is as follows.

Rule 9.3 (basic Chernoff bound) *Let X be a random variable for which the moment-generating function $M_X(t)$ exists. Then,*

$$P(X \geq c) \leq \min_{t>0} \left[e^{-ct} M_X(t)\right] \quad \text{for any constant } c,$$

where the minimum is taken over all $t > 0$ for which $M_X(t)$ is finite.

This is a very useful bound for tail probabilities. Chernoff bounds are widely used for the performance analysis of randomized algorithms in computer science. The proof of Rule 9.3 is based on Markov's inequality. To get the basic Chernoff bound from this inequality, note that

$$P(X \geq c) = P(tX \geq tc) = P(e^{tX} \geq e^{tc}) \quad \text{for any } t > 0.$$

Applying Markov's inequality with $Y = e^{tX}$ and $a = e^{ct} > 0$, we get

$$P(X \geq c) \leq e^{-ct} M_X(t) \quad \text{for any } t > 0,$$

implying the desired result.

Noting that $P(X \leq c) = P(tX \geq tc)$ for $t < 0$, a repetition of the above proof shows that $P(X \leq c) \leq \min_{t<0} \left[e^{-ct} M_X(t)\right]$ for any constant c.

Example 9.2 Suppose that the random variable X has the standard normal distribution. Verify that

$$P(X \geq c) \leq e^{-\frac{1}{2}c^2} \quad \text{for any constant } c > 0.$$

Solution. To get this bound, we use the result $M_X(t) = e^{\frac{1}{2}t^2}$ and apply the basic Chernoff bound. The minimizing value of t in the bound $e^{-ct} e^{\frac{1}{2}t^2}$ follows by putting the derivative of $\frac{1}{2}t^2 - ct$ equal to zero. This gives $t = c$ for any positive value of the constant c. Substituting $t = c$ into the bound yields the desired result.

For an $N(0, 1)$-distributed random variable X, the exponentially decreasing Chernoff bound $e^{-\frac{1}{2}c^2}$ is much sharper than the polynomial Chebyshev bound $\frac{1}{2c^2}$. The latter bound follows by noting that $P(|X| \geq c) = 2P(X \geq c)$ for $c > 0$. For example, $P(X \geq c)$ with $c = 5$ has the exact value 2.87×10^{-7}, where the Chebyshev bound and the Chernoff bound are $\frac{1}{50}$ and 3.73×10^{-6}.

9.1 Bounds and Inequalities

Chernoff bounds are particularly useful to bound the tail probabilities of a sum of independent random variables. There are many versions of Chernoff bounds. A practically useful bound is given in the next rule.

Rule 9.4 *Suppose that $X_1, X_2, \ldots, X_n$ are independent Bernoulli random variables with $P(X_i = 1) = p_i$ and $P(X_i = 0) = 1 - p_i$ for $i = 1, 2, \ldots, n$, where $p_i > 0$ for all i. Let $X = \sum_{i=1}^{n} X_i$ and denote by $\mu = \sum_{i=1}^{n} p_i$ the expected value of X. Then,*

$$P(X \geq c\mu) \leq e^{-(c\ln(c)-c+1)\mu} \quad \text{for } c \geq 1.$$

The proof goes as follows. Fix $c > 1$ (the Chernoff bound is trivially satisfied for $c = 1$). Noting that $P(X \geq c\mu) = P(c^X \geq c^{c\mu})$ when $c > 1$ (why?), Markov's inequality gives

$$P(X \geq c\mu) \leq \frac{E(c^X)}{c^{c\mu}}.$$

The X_i are independent and so $E(c^X) = E(c^{X_1}) \times E(c^{X_2}) \times \cdots \times E(c^{X_n})$. Noting that $E(c^{X_i}) = 1 - p_i + p_i c$, we get

$$E(c^X) = (1 - p_1 + p_1 c) \times (1 - p_2 + p_2 c) \times \cdots \times (1 - p_n + p_n c).$$

Using the inequality $e^x \geq 1 + x$ for $x > 0$, it follows that

$$E(c^X) \leq e^{(c-1)p_1} \times e^{(c-1)p_2} \times \cdots \times e^{(c-1)p_n} = e^{(c-1)\mu}.$$

Next note that $c^{c\mu} = e^{c\mu \ln(c)}$. Putting the pieces together, we get the desired Chernoff bound $P(X \geq c\mu) \leq e^{(c-1)\mu}/e^{c\mu \ln(c)} = e^{-(c\ln(c)-c+1)\mu}$.

Remark 9.2 Taking $c = 1 + \delta$ in the Chernoff bound in Rule 9.4 and using the series expansion $\ln(1 + x) = x - \frac{x^2}{2} + \frac{x^3}{3} - \cdots$ for $|x| < 1$, we obtain the weaker Chernoff bound (verify!)

$$P(X \geq (1 + \delta)\mu) \leq e^{-\frac{1}{3}\delta^2 \mu} \quad \text{when } 0 < \delta < 1.$$

The random variable X from Rule 9.4 also satisfies the Chernoff bound

$$P(X \leq (1 - \delta)\mu) \leq e^{-\frac{1}{2}\delta^2 \mu} \quad \text{when } 0 < \delta < 1.$$

This bound follows from the stronger bound $P(X \leq c\mu) \leq e^{-(c\ln(c)-c+1)\mu}$ for $0 < c < 1$ and the series expansion of $\ln(1 - x)$ for $0 < x < 1$. The stronger bound is obtained by noting that $P(X \leq c\mu) = P(c^X \geq c^{c\mu})$ for $0 < c < 1$ and next mimicking the proof of Rule 9.4.

Chernoff bounds are practically useful in the performance analysis of randomized algorithms in computer science. Many problems arising in the analysis of these algorithms can be described in terms of placing balls into bins. Suppose that m balls are put at random into b bins, one at a time and independently of each other. The load of a bin is the number of balls in the bin. How can we bound the probability that the maximum load over the bins is larger than some particular value? Let us consider the heavy-loaded case with $m \geq 2b \ln(b)$. Define the random variable L_i as the number of balls in bin i. Then,

$$P\left(L_i \geq e\frac{m}{b} \text{ for some } 1 \leq i \leq b\right) \leq \frac{1}{b},$$

where $e = 2.71828\ldots$ is the base of the natural logarithm. To prove this bound, we first apply the union bound

$$P\left(L_i \geq e\frac{m}{b} \text{ for some } 1 \leq i \leq b\right) \leq \sum_{i=1}^{b} P\left(L_i \geq e\frac{m}{b}\right).$$

Fix i and note that $L_i = \sum_{l=1}^{m} X_l$, where the X_l are independent Bernoulli variables with $P(X_l = 1) = \frac{1}{b}$ and $P(X_l = 0) = 1 - \frac{1}{b}$. Thus, $E(L_i) = \frac{m}{b}$. Using Rule 9.4 with $\mu = \frac{m}{b}$ and $c = e$, we get the Chernoff bound

$$P\left(L_i \geq e\frac{m}{b}\right) \leq e^{-(e\ln(e)-e+1)\frac{m}{b}} \leq e^{-2\ln(b)} = \frac{1}{b^2},$$

where the second inequality uses the assumption $m \geq 2b \ln(b)$. Together, this bound and the above union bound give the desired result.

Problems

9.1 Two tasks have to be done sequentially. The time required for each task is a random variable with an expected value of 30 minutes and a standard deviation of 7 minutes for the first task, and an expected value of 15 minutes and a standard deviation of 4 minutes for the second task. The carrying out times of the two tasks are independent of each other. What is a lower bound for the probability that both tasks will be completed in less than one hour?

9.2 You do not know the probability distribution of a random variable X, but you do know its expected value μ and standard deviation σ. How should k be chosen so that $P(X \in (\mu - k\sigma, \mu + k\sigma)) \geq \beta$ for a given β with $0 < \beta < 1$?

9.3 Let $X_1, \ldots, X_n$ be independent random variables that are uniformly distributed on $(-1, 1)$. Use Rule 9.3 to verify the Chernoff bound

$P\left(\frac{1}{n}\sum_{i=1}^{n} X_i \geq c\right) \leq e^{-\frac{3}{2}c^2 n}$ for $c > 0$. Hint: Use the inequality $\frac{1}{2}(e^t - e^{-t}) \leq te^{t^2/6}$ for $t > 0$.

9.4 Let $X_1, \ldots, X_n$ be independent bounded random variables that do not necessarily have the same probability distribution. It is assumed that $E(X_i) = 0$ for all i. Let $B > 0$ be such that $|X_i| \leq B$ for all i. Also, let $\sigma_i^2 = \sigma^2(X_i)$.

(a) Use the series representation $e^x = \sum_{n=0}^{\infty} \frac{x^n}{n!}$ to verify that $E(e^{tX_i})$ is bounded by $1 + \frac{\sigma_i^2}{B^2}(e^{tB} - 1 - tB)$ for $t > 0$. Next use the inequality $1 + x \leq e^x$ for $x > 0$ to conclude that $E(e^{tX_i}) \leq e^{\sigma_i^2(e^{tB} - 1 - tB)/B^2}$ for $t > 0$.

(b) Use Rule 9.3 and the result in (a) to prove the Chernoff bound

$$P\left(\frac{1}{n}\sum_{i=1}^{n} X_i \geq c\right) \leq e^{-\frac{nc^2}{2\sigma^2 + 2Bc/3}} \quad \text{for } c > 0,$$

where $\sigma^2 = \frac{1}{n}\sum_{i=1}^{n} \sigma_i^2$. This bound is known as *Bernstein's inequality*.

9.5 Let the random variable X be the number of successes in n independent Bernoulli trials with success probability p, where $0 < p < 1$. Choose any $\delta > 0$ such that $(1 + \delta)p < 1$. Use Rule 9.3 to verify the Chernoff bound

$$P(X \geq (1 + \delta)np) \leq \left[\left(\frac{p}{a}\right)^a \left(\frac{1-p}{1-a}\right)^{1-a}\right]^n,$$

where $a = (1 + \delta)p$. Next use the representation $\left(\frac{p}{a}\right)^a \left(\frac{1-p}{1-a}\right)^{1-a} = e^{-f(a)}$ with $f(a) = a\ln(\frac{a}{p}) + (1-a)\ln(\frac{1-a}{1-p})$ for $0 < a < 1$ to obtain the more tractable bound

$$P(X \geq (1+\delta)np) \leq e^{-2\delta^2 p^2 n}.$$

9.6 Let the random variable X be the number of successes in n independent heterogeneous Bernoulli trials, where the ith trial has success probability p_i. Verify that the Chernoff bound in Problem 9.5 remains valid when p is taken as $p = \frac{1}{n}\sum_{i=1}^{n} p_i$. Hint: Use the arithmetic–geometric mean inequality $\frac{1}{n}(a_1 + \cdots + a_n) \geq (a_1 \times \cdots \times a_n)^{1/n}$ for $a_1, \ldots, a_n \geq 0$ to get the bound $E(e^{tX}) \leq (pe^t + 1 - p)^n$.

9.2 Strong Law of Large Numbers

In the analysis of success runs in Section 2.3, we have seen that unexpectedly long runs of heads or tails can already occur with relatively few tosses. To see five or six heads or tails in a row in 20 tosses of a fair coin is not exceptional. It is the case that as the number of tosses becomes very large, the fractions of heads and tails should be about equal and "local clusters" of heads or tails are absorbed in the average. This fact is known as the *law of large numbers*. Just as the name implies, this law only says something about the game after a large number of tosses. This law does not imply that the absolute difference between the numbers of heads and tails should oscillate close to zero. It is even typical for the coin-tossing experiment that the absolute difference between the numbers of heads and tails has a tendency to become larger and larger and to grow proportionally with the square root of the number of tosses. This fact can be explained with the help of the central limit theorem, as was done at the end of Section 4.7. The course of a game of chance, although eventually converging in an average sense, is a whimsical process. To illustrate this, a simulation run of 25,000 coin tosses was made. This was done for both a fair coin $\left(p = \frac{1}{2}\right)$ and an unfair coin $\left(p = \frac{1}{6}\right)$. Table 9.1 summarizes the results of this particular simulation study. The statistic $H_n - np$ gives the observed number of heads minus the expected number after n tosses and the statistic f_n gives the observed relative frequency of heads after n tosses. It is worthwhile taking a close look at the results in Table 9.1. You will see that the realizations of the relative frequency, f_n, approach the true value of the probability p in a rather irregular manner.

There are two versions of the law of large numbers. The weak law of large numbers has already been mentioned and proved in Section 9.1. However, for applications, only the strong law of large numbers is of practical importance. The profits of insurance companies and casinos are based on this law and estimating probabilities by simulation also relies on it. We will give a proof of the strong law of large numbers under a weak assumption. The strong

Table 9.1. *Results of coin-toss simulations*

	Fair coin $\left(p = \frac{1}{2}\right)$		Unfair coin $\left(p = \frac{1}{6}\right)$	
n	$H_n - np$	f_n	$H_n - np$	f_n
25	1.5	0.5600	1.83	0.2400
100	2.0	0.5200	3.33	0.2040
500	−2.0	0.4960	4.67	0.1760
1,000	10.0	0.5100	−3.67	0.1630
2,500	12.0	0.5048	−15.67	0.1604
7,500	11.0	0.5015	21.00	0.1695
15,000	40.0	0.5027	−85.00	0.1610
25,000	64.0	0.5026	−30.67	0.1654

law of large numbers was proved for the first time by the famous Russian mathematician A. N. Kolmogorov in the twentieth century. The weak law of large numbers had already been formulated several centuries earlier by the Swiss mathematician Jakob Bernoulli (1654–1705) in his masterpiece *Ars Conjectandi* (The Art of Conjecturing) that was published posthumously in 1713. In that book Bernoulli was the first to make the mathematical connection between the probability of an event and the relative frequency with which the event occurs.

Rule 9.5 (the strong law of large numbers) *Suppose that $X_1, X_2, \ldots$ is a sequence of independent random variables, each having the same probability distribution with finite mean μ and finite standard deviation σ. Then*

$$P\left(\left\{\omega : \lim_{n \to \infty} \frac{1}{n} \sum_{k=1}^{n} X_k(\omega) = \mu\right\}\right) = 1,$$

where the symbol ω represents an outcome in the underlying sample space on which the process $\{X_1, X_2, \ldots\}$ is defined. In words, $\frac{1}{n}\sum_{k=1}^{n} X_k$ converges to μ with probability one.

The proof goes as follows. To begin, assume that $\mu = 0$. Let $S_k = \sum_{j=1}^{k} X_k$. We first verify that the subsequence S_{k^2}/k^2 converges to zero with probability 1. Fix $\epsilon > 0$. Noting that $E(S_{k^2}) = 0$ and $\text{var}(S_{k^2}) = k^2\sigma^2$, it follows from Chebyshev's inequality that

$$P\left(\left|\frac{S_k^2}{k^2}\right| > \epsilon\right) \leq \frac{k^2\sigma^2}{k^4\epsilon^2} = \frac{\sigma^2}{k^2\epsilon^2} \quad \text{for } k = 1, 2, \ldots.$$

This shows that $\sum_{k=1}^{\infty} P(A_k) < \infty$, where the set $A_k = \{\omega : S_{k^2}(\omega)/k^2 > \epsilon\}$. Therefore, $P(\{\omega : \omega \in A_k \text{ for infinitely many values of } k\}) = 0$, by the first Borel–Cantelli lemma (see Section 1.4.2). This holds for any $\epsilon > 0$, and so

$$P\left(\left\{\omega : \lim_{k \to \infty} \frac{S_{k^2}(\omega)}{k^2} = 0\right\}\right) = 1.$$

To get this result for the whole sequence $\{S_n/n\}$, we proceed as follows. Let

$$r_n = \lfloor \sqrt{n} \rfloor \quad \text{for } n = 1, 2, \ldots.$$

That is, r_n is the largest integer less than or equal to $\sqrt{n}$. Note that

$$|S_n| = |S_{r_n^2} + S_n - S_{r_n^2}| \leq |S_{r_n^2}| + |S_n - S_{r_n^2}| \leq |S_{r_n^2}| + D_{r_n},$$

where the random variable D_r is defined by

$$D_r = \max_{r^2 \leq j < (r+1)^2} |S_j - S_{r^2}| \quad \text{for } r = 1, 2, \ldots.$$

Suppose we can show that

$$P\left(\left\{\omega : \lim_{r\to\infty} \frac{D_r(\omega)}{r^2} = 0\right\}\right) = 1.$$

Then it follows from

$$\left|\frac{S_n}{n}\right| \le \left|\frac{S_{r_n^2}}{n}\right| + \frac{D_{r_n}}{n} \le \left|\frac{S_{r_n^2}}{r_n^2}\right| + \frac{D_{r_n}}{r_n^2} \quad \text{for } n = 1, 2, \ldots$$

that the whole sequence S_n/n converges to zero with probability 1. To verify the convergence property of the D_r, we use Kolmogorov's inequality, which can be seen as an extension of Chebyshev's inequality. The famous Kolmogorov inequality reads as follows: if $Y_1, \ldots, Y_n$ are independent random variables with $E(Y_k) = 0$ and $\text{var}(Y_k) < \infty$ for $k = 1, \ldots, n$, then

$$P\left(\max_{1 \le k \le n} |T_k| \ge c\right) \le \frac{1}{c^2} \sum_{k=1}^n \text{var}(Y_k) \quad \text{for any } c > 0,$$

where $T_k = Y_1 + \cdots + Y_k$. An outline of the proof is given in Problem 9.16. Kolmogorov's inequality implies that

$$P\left(\frac{D_r}{r^2} > \epsilon\right) \le \frac{2r\sigma^2}{r^4\epsilon^2} = \frac{2\sigma^2}{r^3\epsilon^2} \quad \text{for } r = 1, 2, \ldots.$$

Note that we cannot use Chebyshev's inequality to obtain such an upper bound, because of the fact that $E(D_r)$ is in general not zero. Next, by the same Borel–Cantelli argument as used above, we obtain the result that D_r/r^2 converges to zero with probability 1. This completes the proof when $\mu = 0$. When $\mu \ne 0$, apply the result for the case of zero mean to the random variables $X_i - \mu$. This gives that $(1/n)\sum_{j=1}^n (X_j - \mu)$ converges to zero with probability 1, or, equivalently, $(1/n)\sum_{j=1}^n X_j$ converges to μ with probability 1.

We can conclude from the strong law of large numbers that, no matter how small $\epsilon > 0$ is chosen, eventually the sample mean $\frac{1}{n}\sum_{k=1}^n X_k$ gets within a distance ϵ of μ and stays within this bandwidth. This conclusion cannot be drawn from the weak law of large numbers. The weak law states only that for any specified large value of n, the random variable $\frac{1}{n}\sum_{k=1}^n X_k$ is likely to be near μ.

The above proof of the strong law of large numbers used the assumption that the X_k are independent and have the same probability distribution with finite first two moments. This assumption can be weakened in two directions. If the independent random variables X_k have the same probability distribution, then the assumption of a finite second moment can be dropped (the assumption of a finite mean can be dropped as well when the X_k are nonnegative). If the independent X_k do not have the same probability distribution, then the strong law of large numbers remains valid when the X_k have the same finite expected value μ and $\sum_{k=1}^{\infty} \frac{\sigma^2(X_k)}{k^2} < \infty$. The proofs of these results are beyond the scope of this book.

Modes of Convergence

The weak law of large numbers and the strong law of large numbers are important instances of convergence in probability and almost-sure convergence, respectively. The general definition of these convergence concepts is as follows. Suppose that the sequence of random variables $X_1, X_2, \ldots$ and the random variable X are defined on the same probability space. The sequence X_n is said to *converge in probability* to X if

$$\lim_{n \to \infty} P(\{\omega : |X_n(\omega) - X(\omega)| \leq \epsilon\}) = 1 \quad \text{for each } \epsilon > 0.$$

The sequence of random variables X_n is said to *converge almost surely* (or to *converge with probability* 1) to X if

$$P(\{\omega : \lim_{n \to \infty} X_n(\omega) = X(\omega)\}) = 1.$$

Almost-sure convergence is the most intuitive notion of convergence. It implies convergence in probability. This can easily be proved by using an equivalent definition of almost-sure convergence. To do so, define for any $\epsilon > 0$ and $n \geq 1$ the sets

$$A_n(\epsilon) = \{\omega : |X_k(\omega) - X(\omega)| \leq \epsilon \text{ for all } k \geq n\},$$
$$B_n(\epsilon) = \{\omega : |X_n(\omega) - X(\omega)| \leq \epsilon\}.$$

The sets $B_n(\epsilon)$ have to do with convergence in probability. The definition of this convergence concept is $\lim_{n \to \infty} P(B_n(\epsilon)) = 1$ for any $\epsilon > 0$. What about the sets $A_n(\epsilon)$? The assertion is that X_n converges almost surely to X if and only if $\lim_{n \to \infty} P(A_n(\epsilon)) = 1$ for any $\epsilon > 0$. To see this, fix $\epsilon > 0$ and note that the sets $A_n(\epsilon)$ are increasing. Therefore, by the continuity of probability measure (see Section 1.4.2), $\lim_{n \to \infty} P(A_n(\epsilon)) = P(\bigcup_{n=1}^{\infty} A_n(\epsilon))$. The set $P(\bigcup_{n=1}^{\infty} A_n(\epsilon))$ is nothing other than the set $\{\omega : |X_k(\omega) - X(\omega)| \leq \epsilon$ for all $k \geq n(\omega)$ for some integer $n(\omega)\}$. This verifies the assertion. The final step is to note that $A_n(\epsilon) \subseteq B_n(\epsilon)$ and so $P(A_n(\epsilon)) \leq P(B_n(\epsilon))$. Therefore, $\lim_{n \to \infty} P(A_n(\epsilon)) = 1$ implies that $\lim_{n \to \infty} P(B_n(\epsilon)) = 1$ for any $\epsilon > 0$, which completes the proof.

Convergence in probability does not necessarily imply almost-sure convergence. A counterexample can be given, but it will not be done here, in which $\{X_n\}$ converges in probability to 0, whereas there is no ω for which $X_n(\omega)$

converges to 0. Convergence in probability in turn implies *convergence in distribution*, that is,

$$\lim_{n\to\infty} P(X_n \le x) = P(X \le x)$$

for any continuity point of $P(X \le x)$. The reader is asked to prove this assertion in Problem 9.15. The converse assertion is not true, as can be seen from the sequence $X_1, X_2, \ldots$ with $X_{2k-1} = X$ and $X_{2k} = -X$, where $P(X = 1) = P(X = -1) = \frac{1}{2}$. An instance of convergence in distribution is provided by the central limit theorem.

Problems

9.7 Let $X_1, X_2, \ldots, X_n$ be independent random variables each having the uniform distribution on $(0, 1)$. Define the random variable P_n as $(X_1 X_2 \cdots X_n)^{1/n}$. Prove that the random variable P_n converges almost surely to $\frac{1}{e}$ as $n \to \infty$.

9.8 Prove that a sequence of random variables $X_1, X_2, \ldots$ converges almost surely to the random variable X if $\sum_{n=1}^{\infty} P(|X_n - X| > \epsilon) < \infty$ for each $\epsilon > 0$. *Hint*: Use the first Borel–Cantelli lemma.

9.9 Suppose that the random variables $X_1, X_2, \ldots$ are independent and converge almost surely to 0. Prove that $\sum_{n=1}^{\infty} P(|X_n| > \epsilon) < \infty$ for any $\epsilon > 0$.

9.10 Consider a chance experiment with sample space $\Omega = \{1, 2, \ldots\}$ and probability measure $p(\omega) = \frac{c}{\omega^2}$ for $\omega \in \Omega$, where $c = \frac{6}{\pi^2}$. For any $k \ge 1$, let the random variable X_k satisfy $X_k(\omega) = 0$ for $1 \le \omega \le k$, and $X_k(\omega) = k$ for $\omega > k$. Show that the sequence $\{X_k\}$ converges almost surely to 0 but $\lim_{k \to \infty} E(X_k) = \infty$.

9.11 A sequence of random variables $X_1, X_2, \ldots$ is said to *converge in mean square* to the random variable X if $\lim_{n \to \infty} E\big((X_n - X)^2\big) = 0$. Verify that convergence in mean square implies convergence in probability. Can you give a counterexample showing that the converse is not necessarily true?

9.12 Suppose that $X_1, X_2, \ldots$ is a sequence of uncorrelated random variables with common expected value μ and common standard deviation σ. Verify that $M_n = \frac{1}{n} \sum_{k=1}^{n} X_k$ converges in mean square to μ.

9.13 Suppose that $X_1, X_2, \ldots$ is a sequence of independent random variables such that $E(X_i) = \mu$ for all i and $\sigma^2(X_i)$ is bounded in i. Let the random variable $Y_k = X_k X_{k+1}$ for $k \ge 1$. Show that $\frac{1}{n} \sum_{k=1}^{n} Y_k$ converges in probability to μ^2.

9.14 Suppose that $Y_1, Y_2, \ldots$ is an infinite sequence of random variables such that $\sum_{n=1}^{\infty} E(Y_n) < \infty$. Prove that Y_n converges almost surely to zero. *Hint*: Use the fact that $E\left(\sum_{n=1}^{\infty} Y_n\right) = \sum_{n=1}^{\infty} E(Y_n)$, by the nonnegativity of the Y_n.

9.15 Prove that convergence in probability implies convergence in distribution. *Hint*: Use the relation $P(X_n \leq x) = P(X_n \leq x, |X_n - X| \leq \epsilon) + P(X_n \leq x, |X_n - X| > \epsilon) \leq P(X \leq x + \epsilon) + P(|X_n - X| > \epsilon)$ and the corresponding relation with the roles of X_n and X interchanged.

9.16 Suppose that $X_1, \ldots, X_n$ are independent random variables with $E(X_k) = 0$ and $\text{var}(X_k) < \infty$ for all k. Prove Kolmogorov's inequality $P\left(\max_{1 \leq k \leq n} |S_k| \geq c\right) \leq \frac{1}{c^2} E(S_n^2)$ for any $c > 0$ and verify next that $E(S_n^2) = \sum_{k=1}^{n} \text{var}(X_k)$, where $S_k = \sum_{j=1}^{k} X_j$ for $1 \leq k \leq n$. *Hint*: Let A_k be the event that $|S_k| \geq c$ but $|S_j| < c$ for $1 \leq j \leq k - 1$ and let the indicator variable $I_k = 1$ if event A_k occurs and $I_k = 0$ otherwise (the A_k are disjoint and $\sum_{k=1}^{n} I_k \leq 1$). Next, noting that $S_n^2 \geq S_n^2 \sum_{k=1}^{n} I_k$, use the relation $E(S_n^2) \geq \sum_{k=1}^{n} E(S_n^2 I_k) = \sum_{k=1}^{n} E[(S_k^2 + 2S_k(S_n - S_k) + (S_n - S_k)^2) I_k]$ together with $P\left(\bigcup_{k=1}^{n} A_k\right) = P\left(\max_{1 \leq k \leq n} |S_k| \geq c\right)$.

9.17 Suppose that $X_1, X_2, \ldots$ are independent random variables each having the one-sided Cauchy density $\frac{2}{\pi(1+x^2)}$ on $(0, \infty)$. Prove that the random variable $M_n = \frac{1}{n} \max(X_1, X_2, \ldots, X_n)$ converges in distribution to a random variable with probability distribution function $e^{-\frac{2}{\pi x}}$ for $x > 0$. *Hint*: Use the relation $\arctg(x) = \frac{\pi}{2} - \arctg(\frac{1}{x})$ and the expansion $\arctg(y) = y - \frac{y^3}{3} + \frac{y^5}{5} - \cdots$ for $|y| \leq 1$.

9.18 Use the result of Problem 9.14 to give an alternative proof of the strong law of large numbers under the assumption that $E(X_i^4) < \infty$. *Hint*: Verify that $\sum_{n=1}^{\infty} \frac{1}{n^4} E(X_1 + \cdots + X_n)^4 < \infty$.

9.2.1 Law of the Iterated Logarithm

The strong law of large numbers and the central limit theorem are the most important and fundamental limit theorems in probability theory. A good candidate for third place is the law of the iterated logarithm. To formulate this law, let $X_1, X_2, \ldots$ be a sequence of independent random variables having a common probability distribution with finite mean μ and finite standard deviation σ. For $n = 1, 2, \ldots$, the partial sum S_n is defined by

$$S_n = \sum_{k=1}^{n} (X_k - \mu).$$

The central limit theorem tells us that suitably normalized sums of independent random variables can be approximated by a normal distribution. That is, the random variable $S_n/(\sigma\sqrt{n})$ is approximately $N(0, 1)$ distributed for n large. The values of $S_n/(\sigma\sqrt{n})$ as a function of n fluctuate, but fluctuate slowly because they are heavily correlated. However, $S_n/(\sigma\sqrt{n})$ as a function of n may take on, and will take on, arbitrarily large positive and negative values, though the occurrence of this is a rare event. The law of the iterated logarithm provides precise bounds on the fluctuations of $S_n/(\sigma\sqrt{n})$ as n grows. The name of this law comes from the iterated logarithm that appears in the expression for the bounds. For almost every possible realization ω of the stochastic process $\{S_n\}$, there is a finite integer $n(\omega)$ so that

$$-\sqrt{2\ln(\ln(n))} < \frac{S_n(\omega)}{\sigma\sqrt{n}} < \sqrt{2\ln(\ln(n))} \quad \text{for all } n \geq n(\omega).$$

Further, for any $c \in (-\sqrt{2}, \sqrt{2})$, the sequence $\{S_n(\omega)/[\sigma\sqrt{2n\ln(\ln(n))}]\}$ takes on infinitely often a value between c and $\sqrt{2}$ and infinitely often a value between c and $-\sqrt{2}$. The set of all realizations ω for which these statements are true has probability one. The function $\sqrt{2\ln(\ln(n))}$ bordering the fluctuations of the $S_n/(\sigma\sqrt{n})$ is not bounded but increases incredibly slowly as n grows. The function has the value 2.292 when $n = 10^6$ and this value has only increased to 2.505 when $n = 10^{10}$. The proof of the law of the iterated logarithm is quite technical and will not be given here. The law provides a beautiful and fascinating result.

As a concrete example, consider the coin-tossing experiment. What can be said about the fluctuations of the random process that describes the proportion of heads obtained when a fair coin is tossed repeatedly? The coin-tossing experiment can be modeled by a sequence of independent random variables $X_1, X_2, \ldots$ with $P(X_i = 0) = P(X_i = 1) = 0.5$ for all i. The expected value and the standard deviation of the X_i are given by $\mu = 0.5$ and $\sigma = 0.5$. The proportion of heads after n tosses is

$$Y_n = \frac{X_1 + \cdots + X_n}{n}.$$

Then, for almost every realization ω of the coin-tossing process, there is a finite integer $n(\omega)$ so that

$$\frac{1}{2} - \frac{1}{2\sqrt{n}}\sqrt{2\ln(\ln(n))} < Y_n(\omega) < \frac{1}{2} + \frac{1}{2\sqrt{n}}\sqrt{2\ln(\ln(n))} \quad \text{for all } n \geq n(\omega),$$

by the law of the iterated logarithm and the fact that $n(Y_n - \mu) = S_n$. As an illustration, Figure 9.1 displays the result of a simulation of the coin-tossing process. The relative frequency of heads is simulated for 1,000,000

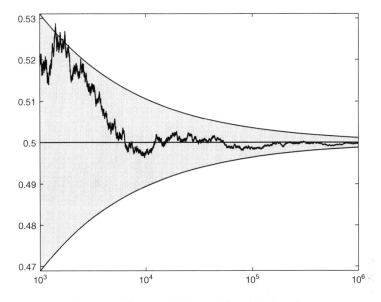

Figure 9.1. The law of the iterated logarithm in action.

tosses of a fair coin. Beginning with the 1,000th toss, the relative frequency of heads is plotted against a log scale for the number of tosses. The simulated sample path nicely demonstrates that the relative frequency stays within the parabolic bounds of the iterated law of the logarithm when the number of tosses increases.

9.3 Kelly Betting System

Suppose that you are playing a game where you have an edge. How should you bet to manage your money in a good way? The idea is always to bet a *fixed proportion* of your present bankroll. When your bankroll decreases you bet less, as it increases you bet more. This strategy is called the *Kelly betting system*, after the American mathematician J. L. Kelly, Jr., who published the system in 1956.[2] The objective of Kelly betting is to maximize the long-run

---

[2] The ideas of Kelly apply not only to gambling scenarios also but to investment scenarios. The Kelly betting system has become a part of mainstream investment theory. It has even been argued that the Kelly criterion should replace the Markowitz criterion as the guide to portfolio selection. The Kelly betting system was popularized by Edward Thorp in his 1962 book *Beat the Dealer*. Since then, several famous investors have made successful use of this betting system.

rate of growth of your bankroll. The optimal value of the fraction to bet can be found by simple arguments based on the strong law of large numbers.

Consider the following model. You are offered a sequence of independent bets, each bet paying out f times your stake with probability p and being a losing proposition with probability $1 - p$. It is assumed that

$$pf > 1.$$

That is, the game is favorable to you. How to gamble, if you must? To avoid trivialities, it is assumed that $0 < p < 1$. Though the expected net payoff of each bet is positive ($pf - 1 > 0$), it is not wise to bet your whole bankroll each time; if you do, you will certainly go bankrupt after a while. Indeed, it is better to bet a fixed proportion of your current bankroll each time. Using the strong law of large numbers, we will show that the betting proportion

$$\alpha^* = \frac{pf - 1}{f - 1}$$

maximizes the long-run growth factor of your bankroll. This is the famous *Kelly formula*. The Kelly betting fraction can be interpreted as the ratio of the expected net payoff for a one-dollar bet and the payoff odds.

It is instructive to give the derivation of the formula for the Kelly betting fraction. Assuming that your starting bankroll is V_0, define the random variable V_n as

$$V_n = \text{the size of your bankroll after } n \text{ bets,}$$

when you bet a fixed fraction α ($0 < \alpha < 1$) of your current bankroll each time. Here it is supposed that winnings are reinvested and that your bankroll is infinitely divisible. Obviously, we have

$$V_n = (1 - \alpha)V_{n-1} + \alpha V_{n-1} R_n \quad \text{for } n = 1, 2, \ldots,$$

where the random variable R_n is equal to the payoff factor f if the nth bet is won and 0 otherwise. By repeated substitution of this equality into itself, we get

$$V_n = (1 - \alpha + \alpha R_1) \times \cdots \times (1 - \alpha + \alpha R_n) V_0 \quad \text{for } n = 1, 2, \ldots,$$

where V_0 is the initial bankroll. Let the growth factor G_n be defined by

$$V_n = V_0 e^{nG_n},$$

or, equivalently, $G_n = \frac{1}{n}\ln\left(\frac{V_n}{V_0}\right)$. Hence, by the above equation for V_n, we get

$$G_n = \frac{1}{n}\Big[\ln(1 - \alpha + \alpha R_1) + \cdots + \ln(1 - \alpha + \alpha R_n)\Big].$$

9.3 Kelly Betting System

The random variables $\ln(1 - \alpha + \alpha R_k)$ are independent and have the same probability distribution. Therefore, by the strong law of large numbers,

$$\lim_{n \to \infty} G_n = p \ln(1 - \alpha + \alpha f) + (1 - p) \ln(1 - \alpha) \quad \text{with probability 1.}$$

Thus, under the strategy of a fixed betting fraction α, the long-run growth factor of your bankroll is given by

$$g(\alpha) = p \ln(1 - \alpha + \alpha f) + (1 - p) \ln(1 - \alpha).$$

It is not difficult to verify that an α_0 with $0 < \alpha_0 < 1$ exists such that $g(\alpha)$ is positive for all α with $0 < \alpha < \alpha_0$ and negative for all α with $\alpha_0 < \alpha < 1$. Choosing a betting fraction between 0 and α_0, your bankroll will ultimately exceed every large level if you simply keep playing for a long enough period of time. Putting the derivative of the function $g(\alpha)$ equal to 0, we find the sought result that the growth factor of your bankroll is maximal for the Kelly betting fraction $\alpha^* = (pf - 1)/(f - 1)$.

What is the asymptotic rate of return for the strategy under which you bet the same fraction α of your bankroll each time? To answer this question, define the random variable γ_n by

$$V_n = (1 + \gamma_n)^n V_0.$$

The random variable γ_n gives the rate of return on your bankroll over the first n bets. It follows from the relationship $V_n = V_0 e^{nG(n)}$ that $\gamma_n = e^{G(n)} - 1$. This implies that $\lim_{n \to \infty} \gamma_n = e^{g(\alpha)} - 1$ with probability 1. Substituting the expression for $g(\alpha)$ and noting that $e^{b \ln(a)} = a^b$, it follows that the asymptotic rate of return for the betting strategy is given by

$$\lim_{n \to \infty} \gamma_n = (1 - \alpha + \alpha f)^p (1 - \alpha)^{1-p} - 1 \quad \text{with probability 1.}$$

For example, for the data $p = 0.4$ and $f = 3$, the Kelly betting fraction $\alpha^* = 0.1$ and the corresponding asymptotic rate of return is 0.0098. In practice, one often uses a fractional Kelly strategy with a betting fraction $\alpha = c\alpha^*$ with $0 < c < 1$, where c is typically in the range from 0.3 to 0.5. The reason is that under fractional Kelly, the fluctuations of your bankroll are less wild than under full Kelly. The increased safety of the fractional Kelly strategy ($\alpha = c\alpha^*$) can be quantified by the approximate relation

$$P(\text{reaching a bankroll of } aV_0 \text{ without falling down first to } bV_0)$$
$$\approx \frac{1 - b^{2/c-1}}{1 - (b/a)^{2/c-1}}$$

for any $0 < b < 1 < a$. Also, the asymptotic rate of return under fractional Kelly is approximately equal to $c(2 - c)$ times the asymptotic rate of return under full Kelly.[3] For example, by betting only half of the Kelly fraction, you give up approximately one-quarter of your maximum growth rate, but you increase the probability of doubling your bankroll without having it halved first from 0.67 to 0.89. More on Kelly betting can be found in Thorp's survey paper, mentioned in the footnote below.

Problems

9.19 You have received a reliable tip that in the local casino the roulette wheel is not exactly fair. The probability of the ball landing on the number 13 is twice what it should be. The roulette table in question will be in use that evening. In that casino, European roulette with 37 slots numbered $0, 1, \ldots, 36$ is played and they allow you to bet any amount you wish. You go with 1,000 euros to the casino and intend to make a lot of bets on number 13. You receive a payoff of 36 times the amount staked if the ball lands on number 13. What betting strategy would you use?

9.20 In a particular game of chance, the payout is f_1 times the amount staked with a probability of p and f_2 times the amount staked with a probability of $1 - p$, where $0 < p < 1$, $f_1 > 1, 0 \leq f_2 < 1$, and $pf_1 + (1 - p)f_2 > 1$. You play this game a large number of times and each time you stake the same fraction α of your bankroll. Verify that the Kelly fraction maximizing the growth rate of your capital is given by the formula $\alpha^* = \min\left(\frac{pf_1+(1-p)f_2-1}{(f_1-1)(1-f_2)}, 1\right)$ with $(1 - \alpha^* + \alpha^* f_1)^p (1 - \alpha^* + \alpha^* f_2)^{1-p} - 1$ as the corresponding rate of return.

9.21 Consider again the Kelly model from Problem 9.20. Use the central limit theorem to approximate the probability distribution of your bankroll after n bets when a fixed fraction α of your bankroll is staked each time. Take the numerical data $p = \frac{1}{2}, f_1 = 1.8, f_2 = 0.4$ and assume an initial bankroll of $V_0 = \$10{,}000$. Use simulation to investigate how good the approximation is when $n = 52$ and $\alpha = \frac{5}{24}$. Can you intuitively explain why it is likely that your bankroll will be very close to zero after 52 bets when the value $\alpha = 1$ is used?

---

[3] These approximate results are taken from Edward O. Thorp, "The Kelly criterion in blackjack, sports betting, and the stock market," revised version 2005, www.edwardothorp.com. Simulation studies reveal that the approximation is very accurate for all cases of practical interest.

9.4 Renewal–Reward Processes

An extremely useful stochastic process in applied probability is the renewal–reward process. This is a random process on which a reward structure is imposed. The most useful result in renewal–reward processes is the so-called renewal–reward theorem. This theorem can be derived directly from the strong law of large numbers and has many applications. It is also a useful result in the long-run analysis of Markov chains in Chapter 10.

Let us first explain the notion of a *stochastic process*, also called a *random process*. A stochastic process $\{S_t, t \in \Gamma\}$ is a collection of random variables S_t, indexed by an ordered time variable $t \in \Gamma$. The process is said to be a continuous-time process if Γ is the set of nonnegative reals and is said to be a discrete-time process if Γ is the set of nonnegative integers. Many stochastic processes have the property of regenerating themselves at certain points in time, so that the continuation of the process after any regeneration epoch is a probabilistic replica of the process starting at time zero and is independent of the behavior of the process before the regeneration epoch. To illustrate this, consider the next example. A light bulb having a continuously distributed lifetime is replaced by a new one upon failure or upon reaching a specified age T, whichever occurs first. For any $t \geq 0$, let the random variable S_t denote the age of the bulb in use at time t. Then the process $\{S_t, t \geq 0\}$ is a continuous-time stochastic process that regenerates itself each time a new bulb is installed.

Let us consider a regenerative stochastic process $\{S_t, t \geq 0\}$. For clarity of presentation, we take a continuous-time process and assume that the process starts in the regeneration state at time zero. The time interval between two successive regeneration epochs is called a *regeneration cycle* (or *cycle*). Let the random variables $L_1, L_2, \ldots$ denote the lengths of the successive cycles. These random variables are independent and have the same probability distribution. It is assumed that the expected value of L_k is finite and positive. Imagine now that rewards are earned or costs are incurred during the process. The reward structure usually consists of reward rates that are earned continuously in time and lump rewards that are earned at specific moments. Let the random variable R_n denote the reward earned during the nth cycle. The assumption is made that the random variables $R_1, R_2, \ldots$ are independent and have the same probability distribution with a finite expected value. For any $t > 0$, define the random variable

$$R(t) = \text{the cumulative reward earned up to time } t.$$

The process $\{R(t), t \geq 0\}$ is called the *renewal–reward process*.

340   9 Additional Topics in Probability

Rule 9.6 (the renewal–reward theorem) *With probability 1,*

$$\lim_{t \to \infty} \frac{R(t)}{t} = \frac{E(R_1)}{E(L_1)}.$$

In words, with probability 1, the long-run average reward per unit time equals the expected reward in a cycle divided by the expected length of the cycle.

The proof of this important theorem goes as follows. For any $t > 0$, define the random variable $N(t)$ as the number of cycles completed up to time t. Since $E(L_1) > 0$, we have that $P(N(t) < \infty) = 1$ for any $t > 0$.[4] A strong law of large numbers applies to $N(t)$:

$$\lim_{t \to \infty} \frac{N(t)}{t} = \frac{1}{E(L_1)} \quad \text{with probability 1.}$$

To prove this, note that $L_1 + \cdots + L_{N(t)} \leq t < L_1 + \cdots + L_{N(t)+1}$, by the definition of $N(t)$. Also, with probability 1, the random variable $N(t)$ tends to ∞ as $t \to \infty$, being a corollary of the fact that $P(L_1 + \cdots + L_n < \infty) = 1$ for all $n \geq 1$. The result $\lim_{t \to \infty} N(t)/t = 1/E(L_1)$ now follows by letting $t \to \infty$ in

$$\frac{L_1 + \cdots + L_{N(t)}}{N(t)} \leq \frac{t}{N(t)} < \frac{L_1 + \cdots + L_{N(t)+1}}{N(t)}$$

and applying the strong law of large numbers to the sequence $\{L_k\}$. The remainder of the proof proceeds as follows. Assuming for a moment that the rewards are nonnegative, we have for any $t > 0$ that

$$\sum_{k=1}^{N(t)} R_k \leq R(t) \leq \sum_{k=1}^{N(t)+1} R_k.$$

This inequality implies

$$\frac{\sum_{k=1}^{N(t)} R_k}{N(t)} \times \frac{N(t)}{t} \leq \frac{R(t)}{t} \leq \frac{\sum_{k=1}^{N(t)+1} R_k}{N(t)} \times \frac{N(t)}{t}.$$

Letting $t \to \infty$, applying the strong law of large numbers to the R_k, and using the relation $\lim_{t \to \infty} N(t)/t = 1/E(L_1)$, we obtain the desired result. If the rewards can take on both positive and negative values, then use the decomposition $R(t) = R^+(t) - R^-(t)$ and $R_n = R_n^+ - R_n^-$, where $R^+(t) = \max(R(t), 0)$, $R^-(t) = -\min(R(t), 0)$, $R_n^+ = \max(R_n, 0)$, and $R_n^- = -\min(R_n, 0)$. Next repeat the above proof for each of the nonnegative renewal–reward processes $\{R^+(t), t \geq 0\}$ and $\{R^-(t), t \geq 0\}$.

---

[4] To see this, note that there is a constant $c > 0$ such that $P(L_1 > c) > 0$. Then, by the second Borel–Cantelli lemma (see Problem 2.37), $P(L_n > c$ for infinitely many $n) = 1$, which implies that $P\left(\sum_{i=1}^{n} L_i \leq t \text{ for all } n \geq 1\right) = 0$ or, equivalently, $P(N(t) = \infty) = 0$.

9.4 Renewal–Reward Processes

The expected-value version of the renewal–reward theorem states that $\lim_{t \to \infty} E[R(t)]/t = E(R_1)/E(L_1)$. Somewhat surprisingly, this result is more difficult to establish than the probability 1 version. The reason is that subtle arguments are needed to justify the interchange of limit and expectation operator.

Example 9.3 A source emitting messages is alternately on and off. Let $U_1, U_2, \ldots$ be the sequence of on-times and $D_1, D_2, \ldots$ be the sequence of off-times. The random variables U_i are independent and have the same probability distribution with expected value μ_{on}. Also, the D_i are independent and have the same probability distribution with expected value μ_{off}. An on-time starts at time 0. What is the long-run fraction of time that the source is on?

Solution. For any $t \geq 0$, let the random variable S_t be equal to 1 if the source is on at time t and 0 otherwise. The stochastic process $\{S_t, t \geq 0\}$ is a regenerative process. The epochs at which an on-period starts can be taken as the regeneration epochs. A cycle consists of an on-period followed by an off-period. Note that it is not needed that the off-times are independent of the on-times. Imagine that a reward at rate 1 is earned when the source is on and no reward is earned otherwise. Then, the long-run fraction of time the source is on is equal to the long-run average reward per unit time. The expected reward earned in a cycle is $E(U_1) = \mu_{on}$ and the expected length of a cycle is $E(U_1 + D_1) = \mu_{on} + \mu_{off}$. Therefore,

$$\text{the long-run average reward per unit time} = \frac{\mu_{on}}{\mu_{on} + \mu_{off}},$$

showing that the long-run fraction of time the system is on is $\frac{\mu_{on}}{\mu_{on}+\mu_{off}}$. The intermediate step of interpreting the long-run fraction of time the process is in a certain state as a long-run average reward per unit time is very helpful in many situations.

Example 9.4 A two-unit reliability system has one operating unit and one identical unit in cold standby. The operating unit has a constant failure rate α. Upon failure of the operating unit, the standby unit is put into operation if available. A failed unit has a fixed repair time $\tau > 0$. Only one unit can be in repair at a time. The system is down when both units have failed. What is the long-run fraction of time the system is down?

Solution. For the stochastic process $\{S_t\}$ describing the status of both units, we take as regeneration epochs the times at which one of the units is put into operation while the other one goes into repair. A cycle is the time between two such consecutive regeneration epochs. The lifetime X of an operating unit is exponentially distributed with parameter α. Under the condition that the lifetime of the operating unit expires during the repair time of the failed unit, the length of the cycle is τ; otherwise, the length of the cycle is τ plus the remaining lifetime of the operating unit. The remaining lifetime has expected value $1/\alpha$, by the memoryless property of the exponential distribution. Hence,

$$E(\text{length of a cycle}) = \tau(1 - e^{-\alpha\tau}) + \left(\tau + \frac{1}{\alpha}\right)e^{-\alpha\tau} = \tau + \frac{1}{\alpha}e^{-\alpha\tau}.$$

The expected downtime during one cycle is $E(\tau - \min(\tau, X))$. This expectation can be evaluated as $\tau - \int_0^\tau x\alpha e^{-\alpha x}dx - \tau\int_\tau^\infty \alpha e^{-\alpha x}dx = \tau - \frac{1}{\alpha}(1 - e^{-\alpha\tau})$. Thus, by the renewal–reward theorem, we have with probability 1 that

$$\text{the long-run fraction of time the system is down} = \frac{\alpha\tau + e^{-\alpha\tau} - 1}{\alpha\tau + e^{-\alpha\tau}}.$$

9.4.1 Renewal Process

Let $X_1, X_2, \ldots$ be a sequence of positive random variables that are independent of each other and have the same probability distribution. Think of the X_i as the interoccurrence times between successive events in some specific situation (for example, arrivals of customers in a queue or replacements of light bulbs). As in the proof of the renewal–reward theorem, the counting process $\{N(t), t \geq 0\}$ is defined by

$$N(t) = \max\left\{n : \sum_{i=1}^n X_i \leq t\right\},$$

where $N(0) = 0$. The random variable $N(t)$ counts the number of events that occur up to time t. The stochastic process $\{N(t), t \geq 0\}$ is called a *renewal process*. The expected value $M(t) = E[N(t)]$ as a function of t is called the *renewal function*. If the interoccurrence times X_i are exponentially distributed with expected value $1/\alpha$, then the renewal process $\{N(t), t \geq 0\}$ is a Poisson process with rate α and the renewal function is given by $M(t) = \alpha t$ for all $t \geq 0$. The Poisson process has already been discussed in Section 4.5. The renewal function $M(t)$ satisfies the so-called *renewal equation*

$$M(t) = F(t) + \int_0^t M(t - x)f(x)\,dx \quad \text{for } t \geq 0$$

when the interoccurrence times X_i are continuous random variables with distribution function $F(x) = P(X_1 \leq x)$ and density function $f(x)$. The derivation of the renewal equation is simple. Note that $N(t)$ is distributed as $1 + N(t - x)$ if $X_1 = x$ with $x \leq t$ and $N(t) = 0$ if $X_1 > t$. Then, by the law of conditional expectation, $E[N(t)] = \int_0^t (1 + E[N(t-x)])f(x)\,dx$, showing that $E[N(t)] = F(x) + \int_0^t E[N(t-x)]f(x)\,dx$. In general, the numerical computation of the renewal function requires a sophisticated method such as a numerical inversion algorithm for Laplace transforms. You are asked to take for granted that the renewal function is uniquely determined by the integral equation.

In Rule 9.6, we proved the strong law of large numbers for $N(t)$:

$$\lim_{t\to\infty} \frac{N(t)}{t} = \frac{1}{E(X_1)} \quad \text{with probability 1.}$$

We also have that $\lim_{t\to\infty} E[N(t)]/t = 1/E(X_1)$. This result is known as the *elementary renewal theorem*. For continuously distributed interoccurrence times, we also have the result

$$\lim_{t\to\infty}\left[E[N(t)] - \frac{t}{\mu}\right] = \frac{\sigma^2 + \mu^2}{2\mu^2} - 1,$$

where $\mu = E(X_1)$ and $\sigma^2 = \text{var}(X_1)$. The proofs of these results will be omitted. Without proof we also state the central limit theorem for $N(t)$:

$$\lim_{t\to\infty} P\left(\frac{N(t) - t/\mu}{\sigma\sqrt{t/\mu^3}} \leq x\right) = \frac{1}{\sqrt{2\pi}} \int_{-\infty}^{x} e^{-\frac{1}{2}z^2}\,dz \quad \text{for all } x.$$

An important random variable in renewal theory is the so-called *excess life* or *residual life*. For any $t > 0$, this random variable is defined by

$$Y(t) = X_1 + \cdots + X_{N(t)+1} - t.$$

Thinking of the X_i as the lifetimes of bulbs, $Y(t)$ is the remaining lifetime of the bulb in use at time t. The expected value of $Y(t)$ is given by

$$E[Y(t)] = \mu\bigl(E[N(t)] + 1\bigr) - t \quad \text{for any } t > 0.$$

To explain this result, note that $N(t) \geq n-1$ if and only if $X_1 + \cdots + X_{n-1} \leq t$. Thus, for every $n \geq 1$, the event $\{N(t) + 1 \geq n\}$ depends only on $X_1, \ldots, X_{n-1}$ and not on $X_n, X_{n+1}, \ldots$. Therefore, we can apply Wald's equation, which is stated in Problem 3.61. Applying this famous equation, we get $E(X_1 + \cdots + X_{N(t)+1}) = E(X_1)E(N(t) + 1)$. Using the asymptotic expansion of $E[N(t)] - t/\mu$, we next find

$$\lim_{t\to\infty} E[Y(t)] = \frac{1}{2}\left(1 + \frac{\sigma^2}{\mu^2}\right)\mu.$$

This is a practically useful formula. It enables us to explain the *waiting-time paradox*. Most people have experienced unexpectedly long waits at bus stops when buses depart irregularly and they arrive at the bus stop at a random moment. If $\sigma/\mu > 1$, then $\lim_{t\to\infty} E[Y(t)] > \mu$. In words, the mean waiting time for the next bus is then larger than the mean interdeparture time. An intuitive explanation of the waiting-time paradox is that it is more likely to hit a long interdeparture time than a short one when arriving at the bus stop at a random moment.

The approximation $E[(Y(t)] \approx \frac{1}{2}\left(1 + \frac{\sigma^2}{\mu^2}\right)\mu$ for t large is very useful in inventory theory. Suppose we have an inventory system in which orders for a certain product arrive randomly, one at a time. The order sizes are independent random variables $X_1, X_2, \ldots$ having the same probability distribution with expected value μ and standard deviation σ. The inventory position of the product is controlled by an (s, S) policy with $0 < s < S$. Each time the stock level falls below s, it is instantaneously replenished to the level S. Then, the average size of a replenishment can be approximated by $S - s + \frac{1}{2}\left(1 + \frac{\sigma^2}{\mu^2}\right)\mu$ when $S - s$ is sufficiently large. This is readily seen by noting that the undershoot of the reorder level s is distributed as the excess life $Y(S - s)$ in a renewal process with the order sizes as interoccurrence times.

Renewal theory has many applications to quite a wide range of practical probability problems. Also, it is a keystone in establishing limit theorems for Markov chains, which are the subject of the next two chapters. A detailed discussion of the beautiful results from renewal theory is beyond the scope of this introductory book.

Problems

9.22 A shuttle regularly goes from Amsterdam Schiphol airport to the Museum Art hotel in the city center and then back. Each trip is done at a fixed speed. Each time, this speed in miles per hour is a sample from the uniform distribution on $(30, 50)$ when the trip is from the airport to the city and is equally likely to be 30 or 50 miles per hour when the trip is from the city center to the airport. The distance from the airport to the hotel is the same as the distance from the hotel to the airport. What stochastic process do you consider and what are the regeneration epochs? Use the renewal–reward theorem to find the long-run proportion of the shuttle's operating time that is spent going to the airport.

9.23 What is the long-run average cost in the replacement example discussed above when there is a fixed cost $c_1 > 0$ for a planned replacement and a fixed cost $c_2 > c_1$ for a failure replacement?

9.24 Customer orders arrive at a production facility according to a Poisson process with rate λ. Processing of the orders starts as soon as N orders have accumulated. All orders are processed simultaneously and the processing time is negligible. A fixed cost of $K > 0$ is incurred for each production setup. Also, there is a holding cost of $h > 0$ per unit time for each order waiting to be processed. What stochastic process do you consider and what are the regeneration epochs? Use the renewal–reward theorem to find the long-run average cost per unit time and verify that the optimal value of N is one of the two integers nearest to $\sqrt{2K\lambda/h}$.

9.25 A communication channel is alternately on and off. An on-time starts at time zero. The on-times are independent random variables having beta density $f(x) = 6x(1 - x)$ for $0 < x < 1$. Any off-time is dependent on the preceding on-time and is equal to $x^2\sqrt{x}$ if the realized value of the on-time is x. What are the regeneration epochs? What is the long-run fraction of time the channel is on?

9.26 Customers asking for a single product arrive according to a Poisson process with rate λ. Each customer asks for one unit of the product. Any demand which cannot be satisfied from stock on hand is lost. Opportunities to replenish the stock occur according to a Poisson process with rate μ, independently of the demand process. A replenishment is only made when the stock on hand is zero and then the stock is raised to the level Q. Use the renewal–reward theorem to verify that the long-run fraction of time the system is out of stock and the long-run fraction of demand that is lost are both equal to $(1/\mu)/(1/\mu + Q/\lambda)$. The finding that these two fractions are the same is a special case of the property "Poisson arrivals see time averages," see also Chapter 11.

9.27 At a processor, jobs arrive according to a Poisson process with rate λ. A job is accepted for processing when the processor is idle; otherwise, the job is rejected and goes elsewhere. The processing time of each job has an expected value of μ. Prove that the long-run fraction of time the processor is idle and the long-run fraction of jobs that are accepted are both equal to $(1 + \lambda\mu)^{-1}$.

9.28 A two-unit reliability system has one operating unit and one identical unit in cold standby. The operating unit has a constant failure rate of α. Upon failure of the operating unit, the standby unit is put into operation if available. The system is down when both units have failed. The system is inspected every T time units. The inspection times provide the only opportunity to replace failed units with new ones. The inspection and replacement times are negligible. What is the long-run fraction of time the system is down?

9.29 A light bulb is replaced only upon failure. The successive lifetimes of the bulbs are independent random variables, each having the same distribution function $F(x)$ with finite mean μ. Use the renewal–reward theorem to show that the long-run fraction of time that the light bulb in use is older than c is $\frac{1}{\mu}\int_c^\infty (1 - F(x))\,dx$ for any $c > 0$. *Note*: This result is in agreement with a basic result in renewal theory, which states that the excess life $Y(t)$ satisfies $\lim_{t\to\infty} P(Y(t) \le x) = \frac{1}{\mu}\int_0^x (1 - F(y))\,dy$ for any $x > 0$ when the interoccurrence times are continuously distributed. This distribution function is called the *equilibrium excess distribution*.

9.30 The lifetime of an operating unit is a random variable X with mean μ and standard deviation σ. Upon failure of the unit, it is immediately replaced by an identical unit. The system incurs a cost at a rate which is equal to the age of the unit in use. Use the renewal–reward theorem to show the long-run average cost per unit time is $\frac{1}{2}\left(1 + \frac{\sigma^2}{\mu^2}\right)\mu$.

9.31 Consider an age-replacement model in which preventive replacements are only possible at special times. Opportunities for preventive replacements occur according to a Poisson process with rate λ. The item is replaced with a new one upon failure or upon a preventive replacement opportunity occurring when the age of the item is T or more, whichever occurs first. The lifetime of the item has a probability density $f(x)$. The cost of replacing the item upon failure is c_0 and the cost of a preventive replacement is c_1 with $0 < c_1 < c_0$. Use the renewal–reward theorem to find the long-run average cost per time unit.

9.32 Verify by substitution into the renewal equation that the renewal function $M(t)$ is equal to $\frac{t}{\mu} - \left(\frac{1-p}{2-p}\right)^2 (1 - e^{-\lambda(2-p)t})$ for $t > 0$ when the interoccurrence times have the density function $f(x) = pe^{-\lambda x} + (1-p)\lambda^2 x e^{-\lambda x}$ for $x > 0$.

9.33 In a communication system, messages requiring transmission arrive according to a Poisson process with a rate of λ per unit time. The messages are temporarily stored in a buffer having ample capacity. Every T time units, the buffer is cleared of all messages present. The buffer is empty at time zero. A fixed cost of $K > 0$ is incurred for each clearing of the buffer. Also, there is a holding cost of $h > 0$ per unit time for each message in the buffer. Use the renewal–reward theorem to verify that the long-run average cost per unit time is minimal for $T = \sqrt{2K/(h\lambda)}$. *Hint*: The contribution of the nth message to the holding costs in the first cycle is $h(S_n)$, where S_n is the arrival time of the nth message and the function $h(x) = h \times (T - x)$ for $0 < x < T$ and $h(x) = 0$ otherwise. The density function of S_n is $\lambda^n \frac{x^{n-1}}{(n-1)!} e^{-\lambda x}$.

9.34 Consider again Problem 9.33. Use the renewal–reward theorem to show that the long-run fraction of time the buffer has i messages is

$\frac{1}{\lambda T}\left(1 - \sum_{k=0}^{i} e^{-\lambda T}\frac{(\lambda T)^k}{k!}\right)$ for any i. *Hint*: Under the condition that a Poisson process has exactly n occurrences of events in a given time interval $(0, T)$, the joint distribution of the n event times is the same as the joint distribution of the order statistics of n independent random variables that are uniformly distributed on $(0, T)$, see Section 5.4.1.

9.35 Suppose that the interoccurrence times X_i of the renewal process $\{N(t)\}$ are Erlang distributed with shape parameter r and scale parameter α. Show that $P(N(t) \leq k) = \sum_{j=0}^{(k+1)r-1} e^{-\alpha t}\frac{(\alpha t)^k}{k!}$ for $k = 0, 1, \ldots$ and any $t > 0$. *Hint*: The Erlang-distributed interoccurence time can be interpreted as the sum of r independent phases, each having the same exponential distribution with scale parameter α.

9.36 For the residual life $Y(t)$ in a renewal process for which the interoccurrence times have probability distribution function $F(x)$ and probability density function $f(x)$, define $Q_t(u) = P(Y(t) > u)$. Use the law of conditional probability to show that $Q_t(u) = 1 - F(t+u) + \int_0^t Q_{t-x}(u)f(x)\,dx$. For the special case that the interoccurrence times are exponentially distributed with parameter λ, verify by substitution that $P(Y(t) > u) = e^{-\lambda u}$ for $u > 0$, regardless of the value of t (*the memoryless property of the Poisson process*).

9.37 Suppose that batches of customers arrive every time unit. The batch sizes are independent of each other and have probability mass function $\{p_k, k = 1, 2, \ldots\}$ with finite mean μ. Can you give an intuitive explanation of the result that the long-run fraction of customers belonging to a batch of size s is sp_s/μ for any $s \geq 1$?

10
Discrete-Time Markov Chains

In previous chapters we have dealt with sequences of independent random variables. However, many random systems evolving in time involve sequences of dependent random variables. Think of the outside weather temperature on successive days, or the price of IBM stock at the end of successive trading days. Many such systems have the property that the current state alone contains sufficient information to give the probability distribution of the next state. The probability model with this feature is called a Markov chain. The concepts of state and state transition are at the heart of Markov chain analysis. The line of thinking through the concepts of state and state transition is very useful to analyze many practical problems in applied probability.

Markov chains are named after the Russian mathematician Andrey A. Markov (1856–1922), who introduced the concept of the Markov chain in a 1906 publication. In a famous paper written in 1913, he used his probability model to analyze the frequencies with which vowels and consonants occur in Pushkin's novel *Eugene Onegin*. Markov showed empirically that adjacent letters in Pushkin's novel are not independent but obey his theory of dependent random variables. Markov's work helped launch the modern theory of stochastic processes (a *stochastic process* is a collection of random variables, indexed by an ordered time variable). The characteristic property of a Markov chain is that its memory goes back only to the most recent state. Knowledge of the current state only is sufficient to describe the future development of the process. A Markov model is the simplest model for random systems evolving in time when the successive states of the system are not independent. But this model is no exception to the rule that simple models are often the most useful models for analyzing practical problems. The theory of Markov chains has applications to a wide variety of fields, including biology, physics, engineering, operations research, and computer science. Markov chains are almost everywhere in science today. A similar method as used by Andrey

Markov to study the alternation of vowels and consonants in Pushkin's novel helps identify genes in DNA. Markovian language models are nowadays used in speech recognition. In physics, Markov chains are used to simulate the macrobehavior of systems made up of many interacting particles. Google's PageRank algorithm is based on a Markov chain whose states are the pages of the World Wide Web. Performance analysis of inventory and queueing systems is based on Markov chains, and so on.

The purpose of this chapter is to introduce you to the exciting world of Markov chains and their applications. Numerous probability problems can be formulated and solved as Markov chains. We first present techniques to analyze the time-dependent behavior of Markov chains. In particular, we give much attention to Markov chains with one or more absorbing states. Such Markov chains have interesting applications in the analysis of success runs, among others. Next we deal with the long-run behavior of Markov chains and give solution methods to answer questions such as: what is the long-run proportion of time that the system will be in any given subset of states? Finally, we discuss the method of Markov chain Monte Carlo simulation, which has revolutionized the field of Bayesian statistics and many other areas of science.

10.1 Markov Chain Model

A Markov chain deals with a collection of random variables, indexed by an ordered time parameter. The Markov model is the simplest conceivable generalization of a sequence of independent random variables. A Markov chain is a sequence of trials having the property that the outcome of each last trial provides enough information to predict the outcome of any future trial. Despite its very simple structure, the Markov model is extremely useful in a wide variety of practical probability problems. The beginning student often has difficulties in grasping the concept of the Markov chain when a formal definition is given. Let's begin with an example that illustrates the essence of what a Markov process is.

Example 10.1 A drunkard wanders about a town square. At each step he no longer remembers anymore the direction of his previous steps. Each step is a unit distance in a randomly chosen direction and has equal probability $\frac{1}{4}$ of going north, south, east, or west, as long as the drunkard has not reached the edge of the square, see Figure 10.1. The drunkard never leaves the square. Should he reach the boundary of the square, his next step is equally likely to be in one of the three remaining directions if he is not at a corner point, and

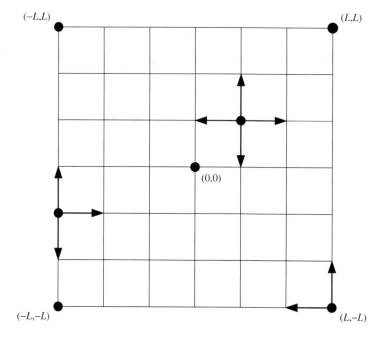

Figure 10.1. The drunkard's walk.

is equally likely to be in one of the two remaining directions otherwise. The drunkard starts in the middle of the square. What stochastic process describes the drunkard's walk?

Solution. To answer this question, we define the random variable X_n as

$X_n =$ the position of the drunkard just after the nth step

for $n = 0, 1, \ldots$ with the convention $X_0 = (0, 0)$. We say that the drunkard is in state (x, y) when the current position of the drunkard is described by the point (x, y). The collection $\{X_0, X_1, \ldots\}$ of random variables is a stochastic process with discrete time parameter and finite state space

$$I = \{(x, y) : x, y \text{ integer and } -L \leq x, y \leq L\},$$

where L is the distance from the middle of the square to its boundary. The successive states of the drunkard are not independent of each other, but the next position of the drunkard depends only on his current position and is not influenced by the earlier positions in his path. That is, the process $\{X_0, X_1, \ldots\}$ has the so-called *Markovian property*, which says that the state at any given time summarizes everything about the past that is relevant to the future.

Many random systems evolving over time can be modeled to satisfy the Markovian property. Having this property introduced informally, we are now ready to give a formal definition of a Markov chain. Let $X_0, X_1, \ldots$ be a sequence of random variables. It is helpful to think of X_n as the state of a dynamic system at time $t = n$. The sequence $X_0, X_1, \ldots$ is called a discrete-time stochastic process. In the sequel, the set of possible values of the random variables X_n is assumed to be *finite* and is denoted by I. The set I is called the *state space* of the stochastic process $\{X_0, X_1, \ldots\}$.

Definition 10.1 *The stochastic process $\{X_n, n = 0, 1, \ldots\}$ with state space I is said to be a discrete-time Markov chain if it possesses the Markovian property, that is, for each time point $n = 0, 1, \ldots$ and all possible values of the states $i_0, i_1, \ldots, i_{n+1} \in I$, the process has the property*

$$P(X_{n+1} = i_{n+1} \mid X_0 = i_0, X_1 = i_1, \ldots, X_{n-1} = i_{n-1}, X_n = i_n)$$
$$= P(X_{n+1} = i_{n+1} \mid X_n = i_n).$$

The term $P(X_{n+1} = i_{n+1} \mid X_0 = i_0, X_1 = i_1, \ldots, X_{n-1} = i_{n-1}, X_n = i_n)$ should be read as follows: it is the conditional probability that the system will be in state i_{n+1} at the *next* time point $t = n + 1$ if the system is in state i_n at the *current* time $t = n$ and has reached the current state i_n via the states $i_0, i_1, \ldots, i_{n-1}$ at the *past* time points $t = 0, 1, \ldots, n - 1$. The Markovian property says that this conditional probability depends only on the current state i_n and is not altered by knowledge of the past states $i_0, i_1, \ldots, i_{n-1}$. The current state summarizes everything about the past that is relevant to the future. At any time $t = n$, the process essentially forgets how it got into the state X_n. It is not true that the state X_{n+1} at the next time point $t = n + 1$ is independent of $X_0, \ldots, X_{n-1}$, but all of the dependency is captured by X_n.

The Markov chain approach is a very powerful tool in applied probability. Using the concept of state and choosing the state in an appropriate way, numerous probability problems can be solved by Markov chain methods.

In Example 10.1 the Markovian property was satisfied in a natural way by choosing the state of the process as the position of the drunkard on the square. However, in other applications the choice of the state variable(s) may require more thought in order to satisfy the Markovian property. To illustrate this, consider Example 10.1 again and assume now that the drunkard never chooses the same direction as was chosen in the previous step. Then we need an extra state variable in order to satisfy the Markovian property. Let's say that the drunkard is in state (x, y, N) when the position of the drunkard on the square is (x, y) and he moved northward in his previous step. Similarly, the states (x, y, E), (x, y, S), and (x, y, W) are defined. Letting X_n be the state of the drunkard after the nth step (with the convention $X_0 = (0,0)$), the stochastic process $\{X_0, X_1, \ldots\}$ satisfies the Markovian property and thus is a Markov chain. The transition probabilities are easy to give. For example, if the current state of the process is (x, y, S) with (x, y) an interior point of the square, then the next state of the process is equally likely to be one of the three states $(x+1, y, E), (x-1, y, W)$, or $(x, y+1, N)$. In the drunkard's walk, the concepts of *state* and *state transition* come up in a natural way. These concepts are at the heart of Markov chain analysis.

In the following, we will restrict our attention to time-homogeneous Markov chains. For such chains the transition probability $P(X_{n+1} = j \mid X_n = i)$ does not depend on the value of the time parameter n and so $P(X_{n+1} = j \mid X_n = i) = P(X_1 = j \mid X_0 = i)$ for all n. We write

$$p_{ij} = P(X_{n+1} = j \mid X_n = i).$$

The probabilities p_{ij} are called the *one-step transition probabilities* of the Markov chain and are the same for all time points n. They satisfy

$$p_{ij} \geq 0 \text{ for } i, j \in I \quad \text{and} \quad \sum_{j \in I} p_{ij} = 1 \quad \text{for all } i \in I.$$

The notation p_{ij} is sometimes confusing for the beginning student: p_{ij} is not a joint probability, but a conditional probability. However, the notation p_{ij} rather than the notation $p(j \mid i)$ has found widespread acceptance.

A Markov chain $\{X_n, n = 0, 1, \ldots\}$ is completely determined by the probability distribution of the initial state X_0 and the one-step transition probabilities p_{ij}. In applications of Markov chains, the art is:

(a) to choose the state variable(s) such that the Markovian property holds;
(b) to determine the one-step transition probabilities p_{ij}.

How to formulate a Markov chain model for a concrete problem is largely an art that is developed with practice. Putting yourselves in the shoes of someone who has to write a simulation program for the problem in question may be helpful in choosing the state variable(s). Once the (difficult) modeling step is done, the rest is simply a matter of applying the theory that will be developed in the next sections. The student cannot be urged strongly enough to try the problems at the end of this section to acquire skills to model new situations. In order to help students develop intuition into how practical situations can be modeled as a Markov chain, we give three examples. The first example deals with the Ehrenfest model for gas diffusion. In physics, the Ehrenfest model resolved at the beginning of the twentieth century a seeming contradiction between the second law of thermodynamics and the laws of mechanics.

Example 10.2 Two compartments A and B together contain r particles. With the passage of every time unit, one of the particles is selected at random and is removed from its compartment to the other. What stochastic process describes the contents of the compartments?

Solution. Let us take as the state of the system the number of particles in compartment A. If compartment A contains i particles, then compartment B contains $r - i$ particles. Define the random variable X_n as

X_n = the number of particles in compartment A after the nth transfer.

By the physical construction of the model with independent selections of a particle, the process $\{X_n\}$ satisfies the Markovian property and thus is a Markov chain. The state space is $I = \{0, 1, \ldots, r\}$. The probability of going from state i to state j in one step is zero, unless $|i - j| = 1$. The one-step transition probability $p_{i,i+1}$ translates into the probability that the randomly selected particle belongs to compartment B and $p_{i,i-1}$ translates into the probability that the randomly selected particle belongs to compartment A. Hence, for $1 \leq i \leq r - 1$,

$$p_{i,i+1} = \frac{r-i}{r} \text{ and } p_{i,i-1} = \frac{i}{r}.$$

Further, $p_{01} = p_{r,r-1} = 1$. The other p_{ij} are zero.

Example 10.3 An absent-minded professor drives every morning from his home to the office and at the end of the day from the office to home. At any given time, his driver's license is located at his home or at the office. If his driver's license is at his location of departure, he takes it with him with probability 0.5. What stochastic process describes whether the professor has the driver's license with him when driving his car home or to the office?

Solution. Your first thought might be to define two states 1 and 0, where state 1 describes the situation that the professor has his driver's license with him when driving his car and state 0 describes the situation that he does not have his driver's license with him when driving his car. However, these two states do not suffice for a Markov model: state 0 does not provide enough information to predict the state at the next drive. In order to give the probability distribution of this next state, you need information about the current location of the driver's license of the professor. You get a Markov model by simply inserting this information into the state description. Let's say that the system is in state 1 if the professor is driving his car and has his driver's license with him, in state 2 if the professor is driving his car and his driver's license is at the point of departure, and in state 3 if the professor is driving his car and his driver's license is at his destination. Define the random variable X_n as

$$X_n = \text{the state at the } n\text{th drive to home or to the office.}$$

The process $\{X_n\}$ has the property that any present state contains sufficient information for predicting future states. Thus, the process $\{X_n\}$ is a Markov chain with state space $I = \{1, 2, 3\}$. Next, we determine the p_{ij}. For example, p_{32} translates into the probability that the professor will not have his driver's license with him at the next drive, given that his driver's license is at the point of departure for the next drive. This gives $p_{32} = 0.5$. Also, $p_{31} = 0.5$. Similarly, $p_{23} = 1$ and $p_{11} = p_{12} = 0.5$. The other p_{ij} are zero.

The third example deals with an inventory problem and the modeling of this problem is more involved than that of the foregoing problems.

Example 10.4 The Johnson hardware shop, a family business since 1888, carries adjustable pliers as a regular stock item. The demand for this tool is stable over time. The total demand during a week has a Poisson distribution with expected value $\lambda = 4$. The demands in the successive weeks are independent of each other. Each demand that occurs when the shop is out of stock is lost. The owner of the shop uses a so-called periodic review (s, S) control rule with $s = 5$ and $S = 10$ for stock replenishment of the item. Under this rule, the inventory position is only reviewed at the beginning of each week. If, upon review, the stock on hand is less than the reorder point s, the inventory is replenished to the order-up point S; otherwise, no ordering is done. The replenishment time is negligible. What stochastic process describes the stock on hand?

Solution. In this application we take as state variable the stock on hand just prior to review (another possible choice would have been the stock on hand just after review). Let the random variable X_n be defined as

$$X_n = \text{the stock on hand at the beginning of the } n\text{th week}$$
$$\text{just prior to review.}$$

It will be immediately clear that the stochastic process $\{X_n\}$ satisfies the Markovian property: the stock on hand at the beginning of the current week and the demand in the coming week determine the stock on hand at the beginning of the next week. It is not relevant how the stock fluctuated in the past. Thus, the process $\{X_n\}$ is a Markov chain. Its state space is finite and is given by $I = \{0, 1, \ldots, S\}$. How do you find the one-step transition probabilities p_{ij}? In any application, the simple but useful advice is to translate

$P(X_{n+1} = j \mid X_n = i)$ in terms of the concrete situation you are dealing with. For example, how to find p_{0j} in the present application? If state 0 is the current state, then the inventory is replenished to level S, and the stock at the beginning of the next week just prior to review is j if and only if the demand in the coming week is $S - j$ for $j \neq 0$. The next state is $j = 0$ if and only if the demand in the coming week is S or more. Armed with this argument, we now specify the p_{ij}. We distinguish between the cases (a) $i < s$ and (b) $i \geq s$. In case (a), the stock on hand just after review is S while in case (b), the stock on hand just after review is i.

Case (a): $i < s$. Then,

$$p_{ij} = P(\text{the demand in the next week will be equal to } S - j)$$

$$= e^{-\lambda} \frac{\lambda^{S-j}}{(S-j)!} \quad \text{for } 1 \leq j \leq S,$$

regardless of the value of $i < s$. Further,

$$p_{i0} = P(\text{the demand in the next week will be } S \text{ or more}) = \sum_{k=S}^{\infty} e^{-\lambda} \frac{\lambda^k}{k!}.$$

Note that the expression for p_{i0} is in agreement with $p_{i0} = 1 - \sum_{j \neq 0} p_{ij}$.

Case (b): $i \geq s$. Then,

$$p_{ij} = P(\text{the demand in the next week will be equal to } i - j)$$

$$= e^{-\lambda} \frac{\lambda^{i-j}}{(i-j)!} \quad \text{for } 1 \leq j \leq i$$

and $p_{ij} = 0$ for $j > i$. Further,

$$p_{i0} = P(\text{the demand in the next week will be } i \text{ or more}) = \sum_{k=i}^{\infty} e^{-\lambda} \frac{\lambda^k}{k!}.$$

Problems

10.1 Two compartments A and B each contain r particles. Of these $2r$ particles, r are of type 1 and r are of type 2. With the passing of every time unit, one particle is selected at random from each of the compartments, and each of these two particles is transferred from its compartment to the other one. Define a Markov chain that describes the numbers of type 1 and type 2 particles in each of the two compartments and specify the one-step transition probabilities.

10.2 Consider the following modification of Example 10.3. In case the driver's license of the professor is at his point of departure, he takes it with him with probability 0.75 when departing from home and with probability 0.5 when departing from the office. Define a Markov chain that describes whether the professor has the driving license with him when driving his car. Specify the one-step transition probabilities.

10.3 You start by rolling a single die. If the outcome is j, you roll j dice. Next you roll as many dice as the largest outcome of the j dice. You continue in this way by rolling each time as many dice as the largest outcome in the previous roll of dice. Describe this process by a Markov chain and specify its one-step transition probabilities.

10.4 Every day, it is either sunny or rainy on Rainbow Island. The weather for the next day depends only on today's weather and yesterday's weather. If the last two days were sunny, it will be sunny on the next day with probability 0.9. This probability is 0.45 if the last two days were rainy. The next day will be sunny with probability 0.7 if today's weather is sunny and yesterday's weather was rainy. If today's weather is rainy and yesterday's weather was sunny, it will be sunny on the next day with probability 0.5. Define a Markov chain describing the weather on Rainbow Island and specify its one-step transition probabilities.

10.5 To improve the reliability of a production system, two identical production machines are connected parallel to one another. For the production process, only one of the two machines is needed. The other machine (if available) takes over when the machine currently in use needs revision. At the end of each production day the used machine is inspected. The probability of an inspection revealing the necessity of a revision of the machine is $\frac{1}{10}$, regardless how long the inspected machine has already been in uninterrupted use. A revision takes exactly two days. There are ample repair facilities, so that the revision of a machine can start immediately. The production process must be stopped when both machines are in revision. Formulate an appropriate Markov chain to describe the functioning of the production system and specify the one-step transition probabilities. *Hint*: Use an auxiliary state variable to indicate the remaining duration of a revision.

10.6 A control device contains two parallel circuit boards. Both circuit boards are switched on. The device operates properly as long as at least one of the circuit boards functions. Each circuit board is subject to random failure. The failure rate increases with the age of the circuit board. The circuit boards are identical and their lifetimes are independent. Let r_i denote the probability of a circuit board failing during the next week if the circuit board has functioned for the past i weeks. Past records of circuit boards give the failure function $r_0 = 0, r_1 = 0.05, r_2 = 0.07, r_3 = 0.12, r_4 = 0.25, r_5 = 0.50$, and $r_6 = 1$. Any failed circuit board is replaced at the beginning of the following week. Also, any six-week-old circuit board is replaced. Formulate an appropriate Markov chain for the failure analysis of the device and specify the one-step transition probabilities.

10.7 A communication channel transmits messages one at a time, and transmission of a message can only start at the beginning of a time slot. The transmission time of any message is one time slot. However, each transmission can fail with a given probability $f = 0.05$. A failed transmission is tried again at the beginning of the next time slot. Newly arriving messages for transmission are temporarily stored in a finite buffer. The buffer has capacity for only $K = 10$ messages (excluding any message in transmission). The number of new messages arriving during any given time slot has a Poisson distribution with mean $\lambda = 5$. If a newly arriving message finds the buffer full, then the message is lost. Formulate an appropriate Markov chain to describe the content of the buffer at the beginning of the time slots and specify its one-step transition probabilities.

10.2 Time-Dependent Analysis of Markov Chains

As said before, a Markov chain $\{X_n, n = 0, 1, \ldots\}$ is completely determined by its one-step transition probabilities p_{ij} and the probability distribution of the initial state X_0. The time-dependent analysis of a Markov chain concerns the calculation of the so-called n-step transition probabilities. The probability of going from state i to state j in the next n transitions of the Markov chain is easily calculated from the one-step transition probabilities. For any $n = 1, 2, \ldots$, the n-step transition probabilities $p_{ij}^{(n)}$ are defined by

$$p_{ij}^{(n)} = P(X_n = j \mid X_0 = i) \quad \text{for } i, j \in I.$$

Note that $p_{ij}^{(1)} = p_{ij}$. A basic result is given in the next rule.

Rule 10.1 (Chapman–Kolmogorov equations) *For any $n \geq 2$,*

$$p_{ij}^{(n)} = \sum_{k \in I} p_{ik}^{(n-1)} p_{kj} \quad \text{for all } i, j \in I.$$

This rule states that the probability of going from state i to state j in n transitions is obtained by summing the probabilities of the mutually exclusive events of going from state i to some state k in the first $n-1$ transitions and then going from state k to state j in the nth transition. A formal proof proceeds as follows. Using the law of conditional probability and the Markovian property, we have that $p_{ij}^{(n)} = P(X_n = j \mid X_0 = i)$ is equal to

$$\sum_{k \in I} P(X_n = j \mid X_0 = i, X_{n-1} = k) P(X_{n-1} = k \mid X_0 = i)$$

$$= \sum_{k \in I} P(X_n = j \mid X_{n-1} = k) P(X_{n-1} = k \mid X_0 = i).$$

By the assumption of time homogeneity, the last expression is equal to $\sum_{k \in I} p_{kj} p_{ik}^{(n-1)}$, proving the desired result.

It is convenient to write the result of Rule 10.1 in terms of matrices. Let

$$\mathbf{P} = (p_{ij})$$

be the matrix having the one-step transition probabilities p_{ij} as entries. If we let $\mathbf{P}^{(n)}$ denote the matrix of the n-step transition probabilities $p_{ij}^{(n)}$, then Rule 10.1 asserts that $\mathbf{P}^{(n)} = \mathbf{P}^{(n-1)} \times \mathbf{P}$ for all $n \geq 2$. By iterating this formula and using the fact that $\mathbf{P}^{(1)} = \mathbf{P}$, we obtain

$$\mathbf{P}^{(n)} = \mathbf{P} \times \mathbf{P} \times \ldots \times \mathbf{P} = \mathbf{P}^n.$$

This gives us the following important result:

Rule 10.2 *The n-step transition probabilities $p_{ij}^{(n)}$ can be calculated as the entries in the matrix product $\mathbf{P}^n$, which is obtained by multiplying the matrix $\mathbf{P}$ by itself n times.*

Example 10.5 On the Island of Hope, the weather each day is classified as sunny, cloudy, or rainy. The next day's weather depends only on today's weather and not on the weather of the previous days. If the present day is sunny, then the next day will be sunny, cloudy, or rainy with probabilities 0.70, 0.10, and 0.20. The transition probabilities for the weather are 0.50, 0.25, and 0.25 when the present day is cloudy and they are 0.40, 0.30, and 0.30 when the present day is rainy. What is the probability that it will be sunny three days from now if it is rainy today? What are the proportions of time the weather will be sunny, cloudy, and rainy over a very long period?

Solution. These questions can be answered by using a three-state Markov chain. Let's say that the weather is in state 1 if it is sunny, in state 2 if it is cloudy, and in state 3 if it is rainy. Define the random variable X_n as the state of the weather on day n. The stochastic process $\{X_0, X_1, \ldots\}$ is then a Markov chain with state space $I = \{1, 2, 3\}$. The matrix $\mathbf{P}$ of one-step transition probabilities is given by

$$\begin{array}{c} \text{from}\backslash\text{to} \\ 1 \\ 2 \\ 3 \end{array} \begin{pmatrix} 1 & 2 & 3 \\ 0.70 & 0.10 & 0.20 \\ 0.50 & 0.25 & 0.25 \\ 0.40 & 0.30 & 0.30 \end{pmatrix}.$$

To find the probability of having sunny weather three days from now, we need the matrix product $\mathbf{P}^3$:

$$\mathbf{P}^3 = \begin{pmatrix} 0.6015000 & 0.1682500 & 0.2302500 \\ 0.5912500 & 0.1756250 & 0.2331250 \\ 0.5855000 & 0.1797500 & 0.2347500 \end{pmatrix}.$$

From this matrix you read off that the probability of having sunny weather three days from now is $p_{31}^{(3)} = 0.5855$ if it is rainy today. What is the probability distribution of the weather on a day far away? Intuitively, you expect that this probability distribution does not depend on the present state of the weather. This is indeed confirmed by the following calculations:

$$\mathbf{P}^5 = \begin{pmatrix} 0.5963113 & 0.1719806 & 0.2317081 \\ 0.5957781 & 0.1723641 & 0.2318578 \\ 0.5954788 & 0.1725794 & 0.2319419 \end{pmatrix}$$

$$\mathbf{P}^{12} = \begin{pmatrix} 0.5960265 & 0.1721854 & 0.2317881 \\ 0.5960265 & 0.1721854 & 0.2317881 \\ 0.5960265 & 0.1721854 & 0.2317881 \end{pmatrix} = \mathbf{P}^{13} = \mathbf{P}^{14} = \cdots.$$

That is, after 12 matrix multiplications the entries agree row-by-row to seven decimal places. You see that the weather on a day far away will be sunny, cloudy, or rainy with probabilities 0.5960, 0.1722, and 0.2318, respectively. It will be clear that these limiting probabilities also give the proportions of time that the weather will be sunny, cloudy, and rainy over a very long period. In this example we have answered the question about the long-run behavior of the weather by computing sufficiently high powers of $\mathbf{P}^n$. A computationally better approach for the long-run behavior will be discussed in Section 10.4.

Many probability problems, which are seemingly unrelated to a Markov chain, can be modeled as a Markov chain with the help of a little imagination. This is illustrated with the next example. This example nicely shows that the line of thinking through the concepts of state and state transition is very useful to analyze the problem (and many other problems in applied probability!).

Example 10.6 What is the probability mass function of the number of different face values that will appear on a roll of six dice?

Solution. To put this problem in the framework of a Markov chain, consider the equivalent problem of repeatedly rolling a single die. Then we can define the state of the process as the number of different face values seen so far. Let the random variable X_n be the state after the nth roll, with the convention $X_0 = 0$. The process $\{X_n\}$ is a Markov chain with state space $I = \{0, 1, \ldots, 6\}$. The one-step transition probabilities are given by $p_{01} = 1$, $p_{ii} = \frac{i}{6}$, and $p_{i,i+1} = 1 - p_{ii}$ for $1 \leq i \leq 5$, $p_{66} = 1$, and the other $p_{ij} = 0$. The probability $p_{0k}^{(6)}$ gives the probability that exactly k different face values will appear on a roll of six dice. Multiplying the matrix of one-step transition probabilities six times by itself yields $p_{01}^{(6)} = 1.28 \times 10^{-4}$, $p_{02}^{(6)} = 0.0199$, $p_{03}^{(6)} = 0.2315$, $p_{04}^{(6)} = 0.5015$, $p_{05}^{(6)} = 0.2315$, and $p_{06}^{(6)} = 0.0154$.

An interesting and useful result is the following:

Rule 10.3 For any two states $i, j \in I$,

$$E(\text{number of visits to state } j \text{ over the time points } t = 1, \ldots, n \mid X_0 = i)$$
$$= \sum_{t=1}^{n} p_{ij}^{(t)} \quad \text{for } n = 1, 2, \ldots.$$

The proof of this result is instructive. Fix $i, j \in I$. For any $t \geq 1$, let

$$I_t = \begin{cases} 1 & \text{if } X_t = j, \\ 0 & \text{otherwise.} \end{cases}$$

The number of visits to state j over the time points $t = 1, \ldots, n$ is then given by the random variable $\sum_{t=1}^{n} I_t$. Using the observation that

$$E(I_t \mid X_0 = i) = 1 \times P(I_t = 1 \mid X_0 = i) + 0 \times P(I_t = 0 \mid X_0 = i)$$
$$= P(X_t = j \mid X_0 = i) = p_{ij}^{(t)},$$

we obtain $E(\sum_{t=1}^{n} I_t \mid X_0 = i) = \sum_{t=1}^{n} E(I_t \mid X_0 = i) = \sum_{t=1}^{n} p_{ij}^{(t)}$, proving the desired result.

As an illustration, consider Example 10.5 again. What is the expected value of the number of sunny days in the coming seven days when it is cloudy today? The answer is that this expected value is equal to $\sum_{t=1}^{7} p_{21}^{(t)}$ days. The value of this sum is calculated as 4.049.

Problems

10.8 A car rental agency rents cars at four locations. A rented car can be returned to any of the four locations. A car rented at location 1 will be returned to location 1 with probability 0.8, to location 2 with probability 0.1, to location 3 with probability 0, and to location 4 with probability 0.1. These probabilities have the values 0.1, 0.7, 0.2, and 0 for cars rented at location 2, the values 0.2, 0.1, 0.5, and 0.2 for cars rented at location 3, and the values 0, 0.2, 0.1, and 0.7 for cars rented at location 4. A particular car is currently at location 3. What is the probability that this car is back at location 3 after being rented out five times? What is the long-run frequency with which any given car is returned to location i for $i = 1, 2, 3$, and 4?

10.9 Consider again Problem 10.4. What is the probability of having sunny weather five days from now if it rained today and yesterday? What is the proportion of time it will be sunny over a very long period? What is the expected number of days it will be sunny in the next 14 days given that it rained the last two days?

10.10 A communication system is either in the on-state (state 1) or the off-state (state 0). Every millisecond, the state of the system may change. An off-state is changed into an on-state with probability α and an on-state is changed into an off-state with probability β, where $0 < \alpha$, $\beta < 1$. Use induction to verify that the n-step transition probabilities of the Markov chain describing the state of the system satisfy

$$p_{00}^{(n)} = \frac{\beta}{\alpha + \beta} + \frac{\alpha(1 - \alpha - \beta)^n}{\alpha + \beta} \quad \text{and} \quad p_{11}^{(n)} = \frac{\alpha}{\alpha + \beta} + \frac{\beta(1 - \alpha - \beta)^n}{\alpha + \beta},$$

where $p_{01}^{(n)} = 1 - p_{00}^{(n)}$ and $p_{10}^{(n)} = 1 - p_{11}^{(n)}$. *Note*: The reader familiar with linear algebra may verify this result from the eigenvalues and eigenvectors of the matrix of one-step transition probabilities.

10.11 A faulty digital video-conferencing system has a clustering error pattern. If a bit is received correctly, the probability of receiving the next bit correctly is 0.999. This probability is only 0.1 if the last bit was received incorrectly. Suppose that the first transmitted bit is received correctly. What is the expected value of the number of incorrectly received bits among the next 5,000 bits?

10.12 Trees in a forest are assumed to fall into four age groups: baby trees (0–10 years of age), young trees (11–20 years of age), middle-aged trees (21–30 years of age), and old trees (older than 30 years of age). The length of one time period is 10 years. In each time period a certain percentage of trees in each age group dies. These percentages are 20%, 5%, 10%, and 25% for the four age groups. Lost trees are replaced by baby trees. Surviving trees enter the next age group, where old trees remain in the fourth age group. Suppose that the forest is newly planted with 10,000 trees. What is the age distribution of the forest after 50 years? What is the age distribution of the forest in the equilibrium situation?

10.13 Let $\{X_n\}$ be any Markov chain. For any i and j, define the random variable $V_{ij}(n)$ as the number of visits to state j over the time points $t = 1, 2, \ldots, n$ if the starting state is i. Verify the result

$$\sigma^2[V_{ij}(n)] = \sum_{t=1}^{n} p_{ij}^{(t)}(1 - p_{ij}^{(t)}) + 2 \sum_{t=1}^{n-1} \sum_{u=t+1}^{n} [p_{ij}^{(t)} p_{jj}^{(u-t)} - p_{ij}^{(t)} p_{ij}^{(u)}].$$

Hint: Use the fact that $P(I_t = 1, I_u = 1 \mid X_0 = i) = p_{ij}^{(t)} p_{jj}^{(u-t)}$ for $u > t \geq 1$, where I_t is defined as in the proof of Rule 10.3.[1] Next apply the result to Example 10.5 to find a normal approximation for the probability of having more than 240 sunny days in the next 365 days, given that it is rainy today.

10.14 Suppose that the results of cricket matches between England and Australia can be modeled by a Markov chain. If England has won the last match, the next match will be a win, a loss, or a draw for England with probabilities 0.44, 0.37, and 0.19. The transition probabilities are 0.28, 0.43, and 0.29 when England has lost the previous match and they are 0.27, 0.30, and 0.43 when the last match was a draw. What is the expected number of wins for England in the next three matches when the last match was a draw?

10.15 Consider again Problem 10.14. Calculate the probability mass function of the number of wins for England in the next three matches, given that the last match was a draw.

10.3 Absorbing Markov Chains

Markov chains can also be used to analyze systems in which some states are "absorbing." Once the system reaches an absorbing state, it remains in that

---

[1] It can be shown that for any i and j, the random variable $V_{ij}(n)$ is approximately normally distributed for n sufficiently large when the Markov chain has the property that any state is accessible from any other state. A state k is said to be *accessible* from another state j if $p_{jk}^{(n)} > 0$ for some $n \geq 1$.

10.3 Absorbing Markov Chains

state permanently. The Markov chain model with absorbing states has many interesting applications. Examples include stochastic models of biological populations, where the absorbing state is extinction and gambling models, where the absorbing state is ruin.

Let $\{X_n\}$ be a Markov chain with one-step probabilities p_{ij}. State i is said to be an *absorbing* state if $p_{ii} = 1$. The Markov chain $\{X_n\}$ is said to be an absorbing Markov chain if it has one or more absorbing states and the set of absorbing states is accessible from the other states. Interesting questions are:

- What is the probability distribution of the time until absorption?
- If there are multiple absorbing states, what is the probability that the system will end up in a specific absorbing state?

We address these questions in the examples below. The first example deals with an instance of the coupon-collector problem.

Example 10.7 A fair die is rolled until each of the six possible outcomes $1, 2, \ldots, 6$ has appeared. How can we calculate the probability mass function of the number of rolls needed?

Solution. Let's say that the system is in state i if i different outcomes have appeared so far. Define the random variable X_n as the state of the system after the nth roll. State 6 is taken as an absorbing state. The process $\{X_n\}$ is an absorbing Markov chain with state space $I = \{0, 1, \ldots, 6\}$. The matrix $\mathbf{P} = (p_{ij})$ of one-step transition probabilities is given by

$$p_{01} = 1, \quad p_{ii} = \frac{i}{6}, \quad \text{and } p_{i,i+1} = 1 - \frac{i}{6} \quad \text{for } i = 1, \ldots, 5, \quad p_{66} = 1,$$

and $p_{ij} = 0$ otherwise. The starting state of the process is state 0. Let the random variable R denote the number of rolls of the die needed to obtain all of the six possible outcomes. The random variable R takes on a value larger than r if and only if the Markov chain has not visited the absorbing state 6 in the first r transitions. Therefore,

$$P(R > r) = P(X_k \neq 6 \text{ for } k = 1, \ldots, r \mid X_0 = 0).$$

However, since state 6 is absorbing, we have that $X_k \neq 6$ for any $k < r$ if $X_r \neq 6$. Therefore,

$$P(X_k \neq 6 \text{ for } k = 1, \ldots, r \mid X_0 = 0) = P(X_r \neq 6 \mid X_0 = 0).$$

Noting that $P(X_r \neq 6 \mid X_0 = 0) = 1 - P(X_r = 6 \mid X_0 = 0)$, we obtain

$$P(R > r) = 1 - p_{06}^{(r)} \quad \text{for } r = 1, 2, \ldots.$$

Since $p_{06}^{(r)}$ is the probability of reaching the absorbing state 6 in r or less steps, the probability that exactly r rolls are needed to get all six possible outcomes is $p_{06}^{(r)} - p_{06}^{(r-1)}$ for $r \geq 6$, where $p_{06}^{(5)} = 0$. The probability $P(R > r)$ is calculated by multiplying the matrix **P** by itself r times. For example, $P(R > r)$ has the values 0.7282, 0.1520, and 0.0252 for $r = 10, 20$, and 30.

Absorbing Markov chains are very useful to analyze success runs. This is illustrated with the following two examples, see also Problems 10.20 and 10.26 for success runs in Bernoulli trials.

Example 10.8 Chess grandmaster Boris Karparoff has entered a match with chess computer Deep Blue. The match will continue until either Karparoff or Deep Blue has won two consecutive games. For the first game as well as any game ending in a draw, we have that the next game will be won by Karparoff with probability 0.4, by Deep Blue with probability 0.3, and will end in a draw with probability 0.3. After a win by Karparoff the probabilities of these outcomes for the next game will have the values 0.5, 0.25, and 0.25, while after a loss by Karparoff the probabilities will have the values 0.3, 0.5, and 0.2. What is the probability that the match will last for longer than 10 games? What is the probability that Karparoff will be the final winner, and what is the expected value of the duration of the match?

Solution. To answer these questions, we use an absorbing Markov chain with two absorbing states. Let's say that the system is in state $(1, K)$ if Karparoff has won the last game but not the game before, and in state $(2, K)$ if Karparoff has won the last two games. Similarly, the states $(1, D)$ and $(2, D)$ are defined. The system is said to be in state 0 if the match is about to begin or the last game is a draw. We take the states $(2, K)$ and $(2, D)$ as absorbing states. Define the random variable X_n as the state of the system after the nth game. The process $\{X_n\}$ is an absorbing Markov chain with five states. Its matrix **P** of one-step transition probabilities is given by

from /to	0	$(1, K)$	$(1, D)$	$(2, K)$	$(2, D)$
0	0.3	0.4	0.3	0	0
$(1, K)$	0.25	0	0.25	0.5	0
$(1, D)$	0.2	0.3	0	0	0.5
$(2, K)$	0	0	0	1	0
$(2, D)$	0	0	0	0	1

Let the random variable L denote the duration of the match. The random variable L takes on a value larger than r only if the Markov chain does not visit either of the states $(2, K)$ and $(2, D)$ in the first r steps. Thus,

10.3 Absorbing Markov Chains

$$P(L > r) = P(X_k \neq (2,K), (2,D) \text{ for } k = 1, \ldots, r \mid X_0 = 0)$$
$$= P(X_r \neq (2,K), (2,D) \mid X_0 = 0),$$

where the last equality uses the fact that the states $(2,K)$ and $(2,D)$ are absorbing, so that $X_k \neq (2,K), (2,D)$ for any $k < r$ if $X_r \neq (2,K), (2,D)$. Thus we obtain

$$P(L > r) = 1 - p^{(r)}_{0,(2,K)} - p^{(r)}_{0,(2,D)}.$$

The value of $P(L > r)$ can be calculated by multiplying the matrix $\mathbf{P}$ by itself r times. We give the matrix product $\mathbf{P}^r$ for $r = 10$ and 30:

$$\mathbf{P}^{10} = \begin{pmatrix} 0.0118 & 0.0109 & 0.0092 & 0.5332 & 0.4349 \\ 0.0066 & 0.0061 & 0.0051 & 0.7094 & 0.2727 \\ 0.0063 & 0.0059 & 0.0049 & 0.3165 & 0.6663 \\ 0.0000 & 0.0000 & 0.0000 & 1.0000 & 0.0000 \\ 0.0000 & 0.0000 & 0.0000 & 0.0000 & 1.0000 \end{pmatrix},$$

$$\mathbf{P}^{30} = \begin{pmatrix} 0.0000 & 0.0000 & 0.0000 & 0.5506 & 0.4494 \\ 0.0000 & 0.0000 & 0.0000 & 0.7191 & 0.2809 \\ 0.0000 & 0.0000 & 0.0000 & 0.3258 & 0.6742 \\ 0.0000 & 0.0000 & 0.0000 & 1.0000 & 0.0000 \\ 0.0000 & 0.0000 & 0.0000 & 0.0000 & 1.0000 \end{pmatrix} = \mathbf{P}^{31} = \cdots.$$

In particular, $P(L > 10) = 1 - 0.5332 - 0.4349 = 0.0319$. The numerical calculations show that by $r = 30$, all of the entries of the matrix product $\mathbf{P}^r$ have converged up to four decimal places. The probability that the system will ultimately be absorbed in state $(2, K)$ is given by $\lim_{r \to \infty} p^{(r)}_{0,(2,K)}$ (why?). Thus, we can read off from the matrix $\mathbf{P}^{30}$ that with probability 0.5506, Karparoff will be the final winner.

Instead of computing the absorption probability by calculating sufficiently high powers of $\mathbf{P}^r$, it can be more efficiently computed by solving a system of linear equations. To write down these equations, we use parametrization. Define f_s as the probability that Karparoff will be the final winner when the starting point is state s. The probability f_0 is of main interest, but we need the other probabilities f_s to write down the linear equations. Obviously, $f_{(2,K)} = 1$ and $f_{(2,D)} = 0$. In general, how do we find f_s? Either the absorbing state $(2, K)$ is reached directly from state s, or it is reached from some other state v. The probability of the event of passing from state s to state v and then proceeding from state v to the absorbing state $(2, K)$ is $p_{sv} f_v$. Thus, by the law of conditional probability, we have the generally applicable formula $f_s = \sum_v p_{sv} f_v$. In this way, we find

$$f_0 = 0.3f_0 + 0.4f_{(1,K)} + 0.3f_{(1,D)},$$
$$f_{(1,K)} = 0.25f_0 + 0.25f_{(1,D)} + 0.5f_{(2,K)},$$
$$f_{(1,D)} = 0.2f_0 + 0.3f_{(1,K)} + 0.5f_{(2,D)},$$

where $f_{(2,K)} = 1$ and $f_{(2,D)} = 0$. The solution of this system of three linear equations in three unknowns is given by $f_0 = 0.5506, f_{(1,K)} = 0.7191$, and $f_{(1,D)} = 0.3258$, in agreement with the entries of the matrix $\mathbf{P}^r$ for r large.

The expected value of the duration of the match can be calculated from $E(L) = \sum_{r=0}^{\infty} P(L > r)$. However, a more elegant approach is to set up a system of linear equations through first-step analysis. Define μ_s as the expected value of the remaining duration of the match when the starting point is state s. The goal is to find μ_0. Given that the system begins in state s, the system will be in state v after the first step with probability p_{sv}, and the additional number of steps from state v until the process enters an absorbing state has expected value μ_v. Thus, by the law of conditional expectation, we have the generally applicable formula $\mu_s = \sum_v (1 + \mu_v) p_{sv}$. This leads to the linear equations

$$\mu_0 = 1 + 0.3\mu_0 + 0.4\mu_{(1,K)} + 0.3\mu_{(1,D)},$$
$$\mu_{(1,K)} = 1 + 0.25\mu_0 + 0.25\mu_{(1,D)} + 0.5\mu_{(2,K)},$$
$$\mu_{(1,D)} = 1 + 0.2\mu_0 + 0.3\mu_{(1,K)} + 0.5\mu_{(2,D)},$$

where $\mu_{(2,K)} = \mu_{(2,D)} = 0$. The solution of these three linear equations in three unknowns is $\mu_0 = 4.079$, $\mu_{(1,K)} = 2.674$, and $\mu_{(1,D)} = 2.618$. In this way we find that the expected value of the duration of the match is 4.079 games. Isn't this approach much more elegant and simpler than the approach of calculating μ_0 as $\sum_{r=0}^{\infty} P(L > r)$?

Example 10.9 On May 23, 2009, Patricia DeMauro set a craps record in a casino in Atlantic City by rolling the dice 154 consecutive times in a single turn before she "sevened out" by rolling a seven. She rolled the dice for 4 hours and 18 minutes, beating the previously reported record set in 1989 by Stanley Fujitake who rolled the dice 119 times at a Las Vegas casino for 3 hours and 6 minutes before his turn finally ended. What is the probability of 154 or more craps rolls in the shooter's turn? What is the expected number of craps rolls in the shooter's turn?

Solution. The rules for the shooter rolling the two dice in craps are as follows. The shooter's turn begins with come-out rolls. These rolls continue until the dice add up to 4, 5, 6, 8, 9, or 10, which establishes the shooter's "point." Once this point is established, the game enters a second stage. The shooter rolls until throwing either the "point," which ends the second stage and begins

10.3 Absorbing Markov Chains

a new set of come-out rolls, or a seven, which ends the shooter's turn. Note that it takes a seven to end the shooter's turn, but the turn cannot end during a come-out roll. Some reflections show that the process of rolling the dice in the shooter's turn can be described by an absorbing Markov chain $\{X_n\}$. The probability a_k of getting a dice total of k in a single roll of two dice is the same as the probability a_{14-k} of getting a dice total of $14-k$ for $k = 2, \ldots, 7$, where $a_k = \frac{k-1}{36}$ for $k = 2, \ldots, 7$. Therefore, we can take a Markov chain with five states rather than eight states. State 0 means that the shooter is in the stage of come-out rolls, state 1 means that the shooter's point is 4 or 10, state 2 means that the shooter's point is 5 or 9, state 3 means that the shooter's point is 6 or 8, and state 4 means that the shooter's turn has ended by sevening out. State 4 is absorbing. Let X_n denote the state after the nth roll of the two dice. It readily follows from the rules for the shooter that the matrix $\mathbf{P} = (p_{ij})$ of the one-step transition probabilities of the Markov chain $\{X_n\}$ is (verify!)

	0	1	2	3	4
0	12/36	6/36	8/36	10/36	0
1	3/36	27/36	0	0	6/36
2	4/36	0	26/36	0	6/36
3	5/36	0	0	25/36	6/36
4	0	0	0	0	1

The probability of not having sevened out after 153 rolls can be computed as $1 - p_{04}^{(153)}$ and is equal to 1.789×10^{-10}. This is a probability of 1 in 5.59 billion. To find the expected number of dice rolls before sevening out, let μ_i be the expected number of remaining dice rolls before sevening out when the current state of the process is i. By the law of conditional expectation, $\mu_i = 1 + \sum_{j=0}^{3} \mu_j p_{ij}$ for $i = 0, 1, 2$, and 3. The solution of these linear equations is given by $\mu_0 = 8.526$, $\mu_1 = 6.842$, $\mu_2 = 7.010$, and $\mu_3 = 7.148$. Thus, the expected number of craps rolls in the shooter's turn is 8.526.

The next example deals with a basic version of the famous Wright–Fisher model from biology.

Example 10.10 Suppose there is a fixed population of N genes that can be one of two types: A or a. These types correspond to two different pieces of genetic information. At each time unit, each gene randomly chooses one of the genes (possibly itself) and adopts the type of the chosen gene at the next time unit, independently of the other genes. What is the probability that all genes will ultimately have type A when there are initially i genes of type A?

Solution. Let X_n be the number of type-A genes at time n. Then $\{X_n\}$ is a Markov chain with state space $I = \{0, 1, \ldots, N\}$. The states 0 and

N are absorbing, that is, $p_{00} = p_{NN} = 1$. The other one-step transition probabilities are

$$p_{jk} = \binom{N}{k}\left(\frac{j}{N}\right)^k\left(1 - \frac{j}{N}\right)^{N-k}.$$

Let f_i be the probability of absorption in state N when starting in i. The absorption probabilities f_i are the unique solution to the equations

$$f_i = \sum_{j=0}^{N} p_{ij} f_j \quad \text{for } i = 1, \ldots, N-1$$

with the boundary conditions $f_0 = 0$ and $f_N = 1$. The equation for f_i follows by conditioning on the next state and applying the law of conditional probability. An explicit expression for the absorption probability f_i can be obtained by the following argument. For fixed j, the p_{jk} represent probabilities from the binomial distribution with expected value $N \times \frac{j}{N} = j$, which implies $E(X_n \mid X_{n-1} = j) = j$. Thus, by conditioning on X_{n-1}, we get

$$E(X_n \mid X_0 = i) = \sum_{j=0}^{N} E(X_n \mid X_0 = i, X_{n-1} = j) P(X_{n-1} = j \mid X_0 = i)$$

$$= \sum_{j=0}^{N} j P(X_{n-1} = j \mid X_0 = i) = E(X_{n-1} \mid X_0 = i)$$

for all n. Iterating the equality $E(X_n \mid X_0 = i) = E(X_{n-1} \mid X_0 = i)$, we find that $E(X_n \mid X_0 = i) = E(X_0 \mid X_0 = i)$ and so

$$E(X_n \mid X_0 = i) = i \quad \text{for all } n.$$

The absorbing states N and 0 satisfy $\lim_{n\to\infty} p_{iN}^{(n)} = f_i$ and $\lim_{n\to\infty} p_{i0}^{(n)} = 1 - f_i$, while $\lim_{n\to\infty} p_{ij}^{(n)} = 0$ for all $j \neq 0, N$. This implies that

$$\lim_{n\to\infty} E(X_n \mid X_0 = i) = 0 \times (1 - f_i) + N \times f_i \quad \text{for all } i.$$

We can now conclude that $N \times f_i = i$ and so

$$f_i = \frac{i}{N} \quad \text{for all } i.$$

As a check, $f_i = \frac{i}{N}$ satisfies indeed the equations for the f_i.

Tabu Probability

An absorbing Markov chain may also be useful to calculate so-called tabu probabilities. A *tabu probability* is the probability of avoiding some given set of states during a certain number of transitions. To illustrate this, we consider

10.3 Absorbing Markov Chains

again Example 10.5 and ask the following question. What is the probability of no rain in the next five days given that it is sunny today? The trick is to make state 3 (rainy weather) absorbing. The Markov matrix **P** in Example 10.5 is adjusted by replacing the third row corresponding to state 3 by the row vector $(0, 0, 1)$. This gives the Markov matrix

$$\mathbf{Q} = \begin{pmatrix} 0.70 & 0.10 & 0.20 \\ 0.50 & 0.25 & 0.25 \\ 0 & 0 & 1 \end{pmatrix}.$$

Some reflection shows that the probability of no rain in the next five days, given that today is rainy, equals $1 - q_{13}^{(5)}$. The matrix product $\mathbf{Q}^5$ is

$$\mathbf{Q}^5 = \begin{pmatrix} 0.2667 & 0.0492 & 0.6841 \\ 0.2458 & 0.0454 & 0.7087 \\ 0 & 0 & 1 \end{pmatrix}.$$

Hence, $1 - q_{13}^{(5)} = 1 - 0.6841 = 0.3159$. Suppose that we had asked the question of what is the probability of no rain during the coming five days given that it is rainy today. The answer to this question requires the matrix product $\mathbf{Q}^4$ rather than $\mathbf{Q}^5$. By conditioning on the state of tomorrow's weather, it is readily seen that the probability called for is given by $p_{31}(1 - q_{13}^{(4)}) + p_{32}(1 - q_{23}^{(4)})$. This probability has the value 0.2698.

Problems

10.16 Suppose that deuce has been reached in a tennis game between John and Peter. The player winning the next point has advantage. On the following point, the player having advantage either wins the game or the game returns to deuce. At deuce, John has probability 0.55 of winning the next point and Peter has probability 0.45. John wins the game with probability 0.60 when John has advantage and Peter wins with probability 0.50 when Peter has advantage. What is the probability that John will win the game and how long is the game expected to last?

10.17 A particle performs a random walk on the positive integers. If the particle is in position k, it moves to position $k - \lfloor k/2 \rfloor$ with probability 0.5 and to position $2k - \lfloor k/2 \rfloor$ with probability 0.5, where $\lfloor x \rfloor$ is the largest integer less than or equal to x. The particle starts in position 1. What is the probability that the particle will reach a position beyond position 10 in the next 25 steps?

10.18 A jar initially contains three white and two red balls. Each time, you take one ball at random out of the jar and insert instead a ball of the opposite color. How would you compute the probability that more than n picks are needed to have only red balls in the jar? What is the value of this probability for $n = 5, 10, 25, 50$, and 100? What is the expected value of the number of picks until there are only red balls in the jar? *Note*: This problem can be seen as a drunkard's walk with state-dependent move probabilities and a reflecting barrier at zero.

10.19 You repeatedly roll two fair dice. What is the probability of two consecutive totals of 7 appearing before a total of 12?

10.20 Use absorbing Markov chains to answer the following questions about success runs.

(a) What is the probability of a run of five or more heads in 20 tosses of a fair coin?

(b) What is the probability of a run of either five or more heads or five or more tails, or both, in 20 tosses of a fair coin?

(c) Player A tosses a fair coin until he gets two consecutive heads, while player B tosses another fair coin until he gets three consecutive heads. What is the probability that player A will be the first player to reach his goal? What is the probability that the two players will reach their goal at the same time?

10.21 In the Italian regional lottery, five numbers are picked from the numbers 1 to 90 in each of 10 cities across the country. While the number 53 had come up in other cities, it did not appear in 182 consecutive draws between 2003 and 2005 in the Venice lottery. Huge amounts were gambled on 53 turning up, leading to incidences of assaults, suicides, and even murder after entire family savings were gambled away. In two years the equivalent of more than US\$4 billion was spent on the number 53. Use an absorbing Markov chain to approximate the probability that in a period of 10 years with 1,040 draws there is *some* number that will not appear in *some* of the 10 Italian city lotteries during 182 or more consecutive draws. *Hint*: Calculate first the exact value of the probability that in the next 1,040 draws of the lottery in a specific city there is a particular number that will not appear during 182 or more consecutive draws and use this probability to approximate the sought probability.

10.22 In each drawing of the 6/45 lottery, six different numbers are drawn from the numbers $1, 2, \ldots, 45$. Calculate for $r = 15, 25, 35$, and 50 the probability that more than r drawings are needed until each of the numbers $1, 2, \ldots, 45$ has been drawn.

10.23 The Bubble Company offers a picture of one of 25 pop stars in a pack of chewing gum. John and Peter each buy one pack every week. They pool the pictures of the pop stars. Assuming equal chances of getting any of the 25 pictures with one purchase, denote by the random variable N the number of weeks until John and Peter have collected two complete sets of 25 pictures. How could you compute $P(N > n)$ and $E(N)$ for this generalized coupon-collector's problem?

10.24 The Penney Ante game is a coin-tossing game between two players A and B. Each of the players chooses beforehand a sequence of heads and tails of length 3. The coin is tossed repeatedly until one of the two sequences appears. The winner is the person whose sequence appears first. Suppose that player A is the first to choose. Player B parries the choices *HHH* and *HHT* with *THH*, the choices *HTH* and *HTT* with *HHT*, the choices *TTT* and *TTH* with *HTT*, and the choices *THT* and *THH* with *TTH* (can you explain the idea behind the response of player B?). Verify that the win probability of player B is $\frac{7}{8}$ against *HHH* and *TTT*, $\frac{3}{4}$ against *HHT* and *TTH*, and $\frac{2}{3}$ otherwise.

10.25 Joe Dalton desperately wants to raise his current bankroll of \$800 to \$1,000 in order to pay his debts before midnight. He enters a casino and decides to play European roulette and to bet on red each time using bold play. Bold play means that Joe bets either his entire bankroll or the amount needed to reach the target bankroll, whichever is smaller. The stake is \$200 if his bankroll is \$200 or \$800 and the stake is \$400 if his bankroll is \$400 or \$600. Joe quits as soon as he has either reached his goal or lost everything.

(a) What is the probability that Joe places more than n bets? Calculate this probability for $n = 2, 3, 5$, and 7.
(b) What is the probability that Joe will reach his goal?
(c) What is the expected value of the total amount that will be staked during the game?

10.26 On August 18, 1913, black came up 26 times in a row on a roulette wheel in a Monte Carlo casino. What is the probability that in 5 million spins of a European roulette wheel the same color will come up 26 or more times in a row? In European roulette the wheel is divided into 37 sections, numbered $1, 2, \ldots, 36$ and 0. Of the sections numbered from 1 to 36, 18 are red and 18 are black. At Caesars Palace in Las Vegas on July 14, 2000, the number 7 came up six times in a row at roulette wheel #211. What is the probability that in 5 million spins of an American roulette wheel any of the numbers $1, 2, \ldots, 36$ will come up six or more times in a row? In American roulette the wheel is divided into 38 sections, numbered $1, 2, \ldots, 36, 0$ and 00.

10.27 You are fighting a dragon with three heads. Each time you swing at the dragon with your sword, there is a 70% chance of knocking off one head and a 30% chance of missing. If you miss, either one additional head or two additional heads will grow immediately, before you can swing again at the dragon. The probability of one additional head is p and of two additional heads is $1 - p$. You win if you have knocked off all of the dragon's heads, but you must run for your life if the dragon has seven or more heads. What is your chance of winning? Calculate this probability for $p = 0, 0.5$, and 1.

10.28 A fair die is rolled repeatedly. What is the expected value of the number of rolls until a run of six different outcomes appears and what is the probability of getting such a run within 100 rolls of the die? What are the answers when restricting to the run 123456?

10.29 In playing the game Yahtzee, what is the probability of rolling five of a kind within three rolls of the dice if your only goal is to get Yahtzee?

10.30 Three equally matched opponents decide to have a ping-pong tournament. Two people play against each other in each game. Drawing lots determines who plays the first game. The winner of a game stays on and plays against the person not active in that game. The games continue until somebody has won two games in a row. What is the probability that the person not active in the first game is the ultimate winner?

10.31 You play the following drinking game with several friends. On a table you have put three whiskey glasses numbered 1, 2, and 3. A fair die is rolled repeatedly. If the outcome of the die is 1 or 6, then glass 1 is emptied when it is full and is filled when it is empty. This same procedure is followed with glass 2 when a 2 or 5 is rolled and with glass 3 when a 3 or 4 is rolled. Initially the three glasses are full. Let N be the number of rolls of the die until all of the three glasses are empty. Calculate $E(N)$ and $P(N > n)$ for $n = 5, 10, 15, 20$, and 25.

10.32 A queue of 50 people is waiting at a box office in order to buy a ticket. The tickets cost five dollars each. For any person, there is a probability of $\frac{1}{2}$ that she/he will pay with a five-dollar note and a probability of $\frac{1}{2}$ that she/he will pay with a ten-dollar note. When the box office opens there is no money in the till. If each person just buys one ticket, what is the probability that none of them will have to wait?

10.33 A fair die is rolled four times. What is the probability that each of the last three rolls is at least as large as the roll preceding it and what is the probability that each of the last three rolls is larger than the roll preceding it?

10.34 Consider again Problem 10.4. What is the probability of having no rain on two consecutive days during the next seven days, given that it was sunny during the last two days? What is the value of this probability if the last two days were rainy?

10.35 Consider again Problem 10.8. A certain car is now at location 4. As soon as this car returns to location 1, it will be overhauled. What is the probability that the car will be rented out more than five times before it returns to location 1? What is this probability if the car is originally at location 1?

10.4 Long-Run Analysis of Markov Chains

In Example 10.5, the long-run behavior of a Markov chain describing the state of the weather was analyzed by taking sufficiently high powers of the matrix of one-step transition probabilities. It was found empirically that the n-step transition probabilities $p_{ij}^{(n)}$ have a limit as n becomes very large. Moreover, it turned out that the limit was independent of the starting state i. The limiting probabilities in the weather example also had a natural interpretation in terms of long-run frequencies. In this section these results will be put into a general framework. In particular, it will be seen that the long-run behavior of a Markov chain can be analyzed more efficiently than by taking high powers of the matrix of one-step transition probabilities.

The long-run (or equilibrium) analysis of Markov chains only makes sense for Markov chains without absorbing states. In the sequel we restrict ourselves to Markov chains with no two or more disjoint closed sets of states. A closed set of states is naturally defined as follows:

Definition 10.2 *A nonempty set C of states is said to be a closed set for the Markov chain* $\{X_n\}$ *if*

$$p_{ij} = 0 \quad \text{for } i \in C \text{ and } j \notin C,$$

that is, the process cannot leave the set C once the process is in the set C. A closed set C is said to be irreducible if it contains no smaller closed set.

The set of all states is always closed by definition. If the set of all states contains no smaller closed set, then the Markov chain is said to be *irreducible*. To illustrate, the three-state Markov chain with $p_{ij} > 0$ for $i,j = 1,2,3$ is irreducible, but the three-state Markov chain with $p_{ij} > 0$ for $i,j = 1,2, p_{3j} > 0$

for $j = 1, 2, 3$, and the other $p_{ij} = 0$ is not irreducible. The state space of the latter Markov chain contains the smaller closed set $C = \{1, 2\}$ (the process is ultimately absorbed in the closed set C when the starting state is 3).

The assumption of no two or more disjoint closed sets is necessary in order to produce the situation in which the effect of the starting state fades away after a sufficiently long period of time. To illustrate this, consider the Markov chain with state space $I = \{1, 2, 3, 4, 5\}$ whose one-step transition probabilities satisfy $p_{ij} > 0$ for $i, j = 1, 2$, $p_{ij} > 0$ for $i, j = 3, 4$, $p_{5j} > 0$ for $j = 1, \ldots, 5$, and the other $p_{ij} = 0$. In this example, the Markov chain has two disjoint closed sets $C_1 = \{1, 2\}$ and $C_2 = \{3, 4\}$, and the effect of the starting state lasts forever. This is best visualized by making a transition diagram in which the nodes are the states and an arrow from node i to node j is only drawn when $p_{ij} > 0$.

In almost any practical application of Markov chains, the assumption of no two or more disjoint closed sets is naturally satisfied. In a finite-state Markov chain having no two or more disjoint closed sets, there is always a state that can be reached from any other state in a finite expected time. The proof of this obvious result is not given here, see Problem 10.49. In the following analysis, the basic assumption that the system has a *finite* state space I is important. The long-run analysis of infinite-state Markov chains involves subtleties, which are beyond the scope of this book.

Rule 10.4 *Suppose that the n-step transition probability $p_{ij}^{(n)}$ of the Markov chain $\{X_n\}$ has a limit as $n \to \infty$ for all $i, j \in I$ such that for each $j \in I$ the limit is independent of the starting state i. Denote the limit by $\pi_j = \lim_{n \to \infty} p_{ij}^{(n)}$ for any $j \in I$. Then the limiting probabilities π_j are the unique solution to the linear equations*

$$\pi_j = \sum_{k \in I} \pi_k p_{kj} \text{ for } j \in I \quad \text{and} \quad \sum_{j \in I} \pi_j = 1.$$

The proof of Rule 10.4 is based on the Chapman–Kolmogorov equations in Rule 10.1. Letting n tend to infinity in these equations, we obtain

$$\pi_j = \lim_{n \to \infty} p_{ij}^{(n)} = \lim_{n \to \infty} \sum_{k \in I} p_{ik}^{(n-1)} p_{kj} = \sum_{k \in I} \lim_{n \to \infty} p_{ik}^{(n-1)} p_{kj} = \sum_{k \in I} \pi_k p_{kj}.$$

The interchange of the order of limit and summation in the third equality is justified by the finiteness of the state space I. Letting $n \to \infty$ in $\sum_{j \in I} p_{ij}^{(n)} = 1$, we obtain $\sum_{j \in I} \pi_j = 1$. It remains to prove that the above system of linear equations has a unique solution. To verify this, let $(x_j, j \in I)$ be any solution to the linear equations $x_j = \sum_{k \in I} x_k p_{kj}$. It is helpful to use matrix notation. Define the row vector $\mathbf{x} = (x_j)$ and the matrix $\mathbf{P} = (p_{ij})$. Then $\mathbf{x} = \mathbf{xP}$. Multiplying both sides of this equation by $\mathbf{P}$, we obtain $\mathbf{xP} = \mathbf{xP}^2$. Thus, by $\mathbf{xP} = \mathbf{x}$, we have $\mathbf{x} = \mathbf{x}\mathbf{P}^2$. Applying this argument repeatedly, we find $\mathbf{x} = \mathbf{x}\mathbf{P}^n$ for all $n = 1, 2, \ldots$. Componentwise, for each $j \in I$,

$$x_j = \sum_{k \in I} x_k p_{kj}^{(n)} \quad \text{for all } n = 1, 2, \ldots.$$

This implies that $x_j = \lim_{n \to \infty} \sum_{k \in I} x_k p_{kj}^{(n)}$. Interchanging the order of limit and summation, we obtain $x_j = \sum_{k \in I} x_k \pi_j = \pi_j (\sum_{k \in I} x_k)$ for all $j \in I$. Hence, $x_j = c\pi_j$ for all $j \in I$ with the constant $c = \sum_{k \in I} x_k$. If the x_k satisfy $\sum_{k \in I} x_k = 1$, we get $c = 1$, proving the uniqueness result.

The limiting probabilities π_j in Rule 10.4 constitute a probability distribution, that is, $\pi_j \geq 0$ for all j and $\sum_{j \in I} \pi_j = 1$. This is not always true for an infinite-state Markov chain. In the counterexample with $I = \{1, 2, \ldots\}$ and $p_{i,i+1} = 1$ for all i, we have $\lim_{n \to \infty} p_{ij}^{(n)} = 0$ for all i, j.

The result of Rule 10.4 motivates the concept of an *equilibrium distribution*.

Definition 10.3 *A probability distribution $\{\eta_j, j \in I\}$ is called an equilibrium distribution of the Markov chain $\{X_n\}$ if*

$$\eta_j = \sum_{k \in I} \eta_k p_{kj} \quad \text{for all } j \in I.$$

The terms *invariant distribution* and *stationary distribution* are also often used. The name equilibrium distribution can be explained as follows. If $P(X_0 = j) = \eta_j$ for all $j \in I$, then, for any time point $n \geq 1$, $P(X_n = j) = \eta_j$ for all $j \in I$. This result should be understood as follows. Suppose that you are going to inspect the state of the process at any time $t = n$ having *only* the information that the starting state of the process was determined according to the probability distribution $\{\eta_j\}$. Then the probability of finding the process in state s is η_s for any $s \in I$. The proof is simple. Suppose it has been verified for $t = 0, 1, \ldots, n-1$ that $P(X_t = j) = \eta_j$ for all $j \in I$. Then,

$$P(X_n = j) = \sum_{k \in I} P(X_n = j \mid X_{n-1} = k) P(X_{n-1} = k)$$

gives that $P(X_n = j) = \sum_{k \in I} p_{kj} \eta_k = \eta_j$ for all $j \in I$, as was to be verified.

It will be seen below that a Markov chain without two or more disjoint closed sets has a unique equilibrium distribution. Such a Markov chain is said to have reached *statistical equilibrium* if its state is distributed according to the equilibrium distribution. Under the assumption that $\lim_{n \to \infty} p_{ij}^{(n)}$ exists for all $i, j \in I$ and is independent of the starting state i, Rule 10.4 states that the Markov chain has a unique equilibrium distribution. The limiting probabilities $\pi_j = \lim_{n \to \infty} p_{ij}^{(n)}$ then constitute the equilibrium probabilities. Three obvious questions are:

- Does $\lim_{n\to\infty} p_{ij}^{(n)}$ always exist?
- Does any Markov chain have an equilibrium distribution?
- If an equilibrium distribution exists, is it unique?

It will be seen below that the answer to the last two questions is positive if the Markov chain has no two or more disjoint closed sets. The answer, however, to the first question is negative. A counterexample is provided by the two-state Markov chain with state space $I = \{1, 2\}$ and one-step transition probabilities $p_{12} = p_{21} = 1$ and $p_{11} = p_{22} = 0$. In this example the system alternates between the states 1 and 2. This means that, as a function of the time parameter n, the n-step transition probability $p_{ij}^{(n)}$ is alternately 0 and 1 and thus has no limit as n becomes very large. The periodicity of the Markov chain is the reason that $\lim_{n\to\infty} p_{ij}^{(n)}$ does not exist in this example. Periodicity of a Markov chain is defined as follows:

Definition 10.4 *A Markov chain $\{X_n\}$ is said to be periodic with period $d \geq 2$ if there are multiple disjoint sets $R_1, \ldots, R_d$ such that a transition from a state in R_k always occurs to a state in R_{k+1} for $k = 1, \ldots, d$ with $R_{d+1} = R_1$. Otherwise, the Markov chain is said to be aperiodic.*

In general, the existence of $\lim_{n\to\infty} p_{ij}^{(n)}$ requires an aperiodicity condition. However, it is not necessary to impose an aperiodicity condition on the Markov chain in order to have the existence of an equilibrium distribution. To work this out, we need the concept of a Cesàro limit. A sequence $(a_1, a_2, \ldots)$ of real numbers is said to have a *Cesàro limit* if $\lim_{n\to\infty} \frac{1}{n} \sum_{k=1}^n a_k$ exists. The Cesàro limit is more general than the ordinary limit. A basic result from calculus is that $\lim_{n\to\infty} \frac{1}{n} \sum_{k=1}^n a_k$ exists and is equal to $\lim_{n\to\infty} a_n$ if the latter limit exists. A beautiful and useful result from Markov chain theory is that

$$\lim_{n\to\infty} \frac{1}{n} \sum_{k=1}^n p_{ij}^{(k)}$$

always exists! An intuitive explanation of this result is as follows. Fix state $j = r$ and imagine that a reward 1 is earned each time the process makes a transition to state r and a reward 0 is earned in any other state. Then, by Rule 10.3, $\sum_{k=1}^n p_{ir}^{(k)}$ is the total expected reward earned up to time n when the starting state is i. Using the renewal–reward theorem (see Section 9.4), it can be shown that the long-run average expected reward per unit time is well-defined. In other words, $\lim_{n\to\infty} \frac{1}{n} \sum_{k=1}^n p_{ir}^{(k)}$ exists. This limit gives also the long-run expected frequency at which the process visits state r.

We now come to the main result of this section. This result will be stated without proof.

10.4 Long-Run Analysis of Markov Chains

Rule 10.5 *Let $\{X_n\}$ be a finite-state Markov chain with no two or more disjoint closed sets. The Markov chain then has a unique equilibrium distribution $\{\pi_j\}$.*

(a) *The equilibrium probabilities π_j are given by*

$$\pi_j = \lim_{n\to\infty} \frac{1}{n} \sum_{k=1}^{n} p_{ij}^{(k)} \quad \text{for all } j \in I,$$

with the averaging limit being independent of the starting state i.

(b) *The π_j are the unique solution to the linear equations*

$$\pi_j = \sum_{k \in I} \pi_k p_{kj} \text{ for } j \in I \text{ and } \sum_{j \in I} \pi_j = 1.$$

(c) *If the Markov chain is aperiodic, then $\lim_{n\to\infty} p_{ij}^{(n)} = \pi_j$ for all $i,j \in I$.*

The equations $\pi_j = \sum_{k \in I} \pi_k p_{kj}$ for $j \in I$ are called the *equilibrium equations*, or *balance equations*, and the equation $\sum_{j \in I} \pi_j = 1$ is called the *normalization equation*. In a similar way as in the proof of Rule 10.4, it can be shown that any solution (x_j) to the equilibrium equations alone is uniquely determined up to a multiplicative constant, that is, for some constant c, $x_j = c\pi_j$ for all $j \in I$. The size of the system of linear equations in part (**b**) of Rule 10.5 is one more than the number of unknowns. However, it is not difficult to see that one of the equilibrium equations is redundant (summing both sides of these equations over j gives "1 = 1" after an interchange of the order of summation). Thus, by deleting one of the equilibrium equations, one obtains a square system of linear equations that uniquely determines the unknowns π_j. Any one of the equilibrium equations can be removed.

An easy way to memorize the equilibrium equations is to note that the equilibrium equations are obtained by multiplying the row vector $\vec{\pi}$ of the equilibrium probabilities by the *column* vectors of the matrix **P** of one-step transition probabilities ($\vec{\pi} = \vec{\pi}\mathbf{P}$).

Example 10.5 (continued) The Markov chain describing the state of the weather has no two or more disjoint closed sets. Thus, the unique equilibrium probabilities of the Markov chain are found from the equations

$$\pi_1 = 0.70\pi_1 + 0.50\pi_2 + 0.40\pi_3,$$
$$\pi_2 = 0.10\pi_1 + 0.25\pi_2 + 0.30\pi_3,$$
$$\pi_3 = 0.20\pi_1 + 0.25\pi_2 + 0.30\pi_3,$$

together with $\pi_1 + \pi_2 + \pi_3 = 1$. One of the equilibrium equations (say, the first one) can be removed to obtain a square system of three linear equations in three unknowns. Solving these equations gives

$$\pi_1 = 0.5960, \ \pi_2 = 0.1722, \ \pi_3 = 0.2318.$$

Noting that the Markov chain is aperiodic, this result agrees with the earlier calculated matrix product $\mathbf{P}^n$ for n sufficiently large. The equilibrium probability π_j can be given two interpretations in this example. First, it can be stated that the weather on a day in the far distant future will be sunny, cloudy, or rainy with probabilities 0.5960, 0.1722, and 0.2318, respectively. Second, these probabilities also give the long-run expected proportions of time during which the weather will be sunny, cloudy, or rainy.

The next example deals with a periodic Markov chain. In such Markov chains one should be careful in interpreting the equilibrium probabilities.

Example 10.2 (continued) What are the equilibrium probabilities in the Ehrenfest model. How can you interpret these probabilities?

Solution. The equilibrium equations for the Ehrenfest model are

$$\pi_j = \frac{r-j+1}{r}\pi_{j-1} + \frac{j+1}{r}\pi_{j+1} \quad \text{for } j = 1, \ldots, r-1$$

with $\pi_0 = \frac{1}{r}\pi_1$ and $\pi_r = \frac{1}{r}\pi_{r-1}$. Intuitively, any marked particle is to be found equally likely in either of the two compartments after many transitions. This suggests the binomial distribution for the equilibrium probabilities. Indeed, by substitution into the equilibrium equations, it is readily verified that $\pi_j = \binom{r}{j}(\frac{1}{2})^r$ for $j = 0, 1, \ldots, r$. The equilibrium distribution is unique, since the Markov chain has no two or more disjoint closed sets. However, the Markov chain is periodic: a transition from any state in the subset of even-numbered states leads to a state in the subset of odd-numbered states, and vice versa. Hence, $\lim_{n\to\infty} p_{ij}^{(n)}$ does not exist and the proper interpretation of π_j is the interpretation as the long-run expected proportion of time during which compartment A contains j particles.

In each of the above two examples, the equilibrium probabilities could be interpreted as long-run expected frequencies. However, a much stronger result for the π_j can be given.

Rule 10.6 *Let $\{\pi_j\}$ be the unique equilibrium distribution of a finite-state Markov chain $\{X_n\}$ that has no two or more disjoint closed sets. Then, for any fixed state $j \in I$,*

$$\lim_{n\to\infty} \frac{1}{n}\sum_{k=1}^{n} X_k = \pi_j \text{ with probability } 1,$$

independently of the starting state $X_0 = i$. In words, the long-run actual proportion of time the process will be in state j is π_j with probability 1.

The term "with probability 1" is subtle and should be interpreted as follows. Let ω represent a possible outcome of the infinite sequence $X_0, X_1, \ldots$. Then, for any state j,

$$P\left(\{\omega : \lim_{n\to\infty} \frac{1}{n} \sum_{k=1}^{n} I_k(\omega) = \pi_j\}\right) = 1.$$

In other words, the set of outcomes ω for which $(1/n)\sum_{k=1}^{n} I_k(\omega)$ does not converge to π_j as $n \to \infty$ has probability zero. Rule 10.6 is the strong law of large numbers for Markov chains. The proof can be based on the renewal–reward theorem, using the fact that the Markov chain can be seen as a regenerative stochastic process with the property that any regeneration state of the process can be reached from any other state in a finite expected time. We omit further details of the proof. Part (**a**) of Rule 10.5 and Rule 10.3 together tell us that the long-run *expected* proportion of time the process will be in state j is π_j for any starting state i, while Rule 10.6 makes the much stronger statement that the long-run *actual* proportion of time the process will be in state j is equal to π_j with probability one.

If the Markov chain is aperiodic, then π_j can also be interpreted as the probability of finding the system in state j at a point of time in the far distant future. One should understand this interpretation as follows: if you inspect the process after it has been running for a very long time and you have *no* information about states visited in the past, then you will find the process in state j with probability π_j. The interpretation of π_j as a long-run frequency is more concrete and is often more useful from a practical point of view.

Also, a physical interpretation can be given to the equilibrium equations. In physical terms, $\pi_k p_{kj}$ is the long-run average rate at which the process goes from state k to state j. Thus, the equation $\pi_j = \sum_{k \in I} \pi_k p_{kj}$ expresses in mathematical terms the physical principle:

> the average rate at which the process makes a transition *from* state j is equal to the average rate at which the process makes a transition *to* state j.

In many applications a cost structure is imposed on a Markov chain. The following useful ergodic theorem holds for Markov chains.

Rule 10.7 *Let $\{\pi_j\}$ be the unique equilibrium distribution of a finite-state Markov chain $\{X_n\}$ that has no two or more disjoint closed sets. Assume that a cost $c(j)$ is incurred at each visit of the Markov chain to state j for any $j \in I$. Then,*

$$\lim_{n\to\infty} \frac{1}{n} \sum_{k=1}^{n} c(X_k) = \sum_{j \in I} c(j)\pi_j \text{ with probability } 1,$$

independently of the starting state $X_0 = i$. *In words, the long-run actual average cost per unit time is* $\sum_{j \in I} c(j) \pi_j$ *with probability* 1.

A formal proof is omitted, but the result is intuitively obvious from the interpretation of the π_j in Rule 10.6.

Page-Ranking Algorithm

The theory of Markov chains has many applications. The most famous application is Markov's own analysis of the frequencies with which vowels and consonants occur in Pushkin's novel *Eugene Onegin*, see also Problem 10.36 below. An important application of more recent date is the application of Markov chains to the ranking of webpages. The page-ranking algorithm is one of the methods Google uses to determine a page's relevance or importance. Suppose that you have n interlinked webpages. Let n_j be the number of outgoing links on page j. It is assumed that $n_j > 0$ for all j. Let α be a given number with $0 < \alpha < 1$. Imagine that a random surfer jumps from his current page by choosing with probability α a random page amongst those that are linked from the current page, and by choosing with probability $1 - \alpha$ a completely random page. Hence, the random surfer jumps around the Web from page to page according to a Markov chain with the one-step transition probabilities

$$p_{jk} = \alpha \, r_{jk} + (1 - \alpha) \frac{1}{n} \quad \text{for } j, k = 1, \ldots, n,$$

where $r_{jk} = \frac{1}{n_j}$ if page k is linked from page j and $r_{jk} = 0$ otherwise. The parameter α was originally set to 0.85. The inclusion of the term $(1 - \alpha)/n$ can be justified by assuming that the random surfer occasionally gets bored and then randomly jumps to any page on the Web. Since the probability of such a jump is rather small, it is reasonable that it does not influence the ranking very much. By the term $(1 - \alpha)/n$ in the p_{jk}, the Markov chain has no two or more disjoint closed sets and is aperiodic. Thus, the Markov chain has a unique equilibrium distribution $\{\pi_j\}$. These probabilities can be estimated by multiplying the matrix $\mathbf{P}$ of one-step transition probabilities by itself repeatedly. Because of the constant $(1 - \alpha)/n$ in the matrix $\mathbf{P}$, things mix up better so that the n-fold matrix product $\mathbf{P}^n$ converges very quickly to its limit. The equilibrium probability π_j gives us the long-run proportion of time that the random surfer will spend on page j. If $\pi_j > \pi_k$, then page j is more important than page k and should be ranked higher.

Concluding Remarks

So far we have paid no attention to state classification in Markov chains, the emphasis was placed on modeling and computational aspects of Markov chains. State classification is needed to fill in the details of the proofs that were skipped in this section on the long-run analysis of Markov chains. In these concluding remarks we briefly discuss the important concept of transient and recurrent states, and a relation between mean recurrence times and equilibrium probabilities. More on state classification in Problem 10.49.

A state j is said to be *recurrent* if

$$\sum_{n=1}^{\infty} p_{jj}^{(n)} = \infty$$

and is said to be *transient* otherwise. Hence, $\sum_{n=1}^{\infty} p_{jj}^{(n)} < \infty$ for a transient state j. The rationale for this definition is the fact that $\sum_{n=1}^{\infty} p_{jj}^{(n)}$ represents the expected value of the number of returns of the process to state j over the time points $n = 1, 2, \ldots$, given that the process starts in state j, see Rule 10.3. Loosely speaking, a recurrent state is one to which the process keeps coming back and a transient state is one which the process eventually leaves forever. Alternatively, a state j can be defined to be recurrent if $P(X_n = j$ for infinitely many $n \mid X_0 = j) = 1$ and transient if $P(X_n = j$ for infinitely many $n \mid X_0 = j) = 0$. A finite-state Markov chain has always at least one recurrent state (why?).

For any recurrent state j, define the mean recurrence time μ_{jj} as the expected value of the number of transitions needed to return from state j to itself. In a finite-state Markov chain, the mean recurrence time μ_{jj} is finite for any recurrent state j, see Problem 10.49. The renewal–reward theorem in Section 9.4 then implies that

$$\pi_j = \frac{1}{\mu_{jj}}$$

for any recurrent state j in a finite-state Markov chain having no two or more disjoint closed sets. To see this, note that the Markov chain is a regenerative stochastic process. Any recurrent state j can serve as a regeneration state. Take as cycle the time interval between two successive visits to the recurrent state j and assume that a reward of 1 is earned each time state j is visited. Then the expected length of one cycle is μ_{jj} and the expected reward earned in one cycle is 1. This gives that the long-run average reward per unit time is equal to $\frac{1}{\mu_{jj}}$ with probability one. Moreover, the long-run average reward per unit time gives the long-run proportion of time the process will be in state j, which is given by π_j.

It is important to note that a finite-state Markov chain with no two or more disjoint closed sets has the property

$$\pi_j > 0 \text{ if and only if state } j \text{ is recurrent.}$$

The result $\pi_j > 0$ for a recurrent state j is a direct consequence of $\pi_j = 1/\mu_{jj}$ and $\mu_{jj} < \infty$. For a transient state j, we have $\sum_{n=1}^{\infty} p_{jj}^{(n)} < \infty$, which implies that $\lim_{n \to \infty} p_{jj}^{(n)} = 0$ and so $\pi_j = 0$.

The above definition of a transient state and a recurrent state applies as well to Markov chains with a countably infinite state space. As noted above, a finite-state Markov chain has the property that the set of recurrent states is not empty and the mean recurrence time μ_{jj} is finite for any recurrent state j. These properties do not necessarily hold for an infinite-state Markov chain. We give two counterexamples:

(i) The Markov chain has state space $I = \{0, 1, 2, \ldots\}$ and $p_{i,i+1} = 1$ for all $i \in I$. Then all states are transient.
(ii) The Markov chain has state space $I = \{0, \pm 1, \pm 2, \ldots\}$ and $p_{i,i-1} = p_{i,i+1} = 0.5$ for all $i \in I$. Then it can be shown that any state j is recurrent with mean recurrence time $\mu_{jj} = \infty$.

The phenomena in (i) and (ii) cannot occur in infinite-state Markov chains satisfying the regularity condition that some state r exists such that state r will ultimately be reached with probability one from any starting state i and the mean recurrence time μ_{rr} is finite. Under this regularity condition it can be shown that the equilibrium results of this section also hold for infinite-state Markov chains, where Rule 10.7 requires the additional condition that the expected value of the cost (in absolute sense) incurred until the first visit to the recurrent state r is finite for each starting state.

Problems

10.36 In a famous paper written in 1913, Andrey Markov analyzed an unbroken sequence of 20,000 letters from the novel Eugene Onegin. He found that the probability of a vowel following a vowel is 0.128, and the probability of a vowel following a consonant is 0.663. Use a Markov chain to estimate the percentages of vowels and consonants in the novel (and check whether the percentages are in agreement with Markov's actual count of 8,638 vowels and 11,362 consonants).

10.37 Consider again Example 10.3. What is the long-run proportion of time the professor has his license with him? Also, answer this question for Problem 10.2.

10.38 In a small student city, there are three restaurants: an Italian restaurant, a Mexican restaurant, and a Thai restaurant. A student eating in the Italian restaurant on a given evening will eat the next evening in the

Italian restaurant with probability 0.10, in the Mexican restaurant with probability 0.35, in the Thai restaurant with probability 0.25, or at home with probability 0.30. The probabilities of switching are 0.4, 0.15, 0.25, and 0.2 when eating in the Mexican restaurant, 0.5, 0.15, 0.05, and 0.3 when eating in the Thai restaurant, and 0.40, 0.35, 0.25, and 0 when eating at home. What proportion of time will the student eat at home?

10.39 A virus can exist in four different strains $i = 1, 2, 3,$ and 4. At each generation the virus stays in the same strain or mutates to another strain that is randomly chosen. The mutation probability is r_i when the virus is in strain i, where the r_i are given by $r_1 = 0.18$, $r_2 = 0.3$, $r_3 = 0.24$, and $r_4 = 0.15$. What is the limiting probability of observing the virus in strain i?

10.40 Consider again Problem 10.4. Calculate the equilibrium probabilities of the Markov chain describing the weather. What is the long-run proportion of days it will be sunny? What is the probability that it will be rainy on a given Sunday many days from now?

10.41 Consider again Problem 10.1. Verify that the equilibrium distribution of the number of type-1 particles in compartment A is a hypergeometric distribution.

10.42 In a long DNA sequence of a particular genome, it has been recorded how often each of the four bases A, C, G, and T is followed by another base. The DNA sequence is modeled by a Markov chain with state space $I = \{A, C, G, T\}$ and one-step transition probabilities

from\to	A	C	G	T
A	0.340	0.214	0.296	0.150
C	0.193	0.230	0.345	0.232
G	0.200	0.248	0.271	0.281
T	0.240	0.243	0.215	0.302

What is the long-run frequency of the base A appearing? What is the long-run frequency of observing base A followed by another A?

10.43 In a certain town, there are four entertainment venues. Both Linda and Bob visit one of these venues every weekend, independently of each other. Each of them visits the venue of the week before with probability 0.4 and chooses otherwise at random one of the other three venues. What is the long-run fraction of weekends that Linda and Bob visit the same venue and what is the limiting probability that they visit the same venue two weekends in a row?

10.44 Consider again Problem 10.3. What is the long-run frequency at which k dice are rolled for $k = 1, \ldots, 6$?

10.45 Let $\{X_n\}$ be a Markov chain with no two or more disjoint closed sets and state space $I = \{1, 2, \ldots, N\}$. Suppose that the Markov chain is *doubly stochastic*, that is, for each column of the matrix of one-step transition probabilities the sum of the column elements is equal to one. Verify that the Markov chain has the discrete uniform distribution $\pi_j = \frac{1}{N}$ for all j as its unique equilibrium distribution.

10.46 Let the random variable S_n be the total score of n rolls of a fair die. Prove that the probability that S_n is divisible by r converges to $1/r$ as $n \to \infty$ for any integer $r \geq 2$. *Hint*: Let $X_n = S_n \pmod{r}$ and use the result of Problem 10.45.

10.47 Let r be a recurrent state of a finite-state Markov chain that has no two or more disjoint closed sets. Use the renewal–reward theorem from Section 9.4 to verify that the equilibrium probabilities π_j of the Markov chain satisfy the relation $\pi_j = \gamma_{jr} \pi_r$ for any state $j \neq r$, where γ_{jr} is the expected number of visits to state j between two successive visits of the Markov chain to state r.

10.48 Let $\{X_n\}$ be a finite-state Markov chain with one-step transition probabilities p_{ij}. Suppose that the Markov chain $\{X_n\}$ has no two or more disjoint closed sets. For fixed τ with $0 < \tau < 1$, define the perturbed Markov chain $\{\overline{X}_n\}$ by the one-step transition probabilities $\overline{p}_{ii} = 1 - \tau + \tau p_{ii}$ and $\overline{p}_{ij} = \tau p_{ij}$ for $j \neq i$. Explain why the Markov chain $\{\overline{X}_n\}$ is aperiodic. Verify that the Markov chain $\{\overline{X}_n\}$ has the same unique equilibrium distribution as the original Markov chain $\{X_n\}$.

10.49 Let $\{X_n\}$ be a finite-state Markov chain and let C be a closed set of states.

(a) Two states i and j with $i \neq j$ are said to communicate if there are integers $r, s \geq 1$ such that $p_{ij}^{(r)} > 0$ and $p_{ji}^{(s)} > 0$. Verify that i and k communicate if i communicates with j and j communicates with k.

(b) Verify that C is irreducible if and only if all states in C communicate with each other. *Hint*: Let $C(i) = \{j : p_{ij}^{(n)} > 0 \text{ for some } n \geq 1\} \cup \{i\}$ for $i \in C$.

(c) Verify that the closed set C always contains a recurrent state.

(d) Verify that a transient state cannot be reached from a recurrent state.

(e) Prove that all states in C are recurrent if the set C is irreducible.

(f) Let C be an irreducible set of states. For any $i, j \in C$, let N_{ij} be the number of transitions until the first return to j when starting in i. Prove that $E(N_{ij}) < \infty$ for all $i, j \in C$. *Hint*: for fixed $j \in I$, show that $P(N_{ij} > kr) \leq (1 - \rho)^k$ for $i \in C$, where $r = \max_{i \in C} r_i$, $\rho = \min_{i \in C} p_{ij}^{(r_i)}$, and r_i is the smallest $n \geq 1$ with $p_{ij}^{(n)} > 0$.

(g) Suppose that the set C is irreducible. The period of any state $i \in C$ is defined as the greatest common divisor of the indices $n \geq 1$ for which $p_{ii}^{(n)} > 0$. Verify that all states in C have the same period.

10.50 Let $\{X_n\}$ be an irreducible Markov chain with finite state space I and unique equilibrium distribution $\{\pi_j, j \in I\}$. Suppose that the Markov chain is periodic with period d. Can you intuitively explain why $\lim_{n \to \infty} p_{ij}^{(nd)} = d\pi_j$ for all $i, j \in R_s$ and $1 \leq s \leq d$, where the sets R_s are as in Definition 10.4?

10.51 Consider again Example 10.4. What is the long-run average stock on hand at the end of the week? What is the long-run average ordering frequency and what is the long-run amount of demand lost per week?

10.52 Consider again Problem 10.4. The local entrepreneur Jerry Woodside has a restaurant on the island. On every sunny day, his turnover (in dollars) has an $N(\mu_1, \sigma_1^2)$ distribution with $\mu_1 = 1,000$ and $\sigma_1 = 200$, while on rainy days his turnover is $N(\mu_2, \sigma_2^2)$ distributed with $\mu_2 = 500$ and $\sigma_2 = 75$. What is the long-run average sales per day?

10.53 Consider again Problem 10.6. Suppose that a cost of \$750 is incurred each time the device fails and that the replacement cost of a circuit board is \$100. What is the long-run proportion of weeks the device operates properly? What is the long-run average weekly cost?

10.54 Suppose that a conveyor belt is running at uniform speed and is transporting items on individual carriers equally spaced along the conveyor. Two work stations, 1 and 2, are placed along the conveyor. Each work station can process only one item at a time and has no storage capacity. The processing time of an item at work station i has an exponential distribution with an expected value of $1/\mu_i$ time units, where $1/\mu_1 = 0.75$ and $1/\mu_2 = 1.25$. An item for processing arrives at fixed times $t = 0, 1, \ldots$. The item is lost when both work stations are occupied and is handled otherwise by an idle work station, where station 1 is the first choice when both stations are idle. What is the long-run fraction of items that are lost?

10.55 Leaky bucket control is a procedure used in telecommunication networks to control the packet input into the network. To achieve this, a token buffer is used. An arriving packet is admitted to the network only if the token buffer is not empty; otherwise, the packet is rejected and is lost. An admitted packet takes one token from the buffer and then enters the network. The buffer cannot contain more than M tokens. A token is generated at the beginning of each time slot. The token is stored in the buffer if there is room in the buffer; otherwise, the token is thrown away. Packets arrive one at a time and the number of packets arriving in any time slot is Poisson distributed with expected value λ, independently of the number of arrivals in other time slots. Specify the equilibrium equations of the Markov chain describing the number of tokens in the pool just before a token is generated. Give an expression for the long-run average number of packets admitted per time slot.

10.56 A transport firm has effected an insurance contract for a fleet of vehicles. The premium payment is due at the beginning of each year. There are four possible premium classes with a premium payment of P_i in class i, where $P_{i+1} < P_i$ for $i = 1, 2, 3$. If no damage is claimed in the just-ended year and the last premium charged is P_i, the next premium payment is P_{i+1} (with $P_5 = P_4$); otherwise, the highest premium P_1 is due. The transport firm has obtained the option to decide only at the end of the year whether the accumulated damage during that year should be claimed or not. If a claim is made, then the insurance company compensates the accumulated damage minus an own risk which amounts to r_i for premium class i. The sizes of the damages in successive years are independent random variables that are exponentially distributed with mean $1/\eta$. The claim strategy of the firm is characterized by four given numbers $\alpha_1, \ldots, \alpha_4$ with $\alpha_i > r_i$ for all i. If the current premium class is i, then the firm claims at the end of the year only damages larger than α_i; otherwise, nothing is claimed. Use an appropriate Markov chain to find the long-run average yearly cost.

10.5 Markov Chain Monte Carlo Simulation

Markov chain Monte Carlo (MCMC) methods are powerful simulation techniques to sample from a multivariate probability density that is known up to a multiplicative constant, where the constant is very difficult to compute. In Bayesian statistics one often encounters the situation that obtaining the posterior density requires the computation of a multiplicative constant in the form of a high-dimensional integral, see also Section 7.5. MCMC methods are very useful for obtaining this constant by random sampling and have greatly enhanced the applicability of Bayesian approaches in practice. MCMC methods are so named because the sequence of simulated values forms a Markov chain that has the required probability distribution as equilibrium distribution. MCMC has revolutionized applied mathematics and is used in many areas of science including statistics, computer science, physics, and biology.

This section begins with a re-examination of classical simulation procedures before moving on to the modern computationally intensive technique of

MCMC simulation. The hit-or-miss method and the acceptance–rejection method can be seen as predecessors of MCMC methods and will be discussed briefly in Section 10.5.1. Ideas from these methods also appear in MCMC methods. Next we introduce in Section 10.5.2 the concept of reversible Markov chains. This concept is key to the construction of a Markov chain that has a given probability distribution as equilibrium distribution. A very versatile MCMC method is the Metropolis–Hastings algorithm, which will be considered in Section 10.5.3. A special case is the Gibbs sampling algorithm. This algorithm is one of the best known MCMC methods and will be discussed in Section 10.5.4.

10.5.1 Classical Simulation Methods

The hit-or-miss method generates random points in a bounded region.[2] It can be used to estimate the volume of a domain in a high-dimensional space. Let G be a bounded region in which it is difficult to generate a random point directly. Suppose that an easy test is available to verify whether a given point belongs to G or not. For ease of presentation, the *hit-or-miss method* is described by using a bounded region G in the two-dimensional plane:

Step 0. Choose a rectangle R that envelopes the region G, see Figure 10.2.
Step 1. Generate a random point $(x, y) \in R : x := a_1 + (a_2 - a_1)u_1$ and $y := b_1 + (b_2 - b_1)u_2$, where u_1 and u_2 are random numbers from $(0, 1)$.
Step 2. If the point (x, y) belongs to G, then accept the point as a random point in G; otherwise, repeat step 1.

The expected number of iterations of the method is equal to the ratio of the area of R and the area of G. Therefore, it is important to choose the rectangle R as minimal as possible. The generalization of the hit-or-miss method to higher-dimensional spaces is obvious.

Acceptance–Rejection Method

The acceptance–rejection method is an extension of the hit-or-miss method. It is a technique to simulate from a general probability density $f(x)$ that is difficult to sample from directly. Instead of sampling directly from the density $f(x)$, one uses an envelope density from which sampling is easier. Let $g(x)$ be a

---

[2] The idea of the hit-or-miss method was introduced to the statistics community by physicists N. Metropolis and S. Ulam in their article "The Monte Carlo method," *Journal of the American Statistical Association* **44** (1949): 335–341. In this article, Metropolis and Ulam give a classic example of finding the volume of a 20-dimensional region within a unit cube when the required multiple integrals are intractable.

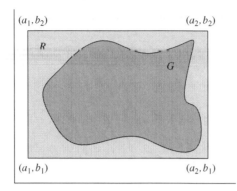

Figure 10.2. Hit-or-miss method.

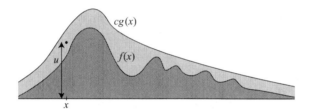

Figure 10.3. Acceptance–rejection method.

probability density that can be simulated by some known method and suppose there is a known constant c such that

$$f(x) \le cg(x) \quad \text{for all } x.$$

Note that c must be at least 1, since both $f(x)$ and $g(x)$ integrate to 1. The acceptance–rejection method goes as follows.

Step 1. Generate a candidate x from $g(x)$ and a random number $u \in (0, 1)$.
Step 2. If $u \le \frac{f(x)}{cg(x)}$, then accept x as a sample from $f(x)$; otherwise, repeat step 1.

In Rule 7.2 we have shown that the method gives a random sample from the target density $f(x)$. Intuitively, the method works because it generates randomly a point $(x, ucg(x))$ in a region covering $f(x)$ and then only keeps points in the required region under the graph of $f(x)$. The acceptance–rejection method is illustrated in Figure 10.3. The particular sample shown in the figure will be rejected. Since the expected number of iterations of steps 1 and 2 to obtain an accepted draw is c, the method is only attractive when c is not too large. The acceptance–rejection method has been described for the probability density of a continuous random variable, but translates literally to the case of a probability mass function of a discrete random variable.

To illustrate the method, let the probability density $f(x)$ be concentrated on a finite interval (a, b) and suppose that $f(x)$ is continuous on (a, b). Choose as candidate density the uniform density $g(x) = 1/(b - a)$ on (a, b). Denoting by m the maximum of $f(x)$ on (a, b), we have $f(x) \leq cg(x)$ for all x with $c = m(b - a)$. The acceptance–rejection method is then as follows. *Step 1*: Generate random numbers u_1 and u_2 from $(0, 1)$, and let $x := a + (b - a)u_1$. *Step 2*: If $u_2 \leq f(x)/m$, then accept x as a sample from $f(x)$; otherwise, return to step 1.

The acceptance–rejection method is a very useful technique for simulating from one-dimensional densities. However, in high-dimensional problems the method is not practical. Apart from the difficulty of finding a high-dimensional candidate density, the bounding constant c will typically be very large.

Importance Sampling

Let $f(x)$ be the probability density of a random variable X and suppose we want to estimate $E[a(X)] = \int a(x)f(x)dx$ for some function $a(x)$. Importance sampling is a useful simulation method for doing this when it is difficult to sample directly from $f(x)$ (or when the variance of the random variable $a(X)$ is very large). This method can also be used when $f(x)$ is only known up to a multiplicative constant. Choose a probability density $g(x)$ that can be simulated by some known method, where $g(x) > 0$ if $f(x) > 0$. Let the random variable Y be distributed according to the density $g(x)$. Then,

$$\int a(x)f(x)dx = \int \left[a(y)\frac{f(y)}{g(y)}\right]g(y)dy = E_Y\left[a(Y)\frac{f(Y)}{g(Y)}\right].$$

This forms the basis for importance sampling. Let $y_1, \ldots, y_n$ be independent samples from $g(y)$, with n large. Then, by the strong law of large numbers,

$$\int a(x)f(x)dx \approx \frac{1}{n}\sum_{i=1}^{n} w_i a(y_i) \quad \text{with } w_i = \frac{f(y_i)}{g(y_i)}.$$

An alternative formulation of importance sampling is to use

$$\int a(x)f(x)dx \approx \frac{\sum_{i=1}^{n} w_i a(y_i)}{\sum_{i=1}^{n} w_i}.$$

Why does this work? The explanation is simple. By the strong law of large numbers, $\frac{1}{n}\sum_{i=1}^{n} w_i \approx \int [f(y)/g(y)]g(y)\,dy = \int f(y)\,dy$ and so $\frac{1}{n}\sum_{i=1}^{n} w_i \approx 1$ for n large. In the alternative formulation it suffices to know the weights w_i up to a multiplicative constant and so this formulation can be used when $f(x)$ is only known up to proportionality.

Importance sampling requires much care in the choice of the proposal density $g(x)$. If the proposal density $g(x)$ is small in a region where $|a(x)f(x)|$ is large, then the estimate of $E[a(X)]$ may be drastically wrong, even after many points y_i have been generated. In high-dimensional problems it is even more difficult to find an appropriate proposal density.

10.5.2 Reversible Markov Chains

Let $\{X_n\}$ be a Markov chain with finite state space I. It is assumed that the Markov chain is irreducible, that is, for any two states i and j there is an $n \geq 1$ such that $p_{ij}^{(n)} > 0$. Any state of an irreducible Markov chain is recurrent, see Problem 10.49. The irreducible Markov chain $\{X_n\}$ has a unique equilibrium distribution $\{\pi_j\}$, see Rule 10.5 in Section 10.4. In this section we also argued that $\pi_j p_{jk}$ can be interpreted as the long-run average number of transitions per unit time from state j to state k. Many Markov chains have the property that the rate of transitions from j to k is equal to the rate of transitions from k to j for any $j, k \in I$. An example of such a Markov chain is a random walk on the set of integers $\{0, 1, \ldots, N\}$, where $p_{ij} = 0$ for $|j - i| > 1$, $p_{i,i+1} > 0$ for $0 \leq i < N$, and $p_{i,i-1} > 0$ for $0 < i \leq N$. In this Markov chain the rate of transitions from $i - 1$ to i is equal to the rate of transitions from i to $i - 1$ for any i because, between any two transitions from $i - 1$ to i, there must be one from i to $i - 1$ and conversely. The Ehrenfest model in Example 10.2 is an illustration of such a random walk.

Definition 10.5 *An irreducible Markov chain $\{X_n\}$ with finite state space I is said to be reversible if the equilibrium probabilities π_j satisfy the so-called detailed balance equations*

$$\pi_j p_{jk} = \pi_k p_{kj} \quad \text{for all } j, k \in I.$$

Why the name of "reversible Markov chain?" Suppose that at time 0 the reversible Markov chain is started according to $P(X_0 = j) = \pi_j$ for $j \in I$. Then, for all $j, k \in I$,

$$P(X_{n-1} = j \mid X_n = k) = P(X_n = j \mid X_{n-1} = k) \quad \text{for all } n \geq 1.$$

In words, in equilibrium the evolution of the process backward in time is probabilistically identical to the evolution of the process forward in time. If you were to watch a reversible process on video, you would not be able to determine whether you were watching the video in forward motion or rewind motion. The proof is simple. Using the fact that any future state of the

10.5 Markov Chain Monte Carlo Simulation

Markov chain has the equilibrium distribution $\{\pi_j\}$ if the initial state is chosen according to this distribution (see below Definition 10.3), it follows that

$$P(X_{n-1} = j \mid X_n = k) = P(X_n = k \mid X_{n-1} = j) \frac{P(X_{n-1} = j)}{P(X_n = k)} = p_{jk} \frac{\pi_j}{\pi_k}.$$

Note that $\pi_k > 0$ for any state k in the irreducible Markov chain. By the reversibility assumption, $p_{jk}\pi_j/\pi_k = p_{kj}$ and so $P(X_{n-1} = j \mid X_n = k) = p_{kj}$. This shows that $P(X_{n-1} = j \mid X_n = k) = P(X_n = j \mid X_{n-1} = k)$.

Why the interest in the detailed balance equations? The answer is that a probability distribution satisfying these equations must be an equilibrium distribution of the Markov chain. This is easy to prove:

Rule 10.8 *Let $\{X_n\}$ be an irreducible Markov chain with finite state space I. If $\{a_j, j \in I\}$ is a probability distribution satisfying*

$$a_j p_{jk} = a_k p_{kj} \quad \text{for all } j, k \in I,$$

then $\{a_j, j \in I\}$ is the unique equilibrium distribution of the Markov chain.

To prove Rule 10.8, sum both sides of the equation $a_j p_{jk} = a_k p_{kj}$ over $k \in I$. Together with $\sum_{k \in I} p_{jk} = 1$, this gives

$$a_j = \sum_{k \in I} a_k p_{kj} \quad \text{for all } j \in I.$$

That is, $\{a_j, j \in I\}$ is an equilibrium distribution of the Markov chain. By Rule 10.5, this equilibrium distribution is unique.

An interesting question is the following. Let $\{a_j, j \in I\}$ be a probability mass function on a finite set of states I with $a_j > 0$ for all j. Can we construct a Markov chain on I with $\{a_j, j \in I\}$ as its unique equilibrium distribution? The answer is yes. The physical construction of the Markov chain proceeds as follows: if the current state of the process is j, then choose at random one of the other states as candidate state for the next state. Suppose the chosen candidate state is k. If $a_k > a_j$, then the next state of the Markov chain is state k. If $a_k \leq a_j$, then the next state is either state k with probability a_k/a_j or the current state j with probability $1 - a_k/a_j$. Thus we have constructed a Markov chain $\{X_n\}$ on I with one-step transition probabilities

$$p_{jk} = \begin{cases} \frac{1}{N-1} \min(\frac{a_k}{a_j}, 1) & \text{for } k \neq j, \\ 1 - \sum_{l \neq j} p_{jl} & \text{for } k = j, \end{cases}$$

where N is the number of states of the Markov chain. It is immediate that the Markov chain $\{X_n\}$ is irreducible (and aperiodic). Also, the Markov chain satisfies the reversibility condition. We have

$$a_j p_{jk} = \frac{1}{N-1}\min(a_k, a_j) = a_k \frac{1}{N-1}\min\left(\frac{a_j}{a_k}, 1\right) = a_k p_{kj}$$

for any $j, k \in I$ with $j \neq k$. It now follows from Rule 10.8 that the Markov chain has $\{a_j\}$ as its unique equilibrium distribution.

A simple but useful extension of the above construction is as follows. For each $i \in I$, choose a set $N(i)$ of neighbors of i with $i \notin N(i)$, where the sets have the property that $k \in N(j)$ if $j \in N(k)$. Letting $M = \max_i |N(i)|$, define a Markov chain by the one-step transition probabilities $p_{jk} = (1/M)\min(a_k/a_j, 1)$ for $k \in N(j)$, $p_{jj} = 1 - \sum_{k \in N(j)} p_{jk}$, and $p_{jk} = 0$ otherwise. Then this Markov chain satisfies the reversibility condition (verify!). Assuming sufficient connectivity between the sets $N(i)$ so that the Markov chain is irreducible, it follows that this Markov chain also has $\{a_j\}$ as its unique equilibrium distribution. In the next section it will be seen that many other constructions are possible for Markov chains, so that the Markov chain has $\{a_j\}$ as equilibrium distribution.

Problems

10.57 A Markov chain model of DNA sequence evolution has a transition matrix **P** of the form

$$\mathbf{P} = \begin{pmatrix} 1-\alpha-\beta-\gamma & \alpha & \beta & \gamma \\ \alpha & 1-\alpha-\beta-\gamma & \gamma & \beta \\ \beta & \gamma & 1-\alpha-\beta-\gamma & \alpha \\ \gamma & \beta & \alpha & 1-\alpha-\beta-\gamma \end{pmatrix},$$

where $\alpha, \beta, \gamma > 0$, and $\alpha + \beta + \gamma < 1$. Verify that the Markov chain is reversible.

10.58 Consider a two-state Markov chain with transition matrix

$$\mathbf{P} = \begin{pmatrix} 1-a & a \\ b & 1-b \end{pmatrix}.$$

For which values of a and b is the Markov chain reversible?

10.59 Consider an undirected graph with a finite number of nodes $i = 1, \ldots, N$, where some pairs of nodes are connected by arcs. It is assumed that the graph is connected, that is, there is a path of arcs between any two nodes. A positive weight w_{ij} is associated with each arc (i, j) in the graph ($w_{ij} = w_{ji}$). A particle is moving from node to node. If the particle is in node i, then it will next move to node j with probability $p_{ij} = w_{ij}/\sum_k w_{ik}$. Verify that the Markov chain describing the position of the particle is reversible and give the equilibrium probabilities. As an application, consider a mouse that is trapped in a closed maze. The maze has 15 rooms. The rooms are positioned according to the three-by-five array below. Doors connect any two neighboring rooms. For example, room 1 is connected with rooms 2 and 6, room 2 with rooms 1, 3, and 7, room 7 with rooms 2, 6, 8, and 12, and so on.

1	2	3	4	5
6	7	8	9	10
11	12	13	14	15

The mouse moves randomly from room to room, where at each time one of the available doors is chosen with equal probability. What is the expected value of the total number of moves needed by the mouse to return to the starting room?

10.60 Let $c(i)$ be a given function on a finite but very large set I. For any $i \in I$, choose a local neighborhood $N(i)$ of state i so that the sets $N(i)$ have the property that $i \notin N(i)$ and $i \in N(k)$ if $k \in N(i)$. For ease, it is assumed that each set $N(i)$ contains the same number of points. The following Markov chain is defined on I. If the current state is i, a candidate state j is chosen at random from $N(i)$. The next state of the process is always j if $c(j) < c(i)$; otherwise, the process either moves to j with probability $e^{-c(j)/T}/e^{-c(i)/T}$ or stays in i with probability $1 - e^{-c(j)/T}/e^{-c(i)/T}$. Here, $T > 0$ is a control parameter. It is assumed that the sets $N(i)$ are such that the Markov chain is irreducible. Prove that the unique equilibrium distribution of the Markov chain is given by $\pi_i = e^{-c(i)/T} / \sum_{k \in I} e^{-c(k)/T}$ for $i \in I$. Also, verify that $\pi_m \to 1$ as $T \to 0$ if the function $c(i)$ assumes its absolute minimum at a unique point m.

10.5.3 Metropolis–Hastings Algorithm

The Metropolis–Hastings algorithm is an example of an MCMC method. It is widely used for high-dimensional problems. In contrast to the acceptance–rejection method, which is only useful for one-dimensional problems and can only generate independent samples, the Metropolis–Hastings algorithm generates a sequence of states from a Markov process. The algorithm is designed to draw samples from a given probability distribution that is known up to a multiplicative constant, where it is not feasible to compute this normalizing constant directly. The algorithm will be explained first for the case of a discrete probability distribution, but the basic idea of the algorithm can be generalized directly to the case of a continuous probability distribution.

Let S be a finite but very large set of states on which a probability mass function $\{\pi(s), s \in S\}$ is given, where $\pi(s) > 0$ for all $s \in S$ and the probability mass function is only known up to a multiplicative constant. The Metropolis–Hastings algorithm generates a sequence of states $(s_0, s_1, \ldots)$ from a Markov chain that has $\{\pi(s), s \in S\}$ as its unique equilibrium distribution. To that end, the algorithm uses a candidate-transition function $q(t \mid s)$ (for clarity of presentation, we use the notation $q(t \mid s)$ rather than p_{st}). This function is to be interpreted as saying that when the current state is s, the candidate for the next state is t with probability $q(t \mid s)$. Thus, one first chooses, for each $s \in S$, a probability mass function $\{q(t \mid s), t \in S\}$. These functions must be chosen in such a way that the Markov matrix with one-step transition probabilities $q(t \mid s)$ is irreducible. The idea is to next adjust these transition probabilities in such a way that the resulting Markov chain has $\{\pi(s), s \in S\}$ as unique equilibrium distribution. The reversibility equations are key to this idea. If the candidate-transition function $q(t \mid s)$ already satisfies the equations

$$\pi(s)q(t \mid s) = \pi(t)q(s \mid t) \quad \text{for all } s, t \in S,$$

then we are done and can conclude from Rule 10.8 that the Markov chain with the probabilities $q(t \mid s)$ as one-step transition probabilities has $\{\pi(s), s \in S\}$ as unique equilibrium distribution. What should one do when the reversibility equations are not fully satisfied? The answer is to modify the transition probabilities by rejecting certain transitions. To work out this idea, fix two states s and t for which the reversibility equation is not satisfied. It is no restriction to assume that $\pi(s)q(t \mid s) > \pi(t)q(s \mid t)$. Otherwise, reverse the roles of s and t. If $\pi(s)q(t \mid s) > \pi(t)q(s \mid t)$, then, loosely speaking, the process moves from s to t too often. How could one restore this? A simple trick to reduce the number of transitions from s to t is to use an acceptance probability $\alpha(t \mid s)$: the process is allowed to make the transition from s to t with probability $\alpha(t \mid s)$, otherwise the process stays in the current state s. The question of how to choose $\alpha(t \mid s)$ remains. The choice of $\alpha(t \mid s)$ is determined by the requirement

$$\pi(s)[q(t \mid s)\alpha(t \mid s)] = \pi(t)[q(s \mid t)\alpha(s \mid t)].$$

Taking an acceptance probability of $\alpha(s \mid t) = 1$ for transitions from t to s, we get

$$\alpha(t \mid s) = \frac{\pi(t)q(s \mid t)}{\pi(s)q(t \mid s)}.$$

10.5 Markov Chain Monte Carlo Simulation

Consequently, for any $s, t \in S$, we define the acceptance probability

$$\alpha(t \mid s) = \min\left[\frac{\pi(t)q(s \mid t)}{\pi(s)q(t \mid s)}, 1\right].$$

Next, the one-step transition probabilities of the Markov chain $\{X_n\}$ we are looking for are defined by

$$q_{MH}(t \mid s) = \begin{cases} q(t \mid s)\alpha(t \mid s) & \text{for } t \neq s, \\ 1 - \sum_{t \neq s} q(t \mid s)\alpha(t \mid s) & \text{for } t = s. \end{cases}$$

The Markov chain $\{X_n\}$ with these one-step transition probabilities satisfies the reversibility condition

$$\pi(s)q_{MH}(t \mid s) = \pi(t)q_{MH}(s \mid t) \quad \text{for all } s, t.$$

Using the assumption that the Markov matrix with the elements $q(t \mid s)$ is irreducible, it is easily verified that the Markov chain $\{X_n\}$ is also irreducible. It now follows from Rule 10.8 that the Markov chain $\{X_n\}$ has $\{\pi(s), s \in S\}$ as its unique equilibrium distribution. It is important to note that for the construction of the Markov chain $\{X_n\}$, it suffices to know the π_s up to a multiplicative constant because the acceptance probabilities involve only the ratios $\pi(s)/\pi(t)$.

Summarizing, the Markov chain $\{X_n\}$ operates as follows. If the current state is s, a candidate state t is generated from the probability mass function $\{q(t \mid s), t \in S\}$. If $t \neq s$, then state t is accepted with probability $\alpha(t \mid s)$ as the next state of the Markov chain; otherwise, the Markov chain stays in state s.

We are now in a position to formulate the Metropolis–Hastings algorithm. The ingredients of the algorithm were derived for a probability mass function $\pi(s)$ on a discrete state space S. However, under suitable regularity conditions, this derivation generalizes to the case of a probability density $\pi(s)$ on a continuous state space S. In the literature, one sometimes uses the terminology *discrete probability density* instead of probability mass function. In the formulation of the Metropolis–Hastings algorithm, the terminology of density will be used to make clear that the algorithm applies both to the case of a discrete distribution and to the case of a continuous distribution. In the continuous case $\{q(t \mid s), t \in S\}$ is a probability density function for any $s \in S$, where in the discrete case $\{q(t \mid s), t \in S\}$ is a probability mass function for any $s \in S$.

Metropolis–Hastings Algorithm

Step 0. For each $s \in S$ choose a density $\{q(t \mid s), t \in S\}$. Let $X_0 := s_0$ for a starting state s_0 and let $n := 1$.

Step 1. Generate a candidate state t_n from the density $\{q(t \mid s_{n-1}), t \in S\}$. Calculate the acceptance probability

$$\alpha = \min\left[\frac{\pi(t_n)q(s_{n-1} \mid t_n)}{\pi(s_{n-1})q(t_n \mid s_{n-1})}, 1\right].$$

Step 2. Generate a random number u from $(0, 1)$. If $u \leq \alpha$, accept t_n and let $s_n := t_n$; otherwise, $s_n := s_{n-1}$.

Step 3. $n := n + 1$. Repeat step 1 with s_{n-1} replaced by s_n.

Note that when the chosen densities $q(t \mid s)$ are symmetric, that is,

$$q(t \mid s) = q(s \mid t) \quad \text{for all } s, t,$$

the acceptance probability α in step 1 reduces to

$$\alpha = \min\left(\frac{\pi(t_n)}{\pi(s_{n-1})}, 1\right).$$

In many applications of the algorithm one wants to estimate $E[h(X)]$ for a given function $h(x)$, where X is a random variable with $\pi(s)$ as its density. The approach is to generate states $s_1, s_2, \ldots, s_m$ by the Metropolis–Hastings algorithm, where m is large. Then $E[h(X)]$ is estimated by

$$\frac{1}{m}\sum_{k=1}^{m} h(s_k),$$

using the ergodic theorem for Markov chains, see Rule 10.7. A practical approach to construct an approximate confidence interval for $E[h(X)]$ is dividing the single very long simulation run into 25 (say) long subruns and treating the averages of the data within the subruns as if they are independent samples from a normal distribution (this is the so-called batch-means method in simulation). How the starting state should be chosen and how many iterations should be done are empirical questions. A simple approach is to use multiple dispersed initial values to start several different chains. One suggestion for an initial value is to start the chain close to the mode of the target density if this density is unimodal. Diagnosing convergence to the target density is an art in itself. To diagnose the convergence of an average $\frac{1}{m}\sum_{k=1}^{m} h(s_k)$, one can look at a plot of this average as a function of the number m of runs.

Implementation Aspects

What are the best options for the proposal densities $q(t \mid s)$? There are two general approaches: independent chain sampling and random walk chain sampling. These approaches will be discussed briefly.[3]

---

[3] For more details the interested reader is referred to S. Chib and E. Greenberg, "Understanding the Metropolis–Hastings algorithm," *The American Statistician* **49** (1995): 327–335.

10.5 Markov Chain Monte Carlo Simulation

(a) In independent chain sampling the candidate state t is drawn independently of the current state s of the Markov chain, that is, $q(t \mid s) = g(t)$ for some proposal density $g(x)$.

(b) In random walk chain sampling the candidate state t is the current state s plus a draw from a random variable Z that does not depend on the current state. In this case, $q(t \mid s) = g(t - s)$ with $g(z)$ the density of the random variable Z. If the proposal density $g(z)$ is symmetric ($g(z) = g(-z)$ for all z), then the acceptance probability $\alpha(t \mid s)$ reduces to $\min(\pi(t)/\pi(s), 1)$.

It is very important to have a well-mixing Markov chain, that is, a Markov chain that explores the state space S adequately and sufficiently fast. In other words, by the choice of the proposal densities one tries to avoid that the Markov chain stays in small regions of the state space for long periods of time. The variance of the proposal density can be thought of as a tuning parameter to get better mixing. It affects both the acceptance probability and the magnitude of the moves of the state. It is a matter of experimentation to find a tradeoff between these two features. In random walk chain sampling, a rule of thumb is to choose the proposal density in such a way that on average about 50% of the candidate states are accepted. In independent chain sampling, it is important that the tail of the proposal density dominates the tail of the target density $\pi(s)$.

The implementation aspects are illustrated with the next example.

Example 10.11 Let the density $\pi(x)$ of the positive random variable X be proportional to $x^{-2.5} e^{-2/x}$ for $x > 0$. Suppose that this target density is simulated by the Metropolis–Hastings algorithm with independent chain sampling, where both the uniform density on $(0, 1000)$ and the Pareto density $1.5 x^{-2.5}$ for $x > 1$ are taken as proposal density. How goes the mixing of the state of the Markov chain? What are the estimates for $E(X)$?

Solution. It is instructive to make a plot of the first 500 (say) states simulated by the Metropolis–Hastings algorithm. Such a plot is given in Figure 10.4, where the left figure corresponds to the uniform proposal density and the right figure to the Pareto proposal density. In each of the simulation studies the starting state was chosen as $s_0 = 1$. In the simulation with the Pareto density, we simulated the shifted density $\pi_{shift}(x) = \pi(x - 1)$ for $x > 1$. It is seen directly from the figure that the mixing of the state is very bad under the uniform density, though this proposal density covers nearly all the mass of the target density. A bad mixing means that the candidate state is often rejected. This happens for the uniform density because the variance of this proposal density is very large, meaning that the candidate state is often far away from the current state and, in turn, implying a very small acceptance probability. The Pareto density exhibits the same tail behavior as the target density $\pi(x)$ and

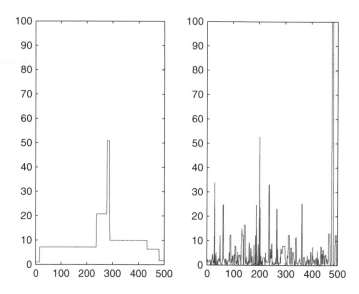

Figure 10.4. Mixing of the state of the Markov chain.

gives an excellent mixing of the state of the Markov chain. The shortcoming of the uniform density as proposal density also appears from the fact that our simulation study with one million runs gave the estimate 3.75 for $E(X)$, where the exact value of $E(X)$ is 4. The simulation study with the Pareto density as proposal density gave the estimate 3.98 for $E(X)$.

Problems

10.61 Let the joint density of the continuous random variables X_1 and X_2 be proportional to $e^{-\frac{1}{2}(x_1^2 x_2^2 + x_1^2 + x_2^2 - 7x_1 - 7x_2)}$ for $-\infty < x_1, x_2 < \infty$. What are the expected value, the standard deviation, and the marginal density of the random variable X_1? Use the Metropolis–Hastings algorithm with random walk chain sampling, where the increments of the random walk are generated from (Z_1, Z_2) with Z_1 and Z_2 independent $N(0, a^2)$ random variables. Experiment with several values of a (say, $a = 0.02$, 0.2, 1, and 5) to see how the mixing in the Markov chain goes and what the average value of the acceptance probability is.

10.62 Simulate the probability distribution $\pi_1 = 0.2$ and $\pi_2 = 0.8$ with the proposal probabilities $q(t \mid s) = 0.5$ for $s, t = 1, 2$ and see how quickly the simulated Markov chain converges.

10.5.4 The Gibbs Sampler

The Gibbs sampler is a special case of Metropolis–Hastings sampling and is used to simulate from a multivariate density whose univariate conditional densities are fully known. This sampler is frequently used in Bayesian statistics. To introduce the method, consider the discrete random vector $(X_1, \ldots, X_d)$ with the multivariate density function

$$\pi(x_1, \ldots, x_d) = P(X_1 = x_1, \ldots, X_d = x_d).$$

The univariate conditional densities of $(X_1, \ldots, X_d)$ are denoted by

$$\pi_k(x \mid x_1, \ldots, x_{k-1}, x_{k+1}, \ldots, x_d)$$
$$= P(X_k = x \mid X_1 = x_1, \ldots, X_{k-1} = x_{k-1}, X_{k+1} = x_{k+1}, \ldots, X_d = x_d).$$

The key to the Gibbs sampler is the assumption that the univariate conditional densities are fully known. The Gibbs sampler generates random draws from the univariate densities and defines a valid Markov chain with $\pi(x_1, \ldots, x_d)$ as its unique equilibrium distribution. The algorithm involves no adjustable parameters and has the feature that the acceptance probability is always equal to 1.

Gibbs Sampling Algorithm

Step 0. Choose a starting state $\mathbf{x} = (x_1, \ldots, x_d)$.
Step 1. Generate a random integer k from $1, \ldots, d$. Simulate a random draw y from the conditional density $\pi_k(x \mid x_1, \ldots, x_{k-1}, x_{k+1}, \ldots, x_d)$. Define state $\mathbf{y}$ by $\mathbf{y} = (x_1, \ldots, x_{k-1}, y, x_{k+1}, \ldots, x_d)$.
Step 2. The new state $\mathbf{x} := \mathbf{y}$. Return to step 1 with $\mathbf{x}$.

This algorithm is a special case of the Metropolis–Hastings algorithm with

$$q(\mathbf{y} \mid \mathbf{x}) = \frac{1}{d} P\left(X_k = y \mid X_j = x_j \text{ for } j = 1, \ldots, d \text{ with } j \neq k\right),$$

for any two states $\mathbf{x}$ and $\mathbf{y}$, where $\mathbf{x} = (x_1, \ldots, x_{k-1}, x_k, x_{k+1}, \ldots, x_d)$ and $\mathbf{y} = (x_1, \ldots, x_{k-1}, y, x_{k+1}, \ldots, x_d)$. In the Gibbs sampler, the acceptance probability $\alpha(\mathbf{y} \mid \mathbf{x})$ is always equal to 1. The proof is simple. Since

$$q(\mathbf{y} \mid \mathbf{x}) = \frac{1}{d} \frac{\pi(\mathbf{y})}{P(X_j = x_j \text{ for } j \neq k)} \quad \text{and} \quad q(\mathbf{x} \mid \mathbf{y}) = \frac{1}{d} \frac{\pi(\mathbf{x})}{P(X_j = x_j \text{ for } j \neq k)},$$

we get $\frac{q(\mathbf{x}|\mathbf{y})}{q(\mathbf{y}|\mathbf{x})} = \frac{\pi(\mathbf{x})}{\pi(\mathbf{y})}$. This shows that $\alpha(\mathbf{y} \mid \mathbf{x}) = \min\left[\frac{\pi(\mathbf{y})\pi(\mathbf{x})}{\pi(\mathbf{x})\pi(\mathbf{y})}, 1\right] = 1$.

Standard Gibbs Sampler

In practice, one usually uses the standard Gibbs sampler, where in each iteration all components of the state vector are adjusted rather than a single component. Letting $\mathbf{x}^{(n)} = (x_1^{(n)}, \ldots, x_d^{(n)})$ denote the state vector obtained in iteration n, the next iteration $n+1$ proceeds as follows:

$x_1^{(n+1)}$ is a random draw from $\pi_1(x \mid x_2^{(n)}, x_3^{(n)}, \ldots, x_d^{(n)})$
$x_2^{(n+1)}$ is a random draw from $\pi_2(x \mid x_1^{(n+1)}, x_3^{(n)}, \ldots, x_d^{(n)})$

.
.
.

$x_d^{(n+1)}$ is a random draw from $\pi_d(x \mid x_1^{(n+1)}, x_2^{(n+1)}, \ldots, x_{d-1}^{(n+1)})$.

This gives the new state vector $\mathbf{x}^{(n+1)} = (x_1^{(n+1)}, \ldots, x_d^{(n+1)})$. The standard Gibbs sampler also generates a sequence of states $\{\mathbf{x}^{(k)}, k = 1, \ldots, m\}$ from a Markov chain having $\pi(x_1, \ldots, x_d)$ as its unique equilibrium distribution. The technical proof is omitted.

The expectation of any function h of the random vector $(X_1, \ldots, X_d)$ can be estimated by $(1/m) \sum_{k=1}^{m} h(\mathbf{x}^{(k)})$ for a Gibbs sequence $\{\mathbf{x}^{(k)}, k = 1, \ldots, m\}$ of sufficient length m. In particular, one might use the Gibbs sequence to estimate the marginal density of any component of the random vector, say the marginal density $\pi_1(x)$ of the random variable X_1. A naive estimate for $\pi_1(x)$ is based on the values $x_1^{(1)}, \ldots, x_1^{(m)}$ from the Gibbs sequence, but a better approach is to use the other values from the Gibbs sequence, together with the explicit expression for the univariate conditional density of X_1. Formally, by the law of conditional expectation, $\pi_1(x) = E[\pi_1(x \mid X_2, \ldots, X_d)]$. Hence, a better estimate for $\pi_1(x)$ is

$$\hat{\pi}_1(x) = \frac{1}{m} \sum_{k=1}^{m} \pi_1(x \mid x_2^{(k)}, \ldots, x_d^{(k)}).$$

This estimate uses more information than the estimate based only on the individual values $x_1^{(1)}, \ldots, x_1^{(m)}$ and will typically be more accurate.

Example 10.12 The bivariate density $\pi(x, y)$ of the random vector (X, Y) is proportional to

$$\binom{r}{x} y^{x+\alpha-1}(1-y)^{r-x+\beta-1} \quad \text{for } x = 0, 1, \ldots, r, \quad 0 \le y \le 1,$$

where r, α, and β are given positive integers. Assuming the data $r = 16$, $\alpha = 2$, and $\beta = 4$, we are interested in estimating the expected value, the variance, and the marginal density of X. The univariate conditional densities of X and Y can be determined explicitly. Therefore, we can use the Gibbs sampler, which method also works for the case of a random vector (X, Y) with a discrete component X and a continuous component Y. How do we find the univariate conditional densities?

Solution. We have that $\pi_1(x \mid y)$ is the ratio of the joint density of (X, Y) and the marginal density of Y. The marginal density of Y is $\sum_{u=0}^{r} \pi(u, y)$. Thus, for any fixed y, it follows from $\pi_1(x \mid y) = \pi(x, y) / \sum_{u=0}^{r} \pi(u, y)$ that

$$\pi_1(x \mid y) = \frac{\binom{r}{x} y^x (1-y)^{r-x}}{\sum_{u=0}^{r} \binom{r}{u} y^u (1-y)^{r-u}} = \binom{r}{x} y^x (1-y)^{r-x}$$

for $x = 0, 1, \ldots, r$. Thus, $\pi_1(x \mid y)$ is the binomial (r, y) density for any fixed y. In the same way, we obtain from $\pi_2(y \mid x) = \pi(x, y) / \int_0^1 \pi(x, u) \, du$ that for any fixed x, the conditional density $\pi_2(y \mid x)$ is proportional to

$$y^{x+\alpha-1}(1-y)^{r-x+\beta-1}$$

for $0 \le y \le 1$. Thus, $\pi_2(y \mid x)$ is the beta$(x + \alpha, r - x + \beta)$ density.

A Gibbs sequence $(x_0, y_0), (x_1, y_1), \ldots, (x_m, y_m)$ of length m is generated as follows, using the standard Gibbs sampler. Choose an integer x_0 between 0 and r. The other elements of the sequence are obtained iteratively by generating alternately a random draw y_j from the beta density $\pi_2(y \mid x_j)$ and a random draw x_{j+1} from the binomial density $\pi_1(x \mid y_j)$. Codes to simulate from the binomial density and the beta density are widely available. In our simulation we have generated $m = 250{,}000$ observations for the state vector (x, y). The first histogram in Figure 10.5 gives the simulated histogram for the marginal density $\pi_1(x)$ of the random variable X. By comparison, the second histogram gives the exact values of $\pi_1(x)$ (a direct computation of

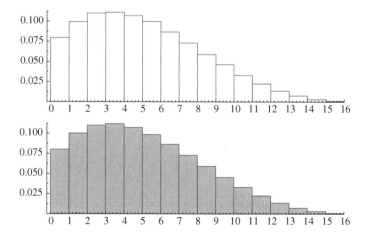

Figure 10.5. Simulated and exact histogram for $\pi_1(x)$.

the proportionality constant for $\pi_1(x)$ is possible). The simulated histogram is based on the estimate

$$\hat{\pi}_1(x) = \frac{1}{m} \sum_{k=1}^{m} \binom{r}{x} y_k^x (1-y_k)^{r-x}$$

for $\pi_1(x)$, using the explicit expression for $\pi_1(x \mid y)$. On the basis of $\hat{\pi}_1(x)$, the estimates 5.35 and 11.20 are found for $E(X)$ and var(X), where the exact values of $E(X)$ and var(X) are 5.33 and 11.17.

Problems

10.63 Consider again Problem 10.61. Use the Gibbs sampler to simulate the marginal density of X_1 together with the expected value and the standard deviation.

10.64 In an actuarial model the random vector (X, Y, N) has a trivariate density $\pi(x, y, n)$ that is proportional to $\binom{n}{x} y^{x+\alpha-1}(1-y)^{n-x+\beta-1} e^{-\lambda} \frac{\lambda^n}{n!}$ for $x = 0, 1, \ldots, n$, $0 < y < 1$, and $n = 0, 1, \ldots$. The random variable N represents the number of policies in a portfolio, the random variable Y represents the claim probability for any policy, and the random variable X represents the number of policies resulting in a claim. First verify that the univariate conditional density functions of X, Y, and N are given by the binomial(n, y) density, the beta$(x + \alpha, n - x + \beta)$ density, and the Poisson$(\lambda(1 - y))$ density shifted to the point x. Assuming the data $\alpha = 2$, $\beta = 8$, and $\lambda = 50$, use the Gibbs sampler to estimate the expected value, the standard deviation, and the marginal density of the random variable X.

11
Continuous-Time Markov Chains

Many random phenomena happen in continuous time. Examples include the occurrence of cell phone calls, the spread of epidemic diseases, stock fluctuations, etc. A continuous-time Markov chain is a very useful stochastic process to model such phenomena. It is a process that goes from state to state according to a Markov chain, but the times between state transitions are continuous random variables having an exponential distribution.

The purpose of this chapter is to give a first introduction to continuous-time Markov chains. The basic concept of the continuous-time Markov chain model is the so-called transition rate function. Several examples will be given to make you familiar with this concept. Then you will learn how to analyze the time-dependent behavior of the process and the long-run behavior of the process. The time-dependent state probabilities can be calculated from linear differential equations, while the limiting state probabilities are obtained by using the appealing flow-rate-equation method. This powerful method has numerous practical applications. The method will be illustrated with examples taken from queueing, inventory, and reliability. The queueing examples include practically important models such as the Erlang loss model and the infinite-server model.

11.1 Markov Chain Model

A continuous-time stochastic process $\{X(t), t \geq 0\}$ is a collection of random variables indexed by a continuous time parameter $t \in [0, \infty)$. The random variable $X(t)$ is called the state of the process at time t. In an inventory problem $X(t)$ might be the stock on hand at time t and in a queueing problem $X(t)$ might be the number of customers present at time t. The formal definition of a continuous-time Markov chain is a natural extension of the definition of a discrete-time Markov chain.

Definition 11.1 *The stochastic process $\{X(t), t \geq 0\}$ with discrete state space I is said to be a continuous-time Markov chain if it possesses the Markovian property, that is, for all time points $s, t \geq 0$ and states $i, j, x(u)$ with $0 \leq u < s$,*

$$P(X(t+s) = j) \mid X(u) = x(u) \text{ for } 0 \leq u < s, X(s) = i)$$
$$= P(X(t+s) = j \mid X(s) = i).$$

In words, the Markovian property says that if you know the present state at time s, then all additional information about the states at times prior to time s is irrelevant for the probabilistic development of the process in the future. All that matters for future states is what the present state is. A continuous-time Markov chain is said to be *time homogeneous* if for any $s, t > 0$ and any states $i, j \in I$,

$$P(X(t+s) = j \mid X(s) = i) = P(X(t) = j \mid X(0) = i).$$

The transition functions $p_{ij}(t)$ are defined by

$$p_{ij}(t) = P(X(t) = j \mid X(0) = i) \quad \text{for } t \geq 0 \text{ and } i, j \in I.$$

In addition to the assumption of time homogeneity, we now make the assumption that the state space I is *finite*. This assumption is made to avoid technical complications involved with a countably infinite state space. Under some regularity conditions, however, the results for the case of a finite state space carry over to the case of a countably infinite state space.

Transition Rates

In continuous time there are no smallest time steps and hence we cannot speak about one-step transition probabilities as in discrete time. In a continuous-time Markov chain we would like to know, for very small time steps of length h, what the probability $p_{ij}(h)$ is of being in a *different* state j at time $t + h$ if the present state at time t is i. This probability is determined by the so-called transition rates.[1] These transition rates are to continuous-time Markov chains what the one-step transition probabilities are to discrete-time Markov chains. Formally, the transition rate q_{ij} can be introduced as the derivative of $p_{ij}(t)$ at $t = 0$. For the case of a finite state space, the $p_{ij}(t)$ are differentiable for any $t \geq 0$. The reader is asked to take this deep result for granted. In particular, the right-hand derivative of $p_{ij}(t)$ at $t = 0$ exists for all i, j. Noting that

$$p_{ij}(0) = \begin{cases} 0 & \text{for } j \neq i, \\ 1 & \text{for } j = i, \end{cases}$$

---

[1] Rate is a measure of how quickly something happens. It can be seen as the frequency with which a repeatable event happens per unit time. That is, an arrival rate λ means that the average frequency of arrivals per unit time is λ.

11.1 Markov Chain Model

we have
$$\frac{p_{ij}(h) - p_{ij}(0)}{h} = \frac{p_{ij}(h)}{h} \text{ for } j \neq i \text{ and } \frac{p_{ii}(h) - p_{ii}(0)}{h} = \frac{p_{ii}(h) - 1}{h}.$$

The right-hand derivatives of the $p_{ij}(t)$ at $t = 0$ are denoted by

$$q_{ij} = \lim_{h \to 0} \frac{p_{ij}(h)}{h} \text{ for } j \neq i \text{ and } v_i = \lim_{h \to 0} \frac{1 - p_{ii}(h)}{h}.$$

In order to make clear the practical meaning of the q_{ij} and v_i, we use the mathematical symbol $o(h)$. This symbol is the generic notation for any function $f(h)$ with the property that

$$\lim_{h \to 0} \frac{f(h)}{h} = 0,$$

that is, $o(h)$ represents some unspecified function that is negligibly small compared with h itself as $h \to 0$. For example, any function $f(h) = h^a$ with $a > 1$ is an $o(h)$-function. A useful fact is that both the sum and the product of a finite number of $o(h)$-functions are again $o(h)$-functions. Since $p_{ij}(h) = q_{ij}h + o(h)$ for $j \neq i$ and $p_{ii}(h) = (1 - v_i)h + o(h)$ as $h \to 0$, we can state the following result.

Rule 11.1 *For any $t \geq 0$ and small $h > 0$,*

$$P(X(t+h) = j \mid X(t) = i) = \begin{cases} q_{ij}h + o(h) & \text{for } j \neq i, \\ 1 - v_i h + o(h) & \text{for } j = i. \end{cases}$$

Hence, q_{ij} gives the rate at which the process tries to enter a different state j when the process is in state i, while v_i gives the rate at which the process tries to leave state i for another state when the process is in state i. The q_{ij} are called the *transition rates* of the continuous-time Markov chain. Since $p_{ii}(h) + \sum_{j \neq i} p_{ij}(h) = 1$, it follows that

$$v_i = \sum_{j \neq i} q_{ij} \quad \text{for all } i \in I.$$

It is important to note that the q_{ij} are rates, not probabilities and, as such, while they must be nonnegative, they are not bounded by 1. However, for very small h, you can interpret $q_{ij}h$ as a probability, namely as the probability that in the next time interval of length h the process will jump to a different state j when the present state is i. Also, it is important to note from Rule 11.1 and the finiteness of the state space that the probability of two or more state changes in such a small time interval is $o(h)$. The transition rates q_{ij} are obtained from the transition functions $p_{ij}(t)$. Conversely, it can be proved that the transition rates q_{ij} uniquely determine the transition functions $p_{ij}(t)$ when the state space is finite.

How should you find the q_{ij} in practical applications? You will see that the exponential distribution is the building block for the transition rates. Before proceeding with the Markov model, we first discuss the most important properties of this distribution and the closely related Poisson process.

11.1.1 Intermezzo: Exponential Distribution and Poisson Process

In the context of continuous-time Markov chains, it is convenient to use the terminology of an exponentially distributed random variable T with *rate* λ when T has the probability density function $f(t) = \lambda e^{-\lambda t}$ for $t > 0$. The cumulative probability distribution function of T is $P(T \leq t) = 1 - e^{-\lambda t}$ for $t \geq 0$, and T has $1/\lambda$ as expected value. The key property of the exponential distribution is its memoryless property. That is,

$$P(T > t + s \mid T > t) = P(T > s) \quad \text{for all } s \geq 0,$$

no matter what the value of t is. Imagining that T represents the lifetime of an item, this property says that the remaining lifetime of the item has the same exponential distribution as the original lifetime, regardless of how long the item has already been in use. This memoryless property has been proved in Section 4.5. The exponential distribution is the only continuous distribution possessing this property. For building and understanding continuous-time Markov chains, it is more convenient to express the memoryless property of the exponential distribution as

$$P(T \leq t + h \mid T > t) = \lambda h + o(h) \quad \text{as } h \to 0,$$

no matter what the value of t is. In other words, the exponential distribution has a constant failure rate. The proof is as follows. By the memoryless property, $P(T \leq t + h \mid T > t) = P(T \leq h) = e^{-\lambda h}$. Expanding out $e^{-\lambda h}$ in a Taylor series, we find

$$P(T \leq t + h \mid T > t) = 1 - \left(1 - \frac{\lambda h}{1!} + \frac{(\lambda h)^2}{2!} - \frac{(\lambda h)^3}{3!} + \cdots\right)$$
$$= \lambda h + o(h) \quad \text{as } h \to 0.$$

Another property of the exponential distribution that will frequently be used in the sequel is that the minimum of r independent random variables having exponential distributions with rates $\mu_1, \ldots, \mu_r$ is exponentially distributed with rate $\mu_1 + \cdots + \mu_r$, see Rule 5.4. Can you explain this property intuitively using the failure rate representation of the exponential distribution?

The Poisson process often appears in applications of continuous-time Markov chains. This process has already been discussed in Section 4.5. Several

11.1 Markov Chain Model

equivalent definitions of the Poisson process can be given. Let $\{N(t), t \geq 0\}$ be a continuous-time stochastic process, where $N(t)$ counts the number of events (e.g., arrivals of customers or jobs) occurring up to time t.

First definition *A counting process $\{N(t), t \geq 0\}$ with $N(0) = 0$ is said to be a Poisson process with rate λ if the interoccurrence times of the events are independent random variables having a common exponential distribution with expected value $1/\lambda$.*

From this definition, the *memoryless property* can be shown for the Poisson process: at each point in time, the waiting time until the next occurrence of an event has the same exponential distribution as the original interoccurrence times, regardless of how long ago the last event occurred, see also Problem 9.36 in Section 9.4. For our purposes in continuous-time Markov chains, the following equivalent definition is more appropriate.

Second definition *A counting process $\{N(t), t \geq 0\}$ with $N(0) = 0$ is said to be a Poisson process with rate λ if*

(a) *Occurrences of events in any time interval $(t, t + h)$ are independent of what happened up to time t.*
(b) *For any $t \geq 0$, the probability $P(N(t+h) - N(t) = 1)$ of one occurrence of an event in a very small time interval $(t, t+h)$ is $\lambda h + o(h)$, the probability $P(N(t+h) - N(t) = 0)$ of no occurrence of an event is $1 - \lambda h + o(h)$, and the probability $P(N(t+h) - N(t) \geq 2)$ of two or more occurrences is $o(h)$ as $h \to 0$.*

The reason for the name "Poisson process" is provided by the third definition.

Third definition *A counting process $\{N(t), t \geq 0\}$ with $N(0) = 0$ is said to be a Poisson process with rate λ if*

(a) *The numbers of events occurring in any two disjoint time intervals are independent of each other.*
(b) *The probability distribution of the number of events occurring in any time interval $(s, s + t)$ depends only on the length of the time interval and has a Poisson distribution with expected value λt.*

A formal proof of the equivalence of the three definitions will not be given, but parts of the proof are already contained in Section 4.5 and Problem 8.7. We give only a rough sketch of an explanation. To make plausible that definition 2 implies definition 3, divide the interval $(0, t)$ into k subintervals of length $h = \frac{t}{k}$. Then, for k large enough, there is only zero or one event per subinterval, and the number of subintervals with one event is binomially distributed with parameters $n = k$ and $p = \frac{\lambda t}{k}$. This binomial distribution tends to the Poisson distribution with expected value λt as $k \to \infty$. To explain that definition 3 implies definition 1, note that the waiting time until the first occurrence of an event is larger than t only if no event occurs in $(0, t)$ and thus has probability $e^{-\lambda t}$, showing that this waiting time has the exponential density $\lambda e^{-\lambda t}$. Repeating this argument, we see that the time between the $(n-1)$th and the nth occurrence of an event has exponential density $\lambda e^{-\lambda t}$. The independence of the interoccurrence times follows from the absence of after-effects in the occurrences of events. Finally, using the failure-rate representation of the exponential density and its memoryless property, it can be explained that definition 1 implies definition 2.

To conclude, we give two important properties of the Poisson process.

(I) (*merging of Poisson processes*) Let $\{N_1(t), t \geq 0\}$ and $\{N_2(t), t \geq 0\}$ be two independent Poisson processes with respective rates λ_1 and λ_2. Then the merged process $\{N(t), t \geq 0\}$ with $N(t) = N_1(t) + N_2(t)$ is a Poisson process with rate $\lambda_1 + \lambda_2$.

(II) (*splitting of a Poisson process*) Let $\{N(t), t \geq 0\}$ be a Poisson process with rate λ. Suppose that any event is classified as a type-1 event with probability p and as a type-2 event with probability $1 - p$, independently of the other events. Let $N_1(t)$ be the number of type-1 events in $(0, t)$ and $N_2(t)$ be the number of type-2 events in $(0, t)$. Then $\{N_1(t), t \geq 0\}$ and $\{N_2(t), t \geq 0\}$ are Poisson processes with respective rates $\lambda_1 = \lambda p$ and $\lambda_2 = \lambda(1-p)$. Moreover, these two Poisson processes are independent of each other.

A sketch of the proof of these properties is as follows. The merging property can be seen by using the second definition of the Poisson process: the probability of one occurrence of an event in a very small time interval $(t, t+h)$ is $\lambda_1 h + \lambda_2 h + o(h) = (\lambda_1 + \lambda_2)h + o(h)$, and the probability of two or more events is $o(h)$ as $h \to 0$. The splitting property follows by using the third definition and mimicking the proof of Rule 3.13.

11.1.2 Alternative Construction of the Markov Process

In this paragraph we give a revealing way to think about a continuous-time Markov chain. The process can be characterized in another way that leads to an understanding of how to simulate the process. When the continuous-time Markov chain $\{X(t)\}$ enters state i at some time, say time 0, the process stays there a random amount of time before making a transition into a different state. Denote this amount of time by T_i. What does Definition 11.1 suggest about the

11.1 Markov Chain Model

random variable T_i? Could we find $P(T_i > t)$ directly? It is not obvious. But, instead, suppose we ask about $P(T_i > t + s \mid T_i > t)$. Then, by the Markov property and the assumption of time homogeneity, $P(T_i > t + s \mid T_i > t)$ equals

$$P(X(u) = i \text{ for } 0 \leq u \leq t + s \mid X(u) = i \text{ for } 0 \leq u \leq t)$$
$$= P(X(u) = i \text{ for } t \leq u \leq t + s \mid X(t) = i)$$
$$= P(X(u) = i \text{ for } 0 \leq u \leq s \mid X(0) = i) = P(T_i > s).$$

This shows that, regardless of the value of t,

$$P(T_i > t + s \mid T_i > t) = P(T_i > s) \quad \text{for all } s \geq 0.$$

In other words, the random variable T_i has the memoryless property and thus has an exponential distribution. It is easily seen from Rule 11.1 that this exponential distribution must have rate v_i. When the process leaves state i after the exponential time T_i, it will enter a different state j with some probability, call it p_{ij}. This probability is given by

$$p_{ij} = \frac{q_{ij}}{v_i} \quad \text{for } j \neq i.$$

To see this, note that the probability of going to state j in the next time interval h when the process is in state i can be represented both by $v_i h \times p_{ij} + o(h)$ and by $q_{ij} h + o(h)$ as $h \to 0$, and so $v_i p_{ij} = q_{ij}$.

Summarizing, a continuous-time Markov chain specified by the transition rates q_{ij} can be simulated as follows.

(a) When the process enters state i, draw from the exponential distribution with rate $v_i = \sum_{j \neq i} q_{ij}$ to determine how long the process stays in i before moving to a different state.
(b) When the process leaves state i, draw the next state j from the probability mass function $p_{ij} = q_{ij}/v_i$ with $j \neq i$.

11.1.3 Illustrative Examples

To illustrate the concepts above, we give a number of specific applied probability problems and show how they can be translated into a continuous-time Markov chain model. As a follow-up to these examples, several other problems are given as modeling exercises.

Example 11.1 A source transmitting messages is alternately on and off. The off-times are independent random variables having a common exponential distribution with rate α and the on-times are independent random variables having a common exponential distribution with rate β. Also, the off-times and on-times are independent of each other. What is an appropriate continuous-time Markov chain model?

Solution. For any $t \geq 0$, let the random variable $X(t)$ be equal to 1 if the source is on at time t and 0 otherwise. Because of the memoryless property of the exponential distribution and the assumption that the off-times are independent of the on-times, the process $\{X(t)\}$ satisfies the Markovian property and thus is a continuous-time Markov chain with state space $I = \{0, 1\}$. The off-time has a constant failure rate α and so the probability that an off-time will expire in $(t, t+h)$ given that the source is off at time t equals $\alpha h + o(h)$ for h small. Thus,

$$P(X(t+h) = 1 \mid X(t) = 0) = \alpha h + o(h) \quad \text{for } h \text{ small}$$

and so the transition rate $q_{01} = \alpha$. In the same way, $q_{10} = \beta$. This completes the specification of the continuous-time Markov chain model.

Example 11.2 An inflammable product is stored in a special tank at a filling station. Customers asking for the product arrive according to a Poisson process with rate λ. Each customer asks for one unit. Any demand that occurs when the tank is out of stock is lost. Opportunities to replenish the stock in the tank occur according to a Poisson process having rate μ and being independent of the demand process. For reasons of security it is only allowed to replenish the stock when the tank is empty. The replenishment time is negligible and the replenishment quantity is Q units of stock. What is an appropriate continuous-time Markov chain model for the stock on hand?

Solution. For any $t \geq 0$, let the random variable $X(t)$ denote how many units of stock are in the tank at time t. The process $\{X(t)\}$ is a continuous-time Markov chain with state space $I = \{0, 1, \ldots, Q\}$. The Markovian property is satisfied as a consequence of the fact that the Poisson processes are memoryless, that is, at any time it is irrelevant for the occurrence of future events how long ago the last event occurred. What about the transition rates? If the process is in state i with $1 \leq i \leq Q$, a transition from i to $i - 1$ occurs when a customer arrives. The probability of one customer arriving in $(t, t + h)$ is $\lambda h + o(h)$ for h small, while the probability of two or more customers arriving in $(t, t + h)$ is $o(h)$. Thus, $P(X(t+h) = i - 1 \mid X(t) = i) = \lambda h + o(h)$ for h small and so

$$q_{i,i-1} = \lambda \quad \text{for } 1 \leq i \leq Q.$$

Also, $q_{ij} = 0$ for $j \neq i - 1$. If the process is in state 0, a transition from 0 to Q occurs when a replenishment opportunity arises. Replenishment opportunities occur at a rate μ and so $q_{0Q} = \mu$. The other q_{0j} are zero.

11.1 Markov Chain Model

Example 11.3 Consider a population of m individuals, some of whom are infected and the others are susceptible. Initially, one individual is infected. The times between contacts between any two individuals of the population are exponentially distributed with rate μ. The times between the various contacts are independent of each other. If a susceptible individual comes into contact with an infected individual, then the susceptible individual becomes infected. Once infected, an individual stays infected. What stochastic process describes the spread of the disease?

Solution. The key to the solution is the fact that the minimum of r independent random variables having exponential distributions with rates $\mu_1, \ldots, \mu_r$ is exponentially distributed with rate $\mu_1 + \cdots + \mu_r$, see Rule 5.4. Denote by the random variable $X(t)$ the number of infected individuals at time t. Then, by the memoryless property of the exponential distribution, the process $\{X(t)\}$ is a continuous-time Markov chain with state space $I = \{1, 2, \ldots, m\}$. How do we find the transition rates? If there are i infected individuals, then the time until one of the $m - i$ susceptible individuals becomes infected is the smallest of $(m - i) \times i$ independent exponentials each having rate μ. Hence, this time has an exponential distribution with rate $(m - i)i\mu$, and so the transition rates $q_{i,i+1}$ are given by

$$q_{i,i+1} = (m - i)i\mu \quad \text{for } i = 1, 2, \ldots, m - 1.$$

The other transition rates q_{ij} are zero, because of the assumption that an infected individual stays infected once it is infected. Note that state m is an absorbing state with leaving rate $\nu_m = 0$.

Example 11.4 An electronic system uses one operating unit but has built-in redundancy in the form of $s - 1$ standby units. The s units are identical. The standby units are not switched on (cold standby). The lifetime of the operating unit is exponentially distributed with rate λ. If the operating unit fails, then it is immediately replaced by a standby unit if available. There are ample repair facilities, so that each failed unit enters repair immediately. The repair time of a failed unit is exponentially distributed with rate μ. A repaired unit is as good as new. What process describes the number of units in repair?

Solution. Let the random variable $X(t)$ denote the number of units in repair at time t. The Markovian property is satisfied for the process $\{X(t)\}$ and so this process is a continuous-time Markov chain with state space $I = \{0, 1, \ldots, s\}$. For the determination of the transition rates, we use again the fact that the smallest of i independent random variables having an exponential distribution with rate μ is exponentially distributed with rate $i\mu$. If at time t there are i units in repair, then the probability of having exactly one repair completed in the time interval $(t, t + h)$ is $i\mu h + o(h)$ for h small. The probability of having two or more repair completions in this interval is $o(h)$. The same is true for the probability that in this interval one or more repairs are completed and at the same time one or more units will fail. Thus, for $1 \leq i \leq s$,

$$P(X(t+h) = i - 1 \mid X(t) = i) = i\mu h + o(h) \quad \text{for } h \text{ small}$$

and so $q_{i,i-1} = i\mu$. Using the fact that a unit in operation at time t will fail in $(t, t + h)$ with probability $\lambda h + o(h)$ for h small, we obtain by similar arguments that

$$P(X(t+h) = i + 1 \mid X(t) = i) = \lambda h + o(h) \quad \text{for } h \text{ small}.$$

This gives $q_{i,i+1} = \lambda$ for $0 \leq i \leq s - 1$. The other q_{ij} are zero.

Problems

11.1 Consider a two-unit reliability system with one operating unit and one unit in warm standby. The operating unit has a constant failure rate of λ, while the unit in warm standby has a constant failure rate of η with $0 \leq \eta < \lambda$. Upon failure of the operating unit, the unit in warm standby is put into operation if available. The time it takes to replace a failed unit is exponentially distributed with rate μ. It is only possible to replace one failed unit at a time. Define an appropriate continuous-time Markov chain and specify its transition rates.

11.2 Cars arrive at a gasoline station according to a Poisson process with an average of λ cars per minute. A car finding upon arrival four other cars present goes elsewhere. The gasoline station has only one pump. The amount of time required to serve a car has an exponential distribution with an expected value of $1/\mu$ minutes. Define an appropriate continuous-time Markov chain and specify its transition rates.

11.3 A barbershop with two barbers has room for at most seven customers. Potential customers arrive according to a Poisson process with a rate of λ customers per hour. A customer who finds upon arrival i other customers present balks with probability $b(i)$, where $b(7) = 1$. The two barbers each work at an exponential rate of μ customers per hour. Formulate a continuous-time Markov chain to describe the number of customers present in the barbershop.

11.4 A ferry travels between two points A and B on the shores of a lake and transports cars from A to B. Cars arrive at point A according to a Poisson

11.1 Markov Chain Model

process with rate λ. A car that finds upon arrival no ferry present does not wait but takes another route. The ferry departs as soon as seven cars are on the ferry. The time the ferry needs to travel to point B is exponentially distributed with expected value $1/\mu_1$. After having arrived at point B, the ferry needs an exponentially distributed time with expected value $1/\mu_2$ to return to point A. Formulate a continuous-time Markov chain to describe the process.

11.5 The so-called sheroot is a familiar sight in Middle East street scenes. It is a seven-seat taxi that drives from a fixed stand in a town to another town. A sheroot leaves the stand as soon as all seven seats are occupied by passengers. Consider a sheroot stand which has room for only one sheroot. Potential passengers arrive at the stand according to a Poisson process with rate λ. A customer who finds upon arrival no sheroot present and seven other passengers already waiting goes elsewhere for transport; otherwise, the customer waits until a sheroot departs. After departure of a sheroot, the time until a new sheroot becomes available has an exponential distribution with mean $1/\mu$. Define an appropriate continuous-time Markov chain and specify its transition rates.

11.6 Ships arrive at a container terminal according to a Poisson process with rate λ. The ships bring loads of containers. The quay has capacity for only four ships. A ship finding upon arrival the quay fully occupied travels through to a neighboring harbor. The terminal has a single unloader that handles only one ship at a time. The unloading time of each ship is exponentially distributed with mean $1/\mu$. The unloader, however, is subject to breakdowns. A breakdown can only occur when the unloader is operating. The operating period of the unloader is exponentially distributed with mean $1/\delta$. The time to repair a broken unloader is exponentially distributed with mean $1/\beta$. Any interrupted unloading of a ship is resumed at the point at which it was interrupted. The unloading times, the operating times, and the repair times are independent of each other. Formulate a continuous-time Markov chain to describe the process and specify its transition rates.

11.7 Messages for transmission arrive at a communication channel according to a Poisson process with rate λ. The channel can transmit only one message at a time. The transmission time is exponentially distributed with mean $1/\mu$. The following access control rule is used. The gate is closed for newly arriving messages if the number of messages in the system (including the message in transmission) reaches the level R. As soon as the number of messages in the system has dropped to r, newly arriving messages are again admitted to the transmission channel.

The control parameters r and R are given integers with $0 \leq r < R$. Formulate a continuous-time Markov chain to describe the process and specify its transition rates.

11.8 Consider a conveyor system at which items for processing arrive according to a Poisson process with rate λ. The amounts of work brought in by the items are independent random variables having a common exponential distribution with mean $1/\mu$. The conveyor system has two work stations 1 and 2 that are placed according to this order along the conveyor. Work station i has a constant processing rate of σ_i units of work per unit time. Each work station can handle only one item at a time. The work stations have no storage capacity and an arriving item is sent elsewhere when both stations are occupied. Station 1 is the first choice for an arriving item when this station is free. Formulate a continuous-time Markov chain to describe the process and specify its transition rates.

11.9 A production facility has 10 operating machines and a buffer of two machines on standby. Machines in operation are subject to breakdowns. The machines work independently of each other and the running time of any machine has an exponential distribution with mean $1/\lambda$. An operating machine that breaks down is replaced by a standby machine if one is available. A failed machine immediately enters repair. There are ample repair facilities so that any number of machines can be repaired simultaneously. The repair time of a failed machine has an exponential distribution with mean $1/\mu$. Formulate a continuous-time Markov chain to describe the process and specify its transition rates.

11.2 Time-Dependent Probabilities

The time-dependent probabilities $p_{ij}(t)$ are defined by

$$p_{ij}(t) = P(X(t) = j \mid X(0) = i) \quad \text{for } t > 0 \text{ and } i, j \in I.$$

The transition functions $p_{ij}(t)$ are the counterpart of the n-step transition probabilities in the discrete-time Markov chain. It will be no surprise that the Chapman–Kolmogorov equations in Rule 10.1 also have a counterpart:

Rule 11.2 *For any $s, t \geq 0$ and $i, j \in I$,*

$$p_{ij}(t+s) = \sum_{k \in I} p_{ik}(t) p_{kj}(s).$$

The proof is essentially identical to the proof of Rule 10.1. By conditioning on $X(t)$ and using the Markovian property together with the assumption of time homogeneity, we have that $P(X(t+s) = j \mid X(0) = i)$ equals

$$\sum_{k \in I} P(X(t+s) = j \mid X(0) = i, X(t) = k) P(X(t) = k \mid X(0) = i)$$
$$= \sum_{k \in I} P(X(s) = j \mid X(0) = k) P(X(t) = k \mid X(0) = i),$$

showing the desired result.

Unlike the discrete-time result in Rule 10.1, the continuous analogue in Rule 11.2 is not directly amenable for numerical computations. How could we compute the $p_{ij}(t)$? To answer this question, let us differentiate both sides of $p_{ij}(t+s) = \sum_{k \in I} p_{ik}(t) p_{kj}(s)$ with respect to s and next set $s = 0$. Since the state space I is finite, the sum can be differentiated term by term. This gives

$$p'_{ij}(t) = \sum_{k \in I} p_{ik}(t) p'_{kj}(0).$$

Next we use the result $p'_{kj}(0) = q_{kj}$ for $k \neq j$ and $p'_{jj}(0) = -v_j$. Then we obtain the so-called *Kolmogorov forward differential equations*:

Rule 11.3 *Assume a finite state space I. Then, for any given state i, the transition functions $p_{ij}(t)$, $j \in I$ satisfy the linear differential equations*

$$p'_{ij}(t) = -v_j p_{ij}(t) + \sum_{k \neq j} q_{kj} p_{ik}(t) \quad \text{for } t > 0 \text{ and } j \in I.$$

Another set of differential equations, known as the *Kolmogorov backward differential equations*, can also be obtained from Rule 11.2. Differentiating both sides of $p_{ij}(t+s) = \sum_{k \in I} p_{ik}(t) p_{kj}(s)$ with respect to t and next setting $t = 0$, we obtain for any $j \in I$ that (verify!)

$$p'_{ij}(s) = -v_i p_{ij}(s) + \sum_{k \neq i} q_{ik} p_{kj}(s) \quad \text{for } s > 0 \text{ and } i \in I.$$

Linear differential equations can be solved numerically by standard codes that are available in numerical software packages. It is only possible in special cases to give an explicit solution.

Example 11.1 (continued) For starting state $i = 1$, the Kolmogorov forward differential equations are given by

$$p'_{10}(t) = -\alpha p_{10}(t) + \beta p_{11}(t) \quad \text{for } t > 0,$$
$$p'_{11}(t) = -\beta p_{11}(t) + \alpha p_{10}(t) \quad \text{for } t > 0.$$

Since $p_{11}(t) = 1 - p_{10}(t)$, the first equation can be written as

$$p'_{10}(t) = \beta - (\alpha + \beta) p_{10}(t) \quad \text{for } t > 0.$$

By the theory of linear differential equations, a solution of the form $p_{10}(t) = c_1 + c_2 e^{-(\alpha+\beta)t}$ is expected. Substitution of this expression leads to

$$-c_2(\alpha + \beta)e^{-(\alpha+\beta)t} = \beta - c_1(\alpha + \beta) - c_2(\alpha + \beta)e^{-(\alpha+\beta)t}.$$

This gives $\beta - c_1(\alpha + \beta) = 0$ and so $c_1 = \beta/(\alpha + \beta)$. The constant c_2 is found by using the boundary condition $p_{10}(0) = 0$. This gives $c_1 + c_2 = 0$ and so $c_2 = -c_1$. We can now conclude that

$$p_{10}(t) = \frac{\beta}{\alpha + \beta} - \frac{\beta}{\alpha + \beta}e^{-(\alpha+\beta)t} \quad \text{and} \quad p_{11}(t) = \frac{\alpha}{\alpha + \beta} + \frac{\beta}{\alpha + \beta}e^{-(\alpha+\beta)t}$$

for all $t \geq 0$. In the same way, for all $t \geq 0$,

$$p_{01}(t) = \frac{\alpha}{\alpha + \beta} - \frac{\alpha}{\alpha + \beta}e^{-(\alpha+\beta)t} \quad \text{and} \quad p_{00}(t) = \frac{\beta}{\alpha + \beta} + \frac{\alpha}{\alpha + \beta}e^{-(\alpha+\beta)t}.$$

In this example the $p_{ij}(t)$ have a limit as $t \to \infty$. Moreover, the limit is independent of the starting state i and is given by $\beta/(\alpha + \beta)$ when $j = 0$ and by $\alpha/(\alpha + \beta)$ when $j = 1$.

First-Passage Time Probabilities

First-passage time probabilities are important in reliability problems, among others. For example, in Example 11.4 one might be interested in the probability distribution of the time until the system goes down for the first time when the down state corresponds to the situation that all units are in repair.

Consider a continuous-time Markov chain $\{X(t)\}$. Let r be any state such that r can be reached from any other state. Define the probability

$$Q_i(t) = P(X(u) \neq r \text{ for all } 0 \leq u \leq t \mid X(0) = i) \quad \text{for } t \geq 0 \text{ and } i \neq r,$$

where $Q_i(0) = 1$ for all $i \neq r$. In words, $Q_i(t)$ is the probability that the time until the first visit of the process to state r is more than t, given that the starting state is i. How could we compute the first-passage time probabilities $Q_i(t)$? The idea is simple. Modify the process $\{X(t)\}$ by making state r absorbing. That is, define a continuous-time Markov chain $\{\overline{X}(t)\}$ by the transition rates $\overline{q}_{ij}$ with $\overline{q}_{ij} = q_{ij}$ for $i \neq r$ and $\overline{q}_{rj} = 0$ for all j. Let $\overline{p}_{ij}(t) = P(\overline{X}(t) = j \mid \overline{X}(0) = i)$. Using the fact that r is an absorbing state for the process $\{\overline{X}(t)\}$, we have (why?)

$$Q_i(t) = 1 - \overline{p}_{ir}(t) \quad \text{for } t > 0 \text{ and } i \neq r.$$

Applying the Kolmogorov backward differential equations for the $\overline{p}_{ir}(t)$ with $\overline{p}_{rr}(t) = 1$ for all t, we find that the probabilities $Q_i(t)$ can be obtained from the linear differential equations (verify!)

$$Q_i'(t) = -v_i Q_i(t) + \sum_{k \neq i, r} q_{ik} Q_k(t) \quad \text{for } t > 0 \text{ and } i \neq r.$$

These differential equations for the first-passage time probabilities $Q_i(t)$ can also be obtained from first principles. To do so, we evaluate $Q_i(t + \Delta t)$ by conditioning on what may happen in $(0, \Delta t)$ for Δt small. This gives

$$Q_i(t + \Delta t) = (1 - \nu_i \Delta t) Q_i(t) + \sum_{k \neq i, r} q_{ik} \Delta t \, Q_k(t) + o(\Delta t).$$

Then the linear differential equations for the $Q_i(t)$ follow by letting $\Delta t \to 0$ in $[Q_i(t + \Delta t) - Q_i(t)]/\Delta t$ and noting that $o(\Delta t)/\Delta t \to 0$ as $\Delta t \to 0$.

Example 11.4 (continued) For the special case of $s = 2$ units, let us derive an explicit expression for $Q_i(t)$, being the probability that the time until the first visit to state 2 (system is down) is more than t when the initial state is i for $i = 0, 1$. The linear differential equations for the $Q_i(t)$ are

$$Q_0'(t) = -\lambda Q_0(t) + \lambda Q_1(t) \qquad \text{for } t > 0,$$
$$Q_1'(t) = -(\lambda + \mu) Q_1(t) + \mu Q_0(t) \qquad \text{for } t > 0,$$

with the boundary condition $Q_0(0) = Q_1(0) = 1$. Using the standard theory for linear differential equations, it can be verified that

$$Q_i(t) = c_i e^{\eta_1 t} + (1 - c_i) e^{\eta_2 t} \qquad \text{for } t > 0$$

for $i = 0, 1$, where $\eta_{1,2} = -\frac{1}{2}(2\lambda + \mu) \pm \frac{1}{2}\sqrt{\mu^2 + 4\lambda\mu}$, $c_0 = \eta_2/(\eta_2 - \eta_1)$, and $c_1 = (\eta_1 + \lambda) c_0/\lambda$. The first-passage time distribution enables us to calculate the expected value of the first-passage time. However, a simpler method for the calculation of the expected values of the first-passage times is to solve a system of linear equations, see Problem 11.16.

Problems

11.10 For a continuous-time Markov chain $\{X(t)\}$, show that $\int_0^T p_{ij}(t) dt$ gives the expected amount of time the process will be in state j during a given time interval $(0, T)$ if it is in state i at time 0. *Hint*: Let the indicator variable $I(t)$ be 1 if $X(t) = j$ and 0 otherwise.

11.11 A particular gene is free (state 1) or repressed by a protein (state 2). The gene stays in state i during an exponentially distributed time with mean $1/\mu_i$ before it moves to the other state for $i = 1, 2$. What is the expected amount of time the gene will be repressed during a given time interval $(0, T)$ if it is free at time 0?

11.12 In mathematical biology the evolution of DNA as a string of four states is often modeled by a continuous-time Markov chain whose matrix of transition rates is of the form

$$\begin{pmatrix} - & \alpha & \beta & \beta \\ \alpha & - & \beta & \beta \\ \beta & \beta & - & \alpha \\ \beta & \beta & \alpha & - \end{pmatrix}.$$

Verify by substitution into the Kolmogorov forward differential equations that the transition probabilities $p_{ij}(t)$ are given by

$$\begin{pmatrix} 1-r(t)-2s(t) & r(t) & s(t) & s(t) \\ r(t) & 1-r(t)-2s(t) & s(t) & s(t) \\ s(t) & s(t) & 1-r(t)-2s(t) & r(t) \\ s(t) & s(t) & r(t) & 1-r(t)-2s(t) \end{pmatrix},$$

where $r(t) = \frac{1}{4}\left(1 + e^{-4\beta t} - 2e^{-2(\alpha+\beta)t}\right)$ and $s(t) = \frac{1}{4}\left(1 - e^{-4\beta t}\right)$ for $t > 0$.

11.13 Consider a continuous-time Markov chain $\{X(t)\}$ with transition rates q_{ij}.

(a) What is the conditional probability that exactly one state change occurred over the time interval $(0, T)$ given that $X(0) = a$ and $X(T) = b$ with $a \neq b$. *Hint*: Use the alternative construction of a continuous-time Markov chain.

(b) What is the expected amount of time that the process has stayed in state j during $(0, T)$ given that $X(0) = a$ and $X(T) = b$? *Hint*: Let the indicator variable $I(t)$ be equal to 1 if $X(t) = j$ and 0 otherwise.

11.14 Consider a two-state continuous-time Markov chain with transition rates $q_{12} = \alpha$ and $q_{21} = \beta$. If the Markov chain is in state i, then events occur according to a Poisson process with rate λ_i for $i = 1, 2$. What is the expected value of the number of events in $(0, t)$ if the Markov process starts in state 1?

11.15 A buffer temporarily stores incoming messages. The messages arrive according to a Poisson process with rate λ. The buffer has capacity for only 10 messages. A message that finds upon arrival the buffer full is lost. Each accepted message stays in the buffer during an exponentially distributed time with expected value $1/\mu$, independently of the sojourn times of the other messages. Let f_i be the probability that the buffer gets full before it becomes empty when there are currently i messages in the buffer. Give a system of linear equations for the f_i.

11.16 Consider a continuous-time Markov chain with finite state space I and transition rates q_{ij}. For a given state r that can be reached from any other state, define μ_i as the expected value of the time until reaching state r when starting from state i. Verify that the μ_i are the unique solution to the linear equations $\mu_i = 1/v_i + \sum_{j \neq i,r} p_{ij} \mu_j$ for $i \neq r$, where $p_{ij} = q_{ij}/v_i$ for $j \neq i$.

11.17 Let $\{J(t), t \geq 0\}$ be a stochastic process with finite state space I in which the state transitions occur at epochs generated by a Poisson process with rate ν. The transitions are governed by a discrete-time Markov chain having $\mathbf{R} = (r_{ij})$ as matrix of one-step transition probabilities and being independent of the Poisson process.

(a) Let $p_{ij}(t) = P(J(t) = j \mid J(0) = i)$ for $t > 0$. Show that

$$p_{ij}(t) = \sum_{n=0}^{\infty} r_{ij}^{(n)} e^{-\nu t} \frac{(\nu t)^n}{n!} \quad \text{for } t > 0 \text{ and } i, j \in I,$$

where the $r_{ij}^{(n)}$ are the elements of the matrix product $\mathbf{R}^n$, with $\mathbf{R}^0$ the identity matrix $\mathbf{I}$.

(b) Suppose that the parameters ν and r_{ij} of the process $\{J(t)\}$ are chosen as $\nu = \max_i v_i$, $r_{ij} = q_{ij}/\nu$ for $j \neq i$, and $r_{ii} = 1 - v_i/\nu$ for $j = i$, where $v_i = \sum_{j \neq i} q_{ij}$ and the q_{ij} are the transition rates of a continuous-time Markov chain $\{X(t)\}$. Use part (a) to verify that the process $\{J(t)\}$ is also a continuous-time Markov chain with transition rates q_{ij}. In other words, the process $\{J(t)\}$ is probabilistically identical to the process $\{X(t)\}$. *Note*: The result of part (a) gives us an alternative method to compute the time-dependent probabilities of the continuous-time Markov chain $\{X(t)\}$. This method is known as the *uniformization method*.

11.18 Consider again Problem 11.1. The system goes down if no standby unit is available upon failure of the operating unit. Let the random variable T denote the time until the system goes down for the first time, given that each of the units is in a good condition at time 0. Take the numerical data $\lambda = 1$, $\eta = 0$, and $\mu = 50$. Calculate $E(T)$ and $P(T > t)$ for $t = 10, 25, 50, 100$, and 200. What are the numerical answers for the case of two standby units with the stipulation that it is only possible to replace one unit at a time?

11.19 The Hubble space telescope carries a variety of instruments, including six gyroscopes to ensure stability of the telescope. The six gyroscopes are arranged in such a way that any three gyroscopes can keep the telescope operating with full accuracy. The operating times of the gyroscopes are independent of each other and have an exponential distribution with an expected value of $1/\lambda$ years. A fourth gyroscope failure triggers the sleep mode of the telescope. In the sleep mode, further observations by the telescope are suspended. It requires an exponential time with an expected value of $1/\mu$ years to turn the telescope into the sleep mode. Once the telescope is in the sleep mode, the base station on earth receives a sleep signal. A shuttle mission to the telescope is next prepared. It takes an exponential time with an expected value of $1/\eta$ years before the repair crew arrives at the telescope and has repaired the stabilizing unit with the gyroscopes. In the mean time, the other two gyroscopes may fail. If this happens, then a crash destroying the telescope will be inevitable. What is the probability that the telescope will crash in the next T years? What is the probability that no shuttle mission will be prepared in the next T years? Compute these probabilities for the numerical data $T = 10$, $\lambda = 0.1$, $\mu = 100$, and $\eta = 5$.

11.20 A drunkard leaves the pub at 22:30 hours. Twenty steps to the left of the pub is a police station and ten steps to the right is his home. Each time it takes the drunkard an exponentially distributed time with an expected value of 30 seconds to make a step either to the left or to the right. Any step is with equal probabilities to the left or to the right, independently of the other steps. The drunkard is thrown in a cell if he arrives at the police station. What is the probability that he is not locked up in a police cell and reaches his home before midnight?

11.3 Limiting Probabilities

In this section the limiting behavior of the state probabilities $p_{ij}(t)$ will be studied. It will be assumed that the continuous-time Markov process $\{X(t)\}$ is irreducible. The process is said to be *irreducible* if for any two states i and j there is some $t > 0$ so that $p_{ij}(t) > 0$. The irreducibility assumption is satisfied in nearly all practical applications.

Rule 11.4 *Let $\{X(t)\}$ be an irreducible continuous-time Markov chain with a finite state space I. Then*

(a) $\lim_{t\to\infty} p_{ij}(t)$ *exists for all i, j and is independent of the initial state i.*
(b) *Letting $p_j = \lim_{t\to\infty} p_{ij}(t)$ for $j \in I$, the limiting probabilities p_j are the unique solution to the equilibrium equations (also called the balance equations)*

$$p_j \sum_{k\neq j} q_{jk} = \sum_{k\neq j} p_k q_{kj} \quad \text{for } j \in I \quad \text{and} \quad \sum_{j \in I} p_j = 1.$$

This rule is the continuous analogue of Rule 10.5. Contrary to discrete-time Markov chains, we have for continuous-time Markov chains that possible

11.3 Limiting Probabilities

periodicity in the state transitions is no issue, because the times between the state transitions are continuous random variables. A formal proof of Rule 11.4 will not be given. The linear equations in part (**b**) can be obtained heuristically from the Kolmogorov forward differential equations. This goes as follows. Since $p_{ij}(t)$ converges to a constant as $t \to \infty$, it will be intuitively clear that $\lim_{t \to \infty} p'_{ij}(t)$ must be equal to zero. Letting $t \to \infty$ on both sides of the differential equations in Rule 11.3, we get $0 = -v_j p_j + \sum_{k \neq j} q_{kj} p_k$ for $j \in I$. Noting that $v_j = \sum_{k \neq j} q_{jk}$, the balance equations in part (**b**) of Rule 11.4 follow. The normalization equation $\sum_{j \in I} p_j = 1$ is a direct consequence of $\sum_{j \in I} p_{ij}(t) = 1$. The normalization equation is needed to ensure the uniqueness of the solution; otherwise, the solution of the balance equations is uniquely determined up to a multiplicative constant.

The limiting probability p_j can be interpreted as the probability that an outside observer finds the system in state j when the observer enters the system after it has run for a very long time and the observer has no information about the past of the process. Another interpretation of p_j is

the long-run proportion of time the process will be in state $j = p_j$

with probability 1, regardless of the starting state. Similarly to the discrete-time Markov model, this result can be proved by using the renewal–reward theorem. Details of the proof will not be given. Also, for any two states j and k with $k \neq j$, we can state that in the long run

the average number of transitions per unit time from j to $k = q_{jk} p_j$

with probability 1, regardless of the starting state. These two results enable us to explain the balance equations in a physical way. To do so, note that in the long run

$$\text{the average number of transitions per unit time out of } j = p_j \sum_{k \neq j} q_{jk},$$

$$\text{the average number of transitions per unit time into } j = \sum_{k \neq j} q_{kj} p_k.$$

Further, in any time interval the number of transitions out of j can differ by at most 1 from the number of transitions into j. Therefore, the long-run average number of transitions per unit time out of j must be equal to the long-run average number of transitions per unit time into j, showing that $p_j \sum_{k \neq j} q_{jk} = \sum_{k \neq j} p_k q_{kj}$. This can be expressed as

$$\text{rate out of state } j = \text{rate into state } j.$$

The principle "rate out of state j = rate into state j" is the *flow-rate-equation method*. To write down the balance equations in specific applications, it is helpful to use the so-called *transition rate diagram* that visualizes the process. The nodes of the diagram represent the states and the arrows in the diagram give the possible state transitions. An arrow from node i to node j is only drawn when the transition rate q_{ij} is positive.

Example 11.2 (continued) How do we compute the long-run average stock on hand, the long-run average number of stock replenishments per unit time, the long-run fraction of time the system is out of stock, and the long-run fraction of demand that is lost?

Solution. The transition rates of the process $\{X(t)\}$ describing the number of units of stock in the tank have already been determined in Section 11.1. The states and the transition rates of the process are displayed in the transition rate diagram of Figure 11.1. Using this diagram and the principle "rate out of state j = rate into state j," it is easy to write down the balance equations. These equations are given by

$$\mu p_0 = \lambda p_1, \quad \lambda p_j = \lambda p_{j+1} \text{ for } 1 \leq j \leq Q-1, \text{ and } \lambda p_Q = \mu p_0.$$

These equations, together with the equation $\sum_{j=0}^{Q} p_j = 1$, have a unique solution. The solution can be given explicitly. It is readily verified that

$$p_0 = \left(1 + Q\frac{\mu}{\lambda}\right)^{-1} \text{ and } p_1 = p_2 = \cdots = p_Q = \frac{\mu}{\lambda} p_0.$$

The limiting probability p_j represents the long-run fraction of time that the number of units of stock in the tank is equal to j. Hence, with probability 1,

$$\text{the long-run average stock on hand} = \sum_{j=0}^{Q} j p_j = \frac{1}{2} Q(Q+1) \frac{\mu}{\lambda + Q\mu},$$

$$\text{the long-run fraction of time the system is out of stock} = p_0 = \frac{1}{1 + Q\mu/\lambda}.$$

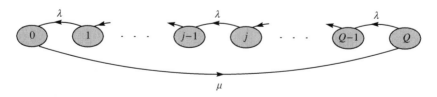

Figure 11.1. The inventory process.

11.3 Limiting Probabilities

A stock replenishment occurs each time the process makes a transition from state 0 to state Q. The long-run average number of transitions per unit time from 0 to Q is $q_{0Q}p_0 = \mu(1 + Q\mu/\lambda)^{-1}$, and so

$$\text{the long-run average frequency of stock replenishments} = \frac{\mu}{1 + Q\mu/\lambda}.$$

In this inventory model, we have the noteworthy result that

> the long-run fraction of demand finding the system out of stock
> = the long-run fraction of time the system is out of stock.

The result that the long-run fraction of demand that is lost is equal to p_0 can be argued directly from the renewal–reward theorem of Chapter 9. The inventory process regenerates itself each time the stock is replenished to the level Q. The expected demand between two successive replenishments is $Q + \frac{\lambda}{\mu}$ and the expected demand that is lost between two successive replenishments is $\frac{\lambda}{\mu}$. Therefore, the long-run fraction of demand that is lost is equal to $\frac{\lambda}{\mu}/(Q + \frac{\lambda}{\mu})$. This expression can be rewritten as $1/(1 + Q\mu/\lambda) = p_0$.

Remark 11.1 The last result in Example 11.2 can be put into a more general framework. Suppose that you have a continuous-time Markov chain $\{X(t)\}$ in which customers/jobs arrive according to a Poisson process, where the arrival process can be seen as an exogenous factor to the system and is not affected by the system itself. Then, for each state j, the long-run fraction of customers/jobs finding upon arrival the process in state j is equal to p_j. This property is known as *Poisson arrivals see time averages*. It can be argued heuristically as follows. Since p_j is the long-run fraction of time the process is in state j, the average number of Poisson arrivals finding the process in state j is λp_j. The average number of arrivals per unit time is λ. Therefore, the long-run fraction of arrivals finding the process in state j is $\frac{\lambda p_j}{\lambda} = p_j$. This property extends beyond continuous-time Markov chains. It also applies to general stochastic systems for which the input process is an exogenous Poisson process that is independent of the evolution of the system in question. The proof of this very useful property is beyond the scope of this book. It has many applications to queueing and inventory systems. In the Problems 9.26 and 9.27 of Chapter 9, the property naturally shows up.

Example 11.4 (continued) Suppose the system is down when all units are in repair. What are the long-run fraction of time the system is down and the long-run average number of units in repair?

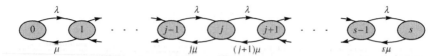

Figure 11.2. The repair process.

Solution. The transition rates of the process $\{X(t)\}$ describing the number of units in repair have already been determined in Section 11.1. The states and the transition rates of the process are displayed in the transition rate diagram of Figure 11.2. Using this diagram and the principle "rate out of state j = rate into state j," we find the balance equations

$$\lambda p_0 = \mu p_1, \quad (\lambda + j\mu) p_j = \lambda p_{j-1} + (j+1)\mu p_{j+1} \text{ for } 1 \leq j \leq s-1,$$
$$s\mu p_s = \lambda p_{s-1}.$$

These linear equations can be reduced to the recursive equations

$$j\mu p_j = \lambda p_{j-1} \quad \text{for } j = 1, 2, \ldots, s.$$

This recursion can be proved by induction (a generally applicable probabilistic proof is discussed in Section 11.3.1). It is true for $j = 1$. Suppose it has been verified for $j = 1, \ldots, k$. Then, for $j = k+1$, it follows from the equation $(\lambda + k\mu) p_k = \lambda p_{k-1} + (k+1)\mu p_{k+1}$ and the induction assumption $k\mu p_k = \lambda p_{k-1}$ that $\lambda p_k = (k+1)\mu p_{k+1}$, which completes the induction proof. The recursive equations lead to an explicit solution for the p_j. Iterating $p_j = \frac{\lambda}{j\mu} p_{j-1}$ gives

$$p_j = \frac{\lambda}{j\mu} p_{j-1} = \frac{\lambda^2}{j\mu(j-1)\mu} p_{j-2} = \cdots = \frac{\lambda^j}{j\mu(j-1)\mu \cdots \mu} p_0 = \frac{(\lambda/\mu)^j}{j!} p_0.$$

Using the normalization equation $\sum_{j=0}^{s} p_j = 1$, it follows next that

$$p_j = \frac{(\lambda/\mu)^j / j!}{\sum_{k=0}^{s} (\lambda/\mu)^k / k!} \quad \text{for } j = 0, 1, \ldots, s.$$

It is interesting to point out that the probabilities p_j can be represented by the truncated Poisson probabilities $e^{-\lambda/\mu} \frac{(\lambda/\mu)^j}{j!} \Big/ \sum_{k=0}^{s} e^{-\lambda/\mu} \frac{(\lambda/\mu)^k}{k!}$.

We can now answer the posed questions. Using the fact that p_j represents the long-run fraction of time that j units are in repair, it follows that the long-run average number of units in repair is $\sum_{j=0}^{s} j p_j$. The long-run fraction of time the system is down is p_s.

The Erlang Loss Model

The reliability model in Example 11.4 is an instance of the famous *Erlang loss model* from queueing theory. In the Erlang loss model customers arrive at a service facility with s identical servers according to a Poisson process with rate λ. A customer finding upon arrival all s servers busy is lost; otherwise, the customer gets assigned one of the free servers. The service times of the customers are independent random variables having a common exponential distribution with mean $1/\mu$. To explain the equivalence of the reliability model with the Erlang loss model, let us formulate the reliability problem in other terms. Instead of assuming an exponentially distributed operating time with mean $1/\mu$, we assume that there is an exogenous process in which catastrophes occur at epochs that are generated by a Poisson process with rate μ. If a catastrophe occurs, the unit in operation (if any) breaks down. This formulation of the reliability problem is equivalent to the original formulation (why?). By identifying units in repair with busy servers, it now follows that the reliability model is an instance of the Erlang loss model. In the context of the loss model, the probability $p_j = \frac{(\lambda/\mu)^j}{j!} / \sum_{k=0}^{s} \frac{(\lambda/\mu)^k}{k!}$ gives the long-run fraction of time that j servers are busy for $0 \leq j \leq s$. The formula for p_s is called the *Erlang loss formula* for the following reason. By the property of Poisson arrivals see time averages, the long-run fraction of customers finding upon arrival all s servers busy is equal to the long-run fraction of time that all servers are busy. Therefore, the probability p_s gives also the long-run fraction of customers that are lost.

The Danish telephone engineer A. K. Erlang (1878–1929) was a genius. He was not only the first person to derive the formula named after him, but also conjectured that the formula for the limiting probability p_j of j busy servers remains valid when the service times are generally distributed with mean $1/\mu$. In particular, this applies to the loss probability. A proof of this important insensitivity result was only given many years after Erlang made his conjecture. More generally, for stochastic service systems in which customers arrive according to a Poisson process and customers never wait in a queue, steady-state performance measures are typically insensitive to the shape of the service-time distribution and require only the expected value of this distribution. Service systems in which customers never queue are loss systems and infinite-server systems. The importance of the insensitivity result cannot be overemphasized.

11.3.1 Birth-and-Death Processes and Queueing Applications

To avoid technicalities, we have assumed so far a finite state space. What about a continuous-time Markov chain with a countably infinite number of states?

One then has to rule out the possibility of an infinite number of state transitions in a finite time interval. For that purpose the assumption is made that the rates of the exponentially distributed holding times in the states are bounded. Under this assumption, Rule 11.3 remains valid for the case of an infinite state space. To ensure that Rule 11.4 is also valid for the case of an infinite state space, a sufficient condition is the existence of some state r so that the expected time until the first transition into state r is finite for any starting state.

An important class of continuous-time Markov chains are the so-called birth-and-death processes. These processes have many applications. A continuous-time Markov chain with states $0, 1, \ldots$ is said to be a *birth-and-death process* if the transition rates satisfy $q_{ij} = 0$ for $|i - j| > 1$, that is, transitions are only possible to adjacent states. Queueing systems can often be modeled as a birth-and-death process in which the population consists of the number of waiting customers. Also, birth-and-death processes are often used in applications in biology. In birth-and-death processes the limiting probabilities p_j can be given explicitly.

Rule 11.5 *Let $\{X(t)\}$ be a birth-and-death process with states $0, 1, \ldots$ and transition rates $\lambda_i = q_{i,i+1}$ and $\mu_i = q_{i,i-1}$, where $\lambda_i > 0$ for all $i \geq 0$ and $\mu_i > 0$ for all $i \geq 1$. Suppose that the so-called ergodicity condition*

$$\sum_{j=1}^{\infty} \frac{\lambda_0 \lambda_1 \cdots \lambda_{j-1}}{\mu_1 \mu_2 \cdots \mu_j} < \infty$$

is satisfied. Then the limiting probabilities p_j exist and satisfy the recursive equation

$$\mu_j p_j = \lambda_{j-1} p_{j-1} \quad \text{for } j = 1, 2, \ldots.$$

An explicit expression for the p_j is given by

$$p_j = \frac{\lambda_0 \lambda_1 \cdots \lambda_{j-1}}{\mu_1 \mu_2 \cdots \mu_j} p_0 \quad \text{for } j = 1, 2, \ldots,$$

where the probability $p_0 = \left(1 + \sum_{j=1}^{\infty} \frac{\lambda_0 \lambda_1 \cdots \lambda_{j-1}}{\mu_1 \mu_2 \cdots \mu_j}\right)^{-1}$.

It is instructive to derive the explicit formula for the p_j. It will appear from the derivation that the ergodicity condition is necessary for the existence of limiting probabilities. The reader has to take for granted that this condition is also sufficient for the existence of the p_j. The key idea for the derivation of the p_j is the observation that for any state $j \neq 0$, the long-run average number of transitions per unit time out of the set $\{j, j+1, \ldots\}$ must be equal to the long-run average number of transitions per unit time into the set $\{j, j+1, \ldots\}$

11.3 Limiting Probabilities

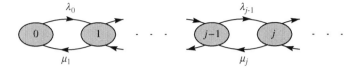

Figure 11.3. The birth-and-death process.

(explain!). This set can only be left by a transition from j to $j-1$ and can only be entered by a transition from $j-1$ to j, see also Figure 11.3. Thus,

$$\mu_j p_j = \lambda_{j-1} p_{j-1} \quad \text{for } j = 1, 2, \ldots.$$

Iterating this recursive equation gives

$$p_j = \frac{\lambda_{j-1}}{\mu_j} p_{j-1} = \frac{\lambda_{j-1}\lambda_{j-2}}{\mu_j \mu_{j-1}} p_{j-2} = \cdots = \frac{\lambda_{j-1}\lambda_{j-2}\cdots\lambda_0}{\mu_j \mu_{j-1}\cdots \mu_1} p_0.$$

Using the fact that $\sum_{j=0}^{\infty} p_j = 1$, the expression for p_0 follows and the derivation of the explicit formula for the p_j has been completed.

Example 11.5 Customers arrive at a service facility according to a Poisson process with rate λ. Each arriving customer immediately gets assigned a free server. There are ample servers. The service times of the customers are independent random variables having a common exponential distribution with mean $1/\mu$. What is the limiting distribution of the number of busy servers? Do you think this limiting distribution is insensitive to the shape of the service-time distribution?

Solution. Let the random variable $X(t)$ be the number of busy servers at time t. The process $\{X(t)\}$ is a birth-and-death process with states $0, 1, \ldots$ and transition rates

$$\lambda_i = \lambda \text{ for } i = 0, 1, \ldots \quad \text{and} \quad \mu_i = i\mu \text{ for } i = 1, 2, \ldots.$$

By Rule 11.5, $p_j = \frac{(\lambda/\mu)^j}{j!} p_0$ for $j \geq 0$. Using the normalization equation $\sum_{j=0}^{\infty} p_j = 1$, it follows next that $p_0 = e^{-\lambda/\mu}$ and so

$$p_j = e^{-\lambda/\mu} \frac{(\lambda/\mu)^j}{j!} \quad \text{for } j = 0, 1, \ldots.$$

In other words, the limiting distribution of the number of busy servers is a Poisson distribution with expected value λ/μ. This limiting distribution is in agreement with the limiting distribution of the number of busy servers in the Erlang loss model with a very large number of servers. By the insensitivity property of the Erlang loss model, it will not come as a surprise that the limiting distribution of the busy servers in the infinite-server model does not change when the service time of a customer has a general distribution with expected value $1/\mu$.

Unlike the Erlang loss model, a rather simple proof can be given for the insensitivity property of the infinite-server model with Poisson arrivals and generally distributed service times. We only give a sketch of a heuristic proof. Suppose first that the service times are deterministic and are equal to the constant $D = 1/\mu$. Fix t with $t > D$. If each service time is a constant D, the only customers present at time t are those customers who have arrived in $(t - D, t]$. Thus, the number of customers present at time t is Poisson distributed with expected value λD, which verifies the insensitivity property for the special case of deterministic service times. Next consider the case that the service time can take on only a finite number of values $D_1, \ldots, D_s$ with respective probabilities $p_1, \ldots, p_s$, where $\sum_{k=1}^{s} p_k D_k = 1/\mu$. Mark the customers with the same fixed service time D_k as type-k customers. The type-k customers arrive according to a Poisson process with rate λp_k. Moreover, the various Poisson arrival processes of the marked customers are independent of each other, see Section 11.1.1. Fix now t with $t > \max_k D_k$. By the above argument, the number of type-k customers present at time t is Poisson distributed with mean $(\lambda p_k) D_k$ and is independent of the other types of customers present at time t. Hence, the total number of customers present at time t has a Poisson distribution with mean $\sum_{k=1}^{s} \lambda p_k D_k = \lambda/\mu$, because the sum of independent random variables with a Poisson distribution is also Poisson distributed. This proves the insensitivity property for the case that the service time has a discrete distribution with finite support. Any service-time distribution can be arbitrarily closely approximated by a discrete distribution with finite support. This makes plausible that the insensitivity result holds for any service-time distribution. The infinite-server queueing model with Poisson arrivals is known as the $M/G/\infty$ queueing model and is a very useful model because of its insensitivity property.

In the next example we analyze the *Erlang delay model*, also known as the $M/M/s$ queueing model.

Example 11.6 Consider a multi-server station with s servers and infinite waiting room. Customers arrive according to a Poisson process with rate λ. If an arriving customer finds a server free, then the customer immediately enters service; otherwise, the customer joins the queue. The service time of a customer has an exponential distribution with rate μ. It is assumed that $\lambda < s\mu$. What is the limiting distribution of the number of customers present? What is the long-run fraction of customers who have to wait? What is the average waiting time per customer?

11.3 Limiting Probabilities

Solution. Letting $X(t)$ denote the number of customers in the system (including those in service), the continuous-time Markov chain $\{X(t)\}$ is a birth-and-death process with states $0, 1, \ldots$. A birth occurs when a new customer arrives and a death occurs when a service is completed. The parameters of this birth-and-death process are given by

$$\lambda_i = \lambda \text{ for all } i, \quad \mu_i = i\mu \text{ for } 1 \le i \le s-1, \text{ and } \mu_i = s\mu \text{ for } i \ge s.$$

The ergodicity condition in Rule 11.5 boils down to the condition $\lambda < s\mu$. This is a natural condition, which states that the arrival rate must be smaller than the total service rate. It is convenient to use the notation

$$\rho = \frac{\lambda}{s\mu}.$$

It is a matter of simple algebra to verify that under the condition $\rho < 1$, the expression for the probabilities p_j in Rule 11.5 reduces to

$$p_j = \frac{(s\rho)^j}{j!} p_0 \text{ for } 0 \le j \le s-1 \text{ and } p_j = \frac{(s\rho)^s \rho^{j-s}}{s!} p_0 \text{ for } j \ge s,$$

where $p_0 = \left(\sum_{k=0}^{s-1} \frac{(s\rho)^k}{k!} + \frac{(s\rho)^s}{s!} \frac{1}{1-\rho} \right)^{-1}$. The limiting probability p_j represents the long-run fraction of time that j customers are present in the system. Of particular interest are the performance measures $P_{delay} = \sum_{j=s}^{\infty} p_j$ and $L_q = \sum_{j=s}^{\infty} (j-s) p_j$, where P_{delay} represents the long-run fraction of time that all s servers are occupied and L_q represents the long-run average number of customers waiting in queue (excluding those in service). It is a matter of simple algebra to verify that

$$P_{delay} = \frac{(s\rho)^s}{s!(1-\rho)} p_0 \text{ and } L_q = \frac{\rho}{1-\rho} P_{delay}.$$

The formula for P_{delay} is often called the *Erlang delay formula*. By the property of Poisson arrivals see time averages, the long-run fraction of customers finding upon arrival s or more other customers present is equal to the long-run fraction of time that s or more customers are present. Thus, P_{delay} also gives the long-run fraction of customers who have to wait in queue. Observe that the average queue size L_q increases in a nonlinear way as a function of the load factor ρ and explodes when ρ approaches 1. Another important quantity is the long-run average waiting time in queue per customer (excluding service time). Denote this quantity by W_q. Observe that service completions occur at a rate of $s\mu$ as long as all s servers are busy. Using this observation together with the fact that the long-run fraction of customers finding upon arrival k other customers present is equal to the long-run fraction of time that k customers are present, it follows that $W_q = \sum_{k=s}^{\infty} \frac{k-s+1}{s\mu} p_k$, assuming service in order of arrival. This leads after some algebra to

$$W_q = \frac{1}{s\mu(1-\rho)} P_{delay}.$$

Comparing the formulas for L_q and W_q, you see that

$$L_q = \lambda W_q.$$

In words, the average number of customers waiting in queue is equal to the average arrival rate of customers times the average waiting time in queue per customer. This result is known as *Little's formula*. This famous formula is valid for almost any queueing system. A heuristic explanation is as follows. Imagine that a cost at rate 1 is incurred for each customer waiting in a queue. Then a cost at rate j is incurred when j customers are waiting in a queue. Therefore, the long-run average cost per unit time is L_q. Moreover, the long-run average cost per unit time is the average number of customers arriving per unit time (= the arrival rate λ) multiplied by the average cost incurred per customer (= the average waiting time W_q per customer).

For the special case of a single server ($s = 1$), the above formulas simplify to

$$p_j = (1-\rho)\rho^j \quad \text{for } j = 0, 1, \ldots,$$
$$P_{delay} = \rho, \ L_q = \rho^2/(1-\rho), \text{ and } W_q = \rho/[\mu(1-\rho)].$$

To illustrate Example 11.6, suppose that the management of a harbor is facing the decision to replace the current unloader. The choice is between one fast unloader and two slow unloaders. The fast unloader can handle on average two ships per eight and a half hours and each of the slow unloaders can handle on average one ship per eight and a half hours. There is activity in the harbor day and night. Ships with cargo arrive according to a Poisson process with a rate of four ships per 24 hours. There is ample berth space for the ships. The unloading times of the ships are independent random variables with the same exponential distribution. What are the delay probability and the average sojourn time per ship for each of the two alternatives? To answer the question, we use the Erlang delay model for $s = 1$ and $s = 2$. Take the hour as unit of time. For the case of one fast unloader ($s = 1$), we have an arrival rate of $\lambda = 4/24$ and a service rate of $\mu = 2/8.5$. This gives a delay probability of $P_{delay} = 0.7083$ and an average sojourn time per ship of $W_q + 1/\mu = 10.32 + 4.25 = 14.57$ hours. For the case of two slow unloaders ($s = 2$), we have $\lambda = 4/24$ and $\mu = 1/8.5$. This gives a delay probability of $P_{delay} = 0.5874$ and an average sojourn time per ship of $W_q + 1/\mu = 8.56 + 8.5 = 17.06$ hours. Assuming that the average sojourn time per ship is the most important criterion, it is better to have one fast unloader.

Problems

11.21 Consider again Problem 11.1. The system is down when all units are in repair. What is the long-run fraction of time the system is down? What is the numerical answer when $\lambda = 0.1$, $\eta = 0.05$, and $\mu = 1$?

11.22 Consider again Problem 11.2 with the data $\lambda = \frac{1}{6}$ and $\mu = \frac{1}{4}$. Solve the balance equations to answer the following questions.

(a) What is the fraction of time that the pump is occupied?
(b) What is the average number of cars waiting in queue?
(c) What is the fraction of cars that cannot enter the station?
(d) What is the average waiting time in a queue of a car entering the gasoline station?

11.23 An assembly line for a certain product has two stations in series. Each station has only room for one unit of the product. If the assembly of a unit is completed at station 1, then it is forwarded immediately to station 2 provided station 2 is idle; otherwise, the unit remains in station 1 and blocks this station until station 2 becomes free. Units for assembly arrive at station 1 according to a Poisson process with rate λ, but a newly arriving unit is only accepted by station 1 when no other unit is present at station 1. Each unit rejected is handled elsewhere. The assembly times at stations 1 and 2 have an exponential distribution with respective means $1/\mu_1$ and $1/\mu_2$. Specify the balance equations and answer the following questions.

(a) What is the long-run fraction of time that station 1 is blocked?
(b) What is the long-run fraction of time that station 2 is busy?
(c) What is the long-run fraction of items that are rejected?

11.24 A small gas station has two pumps, one for gas and the other for LPG. Customers for gas and LPG arrive according to two independent Poisson processes, with an average of 20 customers per hour for gas and an average of 5 customers per hour for LPG. The service times are exponentially distributed with an expected value of two minutes for gas and an expected value of three minutes for LPG. The gas station has an area where two cars can wait until it is their turn to be served. A customer stops at the station only when the desired pump is free or the waiting area is not full.

(a) What is the equilibrium distribution of the total number of customers at the gas station?
(b) What is the average number of customers served per unit time?
(c) What is the long-run fraction of LPG customers who find no place at the station?

11.25 Consider again Problem 11.5. What is the long-run average number of waiting passengers? What is the long-run fraction of potential passengers who go elsewhere for transport?

11.26 In a single-product inventory system, the depletion of stock is due to demand and deterioration. Customers asking for one unit of the product arrive according to a Poisson process with rate λ. The lifetime of each unit of product has an exponential distribution with mean $1/\mu$. Each time the stock drops to zero, a replenishment order for Q units is placed. The lead time of each order is negligible. What are the long-run average stock on hand and the long-run average number of orders placed per time unit?

11.27 Consider again Problem 11.2. It is now assumed that the service time of a car has an Erlang distribution with shape parameter 2 and an expected value of four minutes. Answer again the questions posed in Problem 11.22. *Hint*: Use the trick of imagining that the service time of a car consists of two independent consecutive phases, each having an exponential distribution with an expected value of two minutes. This powerful trick is based on Remark 8.2 in Section 8.3.

11.28 Consider again Problem 11.1. It is now assumed that the repair time of a failed unit has an Erlang distribution with shape parameter 3 and scale parameter β. Use the trick outlined in Problem 11.27 to answer the question of Problem 11.21. What is the numerical answer when $\lambda = 0.1$, $\eta = 0.05$, and $\beta = 3$?

11.29 Let $\{X(t)\}$ be an irreducible continuous-time Markov chain with a finite state space I and transition rates q_{ij}. The process $\{X(t)\}$ is said to be *reversible* if its limiting probabilities p_j satisfy the *detailed balance equations* $p_j q_{jk} = p_k q_{kj}$ for all $j \neq k$.

(a) Explain why, in each of Problems 11.1–11.3, 11.6, and 11.9, the continuous-time Markov chain is reversible.

(b) Suppose that the continuous-time Markov chain $\{X(t)\}$ is reversible. Let $\{\overline{X}(t)\}$ be a truncation of the process $\{X(t)\}$ to the set A with $A \subset I$, that is, $\{\overline{X}(t)\}$ is a continuous-time Markov chain with state space A and transition rates $\overline{q}_{ij} = q_{ij}$ for $i, j \in A$. Assuming that the process $\{\overline{X}(t)\}$ is irreducible, verify that its limiting probabilities $\overline{p}_j$ are given by $\overline{p}_j = p_j / \sum_{k \in A} p_k$ for $j \in A$.

11.30 An operating unit is in a perfect state, in a good state, or in an acceptable state. The unit stays in the perfect state during an exponentially distributed time with mean $1/\mu_1$, in the good state during an exponentially distributed time with mean $1/\mu_2$, and in the acceptable state during an exponentially distributed time with mean $1/\mu_3$. The operating unit goes to the good state or the acceptable state with equal probabilities upon completion of the sojourn time in the perfect state. The unit is replaced by a new one when the sojourn time in the acceptable state expires. What is the average number of replacements per unit time?

11.31 In a factory, type-1 parts and type-2 parts are processed one by one. The processing times of the parts are independent of each other, where the processing time of a type-i part is exponentially distributed with mean $1/\mu_i$. Upon completion of the processing of a type-i part, the next part to be processed is of the same type with probability a_i and of the other type with probability $1 - a_i$. What is the long-run fraction of time that type-i parts are processed? What is the average number of type-i parts processed per unit time? *Hint*: Verify first that the sojourn time in state i is exponentially distributed with expected value $1/[(1 - a_i)\mu_i]$, where the system is said to be in state i when a type-i part is being processed.

11.32 Consider an irreducible continuous-time Markov chain $\{X(t), t \geq 0\}$ with finite state space $I = \{0, 1, \ldots, m\}$.

(a) Let μ_{00} be the expected time between two transitions into state 0. Argue that $\mu_{00} = \frac{1}{v_0 p_0}$, where v_0 is the rate at which the system leaves state 0 and $p_0 = \lim_{t \to \infty} P(X(t) = 0)$.

(b) Let T be the first time that the process has been in state 0 for τ consecutive units of time when $X(0) = 0$. Show that $E(T) = \frac{1}{v_0 p_0}(e^{v_0 \tau} - 1)$.

11.33 Solve again Problem 10.54 when items for processing arrive by a Poisson process with rate $\lambda = 1$ rather than by a deterministic process.

11.34 Consider again Problem 11.9. Give a recursion scheme for the computation of the limiting distribution of the number of machines in repair.

11.35 Suppose that one wishes to determine the capacity of a stockyard at which containers arrive according to a Poisson process with a rate of $\lambda = 1$ per hour. A container finding a full yard upon arrival is brought elsewhere. The time that a container is stored in the yard is exponentially distributed with mean $1/\mu = 10$ hours. Determine the required capacity of the yard so that no more than 1% of the arriving containers find the yard full. How does the answer change when the time that a container is stored in the yard is uniformly distributed between 5 and 15 hours?

11.36 Long-term parkers and short-term parkers arrive at a parking place for cars according to independent Poisson processes with respective rates of $\lambda_1 = 4$ and $\lambda_2 = 6$ per hour. The parking place has room for $N = 10$ cars. Each arriving car which finds all places occupied goes elsewhere. The parking time of long-term parkers is uniformly distributed between 1 and 2 hours, while the parking time of short-term parkers has a uniform distribution between 20 and 60 minutes. Calculate the probability that a car finds all parking places occupied upon arrival.

11.37 Consider a continuous-review inventory system in which customers asking for a certain item arrive according to a Poisson process with rate λ. Each customer asks for one unit. Customer demands occurring when the system is out of stock are lost. The $(S-1, S)$ control rule is used. Under this control rule, the base stock is S and a replenishment order for exactly one unit is placed each time the stock on hand decreases by one unit. The lead time of each replenishment order is a constant $L > 0$. What is the limiting distribution of the stock on hand? What are the long-run average stock on hand and the long-run fraction of demand that is lost?

11.38 Consider again the $(S-1, S)$ inventory model from Problem 11.37, but it is now assumed that demand occurring when the system is out of stock is back-ordered. How do the answers to the questions change?

11.39 Customers asking for a certain product arrive according to a Poisson process with rate λ. Each customer asks for one unit of the product. Any demand that cannot be satisfied directly from stock on hand is back-ordered. The control rule is based on the inventory position, which is defined as the stock on hand minus the amount back-ordered plus the amount on order. Each time the inventory position drops to the reorder level s, a quantity of Q units of the product is ordered. The lead time of any replenishment order is a fixed constant $L > 0$. Use the limiting distribution of the inventory position to find the limiting distribution of the stock on hand.

11.40 A pizzeria at the airport has 20 tables. Each table can be occupied either by one person or two. Customers arrive either singly or in pairs. If a table is free, then a pair is always seated. A single customer is only seated if five or more tables are free. Customers who are not assigned a table leave and go elsewhere. Single customers and pairs arrive according to independent Poisson processes with respective rates of 18 per hour and 12 per hour. A single customer keeps the table occupied during an exponentially distributed time with a mean of 30 minutes and a pair keeps the table occupied during an exponentially distributed time with a mean of 40 minutes. Formulate a continuous-time Markov chain to answer the following questions:

(a) What is the average number of occupied tables?
(b) What are the average number of single customers served per hour and the average number of pairs served per hour?

11.3 Limiting Probabilities

11.41 A machine is continuously processing parts (there is always raw material available). The processing time of a part is exponentially distributed with a mean of 20 seconds. A finished part is transported immediately to an assembly cell by an automatic conveyor system. The transportation time is exactly 3 minutes. What is the long-run fraction of time that there are more than 15 parts on the conveyor line?

11.42 Messages arrive at a communication channel according to a Poisson process with rate λ. The message length (in bits) is exponentially distributed with mean $1/\mu$. An arriving message finding the line idle is provided with service immediately; otherwise, the message waits until access to the line can be given. The communication line is only able to submit one message at a time, but has available two possible transmission rates σ_1 and σ_2 with $0 < \sigma_1 < \sigma_2$. That is, σ_1 bits can be transmitted per unit time under the slower rate and σ_2 bits per unit time under the fast rate. It is assumed that $\lambda/(\sigma_2 \mu) < 1$. At any time the transmission line may switch from one rate to the other. The transmission rate is controlled by a single-critical-number rule. The transmission rate σ_1 is used whenever less than R messages are present, while otherwise the faster transmission rate σ_2 is used. Give a recursion scheme for the computation of the limiting distribution of the number of messages present.

11.43 At a taxi stand, customers and taxis arrive according to independent Poisson processes with respective rates of 10 per hour and 7 per hour. Suppose that a taxi will wait no matter how many other taxis are at the stand. However, an arriving customer who finds no taxi present leaves to find alternative transportation. What is the long-run average number of taxis waiting at the stand? What is the long-run proportion of arriving customers who get a taxi?

11.44 Customers arrive at a service facility with infinite waiting room according to a Poisson process with rate λ. An arriving customer finding upon arrival j other customers present joins the queue with probability $1/(j+1)$ for $j = 0, 1, \ldots$ and goes elsewhere otherwise. There is a single server who can handle only one customer at a time. The service time of each customer is exponentially distributed with mean $1/\mu$. What are the limiting probabilities of the number in the system? What is the long-run fraction of customers who go elsewhere?

11.45 Oil tankers leave Norway with destination Rotterdam according to a Poisson process with an average of two tankers per day, where the sailing time to Rotterdam is gamma distributed with an expected value of two days and a standard deviation of one day. Estimate the probability that more than seven oil tankers are underway from Norway to the harbor of Rotterdam.

11.46 Consider again the multi-server queueing model from Example 11.6. Let W_n be the waiting time in a queue of the nth arriving customer. Under the assumption that customers enter service in order of arrival, verify that $\lim_{n\to\infty} P(W_n > t) = P_{delay}\, e^{-s\mu(1-\rho)t}$ for all $t \geq 0$. *Hint*: Note that the times between service completions have an exponential density with mean $s\mu$ if all servers are busy and use the fact that the sum of k independent exponentials with parameter η is an Erlang-distributed random variable S_k with $P(S_k > x) = \sum_{i=0}^{k-1} e^{-\eta x}(\eta x)^i/i!$ for $x \geq 0$.

11.47 Phone calls with service requests arrive at a call center according to a Poisson process with rate λ. There are s agents to handle the service requests. The service times of the requests are independent random variables having an exponential density with mean $1/\mu$. An arriving request immediately gets assigned a free server if one is available; otherwise, the request joins the queue. The caller becomes impatient if, after an exponentially distributed time with mean $1/\theta$, the service has not yet started and then leaves the system. Give a recursive scheme for the limiting distribution of the number of requests receiving service or waiting in a queue. What is the long-run fraction of balking callers?

11.48 In a tourist village there is a bike rental with 25 bikes. Tourists for a rental bike arrive according to a Poisson process with an average of 10 per hour. A tourist who finds no bike available refrains from renting a bike. The tourists use the bike to make a trip in the village surroundings. The length of a bike trip is exponentially distributed with a mean of 2 hours. The tourists can return the bike either to the bike rental in the village or to a depot outside the village and then take a bus to the village. The probability that a bike is returned to the bike rental is 0.75. The bikes at the depot are transported to the village as soon as there are 10 bikes at the depot. The transportation time is negligible.

(a) Formulate an appropriate continuous-time Markov chain.
(b) What are the average numbers of bikes at the bike rental and the depot?
(c) What is the fraction of tourists who find no bike available?
(d) What is the average number of transports per unit time from the depot to the bike rental?

11.49 Suppose that M independent sources generate service requests for c service channels, where $M \geq c$. A service request that is generated when all c channels are occupied is lost. Each source is alternately on and off. A source is off when it has a service request being served and is on otherwise. A source in the on-state generates a new service request after an exponentially distributed time with mean $1/\alpha$ and the service time of each request is exponentially distributed with mean $1/\mu$. What is the limiting distribution of the number of occupied service channels? What is the long-run fraction of service requests that are lost? *Note*: This probability model is called the *Engset loss model*.

11.50 In an electronic system, there are c elements of a crucial component connected in parallel to increase the reliability of the system. Each component is switched on and the lifetimes of the components have an exponential distribution with mean $1/\alpha$. The lifetimes of the components are independent of each other. The electronic system is working as long as at least one of the components is functioning; otherwise, the system is down. It takes an exponentially distributed time with mean $1/\mu$ to replace a failed component. Only one failed component can be replaced at a time.

(a) What is the long-run fraction of time the system is down? Do you think this performance measure is insensitive to the specific form of the lifetime distribution? *Hint*: Compare the transition-rate diagram with the transition-rate diagram in the Erlang loss model.

(b) Use the results of Problem 11.49 to answer the questions in (a) when there are ample repairmen to replace failed components.

11.51 A national chain of stores allows customers to return bought products for a full refund within a period of one week. Consider the following inventory model for an exclusive product at a particular store. The demands and the returns of the product follow two independent Poisson processes with rates λ and μ. The rule for controlling the inventory level is characterized by two parameters Q and R with $0 < Q < R$. The stock of the product on hand is raised to the level Q if it drops to zero by demand, while the stock on hand is decreased to the level Q when it reaches the value R by return. It is assumed that the adjustments of the inventory level are instantaneous.

(a) Formulate an appropriate continuous-time Markov chain.
(b) What is the average stock on hand?
(c) What are the average number of stock replenishments per unit time and the average number of stock reductions per unit time?

Appendix A: Counting Methods

Many probability problems require counting techniques. In particular, these techniques are extremely useful for computing probabilities in a chance experiment in which all possible outcomes are equally likely. In such experiments, one needs effective methods to count the number of outcomes in any specific event. In counting problems, it is important to know whether the order in which the elements are counted is relevant or not.

In the discussion below, we use the fundamental principle of counting: if there are a ways to do one activity and b ways to do another activity, then there are $a \times b$ ways of doing both. As an example, suppose that you go to a restaurant to get some breakfast. The menu says pancakes, waffles, or fried eggs, while for a drink you can choose between juice, coffee, tea, and hot chocolate. Then the total number of different choices of food and drink is $3 \times 4 = 12$. As another example, how many different license plates are possible when the license plate displays a nonzero digit, followed by three letters, followed by three digits? The answer is $9 \times 26 \times 26 \times 26 \times 10 \times 10 \times 10 = 158,184,000$ license plates.

Permutations

How many different ways can you arrange a number of different objects such as letters or numbers? For example, what is the number of different ways that the three letters A, B, and C can be arranged? By writing out all the possibilities *ABC, ACB, BAC, BCA, CAB,* and *CBA*, you can see that the total number is 6. This brute-force method of writing down all the possibilities and counting them is naturally not practical when the number of possibilities gets large, for example the number of different ways to arrange the 26 letters of the alphabet. You can also determine that the three letters A, B, and C can be written down in six different ways by reasoning as follows. For the first position there are three available letters to choose from, for the second position there are two letters over to choose from, and only one letter for the third position. Therefore the total number of possibilities is $3 \times 2 \times 1 = 6$. The general rule should now be evident. Suppose that you have n distinguishable objects. How many ordered arrangements of these objects are possible? Any ordered sequence of the objects is called a *permutation*. Reasoning similar to that described above shows that there are n ways to choose the first object, leaving $n - 1$ choices for the second object, etc. Therefore, the total number of ways to order n distinguishable objects is $n \times (n-1) \times \cdots \times 2 \times 1$. A convenient shorthand for this product is $n!$ (pronounced: n factorial). Thus, for any positive integer n,

$$n! = 1 \times 2 \times \cdots \times (n-1) \times n.$$

A convenient convention is $0! = 1$. Summarizing,

the total number of ordered sequences (permutations) of n distinguishable objects is $n!$.

Note that $n!$ grows very quickly, since $n! = n \times (n-1)!$. For example, $5! = 720$, $10! = 3,628,800$, and $15! = 1,307,674,368,000$.

Example A.1 A scene from the movie "The Quick and the Dirty" depicts a Russian roulette type of duel. Six identical shot glasses of whiskey are set on the bar, one of which is laced with deadly strychnine. The bad guy and the good guy must drink in turns. The bad guy offers $1,000 to the good guy, if the latter will go first. Is this an offer that should not be refused?

Solution. A handy way to think of the problem is as follows. Number the six glasses from 1 to 6 and assume that the glasses are arranged in a random order after strychnine has been put in one of the glasses. There are 6! possible arrangements of the six glasses. If the glass containing strychnine is in the first position, there remain 5! possible arrangements for the other five glasses. Thus, the probability that the glass in the first position contains strychnine is equal to $5!/6! = 1/6$. By the same reasoning, the glass in each of the other five positions contains strychnine with a probability of $1/6$, before any glass is drunk. It is a fair game. Each of the two "duelists" will drink the deadly glass with a probability of $3 \times (1/6) = 1/2$. The good guy will do well to accept the offer of the bad guy. If the good guy survives the first glass after having drunk it, then the probability that the bad guy will get the glass with strychnine becomes $3 \times (1/5) = 3/5$ (verify!).

Combinations

How many different juries of three persons can be formed from five persons A, B, C, D, and E? By direct enumeration you see that the answer is

10: $\{A,B,C\}$, $\{A,B,D\}$, $\{A,B,E\}$, $\{A,C,D\}$, $\{A,C,E\}$, $\{A,D,E\}$, $\{B,C,D\}$, $\{B,C,E\}$, $\{B,D,E\}$, and $\{C,D,E\}$. In this problem, the order in which the jury members are chosen is not relevant. The answer 10 juries could also have been obtained by a basic principle of counting. First, count how many juries of three persons are possible when attention is paid to the order. Then determine how often each group of three persons has been counted. Thus, the reasoning is as follows. There are 5 ways to select the first jury member, 4 ways to then select the next member, and 3 ways to select the final member. This would give $5 \times 4 \times 3$ ways of forming the jury when the order in which the members are chosen is relevant. However, this order makes no difference. For example, for the jury consisting of persons A, B, and C, it is not relevant which of the 3! ordered sequences ABC, ACB, BAC, BCA, CAB, and CBA has led to the jury. Hence, the total number of ways a jury of 3 persons can be formed from a group of 5 persons is equal to $\frac{5 \times 4 \times 3}{3!}$. This expression can be rewritten as

$$\frac{5 \times 4 \times 3 \times 2 \times 1}{3! \times 2!} = \frac{5!}{3! \times 2!}.$$

In general, you can calculate that the total number of possible ways to choose a jury of k persons out of a group of n persons is equal to

$$\frac{n \times (n-1) \times \cdots \times (n-k+1)}{k!}$$
$$= \frac{n \times (n-1) \times \cdots \times (n-k+1) \times (n-k) \times \cdots \times 1}{k! \times (n-k)!} = \frac{n!}{k! \times (n-k)!}.$$

For nonnegative integers n and k with $k \leq n$, we define

$$\binom{n}{k} = \frac{n!}{k! \times (n-k)!}.$$

The quantity $\binom{n}{k}$ (pronounced: n over k) has the interpretation:

$\binom{n}{k}$ is the total number of ways to choose k different objects out of n distinguishable objects, paying no attention to their order.

In other words, $\binom{n}{k}$ is the total number of combinations of k different objects out of n. The key difference between permutations and combinations is *order*. Combinations are *unordered* selections, permutations are *ordered* arrangements. The numbers $\binom{n}{k}$ are referred to as the *binomial coefficients*. Using the interpretation of $\binom{n}{k}$, we can give a "word proof" of the following two relations:

$$\binom{n}{k} = \binom{n}{n-k} \quad \text{and} \quad \binom{n}{k} = \binom{n-1}{k-1} + \binom{n-1}{k} \quad \text{for } 0 < k < n.$$

The first relation follows by noting that picking k objects out of n is the same as not picking $n-k$ objects out of n. To explain the second relation, it is helpful to imagine that one of the objects has a red label. If you pick k objects out of n, then you can either (i) pick the object with the red label and then pick $k-1$ objects from the remaining $n-1$ objects, or (ii) do not pick the object with the red label and then pick k objects from the remaining $n-1$ objects.

Example A.2 Is the probability of winning the jackpot with a single ticket in the 6/45 lottery larger than the probability of getting 22 heads in a row when tossing a fair coin 22 times?

Solution. In the 6/45 lottery, six different numbers are drawn out of $1, \ldots, 45$. The total number of ways the winning six numbers can be drawn is equal to $\binom{45}{6}$. Hence, the probability of hitting the jackpot with a single ticket is $1/\binom{45}{6} = 1.23 \times 10^{-7}$. This probability is smaller than the probability $\left(\frac{1}{2}\right)^{22} = 2.38 \times 10^{-7}$ of getting 22 heads in a row.

Example A.3 What is the probability of one pair in an ordinary five-card poker hand? One pair means two cards of the same face value plus three cards with different face values.

Solution. The answer to the question requires careful counting to avoid double counting. To count the number of hands with one pair, we proceed as follows. Pick the face value for the pair: $\binom{13}{1}$ choices. Pick two cards from the face value: $\binom{4}{2}$ choices. Pick three other face values: $\binom{12}{3}$ choices. Pick one card from each of the other three face values: $4 \times 4 \times 4$ choices. The number of hands with one pair is $\binom{13}{1} \times \binom{4}{2} \times \binom{12}{3} \times 4^3$ and thus the probability of one pair is

$$\frac{\binom{13}{1} \times \binom{4}{2} \times \binom{12}{3} \times 4^3}{\binom{52}{5}} = 0.4226.$$

Example A.4 What are the probabilities of seven and four matching pairs of socks remaining when six socks are lost during the washing of ten different pairs of socks?

Solution. There are $\binom{20}{6}$ possible ways to choose six socks out of ten pairs of socks. You are left with seven complete pairs of socks only if both socks of three pairs are missing. This can happen in $\binom{10}{3}$ ways. Hence, the probability that you are left with seven complete pairs of socks is equal to

$$\frac{\binom{10}{3}}{\binom{20}{6}} = 0.0031.$$

You are left with four matching pairs of socks only if exactly one sock of each of six pairs is missing. These six pairs can be chosen in $\binom{10}{6}$ ways. There are two possibilities for how to choose one sock from a given pair. This means that there are $\binom{10}{6} \times 2^6$ ways to choose six socks so that four matching pairs of socks are left. Hence, the probability of four matching pairs of socks remaining is equal to

$$\frac{\binom{10}{6} \times 2^6}{\binom{20}{6}} = 0.3467.$$

It is remarkable that the probability of the worst case of four matching pairs of socks remaining is more than a hundred times larger than the probability of the best case of seven matching pairs of socks remaining. When things go wrong, they really go wrong.

In order to test your understanding of the concepts discussed above, you are asked to verify the answers to the following questions.

- How many distinct license plates with three letters followed by three digits are possible? What is the number of possible combinations if the letters and numbers must be different? (answers: 17,576,000 and 11,232,000).
- What is the total number of distinguishable permutations of the 11 letters in the word "Mississippi?" (answer: 34,650).
- Suppose that from ten children, five are to be chosen and lined up. How many different lines are possible? (answer: 30,240).
- Suppose that two teams of five children each have to be formed from ten children. How many different combinations of two teams are possible? (answer: 126).
- A bridge club has 50 members. How many ways can they pick four members to be on a committee to plan a bridge tournament? (answer: 230,300).
- A jury of three women and two men is to be selected from a group of ten women and seven men. In how many ways can this be done? (answer: 2,520).
- In a game of bridge, each of the four players gets 13 cards from a deck of 52 cards. How many deals are possible? (answer: about 5.36×10^{28}).
- How many ordinary five-card poker hands contain exactly three aces? How many ordinary five-card poker hands contain two pairs plus one card with a different face value? (answer: 4,512 and 123,552).

Appendix B: Basics of Set Theory

This appendix discusses a few notions of set theory that are needed for probability theory. A *set* is any well-defined collection of objects without regard to their order. Sets are by definition unordered. The objects of a set are called the *elements* of the set and could refer to numbers, people, or something else. For example, all the students enrolled on a particular probability course can be considered as a set, just as the three natural numbers 1, 2, and 3 can form a set. Sets are commonly represented by listing their contents within curly brackets. For example, $A = \{1, 2, 3\}$ tells us that A is a set whose elements are the first three natural numbers. Sets can also be represented using a rule that identifies the elements of the set. For example, the set $I = \{(x, y) : x^2 + y^2 < 1\}$ represents the interior of the unit circle with radius 1 and centered at the origin $(0, 0)$ in the plane.[1] The set $C = \{(x, y) : x^2 + y^2 = 1\}$ represents the circumference of the unit circle.

Sets are usually denoted by capital letters and their elements by small letters. The notation $s \in A$ means that s is an element of the set A, whereas $s \notin A$ means that s does not belong to the set A. We introduce some further notation.

- The set A is a *subset* of the set B if every element of A is also an element of B. This is denoted as $A \subseteq B$.
- The *union* of sets A and B is the set containing all elements either in A or B or both. It is denoted as $A \cup B$.
- The *intersection* of sets A and B is the set containing all elements in both A and B. It is denoted as $A \cap B$. In probability theory, the abbreviated notation AB is often used for $A \cap B$.

---

[1] The set-builder notation $\{s : s \text{ satisfies property } P\}$ should be read as the set of all elements s such that s satisfies property P. In other words, I is the set of all points (x, y) in the plane such that $x^2 + y^2 < 1$.

- The *difference* of sets A and B is the set containing all elements in A that are not in B. It is denoted as $A \backslash B$.
- Suppose that A is a subset of some universal set U (in probability, the universal set is typically the set of all possible outcomes of some probability experiment). The *complement* of A is the set of all elements in U that are not in A. It is denoted as A^c or $\overline{A}$.
- The set containing no elements is called the *empty set* and is denoted as $\emptyset$.
- Two sets A and B are called *disjoint* (or *mutually exclusive*) if $A \cap B = \emptyset$, that is, if A and B have no element in common.

A few words about the difficult concept of the empty set. The empty set is an important concept in mathematics and its role among sets is conceptually similar to the role of the number 0 in a number system. The empty set is not the same thing as nothing. A set is always something and the empty set is a set with nothing inside it. For example, the set of all real numbers x such that $x^2 + 1 = 0$ is the empty set.

To illustrate the above concepts, take the set $U = \{3, 4, 5, 6, 7, 8, 9, 10, 11\}$ as universal set and consider the subsets $A = \{x \in U : x \text{ is a prime number}\}$, $B = \{x \in U : x \text{ is divisible by } 3\}$, and $C = \{x \in U : x > 15\}$. Then $A = \{3, 5, 7, 11\}$, $B = \{3, 6, 9\}$, and $C = \emptyset$. The union of A and B is $A \cup B = \{3, 5, 6, 7, 9, 11\}$, the intersection of A and B is $A \cap B = \{3\}$, and the difference of A and B is $A \backslash B = \{5, 7, 11\}$. Note that $A \backslash B$ and B are disjoint sets and that their union is equal to $A \cup B$. The complement of A is $A^c = \{4, 6, 8, 9, 10\}$.

It is obvious that for any sets A and B, the following relations apply:

$$A \cup B = B \cup A, \; A \cap B = B \cap A, \; A \cup \emptyset = A, \; \text{and } A \cap \emptyset = \emptyset.$$

Also, it is not difficult to verify that

$$A = (A \cap B) \cup (A \cap B^c)$$

for any two subsets A and B of a universal set U. The proof of this result is based on the fact that $B \cup B^c = U$. Note that the sets $A \cap B$ and $A \cap B^c$ are disjoint.

The following two basic laws from set theory are very useful in probability theory. For any two subsets A and B of some universal set, we have

$$(A \cup B)^c = A^c \cap B^c \quad \text{(De Morgan's first law)},$$
$$(A \cap B)^c = A^c \cup B^c \quad \text{(De Morgan's second law)}.$$

These laws are intuitively clear. The first law states that having an element that is not in A or B or both is the same as having an element that is not in A and not in B. The second law states that having an element that is not in A and not in B is the same as having an element that is not in A or B or both.

Countable and Uncountable Sets

A useful measure of a set is its size, or cardinality. The cardinality of a finite set A is simply the number of distinct elements of A and is denoted by $|A|$. Note that $|A \cup B| = |A| + |B| - |A \cap B|$ for finite sets A and B. A set is said to be *countably infinite* if it can be put into one-to-one correspondence with the set of natural numbers. Put differently, a nonfinite set is countably infinite if and only if it is possible to list all of its elements without repetition as an unending list $\{\omega_1, \omega_2, \ldots\}$. For example, the set of even numbers is countably infinite, just as the set of all perfect square numbers is. Also, $A \cup B$ is countably infinite if A and B are countably infinite. A set is said to be *countable* if it is finite, or countably infinite. A set which is not countable is said to be *uncountable*. The interval $(0, 1)$ of real numbers is the standard example of an uncountable set. Another example of an uncountable set is the set of all possible outcomes of the experiment of tossing a fair coin infinitely often. This set consists of all possible infinite sequences of zeros and ones, where one stands for heads and zero for tails. There is a one-to-one correspondence between these sequences and the numbers in $[0, 1]$ in binary notation. Thus, the set of all possible outcomes of the coin-tossing experiment is uncountable.

Sigma Algebra

Let U be a given universal set. You may think of U as the set of all possible outcomes of a probability experiment. A class $\mathcal{F}$ of subsets of U is called a *sigma algebra* if

(i) $U \in \mathcal{F}$,
(ii) if $A \in \mathcal{F}$, then $A^c \in \mathcal{F}$,
(iii) if $A_n \in \mathcal{F}$ for $n = 1, 2, \ldots$, then $\bigcup_{n=1}^{\infty} A_n \in \mathcal{F}$.

It is important to note that, by (i) and (ii), the empty set $\emptyset$ always belongs to the sigma algebra $\mathcal{F}$. Also, for any finite number of sets $A_1, \ldots, A_r$ from $\mathcal{F}$, we have that $\bigcup_{n=1}^{r} A_n$ belongs to $\mathcal{F}$. This follows by noting that $\bigcup_{n=1}^{r} A_n = \bigcup_{n=1}^{\infty} A_n$ when $A_{r+1} = A_{r+2} = \cdots = \emptyset$. A sigma algebra is not only closed under countable unions, it is also closed under countable intersections. An extension of De Morgan's second law states that, for any infinite sequence of sets A_n,

$$\left(\bigcap_{n=1}^{\infty} A_n \right)^c = \bigcup_{n=1}^{\infty} A_n^c.$$

In probability theory one takes as sigma algebra the collection of all subsets of U when the universal set U is countable. However, when U is uncountable, it is for technical reasons not possible to take the collection of all subsets.

This strange situation is discussed in more detail in advanced probability texts that use measure theory. If U is the set of all real numbers, then one restricts attention to the collection of subsets that can be obtained by starting with all open intervals and their complements, and taking countable unions and intersections of them. This sigma algebra contains all subsets of practical interest but excludes pathological subsets. The excluded subsets are those for which it is impossible to define the notion of length. In the Kolmogorov axioms of probability theory, probabilities are only assigned to the elements of an appropriately chosen sigma algebra. By developing probability theory on a set-theoretical basis, Kolmogorov has not only given a logically consistent foundation for probability theory, but at the same time made it possible to apply highly developed modern branches of mathematics to probability theory.

Appendix C: Some Basic Results from Calculus

This appendix summarizes a number of properties of the famous number e, the exponential function e^x, the natural logarithm function, and the geometric and harmonic series. These concepts play an important role in probability.

Exponential Function

The history of the number e begins with the discovery of logarithms by John Napier in 1614. At this time in history, international trade was experiencing a period of strong growth, and, as a result, there was much attention given to the concept of compound interest. At that time, it was already noticed that $(1+\frac{1}{n})^n$ tends to a certain limit if n is allowed to increase without bound:

$$\lim_{n\to\infty} \left(1 + \frac{1}{n}\right)^n = e,$$

where the famous mathematical constant e is the irrational Euler number[1]

$$e = 2.7182818\ldots.$$

The *exponential function* is defined by e^x, where the variable x runs through the real numbers. This function is one of the most important functions in mathematics.

The number e is positive and so the function e^x is strictly increasing on $(-\infty, \infty)$ with $\lim_{x\to-\infty} e^x = 0$ and $\lim_{x\to\infty} e^x = \infty$. As a consequence, the equation $e^y = c$ has a unique solution y for each fixed $c > 0$. This solution as a function of c is called the *natural logarithm function* and is denoted by $\ln(c)$ for $c > 0$. The natural logarithm function is the inverse function of the exponential function. In other words, $\ln(x)$ is $\log(x)$ with base e.

---

[1] A wonderful account of the number e and its history can be found in E. Maor, *e: The Story of a Number*, Princeton University Press, 1994.

A fundamental property of e^x is that this function itself has as derivative

$$\frac{de^x}{dx} = e^x \quad \text{for all } x.$$

Let's sketch a proof; it gives insight. Consider the function $f(x) = a^x$ for some constant $a > 0$. It then follows from $f(x+h) - f(x) = a^{x+h} - a^x = a^x(a^h - 1)$ that

$$\lim_{h \to 0} \frac{f(x+h) - f(x)}{h} = cf(x)$$

for the constant $c = \lim_{h \to 0}(a^h - 1)/h$. The proof is omitted that this limit always exists. Next, one might wonder for what value of a does the constant $c = 1$, so that $f'(x) = f(x)$. Noting that the condition $(a^h - 1)/h = 1$ can be written as $a = (1 + h)^{1/h}$, it can easily be verified that $\lim_{h \to 0}(a^h - 1)/h = 1$ boils down to $a = \lim_{h \to 0}(1 + h)^{1/h}$, yielding $a = e$.

How do we calculate the function e^x? The generally valid relation

$$\lim_{n \to \infty}\left(1 + \frac{x}{n}\right)^n = e^x \quad \text{for all } x$$

is not useful for that purpose. The calculation of e^x is based on the power series

$$e^x = 1 + x + \frac{x^2}{2!} + \frac{x^3}{3!} + \cdots \quad \text{for all } x.$$

The proof of this power-series expansion requires Taylor's theorem from calculus. The fact that e^x has itself as derivative is crucial in the proof. Note that term-by-term differentiation of the series $1 + x + \frac{x^2}{2!} + \cdots$ leads to the same series, in agreement with the fact that e^x has itself as derivative. The series expansion of e^x leads to the accurate approximation formula

$$e^x \approx 1 + x \quad \text{for } x \text{ close to } 0.$$

This approximation is very useful in probability theory. As an illustration, let's derive an insightful formula for the number of tickets you must fill in so that the probability of matching the winning six numbers in any 6/42 lottery draw is at least α. The odds of matching the winning six numbers with a single ticket is 1 in $\binom{42}{6} = 5{,}245{,}786$. If n tickets are filled in, then the probability that none of these tickets is winning is $(1-p)^n$, with $p = \frac{1}{5{,}245{,}786}$. The value of n must be chosen according to $(1-p)^n \leq 1 - \alpha$. Since p is very close to zero,

$$(1-p)^n \approx e^{-np}.$$

Solving $e^{-np} = 1 - \alpha$ by taking logarithms leads to $n \approx -\frac{1}{p}\ln(1-\alpha)$.

Geometric Series

For any nonnegative integer n,

$$\sum_{k=0}^{n} x^k = \frac{1 - x^{n+1}}{1 - x} \quad \text{for } x \neq 1.$$

This useful result is a direct consequence of

$$(1 - x) \sum_{k=0}^{n} x^k = \left(1 + x + \cdots + x^n\right) - \left(x + x^2 + \cdots + x^n + x^{n+1}\right)$$
$$= 1 - x^{n+1}.$$

The term x^{n+1} converges to 0 for $n \to \infty$ if $|x| < 1$. This leads to

$$\sum_{k=0}^{\infty} x^k = \frac{1}{1 - x} \quad \text{for } |x| < 1.$$

This series is called the *geometric series* and is frequently encountered in probability problems. The series $\sum_{k=1}^{\infty} k x^{k-1}$ can be obtained by differentiating the geometric series $\sum_{k=0}^{\infty} x^k$ term by term and using the fact that the derivative of $1/(1-x)$ is given by $1/(1-x)^2$. The operation of term-by-term differentiation is justified by a general theorem for the differentiation of power series. Thus we have

$$\sum_{k=1}^{\infty} k x^{k-1} = \frac{1}{(1 - x)^2} \quad \text{for } |x| < 1.$$

In the same way, it follows that

$$\sum_{k=2}^{\infty} k(k - 1) x^{k-2} = \frac{2}{(1 - x)^3} \quad \text{for } |x| < 1.$$

Harmonic Series

A surprising result in elementary calculus is that the *harmonic series*

$$\sum_{k=1}^{\infty} \frac{1}{k}$$

diverges and has the value ∞.[2] The partial sum $\sum_{k=1}^{n} \frac{1}{k}$ increases extremely slowly as n gets larger. To illustrate this, the number of terms needed to exceed

---

[2] There are many proofs of this celebrated result. The first proof dates back to about 1350 and was given by the philosopher Nicolas Oresme. His argument was ingenious. Oresme simply observed that $\frac{1}{3} + \frac{1}{4} > \frac{2}{4} = \frac{1}{2}$, $\frac{1}{5} + \frac{1}{6} + \frac{1}{7} + \frac{1}{8} > \frac{4}{8} = \frac{1}{2}$, $\frac{1}{9} + \frac{1}{10} + \cdots + \frac{1}{16} > \frac{8}{16} = \frac{1}{2}$, etc. In general, $\frac{1}{r+1} + \frac{1}{r+2} + \cdots + \frac{1}{2r} > \frac{1}{2}$ for any r, showing that $\sum_{k=1}^{n} \frac{1}{k}$ eventually grows beyond any bound as n gets larger. Isn't it a beautiful argument?

the sum 100 is about 1.509×10^{43} and the number of terms needed to exceed the sum 1,000 is about 1.106×10^{434}. This finding is in agreement with the approximation

$$1 + \frac{1}{2} + \cdots + \frac{1}{n} \approx \ln(n) + \gamma + \frac{1}{2n} \quad \text{for } n \text{ large,}$$

where $\gamma = 0.57721566\ldots$ is the Euler constant. This constant is named after Leonhard Euler (1707–1783), who is considered the most productive mathematician in history. The approximation is very accurate and can be explained from the alternative definition of the natural logarithm function:

$$\ln(y) = \int_1^y \frac{dx}{x} \quad \text{for } y > 0.$$

This fundamental expression for $\ln(y)$ is often used in probability analysis.

To conclude, the sum of the first n positive integers and the sum of their squares are given by

$$\sum_{k=1}^n k = \frac{1}{2}n(n+1) \quad \text{and} \quad \sum_{k=1}^n k^2 = \frac{1}{6}n(n+1)(2n+1).$$

These sums come up in many probability problems and are easily verified by induction.

Appendix D: Basics of Monte Carlo Simulation

In the field of physics, it is quite common to determine the values of certain constants in an experimental way. Computer simulation makes this kind of approach possible in the field of mathematics, too. For example, the value of π can be estimated with the help of some basic principles of simulation (of course, this is not the simplest method for the calculation of π). This is the general idea: take a unit circle (radius 1) with the origin $(0,0)$ as center. In order to generate random points inside the circle,[1] position the unit circle in a square that is described by the four corner points $(-1, 1)$, $(1, 1)$, $(1, -1)$, and $(-1, -1)$. The area of the unit circle is π and the area of the square is equal to 4. Now, generate a large number of randomly chosen points inside the square. This is easily done, as will be seen below. Next, count the number of points that fall within the surface of the unit circle. If you divide this number of points by the total number of generated points, then you get an estimate for $\frac{\pi}{4}$. This method is called the hit-or-miss method.

Monte Carlo simulation is not only often used in simulating physical and mathematical systems in practice, but also a very useful tool for adding an extra dimension to the teaching and learning of probability. It may help students gain a better understanding of probabilistic ideas and overcome common misconceptions about the nature of "randomness." Also, simulation may enable you to get quick answers to specific probability problems. For example, what is the probability that any two adjacent letters are different when the 11 letters of the word "Mississippi" are put in random order? Seemingly a simple probability problem, but it turns out that this combinatorial probability problem is difficult to solve analytically. In computational probability it is

---

[1] What does it mean to choose a "random" point inside a given bounded region in the plane? This means that the probability of the point falling inside some subset of the region must be proportional to the area of the subset.

often beforehand not clear whether a probability problem easily allows for an analytical solution. Many probability problems are too difficult or too time-consuming to solve exactly, while a simulation program is easily written. Monte Carlo simulation can also be used to settle disagreement on the correct answer to a particular probability problem. It is easy to make mistakes in probability, so checking answers is important. Simulation may be helpful for that purpose. Take the famous Monty Hall problem. This probability puzzle raised a lot of discussion about its solution. Paul Erdös, a world-famous mathematician, remained unconvinced about the correct solution of the problem until he was shown a computer simulation confirming the correct result. In the Monty Hall problem, a contestant in a television game show must choose between three doors. An expensive car is behind one of the three doors, and gag prizes are behind the other two. He chooses a door randomly, appealing to Lady Luck. Then the host opens one of the other two doors and, as promised beforehand, takes a door that conceals a gag prize. With two doors remaining unopened, the host now asks the contestant whether he wants to remain with his choice of door, or whether he wishes to switch to the other remaining door. What should the contestant do? Simulating this game is a convincing approach to show that it is better to switch.

Random-Number Generators

How can you conduct a Monte Carlo simulation on your computer? To do so, you need the random-number generator on your computer. A *random-number generator* produces a sequence of numbers that are picked at random between 0 and 1 (excluding the values 0 and 1). It is as if fate falls on a number between 0 and 1 by pure coincidence. Random numbers between 0 and 1 are random samples from the uniform distribution on $(0, 1)$. That is, the probability of the generated number falling into any given subinterval of the unit interval $(0,1)$ equals the length of that subinterval. Any two subintervals of the same length have equal probability of containing the generated number. A random-number generator is the basis of any stochastic simulation. For example, a random-number generator enables us to simulate the outcome of a fair coin toss without actually having to toss the coin. The outcome is heads if the random number lies between 0 and 0.5 (the probability of this is 0.5), and otherwise the outcome is tails.

Producing random numbers between 0 and 1 is not as easily accomplished as it seems, especially when they must be generated quickly, efficiently, and in massive amounts. Even for simple simulation experiments, the required amount of random numbers runs quickly into the tens of thousands or higher. Generating a very large amount of random numbers on a one-time only basis, and storing them up in a computer memory, is practically infeasible. But there is a solution to this kind of practical hurdle that is as handsome as it is practical. Instead of generating *truly* random numbers, a computer can generate so-called *pseudo-random numbers*, and it achieves this through a nonrandom procedure. This procedure is iterative by nature and is determined by a suitably chosen function f. Starting with a seed number z_0, numbers $z_1, z_2, \ldots$ are generated successively by

$$z_1 = f(z_0), z_2 = f(z_1), \ldots, z_n = f(z_{n-1}), \ldots.$$

We refer to the function f as a random-number generator and it must be chosen such that the series $\{z_i\}$ is statistically indistinguishable from a series of truly random numbers. In other words, the output of function f must be able to stand up to a great many statistical tests for "randomness."

The older random-number generators belong to the class of so-called multiplicative congruential generators. Numbers z_n are generated according to

$$z_n = az_{n-1} \text{ (modulo } m\text{),}$$

where a and m are appropriately chosen positive integers and the seed number z_0 is a positive integer. The notation $z_n = az_{n-1}$ (modulo m) means that z_n represents the whole remainder of az_{n-1} after division by m; for example, 17 (modulo 5) = 2. This scheme produces one of the numbers $0, 1, \ldots, m-1$ each time. It takes no more than m steps until some number repeats itself. Whenever z_n takes on a value it has had previously, exactly the same sequence of values is generated again, and this cycle repeats itself endlessly. When the parameters a and m are suitably chosen, the number 0 is not generated and each of the numbers $1, \ldots, m-1$ appears exactly once in each cycle. In this case the parameter m gives the length of the cycle. This explains why a very large integer should be chosen for m. The number z_n determines the random number u_n between 0 and 1 by

$$u_n = \frac{z_n}{m}.$$

We will not delve into the theory behind this. An understanding of the theory is not necessary in order to use the random-number generator on your computer. The quality of the multiplicative congruential generator is strongly dependent on the choice of parameters a and m. A much-used generator is characterized as $a = 16,807$ and $m = 2^{31} - 1$. This generator repeats itself after $m - 1$ values, which is a little over two billion numbers. In the past, this was regarded as plenty. But today, it is not enough for more advanced applications. Nevertheless, the multiplicative congruential generators are still valuable for the simpler applications, despite the fact that n-dimensional strings of the generated numbers do not pass statistical tests on uniformity in the n-dimensional cube for higher values of n.

The newest random-number generators do not use the multiplicative congruential scheme. In fact, they do not involve multiplications or divisions at all. These generators are very fast, have incredibly long periods, and provide high-quality pseudo-random numbers. In software tools such as Matlab you will find not only the so-called Christopher Columbus generator with a cycle length of about 2^{1492} (at 10 million random numbers per second, it will take more than 10^{434} years before the sequence of numbers will repeat!), but also the Mersenne twister generator with a cycle length of $2^{19937} - 1$. This generator would probably take longer to cycle than the entire future existence of humanity. It has passed numerous tests for randomness, including tests for uniformity of high-dimensional strings of numbers. The modern generators are needed in Monte Carlo simulations requiring huge masses of random numbers, as is the case in applications in physics and financial engineering.

Several Basic Methods for Simulation

The simulation of many probability problems can be performed by using a random-number generator together with one of the following basic methods:

- Simulating a random number from a finite interval.
- Simulating a random point inside a bounded region.
- Simulating a random integer from a finite range.
- Simulating a random permutation of integers.
- Simulating a random subset of integers.
- Simulating from a simple probability mass function.

In this appendix, we confine ourselves to these basic methods and give several problems to which these methods can be applied. Methods to simulate random samples from a specific probability distribution function are not discussed here. The basic ideas of generally applicable methods such as the inverse-transformation method and the acceptance–rejection method can be found in Chapters 4 and 10. Matlab and a statistical package such as R contain tailor-made routines for generating random observations from common probability distributions.

- *Simulating a random number from a finite interval.* Suppose that you want to surprise some friends by arriving at their party at a completely random moment in time between 2:30 and 5:00 p.m. How can you determine that moment? To do so, you must blindly choose a random number between 2.5 and 5. How do you choose a random number between two given numbers a

and b with $a < b$? First, you have your computer generate a random number u between 0 and 1. Then, you find a random number between a and b by

$$a + (b - a)u.$$

- *Simulating a random point inside a bounded region.* Let us first explain how to generate a random point inside a rectangle $\{(x, y) : a \leq x \leq b, c \leq y \leq d\}$. To do so, use the random-number generator to generate two random numbers u_1 and u_2 between 0 and 1. Then, $x = a + (b - a)u_1$ and $y = c + (d - c)u_2$ gives you a random point (x, y) inside the rectangle. A random point inside a bounded region in the plane can be generated by taking a rectangle that envelopes the bounded region and generating random points inside the rectangle until a point is obtained that falls inside the region. This is the *hit-or-miss method*. In particular, a random point inside the unit circle $\{(x, y) : x^2 + y^2 \leq 1\}$ is generated by generating random points (x, y) inside the square $\{(x, y) : -1 \leq x, y \leq 1\}$ until $x^2 + y^2 \leq 1$.
- *Simulating a random integer from a finite range.* Suppose that you want to designate one fair prize winner among the 725 people who correctly answered a contest question. You achieve this by numbering the correct entries as $1, 2, \ldots, 725$ and blindly choosing an integer out of the integers $1, 2, \ldots, 725$. How can you choose a random integer out of the integers $1, \ldots, M$? First, have your computer generate a random number u between 0 and 1. Then, using the notation $\lfloor f \rfloor$ for the integer that results by rounding down the number f, the integer

$$1 + \lfloor Mu \rfloor$$

can be considered as a random integer sampled from the integers $1, \ldots, M$. One application is the simulation of the outcome of a roll of a fair die ($M = 6$). For example, the random number $u = 0.428\ldots$ leads to the outcome 3 ($= 1 + \lfloor 6u \rfloor$) of the roll of the die. In general, letting u denote a random number between 0 and 1, a random integer from the integers $a, a+1, \ldots, b$ is given by

$$a + \lfloor (b - a + 1)u \rfloor.$$

It is instructive to illustrate the above procedure with the birthday problem. What is the probability that in a class of 23 children, two or more children have the same birthday? Assume that the year has 365 days and that all possible birthdays are equally likely. In each run of the simulation model, random numbers $u_1, \ldots, u_{23}$ are generated and the birthdays $k_i = 1 + \lfloor 365u_i \rfloor$ are computed. The run is said to be successful if $|k_i - k_j| = 0$ for some $i \neq j$. The sought probability is estimated by the ratio of the number of successful runs and the total number of runs. The exact value can easily be calculated by an analytical approach, see Section 1.5. The analytical approach becomes rather complicated when the problem is changed slightly and one asks for the probability that in a class of 23 children, two or more children have a birthday at most one day from each other. In contrast, the simulation model is hardly affected by the changed problem formulation and needs only a minor adjustment. The test of whether a simulation run is successful is now based on $|k_i - k_j| \leq 1$ or $|k_i - k_j| = 364$ for some $i \neq j$, where $|k_i - k_j| = 364$ corresponds to birthdays on January 1 and December 31.

- *Simulating a random permutation of integers.* Suppose that you want to assign at random a label from the labels numbered 1 to 10 to each of 10 people such that each person gets a different label. This can be done by making a random permutation of the integers $1, \ldots, 10$ and assigning the labels according to the random order in the permutation. The following algorithm generates a random permutation of $1, \ldots, n$ for a given positive integer n:

Step 1. Initialize $t := n$ and $a[j] := j$ for $j = 1, \ldots, n$.
Step 2. Generate a random number u between 0 and 1.
Step 3. Set $k := 1 + \lfloor tu \rfloor$ (random integer from the indices $1, \ldots, t$). Interchange the current values of $a[k]$ and $a[t]$.
Step 4. Let $t := t - 1$. If $t > 1$, return to step 2; otherwise, stop and the desired random permutation $(a[1], \ldots, a[n])$ is obtained.

The idea of the algorithm is first to randomly choose one of the integers $1, \ldots, n$ and to place that integer in position n. Next you randomly choose one of the remaining $n - 1$ integers and place it in position $n - 1$, etc. For the simulation of many probability problems, this is a very useful algorithm.

As an illustration, let us return to the Mississippi problem mentioned at the beginning of this appendix. What is the probability that any two adjacent letters are different in a random permutation of the 11 letters of the word "Mississippi?" A simulation model can be constructed by identifying the letter m with the number 1, the letter i with the number 2, the letter s with the number 3, and the letter p with the number 4. A random permutation of the number array $(1, 2, 3, 3, 2, 3, 3, 2, 4, 4, 2)$ can be simulated by using the algorithm above. Note that the initialization of the algorithm now becomes $a[1] = 1$, $a[2] = 2$, $\ldots$, $a[10] = 4$, $a[11] = 2$. To test whether any two adjacent numbers are different in the resulting random permutation $(a[1], a[2], \ldots, a[11])$, you check whether $a[i+1] - a[i] \neq 0$ for $i = 1, \ldots, 10$. By generating a large number of these random permutations, you obtain an estimate for the sought probability by dividing the number of random permutations in which any two adjacent numbers are different by the total number of random permutations generated. An estimate 0.058 is obtained by making 100,000 simulation runs.

- *Simulating a random subset of integers.* How does a computer generate the "Quick Pick" in the 6/45 lottery, that is, six different integers that are randomly picked from the integers $1, \ldots, 45$? More generally, how does the computer generate randomly r different integers from the integers $1, \ldots, n$? This is accomplished by following the first r iteration steps of the above algorithm for a random permutation until the positions $n, n-1, \ldots, n-r+1$ are filled. The elements $a[n], \ldots, a[n-r+1]$ in these positions constitute the desired random subset. This shows how simple it is to simulate a draw from the lottery. Using the procedure for a random draw, you can easily verify by simulation the remarkable fact that the probability of having two consecutive numbers in a draw of the 6/45 lottery is slightly more than 50%.

- *Simulating from a simple probability mass function.* Suppose that a probability experiment has n possible outcomes $O_1, \ldots, O_n$ with respective probabilities $p_1, \ldots, p_n$. How do you simulate a random outcome? A possible method is the inverse-transformation method. You first generate a random number u between 0 and 1. Then you search for the index l such that $\sum_{k=1}^{l-1} p_k < u \leq \sum_{k=1}^{l} p_k$ and take O_l as random outcome. This method is not practically useful for larger values of n. In the case that the probabilities p_j are fractions with a common denominator, there is a very simple but useful method called the *array method*. To explain this method, assume that each probability p_j can be represented by $k_j/100$ for some integer k_j with $0 < k_j < 100$ for $j = 1, \ldots, n$. You then form an array A[i], $i = 1, \ldots, 100$, by setting the first k_1 elements equal to O_1, the next k_2 elements equal to O_2, etc., and the last k_n elements equal to O_n. Then you generate a random number u between 0 and 1 and calculate the random integer $m = 1 + \lfloor 100u \rfloor$. Next take A[$m$] as the random outcome of the experiment.

Statistical Analysis of the Simulation Output

It is never possible to achieve perfect accuracy through simulation. All you can measure is how likely the estimate is to be correct. When doing a simulation, it is also important to have a probabilistic judgment about the accuracy of the point estimate. Such a probabilistic judgment is provided by the frequentist confidence interval. Simulation of a stochastic system is in fact a statistical experiment in which one or more unknown parameters of the system are estimated from a sequence of observations that are obtained from independent simulation runs of the system.

Let's first consider the situation in which we wish to estimate the unknown expected value $\mu = E(X)$ of a random variable X defined for a given stochastic system (e.g., the expected value of the random time until a complex electronic system fails for the first time). Suppose that n independent repetitions of the chance experiment are performed, where n is large (you should think in terms of hundreds of thousands of runs rather than tens of thousands of runs). The kth performance of the experiment yields the observation X_k of the random variable X. The unknown expected value $\mu = E(X)$ is estimated by the *sample mean*

$$\overline{X}(n) = \frac{1}{n} \sum_{k=1}^{n} X_k.$$

The accuracy of this estimate is expressed by the 95% confidence interval

$$\overline{X}(n) \pm 1.96 \frac{\sqrt{S^2(n)}}{\sqrt{n}},$$

where the statistic $\sqrt{S^2(n)}$ is given by

$$S^2(n) = \frac{1}{n} \sum_{k=1}^{n} \left[X_k - \overline{X}(n)\right]^2.$$

This statistic is called the *sample variance* and is an estimator for $\sigma^2(X)$. The 95% confidence interval should be interpreted as follows: with a probability of about 95%, the statistical procedure will yield an interval in which the true value of μ will be between the two stochastic endpoints of the interval. Each simulation results in another value for $\overline{X}_n$. The percentile 1.96 must be replaced by 1.645 for a 90% confidence interval and by 2.576 for a 99% confidence interval.

The estimator $S^2(n)$ of the unknown variance $\sigma^2(X)$ will not change much after some initial period of the simulation. This means that the width of the confidence interval is nearly proportional to $1/\sqrt{n}$ for n sufficiently large. In other words, the probabilistic error bound decreases as the reciprocal square root of the number of simulation runs. This conclusion leads to a practically important rule of thumb:

To reduce the width of a confidence interval by a factor of two, about four times as many observations are needed.

Appendix D: Basics of Monte Carlo Simulation

By making first a pilot simulation, you may use this rule of thumb to assess how many simulation runs must be done in order to get a simulation result with a prespecified precision.

The goal of many simulation studies is to estimate an unknown probability $P(E)$ for a given event E in a chance experiment. Suppose that n independent simulation runs of the experiment are done. Define the random variable X_i as

$$X_i = \begin{cases} 1 & \text{if event } E \text{ occurs in the } i\text{th run,} \\ 0 & \text{otherwise.} \end{cases}$$

The indicator variables $X_1, \ldots, X_n$ are independent of each other and have the same distribution. Since $P(X_i = 1) = P(E)$, the expected value of the indicator variable X_i is equal to $P(E)$ for any i. Therefore, the sample mean $\overline{X}(n) = \frac{1}{n} \sum_{i=1}^{n} X_i$ provides a point estimate for the unknown probability $P(E)$. The corresponding 95% confidence interval $\overline{X}(n) \pm 1.96 \sqrt{S^2(n)}/\sqrt{n}$ takes the insightful and simple form

$$\overline{X}(n) \pm 1.96 \frac{\sqrt{\overline{X}(n)\left[1 - \overline{X}(n)\right]}}{\sqrt{n}}.$$

The explanation is that $S^2(n) = \overline{X}(n)[1 - \overline{X}(n)]$ for $0-1$ variables X_i.

The confidence interval for an unknown probability also gives insight into the necessary simulation efforts for *very small* probabilities. How large must the number of simulation runs be before the half-width of the 95% confidence interval is no more than one-tenth (say) of the estimate for p? To answer this question, note that $\sqrt{\overline{X}(n)[1-\overline{X}(n)]} \approx \sqrt{p}$ for n very large and p close to zero. The confidence interval for p now gives that the required number n of runs must satisfy $\frac{1.96\sqrt{p}}{\sqrt{n}} \approx 0.1 \times p$, or $n \approx 1.96^2 \times 100/p$. For $p = 10^{-4}$, this means approximately 4 million simulation runs. This result emphasizes that you should be suspicious of simulation studies that consist of only a small number of simulation runs, especially if they deal with small probabilities!

To end this appendix, we give a number of probability problems that can be solved quickly by simulation but that are otherwise not easily amenable to an analytical solution. In solving these problems by simulation, you have to build a mathematical model each time. Simulation is not a simple gimmick! Matlab and Python provide nice programming environments for Monte Carlo simulation.

Simulation Problems

Problem 1. Use simulation to find a quick answer to the following questions. What is the expected value of the distance between two points that are chosen at random inside the unit square? What is the expected value of the distance between two points that are chosen at random inside the unit circle? What are the answers for the unit cube and the unit sphere?

Problem 2. Use simulation to find the expected value of the area of the triangle that is formed by three randomly chosen points inside the unit square. Do the same for three randomly chosen points inside the unit circle.

Problem 3. Use simulation to find the probability that three points chosen at random inside the unit square form an obtuse triangle. What is this probability when the random points are chosen inside the unit circle?

Problem 4. You pick random numbers from (0, 1) until their sum exceeds 1. Use simulation to verify that you need on average $e = 2.71828\ldots$ picks. Also, verify that on average $e^2 - e = 4.67077\ldots$ picks are needed to first exceed 2.

Problem 5. Use simulation to find the probability that the quadratic equation $Ax^2 + Bx + C = 0$ has real roots when A, B, and C are nonzero integers that are chosen at random from the set of integers between -100 and 100.

Problem 6. A random sequence of H's and T's is generated by tossing a fair coin n times. Use simulation to verify experimentally that the length of the longest run of consecutive heads or tails exhibits little variation and has its probability mass concentrated around the value $\log_2(n)$ when n is sufficiently large.

Problem 7. The 52 cards of a well-shuffled deck are turned over one by one. Use simulation to find the probability of a run of r red cards and the probability of a run of either r red cards or r black cards for several values of r.

Problem 8. Use simulation to find the probability that any two adjacent letters are different in a random permutation of the 10 letters of the word "statistics." What is the probability that in a thoroughly shuffled deck of 52 cards, no two adjacent cards are of the same rank?

Problem 9. In the 6/45 lottery, the six winning numbers are picked randomly from 1 to 45. Use simulation to find the probability of two or more consecutive numbers and the probability of three or more consecutive numbers among the winning numbers.

Problem 10. A Christmas party is held for 10 persons. Each person brings a gift to the party for a gift exchange. The gifts are numbered 1 to 10 and each person knows the number of their own gift. Cards with the numbers $1, \ldots, 10$ are put in a hat. The party-goers consecutively pull a card out of the hat in order to determine the present they will receive. If a person pulls out a card corresponding to the number of their own gift, then the card is put back into the hat and that person draws another card. Use simulation to find the probability that the last person gets stuck with his or her own gift.

Problem 11. In a class of n children, each child buys a small Santa Claus present for one of the other children. The gifts are numbered from 1 to n and each child knows the number of their own gift. Also, the children put their names on slips of paper. The n slips of paper and n cards with the numbers of the gifts are put into a box. Each child then draws at random a slip of paper and a card from the box. The child named on the slip gets the gift corresponding to the number on the card. Suppose that none of the children gets his or her own gift. What is the probability that there is at least one pair of children who exchange the gifts they had bought? Use simulation to verify experimentally that a good approximation to this probability is $e^{-0.5}$ for $n \geq 6$, where e is the Euler number 2.71828....

Problem 12. Somebody generates r random numbers from $(0, 1)$ and shows you those numbers, one by one and in random order. Each time he shows you a number, you must decide whether to stop with that number or go on for another one. You cannot go back to any of the previously shown numbers. Use simulation to find, for several values of r, your probability of identifying the overall largest number when you let pass the first $\sqrt{r} - 1$ numbers and then pick the first number exceeding the previous ones. What about the rule that picks a number a exceeding the previous ones only if $a^k \geq 0.5$ when there are still k numbers to come?

Problem 13. Two people are presented, separately, with the same random sequence of 100 numbers. The 100 numbers are generated independently from 1 to 10. Each person chooses at random a number from the first 10 numbers in the sequence. Starting from the position of the chosen number, the person moves forward in the sequence as many numbers as the number chosen. Next, the procedure is repeated for the new number that is hit. The procedure is stopped when a next move would go beyond the last number in the sequence. Use simulation to find the probability that the two people end up at the same number in this so-called Kruskal count.

Problem 14. You pick two different values (an ace and a seven, for example) in a standard deck of cards. Then, the 52 cards are shuffled thoroughly and dealt out face up in a straight line. Use simulation to find the probability that you will find your chosen two values as neighbors.

Problem 15. Suppose that, on two consecutive days, the same n people are seated randomly around a circular table for dinner. What is the probability that no two people sit together at both dinners? Use simulation to verify experimentally that a good approximation to this probability is $e^{-2}\left(1-\frac{4}{n}+\frac{20}{3n^3}\right)$ for $n \geq 7$.

Problem 16. You play against an opponent a game with an ordinary deck of 52 cards that consists of 26 black (*B*) cards and 26 red (*R*) cards. Each player chooses a three-card code sequence of red and black. The cards are laid on the table, face up, one at a time. Each time that one of the two chosen three-card sequences appears, the player of this card sequence gets one point. The cards that were laid down are removed and the game continues with the remaining cards. The player who collects the most points is the winner, while there is a tie if both players have the same number of points. Your opponent chooses first a three-card code sequence. He takes one of the eight possible three-card sequences *BBB*, *BBR*, *BRB*, *BRR*, *RRR*, *RRB*, *RBR*, and *RBB*. You parry the choices *BBB* and *BBR* with *RBB*, the choices *BRB* and *BRR* with *BBR*, the choices *RRR* and *RRB* with *BRR*, and the choices *RBR* and *RBB* with *RRB* (your first color is the opposite of the middle color of your opponent and your second and third colors are the first and second colors of your opponent). Use simulation to find your win probability.

Problem 17. A bowl contains 11 envelopes in the colors red and blue. You are informed that there are four envelopes of one color each containing $100 and seven empty envelopes of the other color, but you do not know the color of the envelopes containing $100. The envelopes are taken out of the bowl, one by one and in random order. Each time an envelope has been taken out, you must decide whether or not to open this envelope. Once you have opened an envelope, the process stops. Your stopping rule is to open the envelope drawn as soon as four or more envelopes of each color have been taken out of the bowl. Use simulation to find the probability of winning $100 under this stopping rule.

Answers to Odd-Numbered Problems

Chapter 1

1.1 Imagine a blue and a red die. Take as sample space the set of ordered pairs (b, r), where b and r are the numbers shown by the blue and the red die. Each of the 36 elements (b, r) is equally likely. There are $2 \times 3 \times 3 = 18$ elements for which exactly one component is odd. Thus, the probability that the sum of the two dice is odd equals $\frac{18}{36} = \frac{1}{2}$. There are $3 \times 3 = 9$ elements for which both components are odd. Thus, the probability that the product of the two numbers rolled is odd equals $\frac{9}{36} = \frac{1}{4}$. Alternatively, you can obtain the probabilities by using as sample space the set consisting of the four equiprobable elements (odd, odd), $(odd, even)$, $(even, even)$, and $(even, odd)$.

1.3 Label the 10 letters of "randomness" as 1 to 10. Take as sample space the set of all permutations of the numbers 1 to 10. All 10! outcomes are equally likely. There are $3 \times 2 \times 8!$ outcomes that begin and end with a vowel and there are $8 \times 3! \times 7!$ outcomes in which the three vowels are adjacent to each other. The probabilities are $3 \times 2 \times 8!/10! = 1/15$ and $8 \times 3! \times 7!/10! = 1/15$.

1.5 Take as sample space the set of all unordered samples of m different numbers. The sample space has $\binom{n}{m}$ equiprobable elements. There are $\binom{n-1}{m-1}$ samples that contain the largest number. Hence, the probability of getting the largest number is $\binom{n-1}{m-1}/\binom{n}{m} = \frac{m}{n}$. Alternatively, you can take as sample space the set of all $n!$ permutations of the integers 1 to n. There are $m \times (n-1)!$ permutations for which the number n is in one of the first m positions. *Note*: More generally, the probability that the largest r numbers are among the m numbers picked is given by both $\binom{n-r}{m-r}/\binom{n}{m}$ and $\binom{m}{r} r! (n-r)!/n!$.

1.7 Imagine that the balls are labeled $1, \ldots, n$. It is no restriction to assume that the two winning balls have the labels 1 and 2. Take as sample space the set of all $n!$ permutations of $1, \ldots, n$. For any k, the number of permutations having either 1 or 2 on the kth place is $(n-1)! + (n-1)!$. Thus, the probability that the kth person picks a winning ball is $\frac{(n-1)!+(n-1)!}{n!} = \frac{2}{n}$ for each k.

1.9 Take as sample space the set of all unordered samples of six different numbers from the numbers 1 to 42. The sample space has $\binom{42}{6}$ equiprobable outcomes. There are $\binom{41}{5}$ outcomes with the number 10. Thus the probability of getting the number 10 is $\binom{41}{5}/\binom{42}{6} = \frac{6}{42}$. The probability that each of the six numbers picked is 20 or more is equal to $\binom{23}{6}/\binom{42}{6} = 0.0192$. Alternatively, the probabilities can be calculated by using the sample space consisting of all ordered arrangements of the numbers 1 to 42, where the numbers in the first six positions are the lottery numbers. This leads to the calculations $6 \times 41!/42!$ and $\binom{23}{6} \times 6! \times 36!/42!$ for the sought probabilities.

1.11 Label the nine socks as $s_1, \ldots, s_9$. The probability model in which the order of selection of the socks is considered relevant has a sample space with $9 \times 8 = 72$ equiprobable outcomes (s_i, s_j). There are $4 \times 5 = 20$ outcomes for which the first sock chosen is black and the second is white, and there are $5 \times 4 = 20$ outcomes for which the first sock is white and the second is black. The sought probability is $40/72 = 5/9$. The probability model in which the order of selection of the socks is not considered relevant has a sample space with $\binom{9}{2} = 36$ equiprobable outcomes. The number of outcomes for which the socks have different colors is $\binom{5}{1} \times \binom{4}{1} = 20$, yielding again the probability $5/9$.

1.13 Take as sample space the set of all unordered pairs of two distinct cards. The sample space has $\binom{52}{2}$ equiprobable outcomes. There are $\binom{1}{1} \times \binom{51}{1} = 51$ outcomes with the ten of hearts, and $\binom{3}{1} \times \binom{12}{1} = 36$ outcomes with hearts and a ten but not the ten of hearts. The sought probability is $(51+36)/\binom{52}{2} = 0.0656$.

1.15 Take as sample space the set of all sequences $(i_1, \ldots, i_k)$, where i_k is the number shown on the kth roll of the die. Each element of the sample space is equally likely. The explanation is that there is a one-to-one correspondence between the elements $(i_1, \ldots, i_k)$ with $\sum_{k=1}^{10} i_k = s$ and the elements $(7 - i_1, \ldots, 7 - i_k)$ with $\sum_{k=1}^{10}(7 - i_k) = 70 - s$.

1.17 Take as sample space the set of all possible samples of three residents. This leads to the value $\binom{4}{1}\binom{4}{1}\binom{4}{1}/\binom{12}{3} = \frac{16}{55}$ for the sought probability.

1.19 Take as sample space the set of the 9! possible orderings of the nine books. The subjects mathematics, physics, and chemistry can be ordered in $3 \times 2 \times 1 = 6$ ways and so the number of favorable orderings is $6 \times 4! \times 3! \times 2!$. The sought probability is $(6 \times 4! \times 3! \times 2!)/9! = 1/210$.

1.21 Label the 11 letters of "Mississippi" as $1, 2, \ldots, 11$ and take as sample space the set of the 11^{11} possible ordered sequences of 11 numbers from $1, \ldots, 11$. There are four positions for a number representing the letter i, four for a number representing the letter s, two for a number representing the letter p, and one for a number representing the letter m. These positions can be chosen in $\binom{11}{4} \times \binom{7}{4} \times \binom{3}{2}$ ways. Therefore, the number of outcomes in which all letters of the word Mississippi are represented is $\binom{11}{4} \times \binom{7}{4} \times \binom{3}{2} \times 4^4 \times 4^4 \times 2^2$. Dividing this number by 11^{11} gives the value 0.0318 for the sought probability.

1.23 Take as sample space the set of all possible combinations of two apartments from the 56 apartments. These two apartments represent the vacant apartments. The sample space has $\binom{56}{2}$ equiprobable elements. The number of elements with no vacant apartment on the top floor is $\binom{48}{2}$. Thus, the sought probability is $[\binom{56}{2} - \binom{48}{2}]/\binom{56}{2} = 0.2675$.

1.25 Take as sample space the set of all ordered pairs (i, j), where i is the first number picked and j is the second number picked. There are n^2 equiprobable outcomes. For $r \leq n+1$, the $r - 1$ outcomes $(1, r-1)$, $(2, r-2), \ldots, (r-1, 1)$ are the only outcomes (i, j) for which $i + j = r$. Thus, the probability that the sum of the two numbers picked is r is $\frac{r-1}{n^2}$ for $2 \leq r \leq n+1$. As a special case, the probability of getting a sum s when rolling two dice is $\frac{s-1}{36}$ for $2 \leq s \leq 7$. By a symmetry argument, the probability of getting a sum s is the same as the probability of getting a sum $14 - s$ for $7 \leq s \leq 12$ (opposite faces of a die always total 7). Thus, the probability of rolling a sum s has the value $\frac{14-s-1}{36}$ for $7 \leq s \leq 12$.

1.27 Take as sample space the square $\{(x, y) : 0 \leq x, y \leq a\}$. The outcome (x, y) refers to the position of the middle point of the coin. The sought probability is the probability that a randomly chosen point inside the square falls within the subset $\{(x, y) : \frac{d}{2} \leq x, y \leq a - \frac{d}{2}\}$ and is equal to $(a - d)^2/a^2$.

1.29 Take as sample space the rectangle $R = \{(x, y) : 0 \leq x \leq 1, \frac{1}{2} \leq y \leq 1\}$, where the outcome (x, y) means that you arrive $60x$ minutes past 7 a.m. and your friend arrives $60y$ minutes past 7 a.m. The probability assigned to each subset of the sample space is the area of the subset divided by the area of the rectangle R. The sought probability is $P(A)$, where the set A is the union of the three disjoint subsets $\{(x, y) : \frac{1}{2} < x, y < \frac{1}{2} + \frac{1}{12}\}$, $\{(x, y) : \frac{1}{2} + \frac{1}{12} < x, y < \frac{3}{4}\}$ and $\{(x, y) : \frac{3}{4} < x, y < 1\}$. This gives $P(A) = \left(\frac{1}{12} \times \frac{1}{12} + \frac{1}{6} \times \frac{1}{6} + \frac{1}{4} \times \frac{1}{4}\right)/\frac{1}{2} = \frac{7}{36}$.

1.31 For $q = 1$, take the square $\{(x,y) : -1 < x, y < 1\}$ as sample space. The sought probability is the probability that a point (x, y) chosen at random in the square satisfies $y \le \frac{1}{4}x^2$ and is equal to $\frac{1}{4}(2+\int_{-1}^{1} \frac{1}{4}x^2 dx) = 0.5417$. For the general case, the sought probability is $\frac{1}{4q^2}(2q^2 + \int_{-q}^{q} \frac{1}{4}x^2 dx) = \frac{1}{2} + \frac{q}{24}$ for $0 < q < 4$ and $\frac{1}{4q^2}(2q^2 + \int_{-2\sqrt{q}}^{2\sqrt{q}} \frac{1}{4}x^2 dx + 2(q - 2\sqrt{q})q) = 1 - \frac{2}{3\sqrt{q}}$ for $q \ge 4$.

1.33 Take as sample space the set $S = \{(y, \theta) : 0 \le y \le \frac{1}{2}D, 0 \le \theta \le \frac{\pi}{2}\}$, where y is the distance from the midpoint of the diagonal of the rectangular card to the closest line on the floor and the angle θ is as described in the figure. It is no restriction to assume that $b \ge a$. Using the figure, it is seen that the card will intersect one of the lines on the floor if and only if the distance y is less than y_θ, where y_θ is determined by $\sin(\alpha + \theta) = y_\theta / (\frac{1}{2}\sqrt{a^2 + b^2})$. Since $\sin(\alpha + \theta) = \sin(\alpha)\cos(\theta) + \cos(\alpha)\sin(\theta)$ with $\sin(\alpha) = \frac{a}{\sqrt{a^2+b^2}}$ and $\cos(\alpha) = \frac{b}{\sqrt{a^2+b^2}}$, we get $y_\theta = \frac{1}{2}(a\cos(\theta) + b\sin(\theta))$. The sought probability is the area under the curve $y = \frac{1}{2}(a\cos(\theta) + b\sin(\theta))$ divided by the area of the set S and is equal to $\frac{1}{(1/4)\pi D} \int_0^{\pi/2} \frac{1}{2}(a\cos(\theta) + b\sin(\theta)) d\theta = \frac{2(a+b)}{\pi D}$.

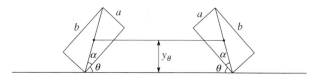

1.35 Take as sample space the unit square $\{(x, y) : 0 \le x, y \le 1\}$. The side lengths $v = x$, $w = y \times (1-v)$, and $1 - v - w$ should satisfy the conditions $v + w > 1 - v - w$, $v + 1 - v - w > w$, and $w + 1 - v - w > v$. These conditions can be translated into $y > \frac{1-2x}{2-2x}$, $y < \frac{1}{2-2x}$, and $x < \frac{1}{2}$. The first probability is given by the area of the shaded region in the first part of the figure and is equal to $\int_0^{0.5} \frac{1}{2-2x} dx - \int_0^{0.5} \frac{1-2x}{2-2x} dx = \ln(2) - 0.5$. To find the second probability, let v be the first random breakpoint chosen on the stick and w be the other breakpoint. The point (v, w) can be represented

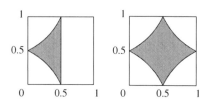

by $v = x$ and $w = y \times (1 - v)$ if $v < \frac{1}{2}$ and by $v = x$ and $w = y \times v$ if $v > \frac{1}{2}$, where (x, y) is a randomly chosen point in the unit square. The second probability is the area of the shaded region in the second part of the figure, and is equal to $2[\ln(2) - 0.5]$.

1.37 The unique chord having the randomly chosen point P as its midpoint is the chord that is perpendicular to the line connecting the point P to the center O of the circle, see the figure. A little geometry tells us that the

chord is longer than the side of the equilateral triangle if and only if the point P falls inside the shaded inner circle. Thus, the sought probability is $\frac{\pi(r/2)^2}{\pi r^2} = \frac{1}{4}$.

1.39 Take as sample the set of all four-tuples $(\delta_1, \delta_2, \delta_3, \delta_4)$, where $\delta_i = 0$ if component i has failed and $\delta_i = 1$ otherwise. The probability $r_1 r_2 r_3 r_4$ is assigned to $(\delta_1, \delta_2, \delta_3, \delta_4)$, where $r_i = f_i$ if $\delta_i = 0$ and $r_i = 1 - f_i$ if $\delta_i = 1$. Let A_0 be the event that the system fails, A_1 be the event that none of the four components fails, and A_i be the event that only component i fails for $i = 2, 3$, and 4. Then $P(A_1) = (1-f_1)(1-f_2)(1-f_3)(1-f_4)$ and $P(A_2) = (1-f_1)f_2(1-f_3)(1-f_4)$, $P(A_3) = (1-f_1)(1-f_2)f_3(1-f_4)$, and $P(A_4) = (1-f_1)(1-f_2)(1-f_3)f_4$. The events A_k are mutually exclusive and their union is the sample space, and so $\sum_{i=0}^{4} P(A_i) = 1$. Hence, the sought probability $P(A_0)$ is $1 - \sum_{i=1}^{4} P(A_i)$.

1.41 Take as sample space the set $\{(1, s), (2, s), \ldots\} \cup \{(1, e), (2, e), \ldots\}$. Assign to the outcomes (i, s) and (i, e) the probabilities $(1 - a_7 - a_8)^{i-1} a_7$ and $(1 - a_7 - a_8)^{i-1} a_8$, where $a_7 = \frac{6}{36}$ and $a_8 = \frac{5}{36}$. The probability of getting a total of 8 before a total of 7 is $\sum_{i=1}^{\infty} (1 - a_7 - a_8)^{i-1} a_8 = a_8/(a_7 + a_8) = \frac{5}{11}$.

1.43 Take as sample space the set $\{(s_1, s_2) : 2 \le s_1, s_2 \le 12\}$, where s_1 and s_2 are the two sums rolled. The probability $p(s_1, s_2) = p(s_1) \times p(s_2)$ is assigned to the outcome (s_1, s_2), where $p(s)$ is the probability of getting the sum s in a roll of two dice. The probabilities $p(s)$ are $p(2) = p(12) = \frac{1}{36}$, $p(3) = p(11) = \frac{2}{36}$, $p(4) = p(10) = \frac{3}{36}$, $p(5) = p(9) = \frac{4}{36}$, $p(6) = p(8) = \frac{5}{36}$, and $p(7) = \frac{6}{36}$. The sought probability is $\sum_{s_1 \ne s_2} p(s_1, s_2) = 1 - \sum_{s=2}^{12} [p(s)]^2 = \frac{575}{648}$.

1.45 The sought probability is at least as large as $1 - P(\bigcap_{n=1}^{\infty} B_n)$. For any $n \geq 1$, $P(B_n) = (1 - (\frac{1}{2})^r)^n$. By the continuity property of probability, $P(\bigcap_{n=1}^{\infty} B_n) = \lim_{n \to \infty} P(B_n)$ and so $1 - P(\bigcap_{n=1}^{\infty} B_n) = 1$.

1.47 Let A be the event that a second-hand car is bought and B be the event that a Japanese car is bought. Noting that $P(A \cup B) = 1 - 0.55$, it follows from $P(A \cup B) = P(A) + P(B) - P(AB)$ that $P(AB) = 0.25 + 0.30 - 0.45 = 0.10$.

1.49 Let A be the event that the truck is used on a given day and B be the event that the van is used on a given day. Then $P(A) = 0.75$, $P(AB) = 0.30$, and $P(A^c B^c) = 0.10$. By De Morgan's first law, $P(A \cup B) = 1 - P(A^c B^c)$. By $P(A \cup B) = P(A) + P(B) - P(AB)$, the probability that the van is used on a given day is $P(B) = 0.90 - 0.75 + 0.30 = 0.45$. Since $P(A^c B) + P(AB) = P(B)$, the probability that only the van is used is $P(A^c B) = 0.45 - 0.30 = 0.15$.

1.51 The probability that exactly one of the events A and B will occur is given by $P((A \cap B^c) \cup (B \cap A^c)) = P(A \cap B^c) + P(B \cap A^c)$. Next note that $P(A \cap B^c) = P(A) - P(A \cap B)$ and $P(B \cap A^c) = P(B) - P(A \cap B)$, yielding the desired result.

1.53 In this "birthday" problem, the probability is $1 - (250 \times 249 \times \cdots \times 221)/250^{30} = 0.8368$.

1.55 This problem is a variant of the birthday problem. One can choose two distinct numbers from the numbers $1, 2, \ldots, 25$ in $\binom{25}{2} = 300$ ways. The desired probability is given by $1 - (300 \times 299 \times \cdots \times 276)/300^{25} = 0.6424$.

1.57 The translation step to the birthday problem is to imagine that each of the $n = 500$ Oldsmobile cars gets assigned a "birthday" chosen at random from $m = 2{,}400{,}000$ possible "birthdays." Using the approximate formula in Problem 1.54, the probability that at least one subscriber gets two or more cars can be calculated as $1 - e^{-\frac{1}{2}n(n-1)/m} = 0.051$.

1.59 (a) The sought probability is $1 - (10{,}000 \times 9{,}999 \times \cdots \times 9{,}990)/10{,}000^{11} = 0.005487$. (b) The sought probability is $1 - (1 - 0.005487)^{300} = 0.8081$.

1.61 Let A_1 be the event that the card of your first favorite team is not obtained and A_2 be the event that the card of your second favorite team is not obtained. The sought probability is $1 - P(A_1 \cup A_2)$. Since $P(A_1 \cup A_2) = \frac{9^5}{10^5} + \frac{9^5}{10^5} - \frac{8^5}{10^5}$, the sought probability is $1 - 0.8533 = 0.1467$.

1.63 Let A (B) be the event that the number 5 (6) is rolled at least once. By the relation $P(A \cap B) = P(A) + P(B) - P(A \cup B) = 1 - P(A^c) + 1 - P(B^c) - (1 - P(A^c \cap B^c))$, we get $P(A \cap B) = 2(1 - \frac{5^6}{6^6}) - (1 - \frac{4^6}{6^6}) = 0.4619$.

1.65 The sought probability is given by $P(\bigcup_{i=1}^{253} A_i)$. For any i, the probability $P(A_i)$ is equal to $(365 \times 1 \times 364 \times 363 \times \cdots \times 344)/365^{23} = 0.0014365$. The events A_i are mutually exclusive and so $P(\bigcup_{i=1}^{253} A_i) = \sum_{i=1}^{253} P(A_i)$. Therefore, the probability that in a class of 23 children exactly two children have the same birthday is $253 \times 0.0014365 = 0.3634$.

1.67 Take as sample space the set of all ordered sequences of the possible destinations of the five winners. The probability that at least one of the destinations A and B will be chosen is $1 - \frac{1}{3^5}$. Let A_k be the event that none of the winners chooses the kth destination. The probability $P(A_1 \cup A_2 \cup A_3)$ is $\frac{2^5}{3^5} + \frac{2^5}{3^5} + \frac{2^5}{3^5} - \frac{1}{3^5} - \frac{1}{3^5} - \frac{1}{3^5} = \frac{31}{81}$.

1.69 Let A be the event that there is no ace among the five cards and B be the event that there is neither a king nor a queen among the five cards. The sought probability is $1 - P(A \cup B)$ and is given by $1 - [\binom{48}{5} + \binom{44}{5} - \binom{40}{5}]/\binom{52}{5} = 0.1765$.

1.71 Take as sample space the set of all 40! possible orderings of the aces and the cards 2 through 10 (the other cards are not relevant). The first probability is $\frac{4 \times 39!}{40!} = \frac{1}{10}$ and the second probability is $\frac{4 \times 3 \times 38!}{40!} = \frac{1}{130}$.

1.73 (a) The number of permutations in which the particular number r belongs to a cycle of length k is $\binom{n-1}{k-1}(k-1)!\,(n-k)!$. The sought probability is $\binom{n-1}{k-1}(k-1)!\,(n-k)!/n! = \frac{1}{n}$. (b) For fixed r, s with $r \neq s$, let A_k be the event that r and s belong to the same cycle with length k. The sought probability is $P(A_2 \cup \cdots \cup A_n) = \frac{1}{n!}\sum_{k=2}^{n}\binom{n-2}{k-2}(k-1)!\,(n-k)!$ and this expression can be simplified to $\frac{1}{2}$.

1.75 In line with the strategy outlined in Problem 1.74, the person with the task of finding the car first opens door 1. This person next opens door 2 if the car key is behind door 1 and next opens door 3 if the goat is behind door 1. The person with the task of finding the car key first opens door 2. This person next opens door 1 if the car is behind door 2 and next opens door 3 if the goat is behind door 2. Under this strategy the probability of winning the car is $\frac{2}{3}$, as is readily verified by considering the six possible configurations of the car, the key, and the goat.

1.77 The probability of correctly identifying five or more wines can be calculated as $\sum_{k=5}^{10} e^{-1} 1^k/k! = 0.00366$ when the person is not a connoisseur and just guesses the names of the wines. This small probability is a strong indication that the person is a connoisseur.

1.79 Let A_i be the event that the ith person gets both the correct coat and the correct umbrella. The sought probability is given by $P(A_1 \cup A_2 \cup A_3 \cup A_4 \cup A_5) = \sum_{k=1}^{5}(-1)^{k+1}\binom{5}{k}\frac{(5-k)!(5-k)!}{5!5!}$ and is equal to 0.1775.

1.81 Let A_k be the event that the three choices of five distinct numbers have number k in common. The sought probability is given by $P(\bigcup_{k=1}^{25} A_k)$ and is calculated as $\sum_{k=1}^{5}(-1)^{k+1}\binom{25}{k}\binom{25-k}{5-k}^3/\binom{25}{5}^3 = 0.1891$.

1.83 Let A_i be the event that the hand does not contain any card of suit i. The probability $P(\bigcup_{i=1}^{4} A_i)$ is $\sum_{k=1}^{3}(-1)^{k+1}\binom{4}{k}\binom{52-13k}{13}/\binom{52}{13} = 0.0511$ for the bridge hand and is $\sum_{k=1}^{3}(-1)^{k+1}\binom{4}{k}\binom{52-13k}{5}/\binom{52}{5} = 0.7363$ for the poker hand.

1.85 The possible paths from node n_1 to node n_4 are the four paths (l_1, l_5), (l_2, l_6), (l_1, l_3, l_6), and (l_2, l_4, l_5). Let A_i be the event that the ith path is functioning. The probability $P(A_1 \cup A_2 \cup A_3 \cup A_4)$ can be evaluated as $p_1p_5 + p_2p_6 + p_1p_3p_6 + p_2p_4p_5 - p_1p_2p_5p_6 - p_1p_3p_5p_6 - p_1p_2p_4p_5 - p_1p_2p_3p_6 - p_2p_4p_5p_6 + p_1p_2p_3p_5p_6 + p_1p_2p_4p_5p_6$. This probability simplifies to $2p^2(1 + p + p^3) - 5p^4$ when $p_i = p$ for all i.

1.87 Let A_i be the event that it takes more than r purchases to get the ith coupon. The sought probability is given by $P(\bigcup_{i=1}^{n} A_i)$ and this probability is equal to $\sum_{k=1}^{n}(-1)^{k+1}\binom{n}{k}\frac{(n-k)^r}{n^r}$. Using this expression for $n = 38, 365$, and 100, we find that the required number of rolls is $r = 13$, the required number of people is $r = 2,287$, and the required number of balls is $r = 497$.

1.89 Let A_i be the event that the ith person does not share his or her birthday with someone else. The sought probability is given by $1 - P(A_1 \cup \cdots \cup A_n)$ and is equal to $1 - \frac{1}{365^n}\sum_{k=1}^{\min(n,365)}(-1)^{k+1}\binom{n}{k}(365) \times \cdots \times (365-k+1) \times (365-k)^{n-k}$. This probability is 0.5008 for $n = 3,064$.

1.91 There are $\binom{10}{2} = 45$ possible combinations of two persons. Let A_k be the event that the two persons from the kth combination have chosen each other's name. The probability $P(A_1 \cup \cdots \cup A_{45})$ can be calculated as $\binom{10}{2}(\frac{1}{9})^2 - \frac{1}{2!}\binom{10}{2}\binom{8}{2}(\frac{1}{9})^4 + \cdots - \frac{1}{5!}\binom{10}{2}\binom{8}{2}\binom{6}{2}\binom{4}{2}\binom{2}{2}(\frac{1}{9})^{10} = 0.4654$, using the fact that $P(A_iA_j) = 0$ for $i \neq j$ when A_i and A_j have a person in common.

1.93 Let A_i be the event that no ball of color i is picked. The sought probability is $1 - P(A_1 \cup A_2 \cup A_3) = 1 - [\sum_{i=1}^{3} P(A_i) - \sum_{i=1}^{3}\sum_{j=1}^{2} P(A_iA_j)]$. If the balls are picked with replacement, then $P(A_1) = 12^5/15^5$, $P(A_2) = 10^5/15^5$, $P(A_3) = 8^5/15^5$, $P(A_1A_2) = 7^5/15^5$, $P(A_1A_3) = 5^5/15^5$, and $P(A_2A_3) = 0$. The sought probability is 0.5237. If the balls are picked without replacement, then $P(A_1) = \sum_{k=1}^{4} \binom{5}{k}\binom{7}{5-k}/\binom{15}{5}$ and $P(A_1A_2) = \binom{7}{5}/\binom{15}{5}$. Similarly for the other terms. The sought probability is 0.6557.

1.95 Let A_i be the event that the four cards of rank i are matched. The sought probability $P(\bigcup_{i=1}^{13} A_i)$ is $\sum_{k=1}^{13}(-1)^{k+1}\binom{13}{k}(4!)^k(52-4k)!/52! = 4.80 \times 10^{-5}$.

Chapter 2

2.1 Take as sample space the set of all ordered pairs (i,j), where the outcome (i,j) represents the two numbers shown on the dice. Each of the 36 possible outcomes is equally likely. Let A be the event that the sum of the two dice is 8 and B be the event that the two numbers shown on the dice are different. There are 30 outcomes (i,j) with $i \neq j$ and four of those outcomes have $i+j = 8$. Therefore, $P(AB) = \frac{4}{36}$ and $P(B) = \frac{30}{36}$, and so $P(A \mid B) = \frac{4/36}{30/36} = \frac{2}{15}$.

2.3 Take as sample space the set of ordered pairs (G,G), (G,F), (F,G), and (F,F), where G stands for a "correct prediction" and F stands for a "false prediction." The probabilities $0.9 \times 0.8 = 0.72$, $0.9 \times 0.2 = 0.18$, $0.1 \times 0.8 = 0.08$, and $0.1 \times 0.2 = 0.02$ are assigned to these elements. Let $A = \{(G,F)\}$ and $B = \{(G,F), (F,G)\}$. The sought probability is $P(A \mid B) = \frac{0.18}{0.26} = \frac{9}{13}$.

2.5 Let A be the event that a randomly chosen household has a cat and B be the event that the household has a dog. Then $P(A) = 0.3$, $P(B) = 0.25$, and $P(B \mid A) = 0.2$. The sought probability $P(A \mid B)$ satisfies $P(A \mid B) = P(AB)/P(B) = P(A)P(B \mid A)/P(B)$ and is $0.3 \times 0.2/0.25 = 0.24$.

2.7 Label the two red balls as R_1 and R_2, the blue ball as B, and the green ball as G. Take as unordered sample space the set consisting of the six equally likely combinations $\{R_1, R_2\}$, $\{R_1, B\}$, $\{R_2, B\}$, $\{R_1, G\}$, $\{R_2, G\}$, and $\{B, G\}$ of two balls. Let C be the event that two non-red balls are grabbed, D be the event that at least one non-red ball is grabbed, and E be the event that the green ball is grabbed. Then $P(CD) = \frac{1}{6}$, $P(D) = \frac{5}{6}$, $P(CE) = \frac{1}{6}$, and $P(E) = \frac{3}{6}$. Thus, $P(C \mid D) = \frac{1}{5}$ and $P(C \mid E) = \frac{1}{3}$.

2.9 Take as unordered sample space the set of all possible combinations of 13 distinct cards. Let A be the event that the hand contains exactly one ace, B be the event that the hand contains at least one ace, and C be the event that the hand contains the ace of hearts. Then $P(A \mid B) = [\binom{4}{1}\binom{48}{12}/\binom{52}{13}]/[1 - \binom{48}{13}/\binom{52}{13}] = 0.6304$ and $P(A \mid C) = \binom{1}{1}\binom{48}{12}/\binom{1}{1}\binom{51}{12} = 0.4388$. The desired probabilities are 0.3696 and 0.5612.

2.11 Let A be the event that each number rolled is higher than all those that were rolled earlier and B be the event that the three different numbers are rolled. Then $P(A) = P(AB)$ and so $P(A) = P(B)P(A \mid B)$. We have $P(B) = \frac{6 \times 5 \times 4}{6^3} = \frac{5}{9}$ and $P(A \mid B) = \frac{1}{3!}$. Thus $P(A) = \frac{20}{36} \times \frac{1}{3!} = \frac{5}{54}$.

2.13 Let A be the event that a randomly chosen student takes Spanish and B be the event that the student takes French. Then $P(A) = 0.35$, $P(B) = 0.15$, and $P(A \cup B) = 0.40$. Thus, $P(AB) = 0.35 + 0.15 - 0.40 = 0.10$ and so $P(B \mid A) = \frac{0.10}{0.35} = \frac{2}{7}$.

2.15 Let A be the event that a randomly chosen voter is a Democrat, B be the event that the voter is a Republican, and C be the event that the voter is in favor of the election issue. **(a)** Since $P(A) = 0.45$, $P(B) = 0.55$, $P(C \mid A) = 0.7$, and $P(C \mid B) = 0.5$, we have $P(AC) = P(A)P(C \mid A) = 0.45 \times 0.7 = 0.315$ and $P(BC) = P(B)P(C \mid B) = 0.55 \times 0.5 = 0.275$. **(b)** Since $P(C) = P(AC) + P(BC)$, we get $P(C) = 0.59$. **(c)** $P(A \mid C) = \frac{0.315}{0.59} = 0.5339$.

2.17 Let A_1 (A_2) be the event that the first (second) card picked belongs to one of the three business partners. Then $P(A_1 A_2) = \frac{3}{5} \times \frac{2}{4} = \frac{3}{10}$.

2.19 Let A be the event that one or more sixes are rolled and B the event that no one is rolled. Then, by $P(AB) = P(B)P(A \mid B)$, we get that the sought probability is equal to $\left(\frac{5}{6}\right)^6 \left(1 - \left(\frac{4}{5}\right)^6\right) = 0.2471$.

2.21 Let A_i be the event that you get a white ball on the ith pick. The probability that you will need three picks is $P(A_1 A_2 A_3) = \frac{3}{5} \times \frac{2}{5} \times \frac{1}{5} = \frac{6}{125}$. Five picks require that one black ball is taken in the first three picks. By the chain rule, the probability that five picks will be needed is $\frac{2}{5} \times \frac{4}{5} \times \frac{3}{5} \times \frac{2}{5} \times \frac{1}{5} + \frac{3}{5} \times \frac{3}{5} \times \frac{3}{5} \times \frac{2}{5} \times \frac{1}{5} + \frac{3}{5} \times \frac{2}{5} \times \frac{4}{5} \times \frac{2}{5} \times \frac{1}{5} = \frac{6}{125}$.

2.23 Let A_i be the event that the ith person in line is the first person matching a birthday with one of the persons in front of him. Then $P(A_2) = \frac{1}{365}$ and $P(A_i) = \frac{364}{365} \times \cdots \times \frac{364-i+3}{365} \times \frac{i-1}{365}$ for $i \geq 3$. The probability $P(A_i)$ is maximal for $i = 20$ and then has the value 0.0323.

2.25 Let A_i be the event that the ith leaving person does not have to squeeze past a still seated person. The sought probability is the same as $P(A_1 A_2 A_3 A_4 A_5) = \frac{2}{7} \times \frac{2}{6} \times \frac{2}{5} \times \frac{2}{4} \times \frac{2}{3} = 0.0127$.

2.27 Under the condition that the events $A_1, \ldots, A_{i-1}$ have occurred, the ith couple can match the birthdays of at most one of the couples 1 to $i - 1$. Thus $P(A_i^c \mid A_1 \cdots A_{i-1}) = \frac{i-1}{365^2}$ and so $P(A_i \mid A_1 \cdots A_{i-1}) = 1 - \frac{i-1}{365^2}$. By the chain rule, the sought probability $1 - P(A_2 A_3 \cdots A_n)$ is $1 - \prod_{i=2}^{n}\left(1 - \frac{i-1}{365^2}\right)$.

2.29 Using the chain rule for conditional probabilities, the sought probability is $\frac{r}{r+b}$ for $k = 1$, $\frac{b}{r+b} \times \frac{r}{r+b-1}$ for $k = 2$, and $\frac{b}{r+b} \times \frac{b-1}{r+b-1} \times \cdots \times \frac{b-(k-2)}{r+b-(k-2)} \times \frac{r}{r+b-(k-1)}$ for $3 \leq k \leq b+1$. This expression can be simplified to $\binom{r+b-k}{r-1}/\binom{r+b}{r}$. The answer to the last question is $\frac{b}{r+b}$, as can be seen directly by a symmetry argument: the probability that the last ball picked is blue is the same as the probability that the first ball picked is blue.

2.31 Since $P(A) = \frac{18}{36}$, $P(B) = \frac{18}{36}$, and $P(AB) = \frac{9}{36}$, we get $P(AB) = P(A)P(B)$. This shows that the events A and B are independent.

2.33 Since A is the union of the disjoint sets AB and AB^c, we have $P(A) = P(AB) + P(AB^c)$. Thus $P(AB^c) = P(A) - P(A)P(B) = P(A)[1 - P(B)]$ and so $P(AB^c) = P(A)P(B^c)$, showing that A and B^c are independent events. Applying this result with A replaced by B^c and B by A, we next get that B^c and A^c are independent events.

2.35 The result follows directly from $P(A_1 \cup \cdots \cup A_n) = 1 - P(A_1^c \cdots A_n^c)$ and the independence of the A_i^c, using $P(A_1^c \cdots A_n^c) = P(A_1^c) \cdots P(A_n^c)$ and $P(A_i^c) = 1 - P(A_i)$.

2.37 The set A can be represented as $\bigcap_{n=1}^{\infty} \bigcup_{k=n}^{\infty} A_k$. Since the sequence of sets $\bigcup_{k=n}^{\infty} A_k$ is decreasing, $P(A) = \lim_{n \to \infty} P(\bigcup_{k=n}^{\infty} A_k)$, by the continuity property of probability. Next use the fact that $P(\bigcup_{k=n}^{\infty} A_k) = 1 - P(\bigcap_{k=n}^{\infty} A_k^c)$. Using the independence of the events A_k^c and the continuity property of probability measure, it is readily verified that $P(\bigcap_{k=n}^{\infty} A_k^c) = \prod_{k=n}^{\infty} P(A_k^c)$ for any $n \geq 1$. By $P(A_k^c) = 1 - P(A_k)$ and $1 - x \leq e^{-x}$, we get $P(\bigcap_{k=n}^{\infty} A_k^c) \leq \prod_{k=n}^{\infty} e^{-P(A_k)}$. Noting that $\sum_{k=n}^{\infty} P(A_n) = \infty$ for any $n \geq 1$, it follows that $\prod_{k=n}^{\infty} e^{-P(A_k)} = e^{-\sum_{k=n}^{\infty} P(A_k)} = 0$ for any $n \geq 1$. This shows that $P(\bigcup_{k=n}^{\infty} A_k) = 1$ for all $n \geq 1$ and so $P(A) = 1$.

2.39 Let A be the event that HAPPY HOUR appears again, B_1 be the event that either the two letters H or the two letters P have been removed, and B_2 be the event that two different letters have been removed. Then $P(B_1) = \frac{2}{9} \times \frac{1}{8} + \frac{2}{9} \times \frac{1}{8}$ and $P(B_2) = 1 - P(B_1)$. Obviously, $P(A \mid B_1) = 1$ and $P(A \mid B_2) = \frac{1}{2}$. Therefore, $P(A) = 1 \times \frac{1}{18} + \frac{1}{2} \times \frac{17}{18} = \frac{19}{36}$, by the law of conditional probability.

2.41 Let A be the event that you ever win the jackpot when buying a single ticket only once. Also, let B be the event that you match the six numbers drawn and C be the event that you match exactly two of these numbers. It follows from $P(A) = P(A \mid B)P(B) + P(A \mid C)P(C)$ that $P(A) = P(B) + P(A)P(C)$. Since $P(B) = 1/\binom{59}{6}$ and $P(C) = \binom{6}{2}\binom{53}{4}/\binom{59}{6}$, we get $P(A) = 1/40{,}665{,}099$.

2.43 Let A be the event of reaching your goal, B_1 be the event of winning the first bet, and B_2 be the event of losing the first bet. Then, by $P(A) = P(A \mid B_1)P(B_1) + P(A \mid B_2)P(B_2)$, we get $P(A) = 1 \times \frac{12}{37} + \frac{9}{37} \times \frac{25}{37}$. Thus, the probability of reaching your goal is 0.4887. This probability is slightly more than the probability 0.4865 of reaching your goal when you use bold play and stake the whole $10,000 on an 18-numbers bet.

2.45 Apply the gambler's ruin formula with $p = 0.3$, $a = 3$, and $b = 7$. The sought probability is 0.0025.

2.47 It is no restriction to assume that the starting point is 1 and the first transition is from point 1 to point 2 (otherwise, renumber the points).

Some reflection shows that the probability of visiting all points before returning to the starting point is nothing other than the probability $\frac{1}{1+10} - \frac{1}{11}$ from the gambler's ruin model.

2.49 Let A be the event that John needs more tosses than Pete and B_j be the event that Pete needs j tosses. Then $P(B_j) = \binom{j-1}{2}(\frac{1}{2})^j$ and $P(A \mid B_j) = \binom{j}{0}(\frac{1}{2})^j + \binom{j}{1}(\frac{1}{2})^j$. Hence, $P(A) = \sum_{j=3}^{\infty} P(A \mid B_j) P(B_j) = 0.1852$.

2.51 Let state i mean that player A's bankroll is i. Also, let E be the event of reaching state k without having reached state $a+b$ when starting in state a and F be the event of reaching state $a+b$ without having reached state $k-1$ when starting in state k. Then, the unconditional probability of player A winning and having k as the lowest value of its bankroll during the game is given by $P(EF) = P(E)P(F \mid E)$. Using the gambler's ruin formula, $P(E) = \frac{b}{a+b-k}$ and $P(F \mid E) = \frac{1}{a+b-k+1}$. Thus, the sought conditional probability is equal to $\frac{b(a+b)}{a(a+b-k)(a+b-k+1)}$ for $k = 1, \ldots, a$. This probability has the values 0.1563, 0.2009, 0.2679, and 0.3750 for $k = 1, 2, 3,$ and 4 when $a = 4$ and $b = 5$.

2.53 Let $p_n(i)$ be the probability of reaching his home no later than midnight without having reached first the police station, given that he is i steps away from his home and he still has time to make n steps before it is midnight. The sought probability is $p_{180}(10)$. By the law of conditional probability, the $p_n(i)$ satisfy the recursion $p_n(i) = \frac{1}{2}p_{n-1}(i+1) + \frac{1}{2}p_{n-1}(i-1)$. The boundary conditions are $p_k(30) = 0$ and $p_k(0) = 1$ for $k \geq 0$, and $p_0(i) = 0$ for $i \geq 1$. Applying the recursion, we find $p_{180}(10) = 0.4572$. In the same way, the value 0.1341 can be calculated for the probability of reaching the police station before midnight. *Note*: As a sanity check, we verified that $p_n(10)$ tends to $\frac{2}{3}$ as n gets large, in agreement with the gambler's ruin formula. The probability $p_n(10)$ has the values 0.5905, 0.6659, and 0.6665 for $n = 360, 1{,}200,$ and 1,440.

2.55 Let A be the event that you choose the bag with one red ball and B be the event that you choose the other bag. Also, let E be the event that the first ball picked is red. The probability that the second ball picked is red is $\frac{1}{4}P(A \mid E) + \frac{3}{4}P(B \mid E)$, by the law of conditional probability. We have $P(A \mid E) = P(AE)/P(E) = P(A)P(E \mid A)/P(E)$ and $P(B \mid E) = 1 - P(A \mid E)$. Since $P(A) = P(B) = \frac{1}{2}$, $P(E) = P(A) \times \frac{1}{4} + P(B) \times \frac{3}{4} = \frac{1}{2}$, and $P(E \mid A) = \frac{1}{4}$, we get $P(A \mid E) = \frac{1}{4}$ and $P(B \mid E) = \frac{3}{4}$. Thus the sought probability is $\frac{1}{4} \times \frac{1}{4} + \frac{3}{4} \times \frac{3}{4} = \frac{5}{8}$.

2.57 Fix j. Label the $c = \binom{7}{j}$ possible combinations of j stops as $l = 1, \ldots, c$. Let A be the event that there will be exactly j stops at which nobody gets off and B_l be the event that nobody gets off at the j stops from combination l. Then $P(A) = \sum_{l=1}^{c} P(A \mid B_l)P(B_l)$. We have that $P(B_l) = (7-j)^{25}/7^{25}$ for all l and $P(A \mid B_l)$ is the unconditional probability that at least one person gets off at each stop when there are $7-j$ stops and 25 persons. Thus $P(A \mid B_l) = 1 - \sum_{k=1}^{7-j}(-1)^{k+1}\binom{7-j}{k}(7-j-k)^{25}/(7-j)^{25}$, using the result of Example 1.18. Next we get, after some algebra, the desired result $P(A) = \sum_{k=0}^{7-j}(-1)^{k}\binom{j+k}{j}\binom{7}{j+k}(7-j-k)^{25}/7^{25}$. More generally, $\sum_{k=0}^{b-j}(-1)^{k}\binom{j+k}{j}\binom{b}{j+k}(b-j-k)^{m}/b^{m}$ gives the probability of exactly j empty bins when $m \geq b$ balls are sequentially placed at random into b bins.

2.59 Let A be the event that both rolls of the two dice show the same combination of two numbers. Also, let B_1 be the event that the first roll of the two dice shows two equal numbers and B_2 be the event that the first roll shows two different numbers. Then $P(B_1) = \frac{6}{36}$ and $P(B_2) = \frac{30}{36}$. Further, $P(A \mid B_1) = \frac{1}{36}$ and $P(A \mid B_2) = \frac{2}{36}$. By the law of conditional probability, $P(A) = \frac{1}{36} \times \frac{6}{36} + \frac{2}{36} \times \frac{30}{36} = \frac{11}{216}$.

2.61 This problem can be seen as a random walk on the integers, where the random walk starts at zero. In the first step the random walk moves from 0 to 1 with probability p and to -1 with probability $q = 1-p$. Take $p < \frac{1}{2}$. Starting from 1, the random walk will ever return to 0 with probability $1 - \lim_{b\to\infty}[1-(q/p)^a]/[1-(q/p)^{a+b}]$ with $a = 1$. This probability is 1. Starting from -1, the random walk will ever return to 0 with probability $1 - \lim_{b\to\infty}[1-(p/q)^a]/[1-(p/q)^{a+b}]$ with $a = 1$. This probability is $\frac{p}{q}$. The sought probability is $p \times 1 + (1-p) \times \frac{p}{1-p} = 2p$ (this result is also valid for $p = \frac{1}{2}$).

2.63 It does not matter what question you ask. To see this, let A be the event that your guess is correct, B_1 be the event that your friend's answer is yes, and B_2 be the event that the answer is no. For the question of whether the card is red, $P(A) = \frac{1}{26} \times \frac{1}{2} + \frac{1}{26} \times \frac{1}{2} = \frac{1}{26}$. For the other question, $P(A) = 1 \times \frac{1}{52} + \frac{1}{51} \times \frac{51}{52} = \frac{1}{26}$.

2.65 Let A be the event that you roll two *consecutive* totals of 7 before a total of 12. Let B_1 be the event that each of the first two rolls results in a total of 7, B_2 be the event that the first roll gives a total of 7 and the second roll a total different from 7 and 12, B_3 be the event that the first roll gives a total different from 7 and 12, B_4 be the event that the first roll gives a total of 7 and the second roll a total of 12, and B_5 be the event that the first roll gives a total of 12. Then $P(A) = 1 \times \frac{6}{36} \times \frac{6}{36} + P(A) \times \frac{6}{36} \times \frac{29}{36} + P(A) \times \frac{29}{36} + 0 \times \frac{6}{36} \times \frac{1}{36} + 0 \times \frac{1}{36}$ and so $P(A) = \frac{6}{13}$.

2.67 The recursion is $p(i,t) = \frac{1}{6-i+1}\sum_{j=0}^{6-i} p(i+1, t-j)$, as follows by conditioning upon the number of tokens you lose at the ith cup. This leads to $p(1,6) = \frac{169}{720}$.

2.69 (a) Define r_n as the probability of getting a run of either r successes or r failures in n trials. Also, define s_n as the probability of getting a run of r successes or r failures in n trials given that the first trial results in a success, and f_n as the probability of getting a run of either r successes or r failures in n trials given that the first trial results in a failure. Then $r_n = ps_n + (1-p)f_n$. The s_n and f_n satisfy the recursive schemes $s_n = p^{r-1} + \sum_{k=1}^{r-1} p^{k-1}(1-p)f_{n-k}$ and $f_n = (1-p)^{r-1} + \sum_{k=1}^{r-1}(1-p)^{k-1} p \, s_{n-k}$ for $n \geq r$, where $s_j = f_j = 0$ for $j < r-1$.

(b) Parameterize the starting state and let $p(r,b,L)$ be the probability that the longest run of red balls will be L or more when the bowl initially contains r red and b blue balls. Fix $r > L$ and $b \geq 1$. Let A be the event that a run of L red balls will occur. To find $P(A) = p(r,b,L)$, let B_L be the conditioning event that the first L balls picked are red and B_{j-1} be the conditioning event that each of the first $j-1$ balls picked is red but the jth ball picked is blue, where $1 \leq j \leq L$. Then $P(B_L) = \frac{r}{r+b} \times \cdots \times \frac{r-(L-1)}{r+b-(L-1)}$ and $P(B_{j-1}) = \frac{r}{r+b} \times \cdots \times \frac{r-(j-2)}{r+b-(j-2)} \times \frac{b}{r+b-(j-1)}$ for $1 \leq j \leq L$. Note that $P(A \mid B_L) = 1$ and $P(A \mid B_{j-1}) = p(r-(j-1), b-1, L)$ for $1 \leq j \leq L$. Then $P(A) = P(B_L) + \sum_{j=1}^{L} P(A \mid B_{j-1})P(B_{j-1})$ gives a recursion scheme for the calculation of the probability $p(r,b,L)$.

2.71 The optimal strategy of player A is to stop after the first spin if this spin gives a score of more than $a_3 = 65$ points. The stopping level a_3 is the largest $a \geq a_2$ for which $C_3(a) = \frac{1}{R} \sum_{k=1}^{R-a} \frac{(a+k)^4}{R^4}$ is still larger than $S_3(a) = \frac{a^4}{R^4}$, where $R = 100$ and $a_2 = 53$ is the optimal switching point for the first player in the two-player game with the numbers $1, 2, \ldots, R$ on the wheel. The overall win probability of player A is $\frac{1}{R} \sum_{a=1}^{a_3} C_3(a) + \frac{1}{R} \sum_{a=a_3+1}^{R} S_3(a) = 0.3123$. The probability that the final score of player A will be a points is given by $\frac{1}{2} a_3(a_3+1)/R^2$ for $a = 0$, $(a-1)/R^2$ for $1 \leq a \leq a_3$, and $1/R + a_3/R^2$ for $a_3 < a \leq R$. The overall win probabilities of players B and C are 0.3300 and 0.3577.

2.73 Let the hypothesis H be the event that oil is present and the evidence E be the event that the test is positive. Then $P(H) = 0.4$, $P(E \mid H) = 0.9$, and $P(E \mid \overline{H}) = 0.15$. Thus the posterior odds are $\frac{0.4}{0.6} \times \frac{0.90}{0.15} = 4$. The posterior probability $P(H \mid E) = \frac{4}{1+4} = 0.8$.

2.75 Let the hypothesis H be the event that the blue coin is unfair and the evidence E be the event that all three tosses of the blue coin show heads. The posterior odds are $\frac{0.2}{0.8} \times \frac{(0.75)^3}{(0.5)^3} = \frac{27}{32}$. The posterior probability $P(H \mid E) = \frac{27}{59} = 0.4576$.

2.77 Let the hypothesis H be the event that both children are boys. **(a)** If the evidence E is the event that at least one child is a boy, then the posterior odds are $\frac{1/4}{3/4} \times \frac{1}{2/3} = \frac{1}{2}$. The posterior probability $P(H \mid E) = \frac{1}{3}$. **(b)** If the evidence E is the event that at least one child is a boy born on a Tuesday, then the posterior odds are $\frac{1/4}{3/4} \times [1 - (\frac{6}{7})^2] / [\frac{1}{3} \times \frac{1}{7} + \frac{1}{3} \times \frac{1}{7} + \frac{1}{3} \times 0] = \frac{13}{14}$. The posterior probability $P(H \mid E) = \frac{13}{27}$. **(c)** If the evidence E is the event that at least one child is a boy born on one of the first k days of the week, then the posterior odds are $\frac{1/4}{3/4} \times [1 - (1 - \frac{k}{7})^2] / [\frac{1}{3} \times \frac{k}{7} + \frac{1}{3} \times \frac{k}{7} + \frac{1}{3} \times 0] = \frac{14-k}{14}$. The posterior probability $P(H \mid E) = \frac{14-k}{28-k}$ for $k = 1, 2, \ldots, 7$.

2.79 Let the hypothesis H be the event that the suspect is guilty and the evidence E be the event that the suspect makes a confession. Then, by Bayes' rule in odds form, $P(H \mid E) > P(H)$ if and only if $P(E \mid H) > P(E \mid \overline{H})$ (use the fact that $p/(1-p) > q/(1-q)$ for $0 < p, q < 1$ if and only if $p > q$).

2.81 Let the hypothesis H be the event that the woman has breast cancer and the evidence E be the event that the test result is positive. Since $P(H) = 0.01$, $P(\overline{H}) = 0.99$, $P(E \mid H) = 0.9$, and $P(E \mid \overline{H}) = 0.1$, the posterior odds are $\frac{0.01}{0.99} \times \frac{0.9}{0.1} = \frac{1}{11}$. Therefore the posterior probability $P(H \mid E) = \frac{1}{12}$. *Note:* As sanity check, the posterior probability can also be obtained by a heuristic but insightful approach. This approach presents the relevant information in terms of frequencies instead of probabilities. Imagine 10,000 (say) women who undergo the test. On average, there are 90 positive tests for the 100 women having the malicious disease, whereas there are 990 false positives for the 9,900 healthy women. Thus, based on the information presented in this way, we find that the sought probability is $90/(90 + 990) = \frac{1}{12}$.

2.83 Let the hypothesis H be the event that you have chosen the two-headed coin and the evidence E be the event that all n tosses result in heads. Then $\frac{P(H|E)}{P(\overline{H}|E)} = \frac{1/10,000}{9,999/10,000} \times \frac{1}{0.5^n}$. This gives $P(H \mid E) = \frac{2^n}{2^n + 9,999}$. The probability $P(H \mid E)$ has the values 0.0929, 0.7662, and 0.9997 for $n = 10$, 15, and 25.

2.85 Let the random variable Θ represent the unknown win probability of Alassi. The prior of Θ is $p_0(0.4) = p_0(0.5) = p_0(0.6) = \frac{1}{3}$. Let E be the event that Alassi wins the best-of-five contest. The likelihood function $L(E \mid \theta)$ is $\theta^3 + \binom{3}{2}\theta^2(1-\theta)\theta + \binom{4}{2}\theta^2(1-\theta)^2\theta$. The posterior probability $p(\theta \mid E)$ is proportional to $p_0(\theta)L(E \mid \theta)$ and has the values 0.2116, 0.3333, and 0.4550 for $\theta = 0.4$, 0.5, and 0.6.

2.87 Let the random variable Θ be 1 if the student is unprepared for the exam, 2 if the student is half prepared, and 3 if the student is well prepared.

The prior of Θ is $p_0(1) = 0.2$, $p_0(2) = 0.3$, and $p_0(3) = 0.5$. Let E be the event that the student has answered correctly 26 out of 50 questions. The likelihood function $L(E \mid \theta)$ is $\binom{50}{26} a_\theta^{26}(1-a_\theta)^{24}$, where $a_1 = \frac{1}{3}$, $a_2 = 0.45$, and $a_3 = 0.8$. The posterior probability $p(\theta \mid E)$ is proportional to $p_0(\theta)L(E \mid \theta)$ and has the values 0.0268, 0.9730, and 0.0001 for $\theta = 1$, 2, and 3.

Chapter 3

3.1 Take as sample space the set consisting of the four outcomes $(0,0)$, $(1,0)$, $(0,1)$, and $(1,1)$, where the first (second) component is 0 if the first (second) student picked has not done the homework and 1 otherwise. The probabilities $\frac{5}{20} \times \frac{4}{19} = \frac{1}{19}$, $\frac{15}{20} \times \frac{5}{19} = \frac{15}{76}$, $\frac{5}{20} \times \frac{15}{19} = \frac{15}{76}$, and $\frac{15}{20} \times \frac{14}{19} = \frac{21}{38}$ are assigned to these four outcomes. The random variable X takes on the value 0 for the outcome $(1,1)$, the value 1 for the outcomes $(0,1)$ and $(1,0)$, and the value 2 for the outcome $(0,0)$. Thus, $P(X = 0) = \frac{21}{38}$, $P(X = 1) = \frac{15}{76} + \frac{15}{76} = \frac{15}{38}$, and $P(X = 2) = \frac{1}{19}$.

3.3 The random variable X can take on the values 0, 1, and 2. By the law of conditional probability, $P(X = 0) = \frac{1}{3} \times 0 + \frac{2}{3} \times \frac{1}{4} = \frac{1}{6}$, $P(X = 1) = \frac{1}{3} \times 0 + \frac{2}{3} \times \frac{1}{2} = \frac{1}{3}$, and $P(X = 2) = \frac{1}{3} \times 1 + \frac{2}{3} \times \frac{1}{4} = \frac{1}{2}$.

3.5 The random variable X can take on the values 2, 3, and 4. Two tests are needed if the first two tests give the depleted batteries, while three tests are needed if the first three batteries tested are not depleted or if a second depleted battery is found at the third test. Thus, by the chain rule for conditional probabilities, $P(X = 2) = \frac{2}{5} \times \frac{1}{4} = \frac{1}{10}$, $P(X = 3) = \frac{3}{5} \times \frac{2}{4} \times \frac{1}{3} + + \frac{2}{5} \times \frac{3}{4} \times \frac{1}{3} + \frac{3}{5} \times \frac{2}{4} \times \frac{1}{3} = \frac{3}{10}$. The probability $P(X = 4)$ is calculated as $1 - P(X = 2) - P(X = 3) = \frac{6}{10}$.

3.7 The random variable X can take on the values 0, 5, 10, 15, 20, and 25. Using the chain rule for conditional probabilities, $P(X = 0) = \frac{4}{7}$, $P(X = 5) = \frac{1}{7} \times \frac{4}{6}$, $P(X = 10) = \frac{2}{7} \times \frac{4}{6}$, $P(X = 15) = \frac{2}{7} \times \frac{1}{6} \times \frac{4}{5} + \frac{1}{7} \times \frac{2}{6} \times \frac{4}{5}$, $P(X = 20) = \frac{2}{7} \times \frac{1}{6} \times \frac{4}{5}$, and $P(X = 25) = \frac{2}{7} \times \frac{1}{6} \times \frac{1}{5} + \frac{2}{7} \times \frac{1}{6} \times \frac{1}{5} + \frac{1}{7} \times \frac{2}{6} \times \frac{1}{5}$.

3.9 $E(X) = 15 \times \frac{1}{4} + 20 \times \frac{1}{4} + 70 \times \frac{1}{4} + 125 \times \frac{1}{4} = 57.5$ and $E(Y) = 15 \times \frac{15}{230} + 20 \times \frac{20}{230} + 70 \times \frac{70}{230} + 125 \times \frac{125}{230} = 88.125$.

3.11 Let the random variable X be the amount of money you win. Then $P(X = m) = \binom{10}{m}/\binom{12}{m}$ and $P(X = 0) = 1 - P(X = m)$. This gives $E(X) = m\binom{10}{m}/\binom{12}{m}$. This expression is maximal for $m = 4$ with $E(X) = \frac{56}{33}$.

3.13 Put for abbreviation $c_k = \binom{52}{k-1}$. Using the second argument from Example 3.2, we get $P(X_2 = k) = \frac{1}{c_k}\binom{48}{k-2}\binom{4}{1} \times \frac{3}{52-(k-1)}$, $P(X_3 = k) = \frac{1}{c_k}$

$\binom{48}{k-3}\binom{4}{2} \times \frac{2}{52-(k-1)}$, and $P(X_4 = k) = \frac{1}{c_k}\binom{48}{k-4}\binom{4}{3} \times \frac{1}{52-(k-1)}$. This leads to $E(X_2) = 21.2$, $E(X_3) = 31.8$, and $E(X_4) = 42.4$.

3.15 Let the random variable X be the total amount staked and the random variable Y be the amount won. The probability p_k that k bets will be placed is $\left(\frac{19}{37}\right)^{k-1}\frac{18}{37}$ for $k = 1, \ldots, 10$ and $\left(\frac{19}{37}\right)^{10}$ for $k = 11$. Thus, $E(X) = \sum_{k=1}^{10}(1+2+\cdots+2^{k-1})p_k + (1+2+\cdots+2^9+1{,}000)p_{11} = 12.583$ dollars. If the round goes to 11 bets, then the player's loss is $23 if the 11th bet is won and $2,023 if the 11th bet is lost. Thus, $E(Y) = 1 \times (1 - p_{11}) - 23 \times p_{11} \times \frac{18}{37} - 2{,}023 \times p_{11} \times \frac{19}{37} = -0.3401$ dollars. The ratio of 0.3401 and 12.583 is in line with the house advantage of 2.70% of the casino.

3.17 Take as sample space the set $0 \cup \{(x, y) : x^2 + y^2 \le 25\}$, where 0 means that the dart has missed the target. The score is a random variable X with $P(X = 0) = 0.25$, $P(X = 5) = 0.75 \times \frac{25-9}{25} = 0.48$, $P(X = 8) = 0.75 \times \frac{9-1}{25} = 0.24$, and $P(X = 15) = 0.75 \times \frac{1}{25} = 0.03$. The expected value of the score is $E(X) = 0.48 \times 5 + 0.24 \times 8 + 0.03 \times 15 = 4.77$.

3.19 Let the random variable X be the payoff when you go for a second spin, given that the first spin showed a score of a points. Then $P(X = a + k) = 1/1{,}000$ for $1 \le k \le 1{,}000 - a$ and $P(X = 0) = a/1{,}000$. Thus, $E(X) = \frac{1}{1{,}000}\sum_{k=1}^{1{,}000-a}(a+k) = \frac{1}{2{,}000}(1{,}000-a)(1{,}000+a+1)$. The largest value of a for which $E(X) > a$ is $a^* = 414$. The optimal strategy is to stop after the first spin if this spin gives a score of more than 414 points.

3.21 Let the random variable X be the payoff of the game. Then, using conditional probabilities, $P(X = 0) = \left(\frac{5}{6}\right)^3$, $P(X = 2) = \binom{3}{1}\frac{1}{6}\left(\frac{5}{6}\right)^2 \times \frac{4}{5}$, $P(X = 2.5) = \binom{3}{1}\frac{1}{6}\left(\frac{5}{6}\right)^2 \times \frac{1}{5}$, $P(X = 3) = \binom{3}{2}\left(\frac{1}{6}\right)^2\frac{5}{6}$, and $P(X = 4) = \left(\frac{1}{6}\right)^3$. This gives $E(X) = 0 \times \frac{125}{216} + 2 \times \frac{60}{216} + 2.5 \times \frac{15}{216} + 3 \times \frac{15}{216} + 4 \times \frac{1}{216} = 0.956$.

3.23 Let the random variable X be the payoff of the game. The probability that X takes on the value k with $k < m$ is the same as the probability that a randomly chosen number from $(0, 1)$ falls into $\left(\frac{1}{k+2}, \frac{1}{k+1}\right)$ or into $\left(1 - \frac{1}{k+1}, 1 - \frac{1}{k+2}\right)$. Thus, $P(X = k) = 2\left(\frac{1}{k+1} - \frac{1}{k+2}\right)$ for $1 \le k \le m-1$ and $P(X = m) = \frac{2}{m+1}$. The stake should be $E(X) = 2\left(\frac{1}{2} + \cdots + \frac{1}{m+1}\right)$.

3.25 Suppose you have rolled a total of $i < 10$ points so far. The expected value of the change of your current total is $\sum_{k=1}^{10-i} k \times \frac{1}{6} - i \times \frac{i-4}{6}$ if you decide to continue for one more roll. This expression is positive for $i \le 5$ and negative for $i \ge 6$. Thus, the one-stage-look-ahead rule prescribes stopping as soon as the total number of points rolled is 6 or more. This rule maximizes the expected value of your reward.

3.27 Suppose that you have gathered a dollars so far. Then there are still $w-a$ white balls in the bowl. If you decide to pick one more ball, then the expected value of the change of your bankroll is $\frac{1}{r+w-a}(w-a) - \frac{r}{r+w-a}a$. This expression is less than or equal to zero for $a \geq \frac{w}{r+1}$. It is optimal to stop as soon as you have gathered at least $\frac{w}{r+1}$ dollars.

3.29 Writing $\sum_{k=0}^{\infty} P(X > k) = \sum_{k=0}^{\infty} \sum_{j=k+1}^{\infty} P(X = j)$ and interchanging the order of summation gives $\sum_{k=0}^{\infty} P(X > k) = \sum_{j=0}^{\infty} \sum_{k=0}^{j-1} P(X = j) = \sum_{j=0}^{\infty} j P(X = j)$ and so $\sum_{k=0}^{\infty} P(X > k) = E(X)$. The interchange of the order of summation is justified by the nonnegativity of the terms involved. An alternative proof can be based on the relation $X = \sum_{n=0}^{\infty} I_n$, where the random variable $I_n = 1$ if $X > n$ and 0 otherwise. Then, by $E(X) = \sum_{n=0}^{\infty} E(I_n)$ and $E(I_n) = P(X > n)$, we get the result. The expected value of the largest of 10 randomly chosen numbers from 1 to 100 is $\sum_{k=0}^{99} \left(1 - \left(\frac{k}{100}\right)^{10}\right) = 91.401$.

3.31 Let $I_k = I_{A_k}$. Then $P(A_1^c \cap \cdots \cap A_n^c) = E[(1 - I_1) \cdots (1 - I_n)]$. We have that $(1 - I_1) \cdots (1 - I_n) = 1 - \sum_{j=1}^{n} I_j + \sum_{j<k} I_j I_k + \cdots + (-1)^n I_1 \cdots I_n$. Taking the expected value of both sides and using the relation $E(I_{i_1} \cdots I_{i_r}) = P(A_{i_1} \cdots A_{i_r})$, we get the inclusion–exclusion formula for $P(\bigcup_{k=1}^{n} A_k) = 1 - P(\bigcap_{k=1}^{n} A_k^c)$.

3.33 Let the indicator variable I_k be 1 if the kth team has a married couple and 0 otherwise. Then $P(I_k = 1) = (12 \times 22)/\binom{24}{3} = \frac{3}{23}$ for any k. The expected value of the number of teams with a married couple is $\sum_{k=1}^{8} E(I_k) = \frac{24}{23}$.

3.35 Let I_k be 1 if two balls of the same color are removed on the kth pick and 0 otherwise. The expected number of times that you pick two balls of the same color is $\sum_{k=1}^{r+b} E(I_k)$. By a symmetry argument, each I_k has the same distribution as I_1. Since $P(I_1 = 1) = \frac{2r}{2r+2b} \times \frac{2r-1}{2r+2b-1} + \frac{2b}{2r+2b} \times \frac{2b-1}{2r+2b-1}$, the expected number of times that you pick two balls of the same color is $[r(2r-1) + b(2b-1)]/(2r+2b-1)$.

3.37 Let the indicator variable I_s be 1 if item s belongs to T after n steps and 0 otherwise. Then $P(I_s = 0) = \left(\frac{n-1}{n}\right)^n$ and so $E(I_s) = 1 - \left(1 - \frac{1}{n}\right)^n$. Thus, the expected value of the number of distinct items in T after n steps is $n\left[1 - \left(1 - \frac{1}{n}\right)^n\right]$. Note that the expected value is about $n\left(1 - \frac{1}{e}\right)$ for n large.

3.39 Label the white balls as $1, \ldots, w$. Let the indicator variable I_k be equal to 1 if the white ball with label k remains in the bag when you stop and 0 otherwise. To find $P(I_k = 1)$, you can discard the other white balls. Therefore $P(I_k) = \frac{1}{r+1}$. The expected number of remaining white balls is $\sum_{k=1}^{w} E(I_k) = \frac{w}{r+1}$.

3.41 Let the indicator variable I_k be equal to 1 if the numbers k and $k+1$ appear in the lottery drawing and 0 otherwise. Then $P(I_k = 1) = \binom{43}{4}/\binom{45}{6}$. The expected number of consecutive numbers is $\sum_{k=1}^{44} E(I_k) = \frac{2}{3}$.

3.43 Let the random variable X_i be equal to 1 if box i contains more than 3 apples and 0 otherwise. Then $P(X_i = 1) = \sum_{k=4}^{25} \binom{25}{k}\left(\frac{1}{10}\right)^k\left(\frac{9}{10}\right)^{25-k}$ and so $E(X_i) = 0.2364$. Thus, the expected value of the number of boxes containing more than 3 apples is $10 \times 0.2364 = 2.364$.

3.45 For any $i \neq j$, let $X_{ij} = 1$ if the integers i and j are switched in the random permutation and $X_{ij} = 0$ otherwise. Then $P(X_{ij} = 1) = \frac{(n-2)!}{n!}$. The expected number of switches is $\sum_{i<j} E(X_{ij}) = \binom{n}{2}\frac{1}{n(n-1)} = \frac{1}{2}$.

3.47 Using the basic sums $\sum_{k=1}^n k = \frac{1}{2}n(n+1)$ and $\sum_{k=1}^n k^2 = \frac{1}{6}n(n+1)(2n+1)$, we get that $E(X)$ and $E(X^2)$ are $\frac{1}{2}(a+b)$ and $\frac{1}{6}(2a^2 + 2ab - a + 2b^2 + b)$, which leads to $\text{var}(X) = \frac{1}{12}(a^2 - 2ab - 2a + b^2 + 2b) = \frac{1}{12}[(b-a+1)^2 - 1]$.

3.49 Let $I_k = 1$ if the kth team has a married couple and $I_k = 0$ otherwise, and let $X = \sum_{k=1}^8 I_k$. Then $E(X^2) = \sum_{k=1}^8 E(I_k^2) + 2\sum_{j=1}^7 \sum_{k=j+1}^8 E(I_j I_k)$. We have $E(I_k^2) = P(I_k = 1) = (12 \times 22)/\binom{24}{3}$ and $E(I_j I_k) = P(I_j = 1, I_k = 1) = (12 \times 11 \times 20 \times 19)/[\binom{24}{3}\binom{21}{3}]$ for $j \neq k$. This gives $E(X) = \frac{24}{23}$, $E(X^2) = \frac{48}{23}$, and $\sigma(X) = 0.9981$.

3.51 We have $E(X) = \sum_{k=1}^{10} k\frac{11-k}{55} = 4$ and $E(X^2) = \sum_{k=1}^{10} k^2 \frac{11-k}{55} = 22$, and so $\sigma(X) = \sqrt{22 - 16} = 2.449$. The number of reimbursed treatments is $Y = g(X)$, where $g(x) = \min(x, 5)$. Then $E(Y) = \sum_{k=1}^4 k\frac{11-k}{55} + \sum_{k=5}^{10} 5\frac{11-k}{55} = \frac{37}{11}$ and $E(Y^2) = \sum_{k=1}^4 k^2\frac{11-k}{55} + \sum_{k=5}^{10} 25\frac{11-k}{55} = \frac{151}{11}$, by applying the substitution rule. The standard deviation $\sigma(Y) = \sqrt{151/11 - (37/11)^2} = 1.553$.

3.53 Let the random variable X be the number of repairs that will be necessary in the coming year and Y be the maintenance costs in excess of the prepaid costs. Then $Y = g(X)$, where the function $g(x) = 100(x - 155)$ for $x > 155$ and $g(x) = 0$ otherwise. By the substitution rule, $E(Y) = \sum_{x=156}^\infty 100(x-155)e^{-150}\frac{150^x}{x!} = 280.995$ and $E(Y^2) = \sum_{x=156}^\infty 100^2 (x-155)^2 e^{-150}\frac{150^x}{x!} = 387{,}929$. The standard deviation of Y is $\sqrt{387{,}929 - 28.995^2} = 555.85$.

3.55 Your probability of winning the contest is $\sum_{k=0}^n \frac{1}{k+1}P(X=k) = E\left(\frac{1}{1+X}\right)$. The random variable X can be written as $X = X_1 + \cdots + X_n$, where X_i is equal to 1 if the ith person in the first round survives this round and 0 otherwise. Since $E(X_i) = \frac{1}{n}$ for all i, we have $E(X) = 1$. Thus, by Jensen's inequality, $E\left(\frac{1}{1+X}\right) \geq \frac{1}{1+E(X)} = \frac{1}{2}$.

3.57 For $x, y \in \{-1, 1\}$, we have $P(X = x, Y = y) = P(X = x \mid Y = y) P(Y = y)$ and $P(X = x \mid Y = y) = P(Z = x/y \mid Y = y)$. Since Y and Z are independent, $P(Z = x/y \mid Y = y) = P(Z = x/y) = 0.5$. This gives $P(X = x, Y = y) = 0.5 \times P(Y = y)$. Also, $P(X = 1) = P(Y = 1, Z = 1) + P(Y = -1, Z = -1)$. Thus, by the independence of Y and Z, we get $P(X = 1) = 0.25 + 0.25 = 0.5$ and so $P(X = -1) = 0.5$. Therefore, the result $P(X = x, Y = y) = 0.5 \times P(Y = y)$ implies $P(X = x, Y = y) = P(X = x)P(Y = y)$, proving that X and Y are independent. However, X is not independent of $Y + Z$. To see this, note that $P(X = 1, Y + Z = 0) = 0$ and $P(X = 1)P(Y + Z = 0) > 0$.

3.59 Noting that $X_i = X_{i-1} + R_i$, we get $X_i = R_2 + \cdots + R_i$ for $2 \leq i \leq 10$. This implies $\sum_{i=2}^{10} X_i = \sum_{k=2}^{10}(11 - k)R_k$. Since $P(R_k = 0) = P(R_k = 1) = \frac{1}{2}$, we have $E(R_k) = \frac{1}{2}$ and $\sigma^2(R_k) = \frac{1}{4}$. The random variables R_k are independent and so, by Rules 3.1 and 3.9, $E\left(\sum_{i=2}^{10} X_i\right) = \sum_{k=2}^{10}(11 - k)E(R_k) = 22.5$ and $\sigma^2\left(\sum_{i=1}^{10} X_i\right) = \sum_{k=2}^{10}(11 - k)^2 \sigma^2(R_k) = 71.25$.

3.61 We have $E\left(\sum_{k=1}^{\infty} X_k I_k\right) = \sum_{k=1}^{\infty} E(X_k I_k)$, since it is always allowed to interchange expectation and summation for nonnegative random variables. Since X_k and I_k are independent, $E(X_k I_k) = E(X_k)E(I_k)$ for any $k \geq 1$. Also, $E(I_k) = P(N \geq k)$. Thus, $E\left(\sum_{k=1}^{\infty} X_k I_k\right) = E(X_1) \sum_{k=1}^{\infty} P(N \geq k) = E(X_1)E(N)$, using the fact that $E(N) = \sum_{n=0}^{\infty} P(N > n)$, see Problem 3.29. *Note*: If the X_k are not nonnegative, the proof still applies in view of the convergence result that $E\left(\sum_{k=1}^{\infty} Y_k\right) = \sum_{k=1}^{\infty} E(Y_k)$ when $\sum_{k=1}^{\infty} E(|Y_k|) < \infty$.

3.63 The probability $P_r = \sum_{k=r}^{6r} \binom{6r}{k}\left(\frac{1}{6}\right)^k \left(\frac{5}{6}\right)^{6r-k}$ of getting at least r sixes in one throw of $6r$ dice has the values 0.6651, 0.6187, and 0.5963 for $r = 1, 2,$ and 3. Thus it is best to throw 6 dice. Pepys believed that it was best to throw 18 dice. *Note*: The probability P_r is decreasing in r and tends to $\frac{1}{2}$ as $r \to \infty$.

3.65 This question can be translated into the question of what the probability is of getting 57 *or less* heads in 199 tosses of a fair coin. A binomial random variable with parameters $n = 199$ and $p = 0.5$ has expected value $199 \times 0.5 = 99.5$ and standard deviation $0.5\sqrt{199} = 7.053$. Thus, the observed number of polio cases in the treatment group is more than six standard deviations below the expected number. Without doing any further calculations, we can say that the probability of this occurring is extremely small (the precise value of the probability is 7.4×10^{-10}). This makes clear that there is overwhelming evidence that the vaccine does work.

3.67 Using the law of conditional probability, we have that the sought probability is given by $\sum_{k=0}^{n} \frac{k}{n}\binom{n}{k}p^k(1-p)^{n-k}$. This probability is nothing other than $\frac{1}{n}E(X)$, where X is binomially distributed with parameters n and p. Thus, the probability that you will be admitted to the program is $\frac{np}{n} = p$.

3.69 Let the random variable X be the number of coins that will be set aside. Then the random variable $100-X$ is binomially distributed with parameters $n = 100$ and $p = \frac{1}{8}$. Therefore, $P(X = k) = \binom{100}{100-k}\left(\frac{1}{8}\right)^{100-k}\left(\frac{7}{8}\right)^k$ for $k = 0, 1, \ldots, 100$.

3.71 Let the random variable X be the number of rounds in which you get cards with an ace. If the cards are well shuffled each time, then X is binomially distributed with parameters $n = 10$ and $p = 1 - \binom{48}{13}/\binom{52}{13}$. The answer to the question should be based on $P(X \leq 2) = \sum_{k=0}^{2}\binom{10}{k}p^k(1-p)^{10-k} = 0.0017$. This small probability is a strong indication that the cards were not well shuffled.

3.73 Let X_i be the number of times that image i shows up in the roll of the five poker dice. Then $(X_1, X_2, \ldots, X_6)$ has a multinomial distribution with parameters $n = 5$ and $p_1 = p_2 = \cdots = p_6 = \frac{1}{6}$. Let X be the payoff to the player for each unit staked. Then $E(X) = 3 \times P(X_1 \geq 1, X_2 \geq 1)$. This gives $E(X) = 3\sum_{x_1=1}^{4}\sum_{x_2=1}^{5-x_1} \frac{5!}{x_1!x_2!(5-x_1-x_2)!}p_1^{x_1}p_2^{x_2}(1-p_1-p_2)^{5-x_1-x_2} = 0.9838$.

3.75 Let the random variable X be the number of king's rolls. Then X has a binomial distribution with parameters $n = 4 \times 6^{r-1}$ and $p = \frac{1}{6^r}$. This distribution converges to a Poisson distribution with expected value $np = \frac{2}{3}$ as $r \to \infty$. The binomial probability $1-(1-p)^n$ tends very fast to the Poisson probability $1-e^{-2/3} = 0.48658$ as $r \to \infty$. The binomial probability has the values 0.51775, 0.49140, 0.48738, 0.48660, and 0.48658 for $r = 1, 2, 3, 5$, and 7.

3.77 The Poisson model is an appropriate model. Using the fact that the sum of two independent Poisson random variables is again Poisson distributed, the sought probability is $1 - \sum_{k=0}^{10} e^{-6.2}6.2^k/k! = 0.0514$.

3.79 Using the Poisson model, an estimate is $1-\sum_{k=0}^{7} e^{-4.2}4.2^k/k! = 0.064$.

3.81 Denote by X the number of weekly winners. An appropriate model for X is the Poisson distribution with expected value 0.25. The standard deviation of this distribution is $\sqrt{0.25} = 0.5$. The observed number of winners lies $\frac{3-0.25}{0.5} = 5.5$ standard deviations above the expected value. Without doing any further calculations, we can say that the probability of three or more winners is quite small ($P(X \geq 3) = 2.2 \times 10^{-3}$).

3.83 Translate the problem into a chance experiment with $\binom{25}{3} = 2{,}300$ trials; a trial for each possible combination of three people. Three people have the same birthday with probability $\left(\frac{1}{365}\right)^2$ and have birthdays falling within one day of each other with probability $7 \times \left(\frac{1}{365}\right)^2$. The Poisson heuristic gives the approximations $1 - e^{-2{,}300 \times (1/365)^2} = 0.0171$ and $1 - e^{-2{,}300 \times 7/(365)^2} = 0.1138$ for the sought probabilities. In a simulation study we found the values 0.016 and 0.103.

3.85 Translate the problem into a chance experiment with n trials. The ith trial is said to be successful if couple i is paired as bridge partners. The success probability of each trial is $p = \frac{1}{2n-1}$. Letting $\lambda = np = \frac{n}{2n-1}$, the Poisson heuristic gives the approximation $1 - e^{-n/(2n-1)}$ for the probability that no couple will be paired as bridge partners. For $n = 10$, the approximate value is 0.4092. This approximate value is quite close to the exact value 0.4088 that can be obtained with the inclusion–exclusion method.

3.87 Translate the problem into a chance experiment with 365 trials. The ith trial is said to be successful if seven or more people have their birthdays on day i. The success probability of each trial is $p = \left[1 - \left(\frac{354}{365}\right)^{75} - 75 \times \frac{1}{365} \times \left(\frac{364}{365}\right)^{74}\right]$. Letting $\lambda = 365p = 6.6603$, the Poisson heuristic gives the approximation $1 - \sum_{k=0}^{6} e^{-\lambda} \lambda^k / k! = 0.499$ for the probability that there are seven or more days so that on each of these days two or more people have their birthday. In a simulation study we found the value 0.516.

3.89 Think of a sequence of $2n$ trials with $n = 5$, where in each trial a person draws at random a card from the hat. The trial is said to be successful if the person draws the card with their own number, or the card with the number of their spouse. The success probability of each trial is $p = \frac{2}{2n}$. The number of successes can be approximated by a Poisson distribution with expected value $\lambda = 2n \times p = 2$. In particular, the probability of no success can be approximated by $e^{-2} = 0.1353$. The exact value is 0.1213 when $n = 5$. This value is obtained from the exact formula $\frac{1}{(2n)!} \int_0^\infty (x^2 - 4x + 2)^n e^{-x} \, dx$, which is stated without proof. *Note*: Using the exact formula, it can be verified experimentally that $e^{-2}(1 - \frac{1}{2n})$ is a better approximation than e^{-2}. In a generalization of the Las Vegas card game, you have two thoroughly shuffled decks of cards, where each deck has $r (= 13)$ types of cards and $s (= 4)$ cards of each type. A match occurs when two cards of the same type occupy the same position in their respective decks. Then the probability of no match can be approximated by e^{-s} for r large, while the exact value of this probability can be calculated from $(-1)^{rs}\frac{(s!)^r}{(rs)!}\int_0^\infty [L_s(x)]^r e^{-x}\,dx$, where $L_s(x) = \sum_{j=0}^s (-1)^j \binom{s}{j}\frac{x^j}{j!}$ is the Laguerre polynomial of degree s.

3.91 Imagine a trial for each person. The trial is said to be successful if the person involved has a lone birthday. The success probability is $p = \left(\frac{364}{365}\right)^{m-1}$. The probability that nobody in the group has a lone birthday is the same as the probability of having no successful trial. Thus, by the Poisson heuristic, the probability that nobody in the group has a lone birthday is approximately equal to $e^{-m(364/365)^{m-1}}$. This leads to the approximate value 3,061 for the minimum number of people that are required in order to have a fifty-fifty probability of no lone birthday. The exact answer is 3,064, see Problem 1.89.

3.93 To approximate the probability of drawing two consecutive numbers, translate the problem into a chance experiment with 44 trials, where there is a trial for any two consecutive numbers from 1 to 45. The probability of drawing two *specific* consecutive numbers is $\binom{43}{4}/\binom{45}{6}$. Thus, letting $\lambda_1 = 44 \times \left[\binom{43}{4}/\binom{45}{6}\right]$, the Poisson heuristic gives the approximation $1 - e^{-\lambda_1} = 0.487$ for the probability of drawing two consecutive numbers. In the same way, letting $\lambda_2 = 43 \times \left[\binom{42}{3}/\binom{45}{6}\right]$, we get the approximation $1 - e^{-\lambda_2} = 0.059$ for the probability of drawing three consecutive numbers. In a simulation study we found the values 0.529 and 0.056 for the two probabilities.

3.95 Let X denote how many numbers you will guess correctly. Then X has a hypergeometric distribution with parameters $R = 5$, $W = 34$, and $n = 5$. Therefore, $P(X = k) = \binom{5}{k}\binom{34}{5-k}/\binom{39}{5}$ for $0 \leq k \leq 5$. Let E be the expected payoff per dollar staked. Then $E = 100{,}000 \times P(X = 5) + 500 \times P(X = 4) + 25 \times P(X = 3) + E \times P(X = 2)$. This gives $E = 0.631$. The house percentage is 36.9%.

3.97 The hypergeometric model with $R = W = 25$ and $n = 25$ is applicable under the hypothesis that the psychologist blindly guesses which 25 persons are left-handed. The probability of identifying correctly 18 or more of the 25 left-handers is then $\sum_{k=18}^{25} \binom{25}{k}\binom{25}{25-k}/\binom{50}{25} = 2.1 \times 10^{-3}$. This small probability provides evidence against the hypothesis.

3.99 Use the hypergeometric model with $R = 8$, $W = 7$, and $n = 10$. The sought probability is equal to the probability of picking 5 or more red balls and this probability is $\sum_{k=5}^{8} \binom{8}{k}\binom{7}{10-k}/\binom{15}{10} = \frac{9}{11}$.

3.101 For a single player, the problem can be translated into the urn model with 24 red balls and 56 white balls. This leads to $Q_k = 1 - \binom{24}{24}\binom{56}{k-24}/\binom{80}{k}$ for $24 \leq k \leq 79$, where $Q_{23} = 1$ and $Q_{80} = 0$. The probability that more than 70 numbers must be called out before one of the players has achieved a full card is given by $Q_{70}^{36} = 0.4552$. The probability that you will be the first player to achieve a full card while no other player has a full card at the same time as you is equal to $\sum_{k=24}^{79}(Q_{k-1} - Q_k) Q_k^{35} = 0.0228$. The probability that you will be among the first players achieving a full card is $\sum_{k=24}^{80} \sum_{a=0}^{35} \binom{35}{a}[Q_{k-1} - Q_k]^{a+1} Q_k^{35-a} = 0.0342$.

3.103 Fix $1 \leq r \leq a$. Let the random variable X be the number of picks needed to obtain r red balls. Then X takes on the value k if and only if $r - 1$ red balls are obtained in the first $k - 1$ picks and another red ball at the kth pick. Thus, for $k = r, \ldots, b + r$, $P(X = k)$ is given by $[\binom{a}{r-1}\binom{b}{k-1-(r-1)} / \binom{a+b}{k-1}] \times \frac{1}{a+b-(k-1)}$. Alternatively, the probability mass function of X can be obtained from the tail probability $P(X > k) = \sum_{j=0}^{r-1} \binom{a}{j}\binom{b}{k-j} / \binom{a+b}{k}$.

3.105 For $0 \leq k \leq 4$, let E_k be the event that a diamond has appeared 4 times and a spade k times in the first $4 + k$ cards, and F_k be the event that the $(4 + k + 1)$th card is a diamond. The sought probability is $\sum_{k=0}^{4} P(E_k F_k) = \sum_{k=0}^{4} P(E_k)P(F_k \mid E_k)$. Thus, the win probability of player A is $\sum_{k=0}^{4}[\binom{8}{4}\binom{7}{k} / \binom{15}{4+k}] \times \frac{4}{11-k} = 0.6224$.

3.107 Let the random variable X be the number of tickets you will win. Then X has a hypergeometric distribution with parameters $R = 100$, $W = 124{,}900$, and $n = 2{,}500$. The sought probability is $1 - P(X = 0) = 0.8675$. Since $R + W \gg n$, the hypergeometric distribution can be approximated by the binomial distribution with parameters $n = 2{,}500$ and $p = \frac{R}{R+W} = 0.0008$. The binomial distribution in turn can be approximated by a Poisson distribution with expected value $\lambda = np = 2$.

3.109 The sought probability is $\sum_{n=15}^{19} \binom{n-1}{14}(\frac{3}{4})^{15}(\frac{1}{4})^{n-15} = 0.4654$, using the negative binomial distribution. As a sanity check, imagine that 19 trials are done. Then the sought probability is the probability of 15 or more successes in 19 trials and equals the binomial probability $\sum_{k=15}^{19} \binom{19}{k}(\frac{3}{4})^k(\frac{1}{4})^{19-k} = 0.4654$.

3.111 Imagine that each player continues rolling the die until one of the assigned numbers of that player appears. Let X_1 be the number of rolls player A needs to get a 1 or 2 and X_2 be the number of rolls player B needs to get a 4, 5, or 6. Then X_1 and X_2 are independent and geometrically distributed with respective parameters $p_1 = \frac{1}{3}$ and $p_2 = \frac{1}{2}$. The probability of player A winning is $P(X_1 \leq X_2) = \sum_{j=1}^{\infty} p_1 (1-p_1)^{j-1}(1-p_2)^{j-1}$ and equals $p_1 / (p_1 + p_2 - p_1 p_2) = \frac{1}{2}$. The length of the game is $X = \min(X_1, X_2)$. The probability $P(X > l)$ equals $(1-p_1)^l (1-p_2)^l = (1-p)^l$, where $p = p_1 + p_2 - p_1 p_2$. The length of the game is geometrically distributed with parameter $p = \frac{2}{3}$.

3.113 Let P_A be the probability of player A winning and P_d be the probability of a draw. By the law of conditional probability, $P_A = a(1-b) + (1-a)(1-b)P_A$ and $P_d = ab + (1-a)(1-b)P_d$. This gives $P_A = \frac{a(1-b)}{a+b-ab}$ and $P_d = \frac{ab}{a+b-ab}$. The length of the game is geometrically distributed with parameter $p = 1 - (1-a)(1-b) = a+b-ab$.

3.115 Suppose the strategy is to stop as soon as you have picked a number larger than or equal to r. The number of trials needed is geometrically distributed with parameter $p = (25-r+1)/25$ and the amount you get paid has a discrete uniform distribution on $r, \ldots, 25$. The expected net payoff is $\frac{1}{25-r+1} \sum_{k=r}^{25} k - \frac{25}{25-r+1} = \frac{1}{2}(25+r) - \frac{25}{25-r+1}$ and has the maximal value 18.4286 for $r = 19$.

3.117 Let X be the number of rounds required for the game. The random variable X is geometrically distributed with parameter $p = \sum_{i=2}^{12} a_i^2$, where a_i is the probability of rolling a dice sum of i and is given by $a_i = \frac{i-1}{36}$ for $2 \le i \le 7$ and $a_i = a_{14-i}$ for $8 \le i \le 12$. The probability of John paying for the beer is $\sum_{k=1}^{5} a_{2k+1}^2 / \sum_{i=2}^{12} a_i^2 = \frac{38}{73}$.

Chapter 4

4.1 Since $c \int_0^{10}(10-x)\,dx$ must be equal to 1, we get $c = \frac{1}{50}$. The probabilities are $P(X \le 5) = \int_0^5 \frac{1}{50}(10-x)\,dx = \frac{3}{4}$ and $P(X > 2) = \int_2^{10} \frac{1}{50}(10-x)\,dx = \frac{16}{25}$.

4.3 The probabilities $P(X \le 15)$, $P(X > 30)$, and $P(20 < X \le 25)$ are equal to $\int_0^{15} 2cxe^{-cx^2}\,dx = 0.3023$, $\int_{30}^{\infty} 2cxe^{-cx^2}\,dx = 0.2369$, and $\int_{20}^{25} 2cxe^{-cx^2}\,dx = 0.1604$, where $c = \frac{1}{625}$. Note that $\int_0^a 2cxe^{-cx^2}\,dx = \int_0^{ca^2} e^{-u}\,du = 1 - e^{-ca^2}$.

4.5 The proportion of pumping engines that will not fail before 10,000 hours of use is $P(X > 10) = \int_{10}^{\infty} 0.02xe^{-0.01x^2}\,dx$ and equals $e^{-0.01u^2}\big|_{10}^{\infty} = e^{-1} = 0.3679$. The other probability is $P(X > 10)/P(X > 5) = e^{-1}/e^{-0.25} = 0.4724$.

4.7 The probability distribution function $F(x) = P(X \le x)$ of the random variable X is $F(x) = 105 \int_0^x y^4(1-y)^2\,dy = x^5(15x^2 - 35x + 21)$ for $0 \le x \le 1$. The solution of the equation $1 - F(x) = 0.05$ is $x = 0.8712$. Thus, the capacity of the storage tank in thousands of gallons should be 0.8712.

4.9 Let the random variable Y be the area of the circle. Then $Y = \pi X^2$. Since $P(X \le x) = x$ for $0 \le x \le 1$ and $P(Y \le y) = P(X \le \sqrt{y/\pi})$, we get $P(Y \le y) = \sqrt{y/\pi}$ for $0 \le y \le \pi$. Differentiation of $P(Y \le y)$ gives that the density function of Y is $1/(2\sqrt{\pi y})$ for $0 < y < \pi$ and 0 otherwise.

4.11 The distribution function of $Y = X^2$ is $P(Y \leq y) = P(X \leq \sqrt{y}) = F(\sqrt{y})$ for $y \geq 0$, where $F(x) = P(X \leq x)$. Differentiation gives that the density function of Y is $\frac{1}{2}f(\sqrt{y})/\sqrt{y}$ for $y > 0$ and 0 otherwise. The distribution function of $W = V^2$ is $P(W \leq w) = P(-\sqrt{w} \leq V \leq \sqrt{w})$ for $w \geq 0$ and so $P(W \leq w) = 2\sqrt{w}/(2a)$ for $0 \leq w \leq a^2$. The density of W is $1/(2a\sqrt{w})$ for $0 < w < a^2$ and 0 otherwise.

4.13 The random variable $V = X/(1-X)$ satisfies $P(V \leq v) = P(X \leq v/(1+v)) = \frac{v}{1+v}$ for $v \geq 0$ and thus has density function $\frac{1}{(1+v)^2}$ for $v > 0$ and 0 otherwise. To get the density of $W = X(1-X)$, note that the function $x(1-x)$ has $\frac{1}{4}$ as its maximal value on $(0, 1)$ and that the equation $x(1-x) = w$ has the solutions $x_1 = \frac{1}{2} - \frac{1}{2}\sqrt{1-4w}$ and $x_2 = \frac{1}{2} + \frac{1}{2}\sqrt{1-4w}$ for $0 \leq w \leq \frac{1}{4}$. Thus, $P(W > w) = P(x_1 \leq X \leq x_2) = \int_{x_1}^{x_2} 1\, dx = \sqrt{1-4w}$. The density function of W is $2/\sqrt{1-4w}$ for $0 < w < \frac{1}{4}$ and 0 otherwise.

4.15 The random variable V takes on a value smaller than or equal to v if and only if the randomly chosen point falls into the set $\{(x, y) : 0 \leq x, y \leq v\}$. Thus the distribution function of V is $P(V \leq v) = v^2$ for $0 \leq v \leq 1$ and so the density function of V is $2v$ for $0 < v < 1$ and 0 otherwise. The random variable W takes on a value larger than w if and only if the randomly chosen point falls into the set $\{(x, y) : w \leq x, y \leq 1\}$. The distribution function of W is then $P(W \leq w) = 1 - (1-w)^2$ for $0 \leq w \leq 1$. The density function of W is $2(1-w)$ for $0 < w < 1$ and 0 otherwise.

4.17 $E(X) = \frac{1}{625}\int_{50}^{75} x(x-50)\, dx + \frac{1}{625}\int_{75}^{100} x(100-x)\, dx = 75$ hundred hours.

4.19 The density function of the distance X thrown by Big John is $f(x) = \frac{x-50}{600}$ for $50 < x < 80$, $f(x) = \frac{90-x}{200}$ for $80 < x < 90$, and $f(x) = 0$ otherwise. This gives $E(X) = \int_{50}^{90} xf(x)\, dx = 73\frac{1}{3}$ meters.

4.21 Let the random variable X be the distance from the randomly chosen point to the base of the triangle. Using a little geometry, we get $P(X > x)$ is the ratio of $\frac{1}{2}[(h-x) \times (h-x)b/h]$ and $\frac{1}{2}h \times b$. Differentiation gives that the density function of X is $2(h-x)/h^2$ for $0 < x < h$ and so $E(X) = \int_0^h x\frac{2(h-x)}{h^2}\, dx = \frac{1}{3}h$.

4.23 Let X be the distance from the point to the origin. Then $P(X \leq a) = \frac{1}{4}\pi a^2$ for $0 \leq a \leq 1$ and $P(X \leq a) = \frac{1}{4}\pi a^2 - 2\int_1^a \sqrt{a^2 - x^2}\, dx = \frac{1}{4}\pi a^2 - a^2 \arccos(\frac{1}{a}) + \sqrt{a^2 - 1}$ for $1 < a \leq \sqrt{2}$. The density function of X is $f(x) = \frac{1}{2}\pi x$ for $0 < x \leq 1$ and $f(x) = \frac{1}{2}\pi x - 2x\arccos(\frac{1}{x})$ for $1 < x < \sqrt{2}$. Numerical integration leads to $E(X) = \int_0^{\sqrt{2}} xf(x)\, dx = 0.765$.

4.25 (a) Differentiating $P(X \leq x \mid X > a) = \int_a^x f(v)\,dv/P(X > a)$ yields that the conditional density of X given that $X > a$ is $f(x)/P(X > a)$ for $x > a$ and 0 otherwise. In the same way, the conditional density of X given that $X \leq a$ is $f(x)/P(X \leq a)$ for $x < a$ and 0 otherwise. (b) By $E(X) = \frac{1}{1-a}\int_a^1 x\,dx$, we get $E(X) = \frac{1}{2}(1+a)$. (c) $E(X \mid X > a) = a + \frac{1}{\lambda}$ and $E(X \mid X \leq a) = \left(1 - e^{-\lambda a} - \lambda a e^{-\lambda a}\right)/\left(\lambda(1 - e^{-\lambda a})\right)$.

4.27 (a) The function $g(x) = \frac{1}{x}$ is convex for $x > 0$. Therefore, by Jensen's inequality, $E\left(\frac{1}{X}\right) \geq \frac{1}{E(X)} = \frac{5}{3}$. (b) $E\left(\frac{1}{X}\right) = \int_0^1 \frac{1}{x} 12x^2(1-x)\,dx = 2$ and $E\left[\left(\frac{1}{X}\right)^2\right] = \int_0^1 \frac{1}{x^2} 12x^2(1-x)\,dx = 6$, and so $\sigma\left(\frac{1}{X}\right) = \sqrt{2}$.

4.29 Let Y be the amount paid by the supplement policy. Then $Y = g(X)$, where $g(x)$ is $\min(500, x - 450)$ for $x > 450$ and 0 otherwise. By the substitution rule, $E(Y) = \int_{450}^{950}(x - 450)\frac{1}{1250}\,dx + 500\int_{950}^{1250} \frac{1}{1250}\,dx = 220$ dollars.

4.31 The net profit is given by $Y = g(X)$, where $g(x) = 2x$ for $0 \leq x \leq 250$ and $g(x) = 2 \times 250 - 0.5(x - 250)$ for $x > 250$. By the substitution rule, $E(Y) = \int_0^{250} 2x f(x)\,dx + \int_{250}^{\infty}[500 - 0.5(x - 250)]f(x)\,dx$, which gives $E(Y) = 194.10$ dollars. The stockout probability is $P(X > 250) = 1 - \int_0^{250} f(x)\,dx = 0.0404$.

4.33 Let U be the random point in $(0, 1)$ and define $g(u) = 1 - u$ if $u < s$ and $g(u) = u$ if $u \geq s$. Then $L = g(U)$ is the length of the subinterval covering the point s. By the substitution rule, $E(L) = \int_0^s (1 - u)\,du + \int_s^1 u\,du = s - s^2 + \frac{1}{2}$.

4.35 The area of the circle is $Y = \pi X^2$. By the substitution rule, we have that $E(Y) = \int_0^1 \pi x^2\,dx = \frac{\pi}{3}$ and $E(Y^2) = \int_0^1 \pi^2 x^4\,dx = \frac{\pi^2}{5}$, and so $\sigma(Y) = \frac{2\pi}{3\sqrt{5}}$.

4.37 (a) $E\left[(X - c)^2\right] = E(X^2) - 2cE(X) + c^2$ and is minimal for $c = E(X)$, as follows by differentiation. The minimal value is the variance of X. (b) $E(|X - c|) = \int_{-\infty}^c (c - x)f(x)\,dx + \int_c^{\infty}(x - c)f(x)\,dx$. The derivative of $E(|X - c|)$ is $2P(X \leq c) - 1$. The minimizing value of c satisfies $P(X \leq c) = \frac{1}{2}$.

4.39 Let Y be the amount of demand that cannot be satisfied from stock. Then $Y = g(X)$, where $g(x) = x - s$ for $x > s$ and $g(x) = 0$ otherwise. By the substitution rule, $E(Y) = \int_s^{\infty}(x - s)\lambda e^{-\lambda x}\,dx = \frac{1}{\lambda}e^{-\lambda s}$ and $E(Y^2) = \int_s^{\infty}(x - s)^2 \lambda e^{-\lambda x}\,dx = \frac{2}{\lambda^2}e^{-\lambda s}$. Also, $\sigma(Y) = \frac{1}{\lambda}[e^{-\lambda s}(2 - e^{-\lambda s})]^{1/2}$.

4.41 The age of the bulb upon replacement is $Y = g(X)$, where $g(x) = x$ for $x \leq 10$ and $g(x) = 10$ for $x > 10$. Then $E(Y) = \int_2^{10} x \frac{1}{10}\,dx + \int_{10}^{12} 10 \frac{1}{10}\,dx$ and $E(Y^2) = \int_2^{10} x^2 \frac{1}{10}\,dx + \int_{10}^{12} 10^2 \frac{1}{10}\,dx$. Thus, $E(Y) = 6.8$ and $\sigma(Y) = 2.613$.

4.43 Since $P(\min(X,Y) > t) = P(X > t, Y > t) = P(X > t)P(Y > t)$, we get $P(\min(X,Y) > t) = e^{-\alpha t}e^{-\beta t} = e^{-(\alpha+\beta)t}$ for all $t > 0$ and so $\min(X,Y)$ is exponentially distributed with parameter $\alpha + \beta$. Using the memoryless property of the exponential distribution, the time T to failure of the reliability system is distributed as $T_1 + T_2 + T_3$, where T_1, T_2, and T_3 are independent and exponentially distributed with respective parameters 5λ, 4λ, and 3λ. Thus, $E(T) = \frac{1}{5\lambda} + \frac{1}{4\lambda} + \frac{1}{3\lambda} = \frac{0.7833}{\lambda}$ and $\sigma^2(T) = \left(\frac{1}{25\lambda^2} + \frac{1}{16\lambda^2} + \frac{1}{9\lambda^2}\right)^{0.5} = \frac{0.4622}{\lambda}$.

4.45 Since the sojourn time of each bus is exactly half an hour, the number of buses on the parking lot at 4 p.m. is the number of buses arriving between 3:30 p.m. and 4 p.m. Taking the hour as unit of time, the buses arrive according to a Poisson process with rate $\lambda = \frac{4}{3}$. Using the memoryless property of the Poisson process, the number of buses arriving between 3:30 p.m. and 4 p.m. is Poisson distributed with expected value $\lambda \times \frac{1}{2} = \frac{2}{3}$.

4.47 The probability that the time between the passings of two consecutive cars is more than c seconds is $p = \int_c^\infty \lambda e^{-\lambda t}\, dt = e^{-\lambda c}$. By the lack of memory of the exponential distribution, p also gives the probability that no car comes around the corner during the c seconds measured from the moment you arrive. The number of passing cars before you can cross the road has the shifted geometric distribution $\{(1-p)^k p, k \geq 0\}$.

4.49 The probability of having a replacement because of a system failure is given by $\sum_{n=0}^\infty P(nT < X \leq (n+1)T - a) = \sum_{n=0}^\infty \left(e^{-\mu nT} - e^{-\mu[(n+1)T-a]}\right)$. This probability is equal to $(1 - e^{-\mu(T-a)})/(1 - e^{-\mu T})$. The expected time between two replacements is $\sum_{n=1}^\infty nTP((n-1)T < X \leq nT)) = T/(1 - e^{-\mu T})$.

4.51 Your win probability is the probability of having exactly one signal in (s, T). This probability is $e^{-\lambda(T-s)}\lambda(T-s)$, by the memoryless property of the Poisson process. Putting the derivative of this expression equal to zero, we get that the optimal value of s is $T - \frac{1}{\lambda}$. The maximal win probability is e^{-1}.

4.53 Take the minute as time unit. Let $\lambda = \frac{8}{60}$ and $T = 10$. The probability that the ferry will leave with two cars is $1 - e^{-\lambda T} = 0.7364$. Let the generic variable X be exponentially distributed with an expected value of $\frac{1}{\lambda} = 7.5$ minutes. The expected value of the time until the ferry leaves is $\frac{1}{\lambda} + E(\min(X, T))$ and equals $\frac{1}{\lambda} + \int_0^T t\lambda e^{-\lambda t}\, dt + T\int_T^\infty \lambda e^{-\lambda t}\, dt = 13.02$ minutes.

4.55 In view of Rule 4.3, we can think of failures occurring according to a Poisson process with a rate of 4 per 1,000 hours. The probability of no more than five failures during 1,000 hours is the Poisson probability $\sum_{k=0}^{5} e^{-4} \frac{4^k}{k!} = 0.7851$. The smallest value of n such that $\sum_{k=0}^{n} e^{-4} \frac{4^k}{k!} \geq 0.95$ is $n = 8$.

4.57 (a) Since the number of goals in the match is Poisson distributed with an expected value of $90 \times \frac{1}{30} = 3$, the answer is $1 - \sum_{k=0}^{2} e^{-3} \frac{3^k}{k!} = 0.5768$. (b) The numbers of goals in disjoint time intervals are independent and so the answer is $e^{-1.5} \frac{1.5^2}{2!} \times e^{-1.5} 1.5 = 0.0840$. (c) Let $a_k = e^{-3 \times (12/25)}(3 \times (12/25))^k/k!$ and $b_k = e^{-3 \times (13/25)}(3 \times (13/25))^k/k!$. Then, using Rule 3.13, the probability of a draw is $\sum_{k=0}^{\infty} a_k \times b_k = 0.2425$ and the probability of a win for team A is $\sum_{k=1}^{\infty} a_k \times \sum_{n=0}^{k-1} b_n = 0.3524$.

4.59 The answer is $\left(1 - \Phi\left(\frac{20}{16}\right)\right) \times 100\% = 10.56\%$.

4.61 An estimate for the standard deviation σ of the demand follows from the formula $50 + \sigma z_{0.95} = 75$, where $z_{0.95} = 1.6449$ is the 95% percentile of the standard normal distribution. This gives the estimate $\sigma = 1.2$.

4.63 By $P(X < 20) = P(X \leq 20) = P\left(\frac{X-25}{2.5} \leq \frac{20-25}{2.5}\right) = \Phi(-2)$, we have $P(X < 20) = 0.0228$. Finding the standard deviation σ of the thickness of the coating so that $P(X < 20) = 0.01$ translates into solving σ from $\Phi\left(\frac{20-25}{\sigma}\right) = 0.01$. The 0.01th percentile of the $N(0, 1)$ distribution is -2.3263, and so $-5/\sigma = -2.3263$, or $\sigma = 2.149$.

4.65 We have $P(|X - \mu| > k\sigma) = P(|Z| > k) = P(Z \leq -k) + P(Z \geq k)$, where Z is $N(0, 1)$ distributed. Since $P(Z \geq k) = P(Z \leq -k)$ and $P(Z \geq k) = 1 - \Phi(k)$, the sought result follows.

4.67 The number of heads in 10,000 tosses of a fair coin is approximately normally distributed with expected value 5,000 and standard deviation 50. The outcome of 5,250 heads lies five standard deviations above the expected value. Without doing any further calculations, we can conclude that the claim is highly implausible $(1 - \Phi(5) = 2.87 \times 10^{-7})$.

4.69 Since $X - Y$ is $N(0, 2\sigma^2)$ distributed, $(X - Y)/(\sigma\sqrt{2})$ is $N(0, 1)$ distributed. Thus, using Problem 4.68, $E(|X - Y|) = \sigma\sqrt{2}\sqrt{2/\pi}$. Also, $E(X + Y) = 2\mu$. The formulas for $E(|X - Y|)$ and $E(X + Y)$ give two equations in $E[\max(X, Y)]$ and $E[\min(X, Y)]$, yielding the sought result. *Note*: $\max(X, Y)$ and $\min(X, Y)$ each have standard deviation $\sigma\sqrt{1 - 1/\pi}$.

4.71 Let X_1 and X_2 be the two measurement errors. Since X_1 and X_2 are independent, $\frac{1}{2}(X_1 + X_2)$ is normally distributed with expected value 0 and standard deviation $\frac{1}{2}\sqrt{0.006^2 l^2 + 0.004^2 l^2} = l\sqrt{52}/2{,}000$. The sought probability is $P(\frac{1}{2}|X_1 + X_2| \leq 0.005l) = P(-0.01l \leq X_1 + X_2 \leq 0.01l)$ and is equal to $\Phi\left(\frac{20}{\sqrt{52}}\right) - \Phi\left(-\frac{20}{\sqrt{52}}\right) = 0.9945$.

4.73 (a) The profit of Joe and his brother after 52 weeks is $\sum_{i=1}^{52} X_i$, where the X_i are independent with $P(X_i = 10) = \frac{1}{2}$ and $P(X_i = -5) = \frac{1}{2}$. The expected value and the standard deviation of the X_i are 2.5 and $\sqrt{62.5 - 2.5^2} = 7.5$ dollars. The sought probability is $P(\sum_{i=1}^{52} X_i \geq 100) \approx 1 - \Phi(\frac{100 - 52 \times 2.5}{7.5\sqrt{52}}) = 0.7105$. (b) Let X_i be the score of the ith roll. The probability $P(\sum_{i=1}^{80} X_i \leq 300)$ can be approximated by $\Phi(\frac{300 - 80 \times 3.5}{1.7078\sqrt{80}}) = 0.9048$.

4.75 (a) The number of sixes in one throw of $6r$ dice is distributed as the binomial random variable $S_{6r} = \sum_{k=1}^{6r} X_k$, where the X_k are independent 0–1 variables with expected value $\mu = \frac{1}{6}$ and standard deviation $\sigma = \frac{1}{6}\sqrt{5}$. The probability $P(S_{6r} \geq r)$ can be written as $P((S_{6r} - 6r\mu)/(\sigma\sqrt{6r}) \geq 0)$. By the central limit theorem, this probability tends to $1 - \Phi(0) = \frac{1}{2}$ as $r \to \infty$. (b) Since the sum of finitely many independent Poisson-distributed random variables is again Poisson distributed, $e^{-n}(1 + \frac{n}{1!} + \frac{n^2}{2!} + \cdots + \frac{n^n}{n!})$ can be interpreted as the probability $P(X_1 + \cdots + X_n \leq n)$, where the X_i are independent and Poisson distributed with parameter 1. Next, repeat the arguments in (a).

4.77 The number of even numbers in any given drawing of the 6/45 lottery has a hypergeometric distribution with expected value $\mu = \frac{132}{45}$ and standard deviation $\sigma = \frac{1}{15}\sqrt{299}$. By the central limit theorem, the total number of even numbers that will be obtained in 52 drawings of the 6/45 lottery is approximately normally distributed with expected value $52\mu = 152.533$ and standard deviation $\sigma\sqrt{52} = 8.313$. The outcome 162 lies $(162 - 152.533)/8.313 = 1.14$ standard deviations above the expected value. This outcome does not cast doubts on the unbiased nature of the lottery drawings.

4.79 The payoff per game has an expected value of $\mu = 12$ dollars and a standard deviation of $\sigma = \sqrt{6,142 - 144} = 77.447$ dollars. By the central limit theorem, the probability of the casino losing money in a given week is approximately equal to $\Phi(-\frac{5,000 \times 3}{77.447\sqrt{5,000}}) = 3.1 \times 10^{-3}$.

4.81 The premium c should satisfy $P(rc - (X_1 + \cdots + X_n) \geq \frac{1}{10}rc)$, where X_i is the amount claimed by the ith policyholder. This probability can be approximated by $\Phi((\frac{9}{10}rc - r\mu)/(\sigma\sqrt{r}))$. Thus, c should be chosen such that $(\frac{9}{10}rc - r\mu)/(\sigma\sqrt{r})$ is equal to the percentile 2.326. Hence, $c \approx \frac{10}{9}(\mu + \frac{2.326\sigma}{\sqrt{r}})$.

4.83 Let X_i be the amount of dollars the casino owner loses on the ith bet. The X_i are independent random variables with $P(X_i = 10) = \frac{18}{37}$ and $P(X_i = -5) = \frac{19}{37}$. Then $E(X_i) = \frac{85}{37}$ and $\sigma(X_i) = \frac{45}{37}\sqrt{38}$. The amount of dollars lost by the casino owner is $\sum_{i=1}^{2,500} X_i$ and is approximately

$N(\mu, \sigma^2)$ distributed with $\mu = 2{,}500 \times \frac{85}{37}$ and $\sigma = 50 \times \frac{45}{37}\sqrt{38}$. The casino owner will lose more than 6,500 dollars with a probability of about $1 - \Phi\left(\frac{6{,}500-\mu}{\sigma}\right) = 0.0218$.

4.85 Let V_n be the bankroll (in dollars) of the gambler after the nth bet. Then $V_n = (1-\alpha)V_{n-1} + \alpha V_{n-1}R_n$, where $\alpha = 0.05$, $V_0 = 1{,}000$, and the R_i are independent random variables with $P(R_i = \frac{1}{4}) = \frac{19}{37}$ and $P(R_i = 2) = \frac{18}{37}$. Iterating this equality gives $V_n = (1-\alpha+\alpha R_1) \times \cdots \times (1-\alpha+\alpha R_n)V_0$. This leads to $\ln(V_n/V_0) = \sum_{i=1}^n \ln(1-\alpha+\alpha R_i)$. The random variables $X_i = \ln(1-\alpha+\alpha R_i)$ are independent with expected value $\mu = \frac{19}{37}\ln(0.9625) + \frac{18}{37}\ln(1.05)$ and variance $\sigma^2 = \frac{19}{37}\ln^2(0.9625) + \frac{18}{37}\ln^2(1.05) - \mu^2$. By the central limit theorem, $\ln(V_{100}/V_0)$ is approximately $N(100\mu, 100\sigma^2)$ distributed (the gambler's bankroll after 100 bets is approximately lognormally distributed). The probability that the gambler will take home more than d dollars is $P(V_n > V_0 + d) = P\bigl(\ln(V_n/V_0) > \ln(1+d/V_0)\bigr)$. This probability is approximately $1 - \Phi\bigl([\ln(1+d/V_0) - 100\mu]/10\sigma\bigr)$ and has the approximate values 0.8276, 0.5494, 0.2581, and 0.0264 for $d = 0$, 500, 1,000, and 2,500.

4.87 The distance can be modeled as $\sqrt{X^2 + Y^2}$, where X and Y are independent $N(0,1)$ random variables. The random variable $X^2 + Y^2$ has chi-squared density $f(v) = \frac{1}{2}e^{-\frac{1}{2}v}$. We have $P(\sqrt{X^2+Y^2} \le r) = P(X^2 + Y^2 \le r^2)$ for $r > 0$. Hence, the probability density of the distance from the center of the target to the point of impact is $2rf(r^2) = re^{-\frac{1}{2}r^2}$ for $r > 0$. The expected value of the distance is $\int_0^\infty r^2 e^{-\frac{1}{2}r^2}\, dr = \frac{1}{2}\sqrt{2\pi}$. The mode of the distance follows by solving the equation $\int_0^x re^{-\frac{1}{2}r^2}\, dr = 0.5$. The mode is $\sqrt{2\ln(2)}$.

4.89 The inverse of the function $e = \frac{1}{2}ms^2$ is $s = \sqrt{\frac{2e}{m}}$, where $\frac{ds}{de} = \sqrt{\frac{1}{2em}}$. An application of Rule 4.6 gives that the probability density of the kinetic energy E is $\frac{2}{c^3}\sqrt{\frac{e}{\pi m}}\, e^{-me/c^2}$ for $e > 0$.

4.91 It follows from $P(Y \le y) = P\bigl(X \le \frac{\ln(y)}{\ln(10)}\bigr)$ that $P(Y \le y) = \frac{\ln(y)}{\ln(10)}$ for $1 \le y \le 10$. Differentiation gives the density of Y.

4.93 Generate n random numbers $u_1, \ldots, u_n$ from the interval $(0,1)$. Then the number $-\frac{1}{\lambda}\bigl[\ln(u_1) + \cdots + \ln(u_n)\bigr] = -\frac{1}{\lambda}\ln(u_1 \times \cdots \times u_n)$ is a random observation from the gamma-distributed random variable.

4.95 The random variable X is with probability p distributed as an exponential random variable with parameter λ_1 and with probability $1-p$ as an exponential random variable with parameter λ_2. Hence, generate two random numbers u_1 and u_2 from $(0,1)$. The random observation is $-(1/\lambda_1)\ln(u_1)$ if $u_1 \le p$ and is $-(1/\lambda_2)\ln(u_2)$ otherwise. For the second part, let V be exponentially distributed with parameter λ. Then the random variable X is with probability p distributed as V and with probability $1-p$ as $-V$. Hence, generate two random numbers u_1 and u_2 from $(0,1)$. The random observation is $-\frac{1}{\lambda}\ln(u_2)$ if $u_1 \leq p$ and $\frac{1}{\lambda}\ln(u_2)$ otherwise. This simulation method is called the composition method.

4.97 Since $P(V > x) = P(X_1 > x, \ldots, X_n > x)$, it follows from the independence of the X_k and the failure rate representation of the reliability function that $P(V > x)$ equals $P(X_1 > x) \cdots P(X_n > x) = e^{-\int_0^x r_1(y)\,dy} \cdots e^{-\int_0^x r_n(y)\,dy}$. This gives that $P(V > x) = e^{-\int_0^x [\sum_{k=1}^n r_k(y)]\,dy}$, proving the desired result.

4.99 Let $F(x) = P(X \leq x)$. Since $\int_0^x r(t)\,dt = \lambda \int_0^x d\ln(1+t^\alpha) = \lambda \ln(1+x^\alpha)$, it follows that $F(x) = 1 - e^{-\lambda \ln(1+x^\alpha)}$. Thus, the reliability function $1 - F(x)$ is equal to $(1+x^\alpha)^{-\lambda}$. The derivative $r'(x) = (\alpha-1-x^\alpha)/(1+x^\alpha)^2$. For the case that $\alpha > 1$, the derivative is positive for $x < (\alpha-1)^{1/\alpha}$ and negative for $x > (\alpha-1)^{1/\alpha}$, showing that $r(x)$ first increases and then decreases.

4.101 The integral $(1/e^{-(1,000/1,250)^{2.1}}) \int_{1,000}^\infty e^{-(y/1,250)^{2.1}}\,dy$ gives the mean residual lifetime $m(1,000)$. By numerical integration, $m(1,000) = 516.70$ hours.

4.103 Form the Lagrange function $F(p_1, \ldots, p_n, \lambda_1, \lambda_2) = -\sum_{i \geq 1} p_i \log p_i + \lambda_1(\sum_{i=1}^n p_i - 1) + \lambda_2(\sum_{i=1}^n p_i E_i - E)$, where λ_1 and λ_2 are Lagrange multipliers. Putting $\partial F/\partial p_i = 0$ results in $-1 - \log p_i + \lambda_1 + \lambda_2 E_i = 0$ and so $p_i = e^{\lambda_1 - 1 + \lambda_2 E_i}$ for all i. Substituting p_i into the constraint $\sum_{i=1}^n p_i = 1$ gives $e^{\lambda_1 - 1} = 1/\sum_{i=1}^n e^{\lambda_2 E_i}$. Thus, $p_i = e^{\lambda_2 E_i}/\sum_{k=1}^n e^{\lambda_2 E_k}$ for all i. Substituting this into $\sum_{i=1}^n p_i E_i = E$ gives the equation $\sum_{i=1}^n E_i e^{\lambda_2 E_i} - E \sum_{k=1}^n e^{\lambda_2 E_k} = 0$ for λ_2. Replacing λ_2 by $-\beta$, we get the desired expression for the p_i^* and the equation for the unknown β.

Chapter 5

5.1 Let X be the low points rolled and Y be the high points rolled. These random variables are defined on the sample space consisting of the 36 equiprobable outcomes (i,j) with $1 \leq i,j \leq 6$, where i is the number shown by the first die and j is the number shown by the second die. For $k < l$, the event $\{X = k, Y = l\}$ occurs for the outcomes (k,l) and (l,k). This gives $P(X = k, Y = l) = \frac{2}{36}$ for $1 \leq k < l \leq 6$. Further, $P(X = k, Y = k) = \frac{1}{36}$ for all k.

5.3 The joint mass function of X and $Y-X$ is $P(X = x, Y-X = z) = P(X = x, Y = z+x) = \frac{e^{-2}}{x!z!}$ for $x, z = 0, 1, \ldots$. Since $\sum_{z=0}^{\infty} \frac{e^{-2}}{x!z!} = \frac{e^{-1}}{x!}$ and $\sum_{x=0}^{\infty} \frac{e^{-2}}{x!z!} = \frac{e^{-1}}{z!}$, the marginal distributions of X and $Y-X$ are Poisson distributions with expected value 1. Noting that $P(X = x, Y-X = z) = P(X = x)P(Y-X = z)$ for all x, z, we have by Rule 3.7 that X and $Y-X$ are independent. Then, by Rule 3.12, the random variable Y is Poisson distributed with expected value 2.

5.5 The sample space for X and Y is the set of $\binom{10}{3} = 120$ combinations of three distinct numbers from 1 to 10. The joint mass function $P(X = x, Y = y)$ is $(y-x-1)/120$ for $1 \le x \le 8$, $x+2 \le y \le 10$. The marginal distributions are $P(X = x) = \sum_{y=x+2}^{10}(y-x-1)/120 = (10-x)(9-x)/240$ for $1 \le x \le 8$ and $P(Y = y) = \sum_{x=1}^{y-2}(y-x-1)/120 = (y-1)(y-2)/240$ for $3 \le y \le 10$. Further, $P(Y-X = k)$ is $\sum_{x=1}^{10-k} P(X = x, Y = x+k) = (k-1)(10-k)/120$ for $2 \le k \le 9$.

5.7 Since $P(X = x, Y = y, N = n) = \frac{1}{6}\binom{n}{x}\left(\frac{1}{2}\right)^n\binom{n}{y}\left(\frac{1}{2}\right)^n$, we get $P(X = x, Y = y) = \frac{1}{6}\sum_{n=1}^{6}\binom{n}{x}\binom{n}{y}\left(\frac{1}{2}\right)^{2n}$ for $0 \le x, y \le 6$. We have $P(X = Y) = \frac{1}{6}\sum_{n=1}^{6}\left(\frac{1}{2}\right)^{2n}\sum_{x=0}^{n}\binom{n}{x}^2$ and so $P(X = Y) = \frac{1}{6}\sum_{n=1}^{6}\binom{2n}{n}\left(\frac{1}{2}\right)^{2n} = 0.3221$.

5.9 The constant c must satisfy $1 = c\int_0^{\infty}\int_0^x e^{-2x}\,dx\,dy = c\int_0^{\infty} xe^{-2x}\,dx$. Since $\int_0^{\infty} 4xe^{-2x}\,dx = 1$, we get $c = 4$. By $P((X, Y) \in C) = \iint_C f(x, y)\,dx\,dy$, we have that $Z = X - Y$ satisfies $P(Z > z) = \int_0^{\infty} dy \int_{y+z}^{\infty} 4e^{-2x}\,dx = \int_0^{\infty} 2e^{-2(y+z)}\,dy = e^{-2z}$ for $z > 0$. Thus, Z has exponential density $2e^{-2z}$.

5.11 Since $c\int_0^1 dx \int_0^1 \sqrt{x+y}\,dy = 1$, we have $c = (15/4)(4\sqrt{2}-2)^{-1}$. Using the basic relation $P((X, Y) \in C) = \iint_C f(x, y)\,dx\,dy$, it follows that $P(X+Y \le z) = c\int_0^z dx \int_0^{z-x} \sqrt{x+y}\,dy$ is $\frac{2c}{3}\int_0^z (z^{3/2}-x^{3/2})\,dx = \frac{2c}{5}z^2\sqrt{z}$ for $0 \le z \le 1$ and $P(X+Y \le z) = c\int_{z-1}^1 dx \int_{z-x}^1 \sqrt{x+y}\,dy$ is $\frac{4c}{15}(2^{5/2}-z^{5/2}) - \frac{2c}{3}(2-z)z^{3/2}$ for $1 \le z \le 2$. The density function of $X+Y$ is $cz\sqrt{z}$ for $0 < z < 1$ and $c(2-z)\sqrt{z}$ for $1 \le z < 2$.

5.13 Since $P(x < X \le x+\Delta x, y < Y \le y+\Delta y) \approx 2\frac{\Delta x}{a} \times \frac{\Delta y}{a}$ for $0 < x < y < a$ when Δx and Δy are small, the joint density function of X and Y is $f(x, y) = \frac{2}{a^2}$ for $0 < x < y < a$ and $f(x, y) = 0$ otherwise. Alternatively, the joint density $f(x, y)$ can be obtained from $P(X > x, Y \le y) = \left(\frac{y-x}{a}\right)^2$ for $0 \le x \le y \le a$. Use next $f(x, y) = \frac{\partial}{\partial x}\left(\frac{\partial}{\partial y} P(X \le x, Y \le y)\right)$ and the formula $P(X \le x, Y \le y) + P(X > x, Y \le y) = P(Y \le y)$.

5.15 Using the basic formula $P((X, Y) \in C) = \iint_C f(x, y)\,dx\,dy$, it follows that $P(X < Y) = \frac{1}{10}\int_5^{10} dx \int_x^{\infty} e^{-\frac{1}{2}(y+3-x)}\,dy = e^{-\frac{3}{2}}$.

5.17 We have $P(\max(X, Y) > a\min(X, Y)) = P(X > aY) + P(Y > aX)$. Thus, by symmetry, $P(\max(X, Y) > a\min(X, Y)) = 2P(X > aY)$. The joint density of X and Y is $f(x, y) = 1$ for $0 < x, y < 1$ and so $P(X > aY) = \int_0^1 dx \int_0^{x/a} dy = \int_0^1 \frac{x}{a} dx = \frac{1}{2a}$. The sought probability is $\frac{1}{a}$.

5.19 The time until both components are down is $T = \max(X, Y)$. Noting that $P(T \leq t) = P(X \leq t, Y \leq t)$, it follows that $P(T \leq t) = \frac{1}{4}\int_0^t dx \int_0^t (2y + 2 - x) dy = 0.125t^3 + 0.5t^2$ for $0 \leq t \leq 1$ and $P(T \leq t) = \frac{1}{4}\int_0^1 dx \int_0^t (2y + 2 - x) dy = 0.75t - 0.125t^2$ for $1 \leq t \leq 2$. The density function of T is $0.375t^2 + t$ for $0 < t < 1$ and $0.75 - 0.25t$ for $1 \leq t < 2$.

5.21 Using the basic formula $P((X, Y) \in C) = \iint_C f(x, y) dx dy$, it follows that the sought probability is $P(B^2 \geq 4A) = \int_0^1 \int_0^1 \chi(a, b) f(a, b) \, da \, db$, where $\chi(a, b) = 1$ for $b^2 \geq 4a$ and $\chi(a, b) = 0$ otherwise. This leads to $P(B^2 \geq 4A) = \int_0^1 db \int_0^{b^2/4} (a + b) \, da = 0.0688$. Similarly, $P(B^2 \geq 4AC) = \int_0^1 \int_0^1 \int_0^1 \chi(a, b, c) f(a, b, c) \, da \, db \, dc$, where $\chi(a, b, c) = 1$ for $b^2 \geq 4ac$ and $\chi(a, b, c) = 0$ otherwise. A convenient order of integration for $P(B^2 \geq 4AC)$ is $\frac{2}{3}\int_0^1 db \int_0^{b^2/4} da \int_0^1 (a + b + c) \, dc + \frac{2}{3}\int_0^1 db \int_{b^2/4}^1 da \int_0^{b^2/(4a)} (a + b + c) \, dc$. This leads to $P(B^2 \geq 4AC) = 0.1960$.

5.23 The marginal densities of X and Y are $f_X(x) = \int_0^x 4e^{-2x} dy = 4xe^{-2x}$ for $x > 0$ and $f_Y(y) = \int_y^\infty 4e^{-2x} dx = 2e^{-2y}$ for $y > 0$.

5.25 The joint density of X and Y is $f(x, y) = 4/\sqrt{3}$ for (x, y) inside the triangle. The marginal density of X is $f_X(x) = \int_0^{x\sqrt{3}} f(x, y) dy = 4x$ for $0 < x < 0.5$ and $f_X(x) = \int_0^{(1-x)\sqrt{3}} f(x, y) dy = 4(1 - x)$ for $0.5 \leq x < 1$. The marginal density of Y is $f_Y(y) = \int_{y/\sqrt{3}}^{1-y/\sqrt{3}} f(x, y) dx = 4/\sqrt{3} - 8y/3$ for $0 < y < \frac{1}{2}\sqrt{3}$.

5.27 We have $P(X \leq x, Y - X \leq z) = \int_0^x dv \int_v^{v+z} e^{-w} dw = (1 - e^{-x})(1 - e^{-z})$ for $x, z > 0$, using the basic formula $P((X, Y) \in C) = \iint_C f(x, y) dx dy$. By partial differentiation, we get that the joint density of X and $Z = Y - X$ is $f(x, z) = e^{-x} e^{-z}$ for $x, z > 0$. The marginal densities of X and Z are $f_X(x) = e^{-x}$ and $f_Z(z) = e^{-z}$. The time until the system goes down is Y. The density function of Y is $\int_0^y e^{-y} dx = ye^{-y}$ for $y > 0$.

5.29 Let X and Y be the packet delays on the two lines. The joint density of X and Y is $f(x, y) = \lambda e^{-\lambda x} \lambda e^{-\lambda y}$ for $x, y > 0$. Using the basic formula $P((X, Y) \in C) = \iint_C f(x, y) dx dy$, we obtain $P(X - Y > v) = \int_0^\infty \lambda e^{-\lambda y} dy \int_{y+v}^\infty \lambda e^{-\lambda x} dx$ for any $v \geq 0$. This leads to $P(X - Y > v) = \frac{1}{2} e^{-\lambda v}$ for $v \geq 0$. For any $v \leq 0$, $P(X - Y \leq v) = P(Y - X \geq -v)$. Thus, by symmetry, $P(X - Y \leq v) = \frac{1}{2} e^{\lambda v}$ for $v \leq 0$. Thus, the density of $X - Y$ is $\frac{1}{2}\lambda e^{-\lambda|v|}$ for $-\infty < v < \infty$, which is the so-called Laplace density.

5.31 We have $P(F \leq c) = P(X + Y \leq c) + P(1 \leq X + Y \leq c + 1)$ for $0 \leq c \leq 1$. Since $P(X + Y \leq c) = \int_0^c dx \int_0^{c-x} dy$ and $P(1 \leq X + Y \leq c + 1) = \int_0^1 dx \int_{1-x}^{\min(c+1-x, 1)} dy$, we get $P(X + Y \leq c) = \frac{1}{2}c^2$ and $P(1 \leq X+Y \leq c+1) = \int_0^c dx \int_{1-x}^1 dy + \int_c^1 dx \int_{1-x}^{c+1-x} dy = \frac{1}{2}c^2 + c(1-c)$. This gives $P(F \leq c) = c$ for all $0 \leq c \leq 1$, proving the desired result.

5.33 The joint density of X and Y is $f(x, y) = 1$ for $0 < x, y < 1$. The distribution function of $Z = X/Y$ is $P(Z \leq z) = \int_0^1 dy \int_0^{\min(1,zy)} dx$. This leads to $P(Z \leq z) = \frac{1}{2}z$ for $0 \leq z \leq 1$ and $P(Z \leq z) = 1 - \frac{1}{2z}$ for $z > 1$. The density of Z is $\frac{1}{2}$ for $0 < z \leq 1$ and $\frac{1}{2z^2}$ for $z > 1$. The probability that the first significant digit of Z is 1 equals $\sum_{n=0}^\infty P(10^n \leq Z < 2 \times 10^n) + \sum_{n=1}^\infty P(10^{-n} \leq Z < 2 \times 10^{-n})$ and is $\frac{5}{18} + \frac{1}{18} = \frac{1}{3}$. In general, the probability that the first significant digit of Z equals k is $\frac{1}{18}\left(\frac{10}{k(k+1)} + 1\right)$ for $1 \leq k \leq 9$.

5.35 We have $P(\max(X, Y) \leq t) = P(X \leq t, Y \leq t) = P(X \leq t)P(Y \leq t)$, and so $P(\max(X, Y) \leq t) = (1 - e^{-\lambda t})^2$ for $t > 0$. Since $P(X + \frac{1}{2}Y \leq t)$ is given by $\int_0^t \lambda e^{-\lambda x} dx \int_0^{2(t-x)} \lambda e^{-\lambda y} dy$, we find $P(X + \frac{1}{2}Y \leq t) = (1 - e^{-\lambda t})^2$.

5.37 Let $X_1, X_2, \ldots$ be a sequence of independent random variables that are uniformly distributed on $(0, 1)$, and let $S_n = X_1 + \cdots + X_n$. The sought probability is given by $P(S_1 > a) + \sum_{n=1}^\infty P(S_n \leq a, a < S_n + X_{n+1} \leq 1)$. Since S_n and X_{n+1} are independent of each other, the joint density $f_n(s, x)$ of S_n and X_{n+1} satisfies $f_n(s, x) = \frac{s^{n-1}}{(n-1)!}$ for $0 < s < 1$ and $0 < x < 1$, using the result (a) of Problem 5.36. Therefore, $P(S_n \leq a, a < S_n + X_{n+1} \leq 1)$ equals $\int_0^a ds \int_{a-s}^{1-s} f_n(s, x) dx = (1-a)\frac{a^n}{n!}$. Thus, the sought probability is $1 - a + \sum_{n=1}^\infty (1-a)\frac{a^n}{n!} = (1-a)e^a$.

5.39 By $P(V > v, W \leq w) = P(v < X \leq w, v < Y \leq w)$ and the independence of X and Y, we have $P(V > v, W \leq w) = P(v < X \leq w) P(v < Y \leq w) = \left(e^{-\lambda v} - e^{-\lambda w}\right)^2$ for $0 \leq v \leq w$. Taking partial derivatives, we get that the joint density of V and W is $f(v, w) = 2\lambda^2 e^{-\lambda(v+w)}$ for $0 < v < w$. It follows from $P(W - V > z) = \int_0^\infty dv \int_{v+z}^\infty 2\lambda^2 e^{-\lambda(v+w)} dw$ that $P(W - V > z) = e^{-\lambda z}$ for $z > 0$, in agreement with the memoryless property of the exponential distribution.

5.41 Define the function $g(x, y)$ as $g(x, y) = T - \max(x, y)$ if $0 \leq x, y \leq T$ and $g(x, y) = 0$ otherwise. The joint density function of X and Y is $e^{-(x+y)}$ for $x, y > 0$. Using the memoryless property of the exponential distribution, the expected amount of time the system is down between two inspections is given by $E[g(X, Y)] = \int_0^T \int_0^T (T - \max(x, y)) e^{-(x+y)} dx dy$. This double integral can be reduced to $2\int_0^T (T - x)(1 - e^{-x})e^{-x} dx = T - 1.5 + 2e^{-T} - \frac{1}{2}e^{-2T}$.

5.43 The expected value of the area of the circle is $\int_0^1 \int_0^1 \pi(x^2 + y^2)(x+y)\,dx\,dy$, by the substitution rule. This integral is equal to $\pi \int_0^1 (x^3 + \frac{1}{2}x^2 + \frac{1}{3}x + \frac{1}{4})\,dx = \frac{5}{6}\pi$.

5.45 The inverse functions $x = a(v,w)$ and $y = b(v,w)$ are $a(v,w) = vw$ and $b(v,w) = v(1-w)$. The Jacobian $J(v,w)$ is equal to $-v$. The joint density of V and W is $f_{V,W}(v,w) = \mu e^{-\mu vw} \mu e^{-\mu v(1-w)} |-v| = \mu^2 v e^{-\mu v}$ for $v > 0$ and $0 < w < 1$. The marginal densities of V and W are $f_V(v) = \int_0^1 \mu^2 v e^{-\mu v}\,dw = \mu^2 v e^{-\mu v}$ for $v > 0$ and $f_W(w) = \int_0^\infty \mu^2 v e^{-\mu v}\,dv = 1$ for $0 < w < 1$. Since $f_{V,W}(v,w) = f_V(v)f_W(w)$ for all v,w, V and W are independent.

5.47 Since Z^2 has the χ_1^2 density $\frac{1}{\sqrt{2\pi}} u^{-\frac{1}{2}} e^{-\frac{1}{2}u}$ when Z is $N(0,1)$ distributed and the random variables Z_1^2 and Z_2^2 are independent, the joint density of $X = Z_1^2$ and $Y = Z_2^2$ is $\frac{1}{2\pi}(xy)^{-\frac{1}{2}} e^{-\frac{1}{2}(x+y)}$ for $x,y > 0$. For the transformation $V = X + Y$ and $W = X/Y$, the inverse functions $x = a(v,w)$ and $y = b(v,w)$ are $a(v,w) = \frac{1}{2}(v+w)$ and $b(v,w) = \frac{1}{2}(v-w)$. The Jacobian $J(v,w)$ is equal to $-\frac{1}{2}$. The joint density of V and W is $f_{V,W}(v,w) = \frac{1}{4\pi^2}(v^2 - w^2)^{-\frac{1}{2}} e^{-\frac{1}{2}v}$ for $v > 0$ and $-\infty < w < \infty$. The random variables V and W are not independent.

5.49 The inverse functions are $a(v,w) = v e^{-\frac{1}{4}(v^2+w^2)}/\sqrt{v^2+w^2}$ and $b(v,w) = w e^{-\frac{1}{4}(v^2+w^2)}/\sqrt{v^2+w^2}$. The Jacobian is $\frac{1}{2} e^{-\frac{1}{2}(v^2+w^2)}$. Since $f_{X,Y}(x,y) = \frac{1}{\pi}$, we get $f_{V,W}(v,w) = \frac{1}{\pi} \times \frac{1}{2} e^{-\frac{1}{2}(v^2+w^2)}$. Noting that $f_{V,W}(v,w) = \frac{1}{\sqrt{2\pi}} e^{-\frac{1}{2}v^2} \times \frac{1}{\sqrt{2\pi}} e^{-\frac{1}{2}w^2}$ for all v,w, it follows that V and W are independent and $N(0,1)$ distributed.

5.51 The joint density of X and Y is $\frac{1}{\Gamma(\alpha)\Gamma(\beta)} x^{\alpha-1} y^{\beta-1} e^{-(x+y)}$. The inverse functions are $a(v,w) = vw$ and $b(v,w) = w(1-v)$. The Jacobian $J(v,w) = w$. Thus, the joint density of V and W is $\frac{1}{\Gamma(\alpha)\Gamma(\beta)}(vw)^{\alpha-1}(w(1-v))^{\beta-1} e^{-w} w$. This density can be rewritten as $\frac{\Gamma(\alpha+\beta)}{\Gamma(\alpha)\Gamma(\beta)} v^{\alpha-1}(1-v)^{\beta-1} \frac{w^{\alpha+\beta-1}}{\Gamma(\alpha+\beta)} e^{-w}$ for all v,w. This shows that V and W are independent, where V has a beta distribution with parameters α and β, and W has a gamma distribution with shape parameter $\alpha + \beta$ and scale parameter 1.

5.53 Since $P(x - \frac{1}{2}\Delta x \le U_{(1)} \le x + \frac{1}{2}\Delta x, y - \frac{1}{2}\Delta y \le U_{(n)} \le y + \frac{1}{2}\Delta y)$ is approximately equal to $\binom{n}{1}\binom{n-1}{1}(y-x)^{n-2} \Delta x \Delta y$ for $0 < x < y < 1$ and $\Delta x, \Delta y$ small, the joint density of $U_{(1)}$ and $U_{(n)}$ is $f(x,y) = \frac{n!}{(n-2)!}(y-x)^{n-2}$ for $0 < x < y < 1$.

5.55 The marginal distributions of X and Y are the Poisson distributions $p_X(x) = e^{-1}/x!$ for $x \ge 0$ and $p_Y(y) = e^{-2} 2^y/y!$ for $y \ge 0$ with $E(X) = \sigma^2(X) = 1$ and $E(Y) = \sigma^2(Y) = 2$. We have $E(XY) = $

$\sum_{x=0}^{\infty}\sum_{y=x}^{\infty} xy \frac{e^{-2}}{x!(y-x)!}$. Noting that $\sum_{y=x}^{\infty} y \frac{e^{-1}}{(y-x)!} = \sum_{y=x}^{\infty}(y-x+x)\frac{e^{-1}}{(y-x)!} = 1+x$, we get $E(XY) = 3$. This gives $\rho(X,Y) = 1/\sqrt{2}$.

5.57 The variance of the portfolio's return is $f^2\sigma_A^2 + (1-f)^2\sigma_B^2 + 2f(1-f)\sigma_A\sigma_B\rho_{AB}$. Putting the derivative of this function equal to zero, it follows that the optimal fraction f is $(\sigma_B^2 - \sigma_A\sigma_B\rho_{AB})/(\sigma_A^2 + \sigma_B^2 - 2\sigma_A\sigma_B\rho_{AB})$.

5.59 (a) Let R_A be the rate of return of stock A and R_B be the rate of return of stock B. Since $R_B = -R_A + 14$, the correlation coefficient is -1. (b) Let $X = fR_A + (1-f)R_B$. Since $X = (2f-1)R_A + 14(1-f)$, the variance of X is minimal for $f = \frac{1}{2}$. Investing $\frac{1}{2}$ of your capital in stock A and $\frac{1}{2}$ in stock B, the portfolio has a guaranteed rate of return of 7%.

5.61 The joint density of (X,Y) is $f(x,y) = \frac{1}{\pi}$ for (x,y) inside the circle C. Then $E(XY) = \iint_C xy \frac{1}{\pi} dx dy$. Since the function xy has opposite signs on the quadrants of the circle, a symmetry argument gives $E(XY) = 0$. Also, by the same argument, $E(X) = E(Y) = 0$. This gives $\rho(X,Y) = 0$, although X and Y are dependent.

5.63 The joint density function $f_{V,W}(v,w)$ of V and W is most easily obtained from the relation $P(v < V < v+\Delta v, w < W < w+\Delta w) = 2\Delta v \Delta w$ for $0 \le v < w \le 1$ and $\Delta v, \Delta w$ small enough. This shows that $f_{V,W}(v,w) = 2$ for $0 < v < w < 1$. Next it follows that $f_V(v) = 2(1-v)$ for $0 < v < 1$ and $f_W(w) = 2w$ for $0 < w < 1$. This leads to $E(VW) = \frac{1}{4}$, $E(V) = \frac{1}{3}$, $E(W) = \frac{2}{3}$, and $\sigma(V) = \sigma(W) = \frac{1}{3\sqrt{2}}$. Thus, $\rho(V,W) = \frac{1}{2}$.

5.65 The joint mass function of X and Y is $P(X=x, Y=y) = \frac{1}{100} \times \frac{1}{x}$ for $1 \le x \le 100$ and $1 \le y \le x$. Hence, $E(XY) = \sum_{x=1}^{100}\sum_{y=1}^{x} xy \times \frac{1}{100x} = 1{,}717$. The marginal distributions are $P(X=x) = \frac{1}{100}$ for $1 \le x \le 100$ and $P(Y=y) = \frac{1}{100}\sum_{x=y}^{100}\frac{1}{x}$ for $1 \le y \le 100$. We have $E(X) = 50.5$, $\sigma(X) = 28.8661$, $E(Y) = 25.75$, and $\sigma(Y) = 22.1402$. This gives $\rho(X,Y) = 0.652$.

5.67 The joint probability mass function $p(x,y) = P(X=x, Y=y)$ is given by $p(x,y) = r^{x-1}p(r+p)^{y-x-1}q$ for $x < y$ and $p(x,y) = r^{y-1}q(r+q)^{x-y-1}p$ for $x > y$. It is a matter of some algebra to get $E(XY) = \frac{p}{q}\frac{1}{(1-r)^2} + p\frac{1+r}{(1-r)^3} + \frac{q}{p}\frac{1}{(1-r)^2} + q\frac{1+r}{(1-r)^3}$. Also, $E(X) = \frac{1}{p}$ and $E(Y) = \frac{1}{q}$. This leads to $\text{cov}(X,Y) = -1/(1-r)$.

5.69 The "if" part follows from the relations $\text{cov}(X, aX+b) = a\text{cov}(X,X) = a\sigma_1^2$ and $\sigma(aX+b) = |a|\sigma_1$. Suppose now that $|\rho| = 1$. Since $\text{var}(V) = \frac{1}{\sigma_2^2}\sigma_2^2 + \frac{\rho^2}{\sigma_1^2}\sigma_1^2 - 2\frac{\rho}{\sigma_1\sigma_2}\text{cov}(X,Y) = 1-\rho^2$, we have $\text{var}(V) = 0$. This result implies that V is equal to a constant and this constant is $E(V) = \frac{E(Y)}{\sigma_2} - \rho\frac{E(X)}{\sigma_1}$. This shows that $Y = aX + b$, where $a = \rho\sigma_2/\sigma_1$ and $b = E(Y) - aE(X)$.

5.71 (a) Suppose that $E(Y^2) > 0$ (if $E(Y^2) = 0$, then $Y = 0$). Let $h(t) = E[(X - tY)^2]$. Then $h(t) = E(X^2) - 2tE(XY) + t^2 E(Y^2)$. The function $h(t)$ is minimal for $t = E(XY)/E(Y^2)$. Substituting this t-value into $h(t)$ and noting that $h(t) \geq 0$, the Cauchy–Schwarz inequality follows. (b) The Cauchy–Schwarz inequality gives $[\text{cov}(X, Y)]^2 \leq \text{var}(X)\text{var}(Y)$ or, equivalently, $\rho^2(X, Y) \leq 1$ and so $-1 \leq \rho(X, Y) \leq 1$. (c) Noting that $E(XY) = E(X)$ and $E(Y^2) = P(X > 0)$, the Cauchy–Schwarz inequality gives $[E(X)]^2 \leq E(X^2)P(X > 0)$. This shows that $P(X > 0) \geq [E(X)]^2/E(X^2)$ and so $P(X = 0) \leq \text{var}(X)/E(X^2)$.

5.73 Since $\text{cov}(X_i, X_j) = \text{cov}(X_j, X_i)$, the matrix $\mathbf{C}$ is symmetric. To prove that $\mathbf{C}$ is positive semi-definite, we must verify that $\sum_{i=1}^{n} \sum_{j=1}^{n} t_i t_j \sigma_{ij} \geq 0$ for all real numbers $t_1, \ldots, t_n$. This property follows from the formula for $\text{var}\left(\sum_{i=1}^{n} t_i X_i\right)$ in Problem 5.72 and the fact that the variance is always nonnegative.

5.75 Let $V = \max(X, Y)$ and $W = \min(X, Y)$. Then $E(V) = 1/\sqrt{\pi}$ and $E(W) = -1/\sqrt{\pi}$, see Problem 4.69. Obviously, $VW = XY$ and so $E(VW) = E(X)E(Y) = 0$, by the independence of X and Y. Thus, $\text{cov}(V, W) = 1/\pi$. We have $\min(X, Y) = -\max(-X, -Y)$. Since the independent random variables $-X$ and $-Y$ are distributed as X and Y, it follows that $\min(X, Y)$ has the same distribution as $-\max(X, Y)$. Therefore, $\sigma^2(V) = \sigma^2(W)$. Also, by $V + W = X + Y$, we have $\sigma^2(V + W) = \sigma^2(X + Y) = 2$. Using the relation $\sigma^2(V + W) = \sigma^2(V) + \sigma^2(W) + 2\text{cov}(V, W)$, we now get $\sigma^2(V) + \sigma^2(W) = 2 - 2/\pi$ and so $\sigma^2(V) = \sigma^2(W) = 1 - 1/\pi$. This leads to $\rho(V, W) = \frac{1/\pi}{1 - 1/\pi} = \frac{1}{\pi - 1}$. *Note:* The result is also true when X and Y are $N(\mu, \sigma^2)$ distributed. To see this, write X and Y as $\sigma \times \left(\frac{X - \mu}{\sigma}\right) + \mu$ and $\sigma \times \left(\frac{Y - \mu}{\sigma}\right) + \mu$ and use the relation $\text{cov}(aV_1 + b, cW_1 + d) = ac\,\text{cov}(V_1, W_1)$.

5.77 The linear least-squares estimate of D_1 given that $D_1 - D_2 = d$ is equal to $E(D_1) + \rho(D_1 - D_2, D_1)\frac{\sigma(D_1)}{\sigma(D_1 - D_2)}[d - E(D_1 - D_2)]$. By the independence of D_1 and D_2, $E(D_1 - D_2) = \mu_1 - \mu_2$, $\sigma(D_1 - D_2) = \sqrt{\sigma_1^2 + \sigma_2^2}$, and $\text{cov}(D_1 - D_2, D_1) = \sigma_1^2$. The linear least squares estimate is thus given by $\mu_1 + \frac{\sigma_1^2}{\sigma_1^2 + \sigma_2^2}(d - \mu_1 + \mu_2)$.

Chapter 6

6.1 It suffices to prove the result for the standard bivariate normal distribution. Let $W = aX + bY$. Let us first assume that $b > 0$. Then $P(W \leq w)$ is given by $\frac{1}{2\pi\sqrt{1-\rho^2}}\int_{-\infty}^{\infty}\left[\int_{-\infty}^{(w-ax)/b}\right.$

$e^{-\frac{1}{2}(x^2-2\rho xy+y^2)/(1-\rho^2)}\,dy\big]\,dx$ for $-\infty < w < \infty$. Thus, $f_W(w) = \frac{1}{2\pi b\sqrt{1-\rho^2}}\int_{-\infty}^{\infty} e^{-\frac{1}{2}[x^2-2\rho x(w-ax)/b+(w-ax)^2/b^2]/(1-\rho^2)}\,dx$ is the density function of W. It is a matter of some algebra to obtain $f_W(w) = (\eta\sqrt{2\pi})^{-1}\exp(-\frac{1}{2}w^2/\eta^2)$, where $\eta = \sqrt{a^2+b^2+2ab\rho}$. This expression also applies for $b \le 0$. To see this, write $W = aX+(-b)(-Y)$ and note that $(X, -Y)$ has the standard bivariate normal density with correlation coefficient $-\rho$. We can now conclude that $aX+bY$ is normally distributed for all a, b if (X, Y) has a bivariate normal distribution. If this distribution has the parameters $(\mu_1, \mu_2, \sigma_1^2, \sigma_2^2, \rho)$, then $X - Y$ is $N(\mu_1 - \mu_2, \sigma_1^2 + \sigma_2^2 - 2\rho\sigma_1\sigma_2)$ distributed and $P(X > Y)$ is $1 - \Phi\big(-(\mu_1-\mu_2)/(\sigma_1^2+\sigma_2^2-2\rho\sigma_1\sigma_2)^{1/2}\big)$.

6.3 We have $P(Z \le z) = \int_0^\infty dy \int_{-\infty}^{yz} f(x,y)\,dx + \int_{-\infty}^0 dy \int_{yz}^\infty f(x,y)\,dx$. The density of Z is $f_Z(z) = \int_0^\infty yf(yz, y)\,dy - \int_{-\infty}^0 yf(yz, y)\,dy = \int_{-\infty}^\infty |y|f(yz, y)\,dy$. Inserting the standard bivariate normal density for $f(x, y)$, the desired result follows after some algebra.

6.5 Any linear combination of V and W is a linear combination of X and Y and thus is normally distributed.

6.7 Since X and V are linear combinations of X and Y, any linear combination of X and V is normally distributed and so (X, V) has a bivariate normal distribution. Noting that $E(V) = 0$ and $\sigma^2(V) = (1 + \rho^2 - 2\rho^2)/(1 - \rho^2) = 1$, the random variable V is $N(0, 1)$ distributed like the random variable X. To prove the independence of X and V, we verify that $\mathrm{cov}(X, V) = 0$. This is immediate from $\mathrm{cov}(X, V) = \big(\mathrm{cov}(X, Y) - \rho\sigma^2(X)\big)/\sqrt{1-\rho^2} = (\rho - \rho)/\sqrt{1-\rho^2} = 0$.

6.9 Any linear combination of $X + Y$ and $X - Y$ is a linear combination of X and Y and thus is normally distributed. This shows that the random vector $(X+Y, X-Y)$ has a bivariate normal distribution. The components $X + Y$ and $X - Y$ are independent if $\mathrm{cov}(X + Y, X - Y) = 0$. Since $\mathrm{cov}(X+Y, X-Y) = \mathrm{cov}(X,X) - \mathrm{cov}(X,Y) + \mathrm{cov}(X,Y) - \mathrm{cov}(Y,Y)$, it follows that $\mathrm{cov}(X + Y, X - Y) = \sigma^2(X) - \sigma^2(Y) = 0$.

6.11 Since S_{n1} and S_{n2} are approximately $N(0, \frac{1}{2}n)$ distributed for n large, it follows from the results in Problem 4.68 that both $|S_{n1}|$ and $|S_{n2}|$ have approximate density $\frac{2}{\sqrt{\pi n}}e^{-u^2/n}$ with $E(|S_{n1}|) = E(|S_{n2}|) \approx \sqrt{n/\pi}$. This gives $E(R_n) \approx 2\sqrt{n/\pi}$. Also, $|S_{n1}|$ and $|S_{n2}|$ are nearly independent for n large. Using the convolution formula in Section 5.2, the density of R_n is approximately $\int_0^r \frac{2}{\sqrt{\pi n}}e^{-u^2/n}\frac{2}{\sqrt{\pi n}}e^{-(r-u)^2/n}\,du$. This integral can be rewritten as $\frac{4}{\sqrt{2\pi n}}e^{-\frac{1}{2}r^2/n}\frac{1}{\sqrt{2\pi}}\int_{-r/\sqrt{n}}^{r/\sqrt{n}}e^{-\frac{1}{2}z^2}\,dz$, showing that the density of R_n is approximately $\frac{1}{\sqrt{2\pi n}}e^{\frac{1}{2}r^2/n}\big[\Phi\big(\frac{r}{\sqrt{n}}\big) - \Phi\big(\frac{-r}{\sqrt{n}}\big)\big]$ for n large.

6.13 The random variables $a_1 X_1 + \cdots + a_n X_n$ and $b_1 Y_1 + \cdots + b_m Y_m$ are normally distributed for any constants $a_1, \ldots, a_n$ and $b_1, \ldots, b_m$. Moreover, these random variables are independent (for any functions f and g, the random variables $f(\mathbf{X})$ and $g(\mathbf{Y})$ are independent if $\mathbf{X}$ and $\mathbf{Y}$ are independent). Since the sum of two independent normally distributed random variables is again normally distributed, $a_1 X_1 + \cdots + a_n X_n + b_1 Y_1 + \cdots + b_m Y_m$ is normally distributed, showing that $(\mathbf{X}, \mathbf{Y})$ has a multivariate normal distribution.

6.15 The value of the chi-square statistic is $\frac{(60{,}179 - 61{,}419.5)^2}{61{,}419.5} + \frac{(55{,}551 - 55{,}475.7)^2}{55{,}475.7} + \cdots + \frac{(60{,}745 - 59{,}438.2)^2}{59{,}438.2} + \frac{(61{,}334 - 61{,}419.5)^2}{61{,}419.5} = 642.46$. The probability that a χ_{11}^2-distributed random variable takes on a value larger than 642.46 is practically zero. This leaves no room at all for doubt about the fact that birth dates are not uniformly distributed over the year.

6.17 The estimate of the parameter of the hypothesized Poisson distribution is $\lambda = \frac{37}{78}$ vacancies per year. The data are divided in three groups: years with 0 vacancies, with 1 vacancy, and with ≥ 2 vacancies. Letting $p_i = e^{-\lambda} \lambda^i / i!$, the expected number of years with 0 vacancies is $78 p_0 = 48.5381$, with 1 vacancy is $78 p_1 = 23.0245$, and with ≥ 2 vacancies is $78(1 - p_0 - p_1) = 6.4374$. The chi-square test statistic with $3 - 1 - 1 = 1$ degrees of freedom has the value 0.055. Since $P(\chi_1^2 > 0.055) = 0.8145$, the Poisson distribution gives an excellent fit.

6.19 The estimate of the parameter of the hypothesized Poisson distribution is $\lambda = 2.25$. The matches with 5 or more goals are aggregated, and so six data groups are considered. The test statistic has approximately a chi-squared distribution with $6 - 1 - 1 = 4$ degrees of freedom and its value is 1.521. The probability $P(\chi_4^2 > 1.521) = 0.8229$. The Poisson distribution gives an excellent fit.

6.21 Think of $n = 98{,}364{,}597$ independent repetitions of a chance experiment with seven possible outcomes, where the outcomes $1, 2, \ldots, 6$ correspond to the prizes $1, 2, \ldots, 6$ and the outcome 7 means that none of these six prizes was won. Letting the constant $\gamma = 1/\binom{45}{6}$, the probabilities p_j of the outcomes are $p_1 = \frac{1}{6}\gamma, p_2 = \frac{5}{6}\gamma, p_3 = \gamma, p_4 = 5\gamma, p_5 = 38\gamma, p_6 = 38 \times 5\gamma$, and $p_7 = 1 - \sum_{j=1}^{6} p_j$. Assuming that the tickets are randomly filled in, the test statistic D has the value 20.848. The probability $P(\chi_6^2 > 20.848) = 0.00195$. This indicates that the tickets are not randomly filled in. People do not choose their numbers randomly, but they often use birth dates, lucky numbers, arithmetical sequences, etc., in order to choose lottery numbers. The same conclusion was reached in a similar study by D. Kadell and D. Ylvisaker entitled "Lotto play: the good, the fair and the truly awful," *Chance* **4** (1991): 22–25.

Chapter 7

7.1 To specify $P(X = x \mid Y = 2)$ for $x = 0, 1$, we use the results from Table 5.1. From this table, we get $P(X = 0, Y = 2) = (1-p)^2, P(X = 1, Y = 2) = 2p(1-p)^2$, and $P(Y = 2) = (1-p)^2 + 2p(1-p)^2$. Therefore, $P(X = 0 \mid Y = 2) = \frac{1}{1+2p}$ and $P(X = 1 \mid Y = 2) = \frac{2p}{1+2p}$.

7.3 Since $P(X = 1, Y = 2) = \frac{1}{6} \times \frac{1}{6}, P(X = x, Y = 2) = \frac{4}{6} \times \frac{1}{6} \times \left(\frac{5}{6}\right)^{x-3} \times \frac{1}{6}$ for $x \geq 3$, and $P(Y = 2) = \frac{5}{6} \times \frac{1}{6}$, it follows that $P(X = 1 \mid Y = 2) = \frac{1}{5}$ and $P(X = x \mid Y = 2) = \frac{4}{30}\left(\frac{5}{6}\right)^{x-3}$ for $x \geq 3$. In the same way, $P(X = x \mid Y = 20)$ is $\left(\frac{4}{6}\right)^{x-1}\frac{1}{6}\left(\frac{5}{6}\right)^{-x}$ for $1 \leq x \leq 19$ and $\left(\frac{4}{6}\right)^{19}\frac{1}{6}\left(\frac{5}{6}\right)^{x-21}$ for $x \geq 21$.

7.5 The joint mass function $P(X = x, Y = y) = \left(\frac{5}{6}\right)^{x-1}\frac{1}{6}\binom{x}{y}\left(\frac{1}{2}\right)^x$ for $0 \leq y \leq x$ and $x \geq 1$. We have $P(Y = y) = \frac{1}{5}\sum_{x=y}^{\infty}\binom{x}{y}\left(\frac{5}{12}\right)^x = \frac{12}{35}\left(\frac{5}{7}\right)^y$ for $y \geq 1$ and $P(Y = 0) = \sum_{x=1}^{\infty}\left(\frac{5}{6}\right)^{x-1}\frac{1}{6}\left(\frac{1}{2}\right)^x = \frac{1}{7}$. For any $y \geq 1$, $P(X = x \mid Y = y)$ is $\frac{7}{12}\binom{x}{y}\left(\frac{5}{12}\right)^x\left(\frac{5}{7}\right)^{-y}$ for $x \geq y$. Further, $P(X = x \mid Y = 0)$ is $\frac{7}{5}\left(\frac{5}{12}\right)^x$.

7.7 The joint mass function $P(X = x, Y = y) = \binom{24}{x}\binom{x}{y}\left(\frac{1}{6}\right)^{x+y}\left(\frac{5}{6}\right)^{24-y}$. Also, $P(Y = y) = \left(\frac{1}{6}\right)^y\left(\frac{5}{6}\right)^{24-y}\sum_{x=y}^{24}\binom{24}{x}\binom{x}{y}\left(\frac{1}{6}\right)^x$. The conditional mass function $P(X = x \mid Y = y) = \binom{24}{x}\binom{x}{y}\left(\frac{1}{6}\right)^x / \sum_{k=y}^{24}\binom{24}{k}\binom{k}{y}\left(\frac{1}{6}\right)^k$ for $y \leq x \leq 24$.

7.9 The marginal densities are $f_X(x) = \int_0^\infty xe^{-x(y+1)}\,dy = e^{-x}$ for $x > 0$ and $f_Y(y) = \int_0^\infty xe^{-x(y+1)}\,dx = \frac{1}{(1+y)^2}$ for $y > 0$. Thus, the conditional densities are $f_X(x \mid y) = (y+1)^2 xe^{-x(y+1)}$ for $x > 0$ and $f_Y(y \mid x) = xe^{-xy}$ for $y > 0$. The probability $P(Y > 1 \mid X = 1)$ is $\int_1^\infty f_Y(y \mid 1)\,dy = \int_1^\infty e^{-y}\,dy = e^{-1}$.

7.11 The marginal densities of X and Y are $f_X(x) = \int_0^x \frac{1}{x}\,dy = 1$ for $0 < x < 1$ and $f_Y(y) = \int_y^1 \frac{1}{x}\,dx = -\ln(y)$ for $0 < y < 1$. Therefore, for any y with $0 < y < 1$, the conditional density $f_X(x \mid y) = -\frac{1}{x\ln(y)}$ for $y \leq x < 1$. For any x with $0 < x < 1$, the conditional density $f_Y(y \mid x) = \frac{1}{x}$ for $0 < y < x$.

7.13 By $f(x, y) = f_X(x)f_Y(y \mid x)$, the joint density of X and Y is $f(x, y) = \frac{1}{x}$ for $0 < x < 1$ and $1 - x < y < 1$. Hence, $P(X + Y > 1.5)$ is $\int_{0.5}^1 dx \int_{1.5-x}^1 \frac{1}{x}\,dy$ and equals $0.5\ln(0.5) + 0.5 = 0.1534$, and $P(Y > 0.5)$ is $\int_0^{0.5} dx \int_{1-x}^1 \frac{1}{x}\,dy + \int_{0.5}^1 dx \int_{0.5}^1 \frac{1}{x}\,dy$ and equals $0.5 - 0.5\ln(0.5) = 0.8466$.

7.15 By $f(x, y) = f_X(x)f_Y(y \mid x)$, the joint density of X and Y is $f(x, y) = 2$ for $0 < y \leq x < 1$. The marginal density of Y is $f_Y(y) = 2(1 - y)$ for $0 < y < 1$. Using again $f(x, y) = f_Y(y)f_X(x \mid y)$, it follows that $f_X(x \mid y) = \frac{1}{1-y}$ for $y < x < 1$, which is the uniform density on $(y, 1)$.

7.17 Let $P_{\Delta y}(x \mid y) = P(X = x \mid y - \frac{1}{2}\Delta y \leq Y \leq y + \frac{1}{2}\Delta y)$. This conditional probability can be written as $P(y - \frac{1}{2}\Delta y \leq Y \leq y + \frac{1}{2}\Delta y \mid X = x) P(X = x)$ divided by $P(y - \frac{1}{2}\Delta y \leq Y \leq y + \frac{1}{2}\Delta y)$. Thus, for continuity points y, we have $P_{\Delta y}(x \mid y) \approx p_X(x)f_Y(y \mid x)\Delta y/[f_Y(y)\Delta y] = p_X(x)f_Y(y \mid x)/f_Y(y)$. Define $p_X(x \mid y)$ as $\lim_{\Delta y \to 0} P_{\Delta y}(x \mid y)$. Then, for fixed y, $p_X(x \mid y)$ as a function of x is proportional to $p_X(x)f_Y(y \mid x)$. The proportionality constant is the inverse of $\sum_x p_X(x)f_Y(y \mid x)$.

7.19 Let Y be the time needed to process a randomly chosen claim. By the relation $f(x, y) = f_Y(y \mid x)f_X(x)$, the joint density function of X and Y satisfies $f(x, y) = 1.5(2 - x)$ for $0 < x < 2$ and $x < y < 2x$. The sought probability is $P((X, Y) \in C)$ with $C = \{(x, y) : 0 \leq x \leq 1, x \leq y \leq \min(2x, 1)\}$ and has the value $\int_0^1 dx \int_x^{\min(2x,1)} 1.5(2 - x)\,dy = 0.5625$.

7.21 We have $P(N = k) = \int_0^1 P(N = k \mid X_1 = u)\,du$, by the law of conditional probability. Thus, $P(N = k) = \int_0^1 u^{k-2}(1 - u)\,du = \frac{1}{k(k-1)}$ for $k \geq 2$, and so $E(N) = \infty$.

7.23 Condition on the unloading time. By the law of conditional probability, the probability of no breakdown is $\int_{-\infty}^{\infty} e^{-\lambda y} \frac{1}{\sigma\sqrt{2\pi}} e^{-\frac{1}{2}(y-\mu)^2/\sigma^2}\,dy = e^{-\mu\lambda + \frac{1}{2}\sigma^2\lambda^2}$.

7.25 Let Y be the outcome of the first roll of the die. Then, by $P(X = k) = \sum_{i=1}^{6} P(X = k \mid Y = i)P(Y = i)$, we get $P(X = k) = \sum_{i=1}^{5} \binom{i}{k}\left(\frac{1}{6}\right)^k\left(\frac{5}{6}\right)^{i-k} \times \frac{1}{6} + \binom{6}{k-1}\left(\frac{1}{6}\right)^{k-1}\left(\frac{5}{6}\right)^{6-k+1} \times \frac{1}{6}$, where $\binom{i}{k} = 0$ for $k > i$ and $\binom{6}{-1} = 0$.

7.27 By the law of conditional probability, the probability of having k red balls among the r selected balls is $\sum_{n=0}^{B} \left[\binom{n}{k}\binom{B-n}{r-k}/\binom{B}{r}\right]\binom{B}{n}p^n(1-p)^{B-n}$. This probability simplifies to $\binom{r}{k}p^k(1-p)^{r-k}$. This result can be seen directly by assuming that the B balls are originally non-colored and next giving each of the r balls chosen the color red with probability p.

7.29 By the law of conditional probability, the probability $P(B^2 \geq 4AC)$ is given by $\int_0^1 P(AC \leq \frac{b^2}{4})\,db = \int_0^1 db\left[\int_0^1 P(C \leq \frac{b^2}{4a})\,da\right]$. This integral can be rewritten as $\int_0^1 db\left[\int_0^{b^2/4} da + \int_{b^2/4}^1 \frac{b^2}{4a}\,da\right] = \int_0^1 db\left[\frac{b^2}{4} - \frac{b^2}{4}\ln\left(\frac{b^2}{4}\right)\right]$. This leads to $P(B^2 \geq 4AC) = \frac{5}{36} + \frac{1}{6}\ln(2) = 0.2544$.

7.31 The expected number of crossings of the zero level during the first n jumps is $\sum_{k=1}^{n-1} E(I_k)$, where $E(I_k) = P(I_k = 1)$. Denote by S_k the position of the particle just before the $(k+1)$th jump. Then, S_k is the sum of k independent standard normally distributed random variables and thus has the density function $\frac{1}{\sqrt{2\pi k}}e^{-\frac{1}{2}x^2/k}$. By conditioning on S_k, we get that $P(I_k = 1)$ is equal to the sum of the integral

$\int_0^\infty \frac{1}{\sqrt{2\pi k}} e^{-\frac{1}{2}x^2/k} dx \int_{-\infty}^{-x} \frac{1}{\sqrt{2\pi}} e^{-\frac{1}{2}y^2} dy$ and the integral $\int_{-\infty}^0 \frac{1}{\sqrt{2\pi k}} e^{-\frac{1}{2}x^2/k} dx \int_{-x}^\infty \frac{1}{\sqrt{2\pi}} e^{-\frac{1}{2}y^2} dy$. Using polar coordinates for these integrals, we get that $P(I_k = 1)$ is equal to $\frac{1}{\pi}\left[\frac{\pi}{2} - \arctg(\sqrt{k})\right] = \frac{1}{\pi}\arctg\left(\frac{1}{\sqrt{k}}\right)$. Therefore, $\sum_{k=1}^{n-1} E(I_k) = \frac{1}{\pi} \sum_{k=1}^{n-1} \arctg\left(\frac{1}{\sqrt{k}}\right)$. Note: An asymptotic expansion for this sum is $\frac{2}{\pi}\sqrt{n} + c + \frac{1}{6\pi\sqrt{n}}$, where $c = -0.68683\ldots$.

7.33 It suffices to find $P(a, b)$ for $a \geq b$. By symmetry, $P(a, b) = 1 - P(b, a)$ for $a \leq b$. For fixed a and b with $a \geq b$, let the random variables S_A and S_B be the total scores of the players A and S_B. Let $f_A(s)$ be the probability density of S_A. Then, by the law of conditional probability, $P(a, b) = \int_0^1 P(A \text{ beats } B \mid S_A = s) f_A(s) ds$. By conditioning on the outcome of the first draw of player A, it follows that $P(S_A \leq s) = \int_0^s (s - u) du$ for $0 < s \leq a$, $P(S_A > s) = 1 - s + \int_0^a (1 - (s - u)) du$ for $a < s \leq 1$, and $P(S_A > s) = \int_{s-1}^a (1 - (s - u)) du$ for $1 < s < 1 + a$. Differentiation gives that the density function $f_A(s)$ of S_A is s for $0 < s \leq a$, $1 + a$ for $a < s \leq 1$, and $1 + a - s$ for $1 < s < 1 + a$. The distribution of S_B follows by replacing a by b in the distribution of S_A. Next it is a matter of tedious algebra to obtain $P(a, b) = \frac{1}{2} - \frac{1}{6}(a - b)(a^2 b + a^2 + ab^2 + b^2 + ab + 3a - 3)$ for $a \geq b$. Also, $P(a, b) = \frac{1}{2} + \frac{1}{6}(b - a)(b^2 a + b^2 + ba^2 + a^2 + ba + 3b - 3)$ for $a \leq b$. Let a_0 be the optimal threshold value of player A. Then $P(a_0, b) \geq 0.5$ for all b with $P(a_0, b) = 0.5$ for $b = a_0$. This leads to the equation $2a_0^3 + 3a_0^2 + 3a_0 - 3 = 0$. This solution of the equation is $a_0 = 0.5634$. If player A uses this threshold value, his win probability is at least 50%, whatever threshold player B uses.

7.35 The optimal stopping level for player A is the solution to the equation $a^4 = \frac{1}{5}(1 - a^5)$ and is given by $a_3 = 0.64865$. The overall win probability of player A is 0.3052. Let $P(a)$ denote the probability that the final score of player A is no more than a. Then $P(0) = \frac{1}{2}a_3^2$, $P(a) = P(0) + \frac{1}{2}a^2$ for $0 \leq a \leq a_3$, and $P(a) = (1 + a_3)a - a_3$ for $a \geq a_3$. The overall win probabilities of the players B and C can be calculated next as 0.3295 and 0.3653.

7.37 Since $f_Y(y \mid s) = \frac{3(s+1)^3}{(s+y)^4}$ for $y > 1$, we have $E(Y \mid X = s) = \int_1^\infty y \frac{3(s+1)^3}{(s+y)^4} dy$. Partial integration results in $E(Y \mid X = s) = 1 + \frac{1}{2}(s+1)$.

7.39 The joint density of X and Y is $f(x, y) = 6(y - x)$ for $0 < x < y < 1$, as follows from $P(x < X \leq x + \Delta x, y < Y \leq y + \Delta y) = 6\Delta x(y - x)\Delta y$ for $\Delta x, \Delta y$ small. This gives $f_X(x) = 3(1 - x)^2$ and $f_Y(y) = 3y^2$. Thus, $f_X(x \mid y) = \frac{6(y-x)}{3y^2}$ for $0 < x < y$ and $f_Y(y \mid x) = \frac{6(y-x)}{3(1-x)^2}$ for $x < y < 1$.

This gives $E(X \mid Y = y) = \int_0^y x \frac{6(y-x)}{3y^2} dx = \frac{1}{3}y$ and $E(Y \mid X = x) = \int_x^1 y \frac{6(y-x)}{3(1-x)^2} dy = \frac{2+x}{3}$.

7.41 Noting that X can be written as $X = \frac{1}{2}(X+Y) + \frac{1}{2}(X-Y)$, it follows that $E(X \mid X+Y = v) = \frac{1}{2}v + \frac{1}{2}E(X-Y \mid X+Y = v)$. By Problem 6.9, $X+Y$ and $X-Y$ are independent and so $E(X-Y \mid X+Y = v) = E(X-Y)$. Also, $E(X-Y) = \mu_1 - \mu_2$. Thus, $E(X \mid X+Y = v) = \frac{1}{2}v + \frac{1}{2}(\mu_1 - \mu_2)$.
Note: The conditional distribution of X given that $X+Y = v$ is the normal distribution with mean $\frac{1}{2}(\mu_1 - \mu_2 + v)$ and variance $\frac{1}{2}\sigma^2(1-\rho)$. This result follows from the relation $P(X \leq x \mid X+Y = v) = P(\frac{1}{2}(X-Y) + \frac{1}{2}v \leq x)$ and the fact that $X - Y$ is $N(\mu_1 - \mu_2, 2\sigma^2(1-\rho))$ distributed.

7.43 Let X be the number of trials until the first success and N be the number of successes in the first n trials. Then, for $1 \leq r \leq n$ and $1 \leq j \leq n-r+1$, we have $P(X = j, N = r) = (1-p)^{j-1}p \binom{n-j}{r-1} p^{r-1}(1-p)^{n-j-(r-1)}$. Since $P(N = r) = \binom{n}{r} p^r (1-p)^{n-r}$, we get $P(X = j \mid N = r) = \binom{n-j}{r-1}/\binom{n}{r}$. Thus, by $E(X \mid N = r) = \sum_{j=1}^{n-r+1} j P(X = j \mid N = r)$, we obtain that $E(X \mid N = r) = \frac{n+1}{r+1}$.

7.45 Denote by X and Y the zinc content and the iron content. The marginal density of Y is $f_Y(y) = \int_2^3 \frac{1}{75}(5x + y - 30) dx = \frac{1}{75}(y - 17.5)$ and so $f_X(x \mid y) = \frac{5x+y-30}{y-17.5}$. Thus, $E(X \mid Y = y) = \int_2^3 x f_X(x \mid y) dx = \frac{15y-260}{6(y-17.5)}$.

7.47 (a) Since $P(X \leq x \mid a < Y < b) = \frac{1}{P(a<Y<b)} \int_{-\infty}^x dv \int_a^b f(v, w) dw$, differentiation yields that $\int_a^b f(x, w) dw / P(a < Y < b)$ is the conditional density of X given that $a < Y < b$. In the same way, $\int_{-\infty}^x f(x, w) dw / P(X > Y)$ is the conditional density of X given that $X > Y$. (b) The formula for $E(X \mid a < Y < b)$ is obvious from (a). Since $X = \frac{1}{2}(X + Y) + \frac{1}{2}(X - Y)$, $E(X \mid X > Y)$ equals $\frac{1}{2}E(X+Y \mid X-Y > 0) + \frac{1}{2}E(X-Y \mid X - Y > 0)$. By the independence of $X + Y$ and $X - Y$ (see Problem 6.9), $E(X + Y \mid X - Y > 0) = E(X+Y) = 0$. Since $V = X - Y$ is $N(0, \sigma^2)$ distributed with $\sigma^2 = 2(1 - \rho)$, we have $E(V \mid V > 0) = \frac{1}{P(V>0)} \frac{1}{\sigma\sqrt{2\pi}} \int_0^\infty v e^{-\frac{1}{2}v^2/\sigma^2} dv$, which yields $E(X - Y \mid X - Y > 0) = \sqrt{(1-\rho)\pi}$.

7.49 By the law of conditional probability, the probability of running out of oil is given by $\frac{2}{3}P(X_1 > Q) + \frac{1}{3}P(X_2 > Q)$, where X_i is $N(\mu_i, \sigma_i^2)$ distributed. The stockout probability can be evaluated as $\frac{2}{3}(1 - \Phi(\frac{Q-\mu_1}{\sigma_1})) + \frac{1}{3}(1 - \Phi(\frac{Q-\mu_2}{\sigma_2}))$. By the law of conditional expectation, the expected value of the shortage is $\frac{2}{3}E[(X_1 - Q)^+] + \frac{1}{3}E[(X_2 - Q)^+]$, where $x^+ = \max(x, 0)$. The expected value can be evaluated as $\frac{2}{3}\sigma_1 I(\frac{Q-\mu_1}{\sigma_1}) + \frac{1}{3}$

$\sigma_2 I\left(\frac{Q-\mu_2}{\sigma_2}\right)$, where $I(k)$ is the normal loss integral $(1/\sqrt{2\pi})\int_k^\infty (x-k)$ $e^{-\frac{1}{2}x^2}\,dx$, which can be written as $I(k) = (1/\sqrt{2\pi})e^{-\frac{1}{2}k^2} - k[1 - \Phi(k)]$. The expected value of the number of gallons left over equals the expected value of the shortage minus $\frac{2}{3}\mu_1 + \frac{1}{3}\mu_2 - Q$.

7.51 Let the geometrically distributed random variable Y be the number of messages waiting in the buffer. Under the condition that $Y = y$, the random variable X is uniformly distributed on $0, 1, \ldots, y-1$. Therefore, $E(X \mid Y = y) = \frac{1}{2}(y-1)$ and $E(X^2 \mid Y = y) = \frac{1}{12}(y^2-1)$. By the law of conditional expectation, $E(X^k) = \sum_{y=1}^\infty E(X^k \mid Y = y)p(1-p)^{y-1}$ for $k = 1, 2$. We find after some algebra that $E(X) = \frac{1}{2p}(1-p)$, $E(X^2) = \frac{2}{3p^2} - \frac{1}{6p} - \frac{1}{2}$, and $\sigma(X) = \frac{1}{2\sqrt{3}p}\sqrt{(1-p)(5+9p)}$.

7.53 For fixed n, let $u_k(i) = E[X_k(i)]$. To find $u_n(0)$, we apply the recursion $u_k(i) = \frac{1}{2}u_{k-1}(i+1) + \frac{1}{2}u_{k-1}(i)$ for i satisfying $\frac{i}{n-k} \leq \frac{1}{2}$ and use the boundary conditions $u_0(i) = \frac{i}{n}$ and $u_k(i) = \frac{i}{n-k}$ for $i > \frac{1}{2}(n-k)$ and $1 \leq k \leq n$. The sought probability $u_n(0)$ has the values 0.7083, 0.7437, 0.7675, and 0.7761 for $n = 5, 10, 25$, and 50. *Note*: $u_n(0)$ tends to $\frac{\pi}{4}$ as n increases without bound, see also Example 8.4.

7.55 Let X_a be your end score when you continue for a second spin after having obtained a score of a in the first spin. Then, by the law of conditional expectation, $E(X_a) = \int_0^{1-a}(a+x)\,dx + \int_{1-a}^1 0\,dx = a(1-a) + \frac{1}{2}(1-a)^2$. The solution of $a(1-a) + \frac{1}{2}(1-a)^2 = a$ is $a^* = \sqrt{2} - 1$. The optimal strategy is to stop after the first spin if this spin gives a score larger than $\sqrt{2} - 1$. Your expected payoff is \$609.48.

7.57 In each of the problems, define v_i as the expected reward that can be achieved when your current total is i points. A recursion scheme for the v_i is obtained by applying the law of conditional expectation. For Problem 3.24, use the recursion $v_i = \frac{1}{6}\sum_{k=2}^6 v_{i+k}$ for $0 \leq i \leq 19$, where $v_j = j$ for $j \geq 20$. The expected reward is $v_0 = 8.5290$. For Problem 3.25, use the recursion $v_i = \frac{1}{6}\sum_{k=1}^6 v_{i+k}$ for $0 \leq i \leq 5$, where $v_j = j$ for $6 \leq j \leq 10$ and $v_{11} = 0$. The expected reward is $v_0 = 6.9988$.

7.59 Define E_i as the expected value of the remaining duration of the game when the current capital of John is i dollars. Then, by a conditioning argument, $E_i = 1 + pE_{i+1} + qE_{i-1}$ for $1 \leq i \leq a+b-1$, where $E_0 = E_{a+b} = 0$. By substitution, it can be verified that the solution of this linear difference equation is $E_i = \frac{i}{q-p} - \frac{a+b}{q-p}\frac{1-(q/p)^i}{1-(q/p)^{a+b}}$ if $p \neq q$ and $i(a+b-i)$ if $p = q$. Taking $p = \frac{18}{37}, q = \frac{19}{37}, a = 2$, and $b = 8$, we get that the expected value of the number of bets is 15.083.

7.61 For fixed n, let $F(i, k)$ be the maximal expected payoff that can be achieved when k tosses can still be done and heads have turned up i times so far. The recursion is $F(i,k) = \max\{\frac{1}{2}F(i+1, k-1) + \frac{1}{2}F(i-1, k-1),\frac{i}{n-k}\}$ for $k = 1,\ldots,n$ with $F(i,0) = \frac{i}{n}$. The maximal expected payoff $F(0,n)$ has the values 0.7679, 0.7780, 0.7834, and 0.7912 for $n = 25$, 50, 100, and 1,000.

7.63 Let state $(l,r,1)$ $((l,r,0))$ mean that r numbers have been taken out of the hat, l is the largest number seen so far, and l was obtained (not obtained) at the last pick. Define $F_r(s)$ as the maximal probability of obtaining the largest number starting from state s for which r numbers have been taken out of the hat. The maximal success probability is $\sum_{l=1}^{N} f_1(l,1)\frac{1}{N}$. The optimality equations are $F_r(l,0) = F_{r+1}(l,0)\frac{l-r}{N-r} + \sum_{j=l+1}^{N} F_{r+1}(j,1)\frac{1}{N-r}$ and $F_r(l,1) = \max\left[\binom{l-r}{n-r}/\binom{N-r}{n-r}, F_{r+1}(l,0)\frac{l-r}{N-r} + \sum_{j=l+1}^{N} F_{r+1}(j,1)\frac{1}{N-r}\right]$ for $l = r,\ldots,N$, where $\binom{l-r}{n-r} = 0$ for $l < n$ and the boundary conditions are $F_n(l,0) = 0$ and $F_n(l,1) = 1$ for $l = n,\ldots,N$. For $n = 10$ and $N = 100$, the maximal success probability is 0.6219 and the optimal stopping rule is characterized by $l_1 = 93$, $l_2 = 92$, $l_3 = 91$, $l_4 = 89$, $l_5 = 87$, $l_6 = 84$, $l_7 = 80$, $l_8 = 72$, and $l_9 = 55$. This rule prescribes stopping in state $(l,r,1)$ if $l \geq l_r$ and continuing otherwise. *Note:* We verified experimentally that l_r is the smallest value of l such that $Q_s(l,r) \geq Q_c(l,r)$, where $Q_s(l,r) = \binom{l-r}{10-r}/\binom{100-r}{10-r}$ is the probability of having obtained the overall largest number when stopping in state (l,r) and $Q_c(l,r) = \sum_{k=1}^{10-r}\frac{1}{k}\binom{100-l}{k}\binom{l-r}{10-r-k}/\binom{100-r}{10-r}$ is the probability of getting the overall largest number when continuing in state (l,r) and stopping as soon as you pick a number larger than l.

7.65 Define the value function $v(i_0,i_1)$ as the maximal expected net winnings you can still achieve starting from state (i_0,i_1), where state (i_0,i_1) means that there are i_0 empty bins and i_1 bins with exactly one ball. The desired expected value $v(b,0)$ can be obtained from the optimality equation $v(i_0,i_1) = \max\left[i_1 - \frac{1}{2}(b-i_0-i_1), \frac{i_0}{b}v(i_0-1,i_1+1) + \frac{i_1}{b}v(i_0,i_1-1) + \frac{b-i_0-i_1}{b}v(i_0,i_1)\right]$ with the boundary condition $v(0,i_1) = i_1 - \frac{1}{2}(b-i_1)$. This equation can be solved by backwards calculations. First, calculate $v(1,i_1)$ for $i_1 = 0,\ldots,b-1$. Next, calculate $v(2,i_1)$ for $i_1 = 0,\ldots,b-2$. Continuing in this way, the desired $v(b,0)$ is obtained. Numerical investigations lead to the conjecture that the optimal stopping rule has the following simple form: you stop only in the states (i_0,i_1) with $i_1 \leq a$, where a is the smallest integer larger than or equal to $2i_0/3$. For $b = 25$, we find that the maximal expected net winnings is \$7.566. The one-stage-look-ahead rule prescribes stopping in the states (i_0,i_1) with $i_0 \leq (1+0.5)i_1$ and continuing otherwise. This stopping rule has an expected net winnings of \$7.509.

7.67 Imagine that the balls are placed into the bins at times generated by a Poisson process with rate 1. Then, for any j, a Poisson process with rate $\frac{1}{b}$ generates the times at which the jth bin receives a ball. Using the independence of these Poisson processes and conditioning upon the time that the ith bin receives its first ball, we get $P(A_i) = \int_0^\infty \left(\sum_{k=1}^m e^{-\frac{1}{b}t} \frac{(t/b)^k}{k!} \right)^{b-1} \frac{1}{b} e^{-\frac{1}{b}t} dt$. The sought probability is $\sum_{i=1}^b P(A_i)$. As a sanity check, this probability is $\frac{b!}{b^b}$ for $m = 1$.

7.69 Imagine that rolls of the two dice occur at epochs generated by a Poisson process with rate 1. Let N be the number of rolls needed to remove all tokens and T be the first epoch at which all tokens have been removed. Then $E(N) = E(T)$ and $E(T) = \int_0^\infty P(T > t) dt$. Also, $T = \max_{2 \le j \le 12} T_j$, where T_j is the first epoch at which all tokens in section j have been removed. The rolls resulting in a dice total of k occur according to a Poisson process with rate p_k for each k and these Poisson processes are independent of each other. The p_k are given by $p_k = \frac{k-1}{36}$ for $2 \le k \le 7$ and $p_k = p_{14-k}$ for $8 \le k \le 12$. By the independence of the T_k, $P(T \le t) = P(T_2 \le t) \cdots P(T_{12} \le t)$. Also, $P(T_k > t) = \sum_{j=0}^{a_k - 1} e^{-p_k t} (p_k t)^j / j!$. Putting the pieces together and using numerical integration, we find $E(N) = 31.922$.

7.71 Imagine that the rolls of the die occur at epochs generated by a Poisson process with rate 1. Then the times at which an odd number is rolled are generated by a Poisson process with rate $\frac{1}{2}$ and the times at which the even number k is rolled are generated by a Poisson process with rate $\frac{1}{6}$ for $k = 2, 4, 6$. These Poisson processes are independent of each other. By conditioning on the first epoch at which an odd number is rolled, we find that the sought probability is $\int_0^\infty (1 - e^{-t/6})^3 \frac{1}{2} e^{-\frac{1}{2}t} dt = 0.05$.

7.73 Imagine that the rolls of the die occur at epochs generated by a Poisson process with rate 1. Then independent Poisson processes each having rate $\mu = \frac{1}{6}$ describe the moves of the horses. The density of the sum of r independent interoccurrence times, each having an exponential distribution with expected value $1/\mu$, is the Erlang density $\mu^r \frac{t^{r-1}}{(r-1)!} e^{-\mu t}$, see Section 5.4.3. Thus, the win probability of horse 1 with starting point $s_1 = 0$ is $\int_0^\infty \left(\sum_{k=0}^4 e^{-\frac{t}{6}} \frac{(t/6)^k}{k!} \right)^2 \left(\sum_{k=0}^3 e^{-\frac{t}{6}} \frac{(t/6)^k}{k!} \right)^2 \left(\sum_{k=0}^5 e^{-\frac{t}{6}} \frac{(t/6)^k}{k!} \right) \left(\frac{1}{6} \right)^6 \frac{t^5}{5!} e^{-\frac{t}{6}} dt$. This gives the win probability 0.06280 for horses 1 and 6. In the same way, we get the win probability 0.13991 for horses 2 and 5, and the win probability 0.29729 for horses 3 and 4. The expected duration of the game is $\int_0^\infty \left(\sum_{k=0}^5 e^{-t/6} \frac{(t/6)^k}{k!} \right)^6 dt = 19.737$ when each horse starts at panel 0.

7.75 Imagine that cards are picked at epochs generated by a Poisson process with rate 1. Let N be the number of picks until each card of some of the suits has been obtained and let T be the epoch at which this occurs. Then $E(T) = E(N)$. The epochs at which any specific card is picked are generated by a Poisson process with rate $\frac{1}{20}$. These Poisson processes are independent of each other. Let T_i be the time until all cards of the ith suit have been picked. The T_i are independent random variables and $T = \min(T_1, T_2, T_3, T_4)$. Since $P(T_i > t) = 1 - (1 - e^{-t/20})^5$ and $P(T > t) = P(T_1 > t) \cdots P(T_4 > t)$, we get $E(T)$ is $\int_0^\infty [1 - (1 - e^{-t/20})^5]^4 \, dt = 24.694$ and so $E(N) = 24.694$.

7.77 Writing $x_i - \theta = x_i - \bar{x} + \bar{x} - \theta$, it follows that $\sum_{i=1}^n (x_i - \theta)^2$ can be represented as $n(\bar{x} - \theta)^2 + \sum_{i=1}^n (x_i - \bar{x})^2 - 2(\bar{x} - \theta) \sum_{i=1}^n (x_i - \bar{x})$. Noting that $\sum_{i=1}^n (x_i - \bar{x}) = 0$, it follows that $L(\mathbf{x} \mid \theta) = (\sigma \sqrt{2\pi})^{-n} e^{-\frac{1}{2} \sum_{i=1}^n (x_i - \theta)^2 / \sigma^2}$ is proportional to $e^{-\frac{1}{2} n(\theta - \bar{x})^2 / \sigma^2}$. Thus, the posterior density $f(\theta \mid \mathbf{x})$ is proportional to $e^{-\frac{1}{2} n(\theta - \bar{x})^2 / \sigma^2} f_0(\theta)$, where $f_0(\theta) = \frac{1}{\sigma\sqrt{2\pi}} e^{-\frac{1}{2}(\theta - \mu_0)^2 / \sigma_0^2}$. Next it is a matter of some algebra to find that the posterior density is proportional to $e^{-\frac{1}{2}(\theta - \mu_p)^2 / \sigma_p^2}$, where $\mu_p = [\sigma_0^2 \bar{x} + (\sigma^2/n)\mu_0]/[\sigma_0^2 + \sigma^2/n]$ and $\sigma_p^2 = [\sigma_0^2(\sigma^2/n)]/[\sigma_0^2 + \sigma^2/n]$. In other words, the posterior density is the $N(\mu_p, \sigma_p^2)$ density. Inserting the data $n = 10$, $\sigma = \sqrt{2}$, $\mu_0 = 73$, and $\sigma_0 = 0.7$, it follows that the posterior density is maximal at $\theta = \mu_p = 73.356$. Using the 0.025 and 0.975 percentiles of the standard normal density, a 95% Bayesian credible interval for θ is $(\mu_p - 1.960\sigma_p, \mu_p + 1.960\sigma_p) = (72.617, 74.095)$.

7.79 The posterior density is proportional to $e^{-\frac{1}{2}(t_1 - \theta)^2 / \sigma^2} \times e^{-\frac{1}{2}(\theta - \mu_0)^2 / \sigma_0^2}$, where $t_1 = 140$, $\sigma = 20$, $\mu_0 = 150$, and $\sigma_0 = 25$ light years. Next, a little algebra shows that the posterior density is proportional to $e^{-\frac{1}{2}(\theta - \mu_p)^2 / \sigma_p^2}$, where $\mu_p = [\sigma_0^2 t_1 + \sigma^2 \mu_0]/[\sigma_0^2 + \sigma^2]$ and $\sigma_p^2 = [\sigma_0^2 \sigma^2]/[\sigma_0^2 + \sigma^2]$. This gives that the posterior density is a normal density with an expected value of $\mu_p = 143.902$ light years and a standard deviation of $\sigma_p = 15.617$ light years. The posterior density is maximal at $\theta = 143.902$. A Bayesian 95% credible interval for the distance is $(\mu_p - 1.960\sigma_p, \mu_p + 1.960\sigma_p) = (113.293, 174.512)$.

7.81 The prior density of the parameter of the exponential lifetime of a light bulb is $f_0(\theta) = c\theta^{\alpha-1} e^{-\lambda \theta}$, where c is a normalization constant. Let E be the event that light bulbs have failed at times $t_1 < \cdots < t_r$ and $m - r$ light bulbs are still functioning at time T. The likelihood function $L(E \mid \theta)$ is defined as $\binom{m}{r} r! \, \theta^r e^{-[t_1 + \cdots + t_r + (m-r)T]\theta}$ (the rationale of this definition is the probability that one light bulb fails in each of the infinitesimal intervals $(t_i - \frac{1}{2}\Delta, t_i + \frac{1}{2}\Delta)$ and $m - r$ light bulbs are still functioning at time T). The posterior density $f(\theta \mid E)$ is proportional to $\theta^{\alpha+r-1} e^{-[\lambda + t_1 + \cdots + t_r + (m-r)T]\theta}$. In other words, the posterior density $f(\theta \mid E)$ is a gamma density with shape parameter $\alpha + r$ and scale parameter $\lambda + \sum_{i=1}^{r} t_i + (m-r)T$.

Chapter 8

8.1 (a) The binomial random variable X can be represented as $X = X_1 + \cdots + X_n$, where the X_i are independent with $P(X_i = 0) = 1 - p$ and $P(X_i = 1) = p$. Since $E(z^{X_i}) = 1 - p + pz$, Rule 8.2 gives that $G_X(z) = (1 - p + pz)^n$. The negative binomial random variable Y can be represented as $Y = Y_1 + \cdots + Y_r$, where the X_i are independent with $P(Y_i = k) = p^{k-1}(1-p)$ for $k \geq 1$. Since $E(z^{Y_i}) = pz/(1 - (1-p)z)$, Rule 8.2 gives that $G_Y(z) = [pz/[1 - (1-p)z]]^r$. (b) For the binomial distribution, $G'_X(1) = np$ and $G''_X(1) = n(n-1)p^2$, implying that $E(X) = np$ and $\sigma^2(X) = np(1-p)$. For the negative binomial distribution, $G'_Y(1) = r/p$ and $G''_Y(1) = (r(1-2p) + r^2)/p^2$, and thus $E(Y) = \frac{r}{p}$ and $\sigma^2(Y) = \frac{r(1-p)}{p^2}$.

8.3 By Rule 8.2, the generating function of the total score S is given by $G_S(z) = (\frac{1}{3}z + \frac{2}{15}z^2 + \frac{2}{15}z^3 + \frac{2}{15}z^4 + \frac{2}{15}z^5)^n$. Since $G_S(-1) = (-\frac{1}{3})^n$, the sought probability is $\frac{1}{2}[(-\frac{1}{3})^n + 1]$.

8.5 We have $G_{X+Y}(z) = e^{-\mu(1-z)}$, where $\mu = E(X + Y)$. By the independence of X and Y, $G_{X+Y}(z) = G_X(z)G_Y(z)$. Since X and Y have the same distribution, $G_X(z) = G_Y(z)$. Therefore, $[G_X(z)]^2 = e^{-\mu(1-z)}$ and so $G_X(z) = e^{-\frac{1}{2}\mu(1-z)}$. The generating function uniquely determines the probability mass function. Thus, X and Y are Poisson distributed with expected value $\frac{1}{2}\mu$.

8.7 Since $E(z^{N(t+h)}) = E(z^{N(t+h)-N(t)}z^{N(t)})$ and $N(t+h) - N(t)$ is independent of $N(t)$, we have $g_z(t+h) = E(z^{N(t+h)-N(t)})g_z(t)$. We can evaluate $E(z^{N(t+h)-N(t)})$ as $(1 - \lambda h) + \lambda hz + o(h)$ for $h \to 0$. This leads to $(g_z(t+h) - g_z(t))/h = -\lambda(1-z)g_z(t) + o(h)/h$ as $h \to 0$, and so $\frac{\partial}{\partial t}g_z(t) = -\lambda(1-z)g_z(t)$ for $t > 0$. Together with $g_z(t) = 1$ for $z = 1$, this gives $g_z(t) = e^{-\lambda t(1-z)}$, showing that the generating function of $N(t)$ is given by the generating function of a Poisson-distributed random variable with expected value λt. By the uniqueness property of the generating function, it follows that $N(t)$ is Poisson distributed with expected value λt.

8.9 Using first-step analysis, we get $E(z^X) = pzE(z^X) + q + rE(z^X)$. This leads to $\sum_{k=0}^{\infty} P(X=k)z^k = q/(1 - pz - r)$. Writing $q/(1 - pz - r)$ as

$(q/(1-r))/(1-pz/(1-r))$ and using the expansion $1/(1-pz/(1-r)) = \sum_{k=0}^{\infty}(\frac{p}{1-r})^k z^k$, we obtain by equating terms that $P(X = k) = \frac{q}{1-r}(\frac{p}{1-r})^k$ for all $k \geq 0$.

8.11 Conditioning on the outcome of the first toss, we get $E(z^X) = \frac{1}{2}zE(z^{X_1}) + \frac{1}{2}zE(z^{X_2})$. The random variable X_1 is equal to $r-1$ if the next $r-1$ tosses give heads and X_1 is distributed as $k + X_2$ for $1 \leq k \leq r-1$ if the next $k-1$ tosses give heads and are followed by tails. Therefore, $E(z^{X_1})$ is $(\frac{1}{2})^{r-1} z^{r-1} + \sum_{k=1}^{r-1}(\frac{1}{2})^k z^k E(z^{X_2})$. Since $E(z^{X_2}) = E(z^{X_1})$, we have $E(z^{X_1}) = (\frac{1}{2}z)^{r-1}/[1 - \sum_{k=1}^{r-1}(\frac{1}{2}z)^k]$. This leads to $E(z^X) = 2(\frac{1}{2}z)^r/[1 - \sum_{k=1}^{r-1}(\frac{1}{2}z)^k]$. Taking the derivatives of $E(z^X)$ and putting $z = 1$, we get $E(X) = 2^r - 1$ and $\text{var}(X) = 2^r(2^r - 2r + 1) - 2$.

8.13 The extinction probability is the smallest root of the equation $u = P(u)$, where the generating function $P(u)$ is given by $P(u) = \frac{p}{1-(1-p)u}$ for $|u| \leq 1$. The equation $u = P(u)$ has the two roots $u = \frac{p}{1-p}$ and $u = 1$. The extinction probability is $\frac{p}{1-p}$ if $p < \frac{1}{2}$ and 1 otherwise.

8.15 The generating function of the offspring distribution is $P(u) = \frac{1}{3} + \frac{2}{3}u^2$. To find u_3, iterate $u_n = P(u_{n-1})$ starting with $u_0 = 0$. This gives $u_1 = P(0) = \frac{1}{3}$, $u_2 = P(\frac{1}{3}) = \frac{1}{3} + \frac{2}{3}(\frac{1}{3})^2 = \frac{11}{27}$, and $u_3 = P(\frac{11}{27}) = \frac{1}{3} + \frac{2}{3}(\frac{11}{27})^2 = 0.4440$. The equation $u = \frac{1}{3} + \frac{2}{3}u^2$ has roots $u = 1$ and $u = \frac{1}{2}$. Therefore, the probability $u_\infty = \frac{1}{2}$. The probabilities for two individuals are $u_3^2 = 0.1971$ and $u_\infty^2 = 0.25$.

8.17 Since the uniform random variable on $(0, 1)$ has the moment-generating function $\int_0^1 e^{tu} du = \frac{e^t - 1}{t}$, it is plausible that X takes on the value 0 with probability $\frac{1}{2}$ and is uniformly distributed on $(0, 1)$ with probability $\frac{1}{2}$. Indeed, for such a mixed random variable X, we have $E(e^{tX}) = \frac{1}{2} + \frac{1}{2}\frac{e^t - 1}{t}$.

8.19 Since the random variables X_i are independent, the moment-generating function of $\sum_{i=1}^n X_i$ is $(\frac{\lambda}{\lambda-t})^{\alpha_1} \cdots (\frac{\lambda}{\lambda-t})^{\alpha_n} = (\frac{\lambda}{\lambda-t})^{\alpha_1 + \cdots + \alpha_n}$. This proves the desired result by the uniqueness property of the moment-generating function.

8.21 The definition $M_X(t) = \int_{-\infty}^{\infty} e^{tx} e^x/(1+e^x)^2 dx$ reveals that $M_X(t)$ is finite only for $-1 < t < 1$. Using the change of variable $u = 1/(1 + e^x)$ and noting the relations $\frac{du}{dx} = -e^x/(1+e^x)^2$ and $e^x = \frac{1-u}{u}$, it follows readily that $M_X(t) = \int_0^1 (\frac{1-u}{u})^t du$ for $-1 < t < 1$. Since $a^t = e^{t \ln(a)}$ has as derivative $\ln(a)e^{at}$, we get from the integral representation of $M_X(t)$ that $M_X'(t) = \int_0^1 \ln(\frac{1-u}{u})(\frac{1-u}{u})^t du$. Thus, $M_X'(0) = \int_0^1 [\ln(1-u) - \ln(u)] du = 0$. In the same way, we find $M_X''(0) = \int_0^1 [\ln(1-u) - \ln(u)]^2 du = \frac{\pi^2}{3}$, showing that $E(X) = 0$ and $\sigma^2(X) = \frac{\pi^2}{3}$.

8.23 (a) Using the decomposition formula for the standard bivariate normal density function $f(x, y)$ in Section 6.1 and the substitution rule, the moment-generating function $E(e^{sX+tY})$ can be evaluated as $\int_{-\infty}^{\infty} e^{sx} \frac{1}{\sqrt{2\pi}} e^{-\frac{1}{2}x^2} dx \int_{-\infty}^{\infty} e^{ty} \frac{1}{\sqrt{2\pi}} \frac{1}{\sqrt{1-\rho^2}} e^{-\frac{1}{2}(y-\rho x)^2/(1-\rho^2)} dy$. The inner integral can be interpreted as $E(e^{tW})$ with $N(\rho x, 1 - \rho^2)$-distributed W and so, using Example 8.5, the inner integral reduces to $e^{\rho xt + \frac{1}{2}(1-\rho^2)t^2}$. Thus, $E(e^{sX+tY})$ is given by $e^{\frac{1}{2}(1-\rho^2)t^2} \int_{-\infty}^{\infty} e^{(s+\rho t)x} \frac{1}{\sqrt{2\pi}} e^{-\frac{1}{2}x^2} dx$. The latter integral can be interpreted as $E(e^{(s+\rho t)Z})$ with $N(0, 1)$-distributed Z and is thus equal to $e^{\frac{1}{2}(s+\rho t)^2}$. Putting the pieces together, we have $E(e^{sX+tY}) = e^{\frac{1}{2}(s^2 + 2\rho st + t^2)}$, as was to be verified. (b). It suffices to verify the assertion for $N(0, 1)$-distributed X and Y. Let $\rho = \rho(X, Y)$ ($= \text{cov}(X, Y)$). Using the assumption and Rule 5.12, the random variable $aX + bY$ is $N(0, a^2 + 2ab\rho + b^2)$ distributed for any constants a, b. Then, by Example 8.5 with $t = 1$, $E(e^{aX+bY})$ is $e^{\frac{1}{2}(a^2 + 2ab\rho + b^2)}$ for all a, b. This proves the desired result with an appeal to the result of part (**a**) and the uniqueness property of the moment-generating function.

Chapter 9

9.1 Let X be the total time needed for both tasks. Then $E(X) = 45$ and $\sigma^2(X) = 65$. Write the probability $P(X < 60)$ as $1 - P(X \geq 45 + 15)$. The one-sided Chebyshev's inequality gives $P(X < 60) \geq 1 - 65/(65 + 225) = 0.7759$.

9.3 The moment-generating function of a random variable X that is uniformly distributed on $(-1, 1)$ is $M_X(t) = \frac{1}{2t}(e^t - e^{-t})$ for all t. Put for abbreviation $\overline{X}_n = \frac{1}{n}(X_1 + \cdots + X_n)$. Since the X_i are independent, $M_{\overline{X}_n}(t)$ is $\left[\frac{1}{2t/n}(e^{t/n} - e^{-t/n})\right]^n$. By Chernoff's bound, $P(\overline{X}_n \geq c) \leq \min_{t>0} e^{-ct} M_{\overline{X}_n}(t)$. Using the inequality $\frac{1}{2}(e^u - e^{-u}) \leq ue^{u^2/6}$ for $u > 0$, we get $P(\overline{X}_n \geq c) \leq \min_{t>0} e^{-ct} e^{t^2/6n}$. The function $e^{-(ct - t^2/6n)}$ is minimal for $t = 3cn$, which gives the desired bound.

9.5 The random variable X is distributed as $\sum_{i=1}^{n} X_i$, where the X_i are independent with $P(X_i = 1) = p$ and $P(X_i = 0) = 1 - p$. This gives $M_X(t) = (pe^t + 1 - p)^n$. By Chernoff's bound, $P(X \geq np(1+\delta)) \leq \min_{t>0} [e^{-np(1+\delta)t}(pe^t + 1 - p)^n]$. Let $g(t) = e^{-p(1+\delta)t}(pe^t + 1 - p)$. The function $g(t)$ takes on its absolute minimum for $t = \ln(\gamma)$ with $\gamma = \frac{(1-p)(1+\delta)}{1-p(1+\delta)}$. This leads to the upper bound $\left(\frac{p\gamma + 1 - p}{\gamma^{p(1+\delta)}}\right)^n$ for $P(X \geq np(1+\delta))$. Next it is a matter of some algebra to obtain the first bound. To get the other bound, note that $f'(a) = \ln\left(\frac{a}{p}\right) - \ln\left(\frac{1-a}{1-p}\right)$ and $f''(a) = \frac{1}{a(1-a)}$ for $0 \leq a < 1$. Next use Taylor's formula $f(a) = f(p) + (a-p)f'(p) + \frac{1}{2!}(a-p)^2 f''(\eta_a)$ for some η_a with $p < \eta_a < a$. Since $f(p) - f'(p) = 0$ and $\eta(1-\eta) \leq \frac{1}{4}$ for $0 < \eta < 1$, we obtain $f(a) \geq 2(a-p)^2 = 2\delta^2 p^2$ for $p < a < 1$, which gives the second bound.

9.7 By $\ln(P_n) = \frac{1}{n}\sum_{k=1}^{n} \ln(X_k)$ and $E(\ln(X_k)) = \int_0^1 \ln(x)dx = -1$, the strong law of large numbers gives $P(\{\omega : \lim_{n \to \infty} \ln(P_n(\omega)) = -1\}) = 1$. This implies that $P(\{\omega : \lim_{n \to \infty} P_n(\omega) = e^{-1}\}) = 1$.

9.9 Fix $\epsilon > 0$. Let $A_n = \{\omega : |X_n(\omega)| > \epsilon\}$. Suppose to the contrary that $\sum_{n=1}^{\infty} P(A_n) = \infty$. Then, by the second Borel–Cantelli lemma, $P(B) = 1$ for $B = \{\omega : \omega \in A_n \text{ for infinitely many } n\}$. This contradicts the assumption $P(\{\omega : \lim_{n \to \infty} X_n(\omega) = 0\}) = 1$. Thus $\sum_{n=1}^{\infty} P(A_n) < \infty$.

9.11 By Markov's inequality, $P(|X_n - X| > \epsilon) = P(|X_n - X|^2 > \epsilon^2) \leq \frac{E(|X_n - X|^2)}{\epsilon^2}$ for each $\epsilon > 0$. Therefore, $\lim_{n \to \infty} P(|X_n - X| > \epsilon) = 0$ for each $\epsilon > 0$, showing convergence in probability. A counterexample is provided by Problem 9.10. *Note*: Mean-square convergence is neither stronger nor weaker than almost-sure convergence.

9.13 Since the X_i are independent, $E(Y_k) = \mu^2$ and so $E\left(\frac{1}{n}\sum_{k=1}^{n} Y_k\right) = \mu^2$. By Chebyshev's inequality, $P(|\frac{1}{n}\sum_{k=1}^{n} Y_k - \mu^2| > \epsilon) \leq \sigma^2\left(\sum_{k=1}^{n} Y_k\right)/(n^2\epsilon^2)$ for each $\epsilon > 0$. By the independence of the X_k, we have $\text{cov}(Y_i, Y_j) = \mu^4 - \mu^4 = 0$ for $j > i+1$. Thus, $\sigma^2\left(\sum_{k=1}^{n} Y_k\right) = \sum_{k=1}^{n} \sigma^2(Y_k) + 2\sum_{i=1}^{n-1} \text{cov}(Y_i, Y_{i+1})$. We have $\sigma^2(Y_k) = E(X_k^2)E(X_{k+1}^2) - \mu^4$ and $\text{cov}(Y_i, Y_{i+1}) = \mu^2 E(X_{i+1}^2) - \mu^4$. Thus, by the boundedness of the $\sigma^2(X_i)$, there is a constant $c > 0$ such that $\sigma^2\left(\sum_{k=1}^{n} Y_k\right) \leq nc$. Then we get that $P(|\frac{1}{n}\sum_{k=1}^{n} Y_k - \mu^2| > \epsilon)$ tends to 0 as $n \to \infty$, as was to be proved.

9.15 Letting $n \to \infty$ in $P(X_n \leq x) \leq P(X \leq x+\epsilon) + P(|X_n - X| > \epsilon)$, we get $\lim_{n \to \infty} P(X_n \leq x) \leq P(X \leq x+\epsilon)$ for any $\epsilon > 0$. Next, letting $\epsilon \to 0$, we obtain $\lim_{n \to \infty} P(X_n \leq x) \leq P(X \leq x)$ when x is a continuity point of $P(X \leq x)$. Interchanging the roles of X_n and X, we get in a similar way that $P(X \leq x) \leq \lim_{n \to \infty} P(X_n \leq x+\epsilon)$. Replacing x by $x - \epsilon$, we obtain $\lim_{n \to \infty} P(X_n \leq x) \geq P(X \leq x-\epsilon)$ for any $\epsilon > 0$. Therefore, $\lim_{n \to \infty} P(X_n \leq x) \geq P(X \leq x)$ when x is a continuity point of $P(X \leq x)$. This completes the proof.

9.17 Since the X_i are independent, $P\left(\frac{M_n}{n} \leq x\right) = P(X_1 \leq nx) \cdots P(X_n \leq nx)$. Therefore, $P\left(\frac{M_n}{n} \leq x\right)$ equals $\left(\frac{2}{\pi}\text{arctg}(nx)\right)^n = \left(1 - \frac{2}{\pi}\text{arctg}(\frac{1}{nx})\right)^n$ for $x > 0$. For any fixed x, we have that $|\frac{1}{nx}| < 1$ for n large enough. Using the power-series expansion of $\text{arctg}(y)$ for $|y| < 1$, it follows that $\lim_{n \to \infty} P\left(\frac{M_n}{n} \leq x\right)$ is equal to $\lim_{n \to \infty} \left(1 - \frac{2}{\pi x n}\right)^n = e^{-\frac{2}{\pi x}}$ for any $x > 0$.

9.19 The Kelly betting fraction suggests staking 2.7% of your current bankroll each time.

9.21 Denote by V_k your bankroll after k bets. Then, by the same arguments as used in the derivation of the Kelly betting fraction, $\ln(V_n/V_0) = \sum_{i=1}^{n} \ln(1 - \alpha + \alpha R_i)$, where the R_i are independent random variables with $P(R_i = f_1) = p$ and $P(R_i = f_2) = 1 - p$. By the central limit theorem, $\ln(V_n/V_0)$ is approximately $N(n\mu_\alpha, n\sigma_\alpha^2)$ distributed for n large enough, where $\mu_\alpha = p\ln(1 - \alpha + f_1\alpha) + (1 - p)\ln(1 - \alpha + f_2\alpha)$ and $\sigma_\alpha^2 = p\ln^2(1 - \alpha + f_1\alpha) + (1 - p)\ln^2(1 - \alpha + f_2\alpha) - \mu_\alpha^2$. This leads to $P(V_n > x) \approx 1 - \Phi(\frac{\ln(x/V_0) - n\mu_\alpha}{\sigma_\alpha\sqrt{n}})$. For $p = 0.5$, $f_1 = 1.8$, $f_2 = 0.4$, $n = 52$, and $\alpha = \frac{5}{24}$, the normal approximation yields the values 0.697, 0.440, and 0.150 for $x/V_0 = 1, 2$, and 5. A simulation study with 1 million runs yields the values 0.660, 0.446, and 0.167. For $\alpha = 1$, the probability is about 0.5 that your bankroll will be no more than $1.8^{26} \times 0.4^{26} \times 10,000 = 1.95$ dollars after 52 weeks. The intuitive explanation is that in the most likely scenario a path will unfold in which the stock price rises during half of the time and falls during the other half of the time.

9.23 The stochastic process $\{S_t\}$ describing the age of the light bulb in use is regenerative. It regenerates itself each time a new bulb is installed. Let the generic random variable X be distributed as the lifetime of a bulb. Let $F(x) = P(X \leq x)$ and let $f(x)$ be the probability density of X. Then the expected length of a cycle is $\int_0^T tf(t)\,dt + \int_T^\infty Tf(t)\,dt = \int_0^T (1 - F(t))\,dt$. The expected cost in one cycle is $c_2 P(X \leq T) + c_1 P(X > T) = c_1 + (c_2 - c_1)F(T)$, so the long-run average cost per unit time is $[c_1 + (c_2 - c_1)F(T)]/\int_0^T (1 - F(t))\,dt$.

9.25 Let S_t be equal to 1 if the channel is on at time t and be equal to 0 otherwise. The stochastic process $\{S_t\}$ is regenerative. Take the epochs at which an on-time starts as the regeneration epochs. Let μ_{on} be the expected length of the on-time X. Then $\mu_{on} = \int_0^1 x\,6x(1-x)\,dx = 0.5$. Let L be the length of a cycle. By the law of conditional expectation, $E(L) = \int_0^1 (x + x^2\sqrt{x})\,6x(1-x)\,dx = \mu_{on} + \mu_{off}$, where $\mu_{off} = \int_0^1 x^2\sqrt{x}\,6x(1-x)\,dx = 0.2424$. Thus, the long-run fraction of time the system is on equals $\mu_{on}/(\mu_{on} + \mu_{off}) = 0.673$.

9.27 The process describing the status of the processor is regenerative. Take as cycle the time interval between two successive epochs at which an arriving job finds the processor idle. Using the memoryless property of the Poisson process, the expected length of a cycle is $\mu + \frac{1}{\lambda}$ and the expected amount of idle time in one cycle is $\frac{1}{\lambda}$. Thus, the long-run fraction of time the server is idle is $(1/\lambda)/(\mu + 1/\lambda) = 1/(1 + \lambda\mu)$. Let N be the number of jobs arriving during the processing time X. Then $E(N \mid X = x) = \lambda x$ and so $E(N) = \lambda\mu$. Thus, the expected number of arrivals during one cycle is $1 + \lambda\mu$. The number of jobs accepted in one cycle is 1, and so the long-run fraction of jobs that are accepted is $1/(1 + \lambda\mu)$.

9.29 Let a cycle be the time interval between two successive replacements of the bulb. Denote by the generic variable X the length of a cycle. Imagine that a cost at rate 1 is incurred if the age of the bulb is greater than c and a cost at rate 0 is incurred otherwise. Then, the cost incurred during one cycle is $\max(X - c, 0)$. The long-run average cost per unit time is $E[\max(X - c, 0)]/E(X)$. Since $E(V) = \int_0^\infty P(V > v)\, dv$ for any nonnegative random variable V (see Problem 4.26), $E[\max(X - c, 0)] = \int_0^\infty P(\max(X - c, 0) > v)\, dv = \int_0^\infty P(X > v + c)\, dv$. Thus, $E[\max(X - c, 0)] = \int_c^\infty (1 - F(x))\, dx$ and the desired result follows.

9.31 Define a cycle as the time elapsed between two consecutive replacements of the item. By conditioning on the lifetime X of the item, the expected length of a cycle is $\int_0^T xf(x)\, dx + \int_T^\infty (T + a(x))f(x)\, dx$, where $a(x) = E[\min(x - T, V)]$ for $x > T$ and V is the length of the interval between T and the first preventive replacement opportunity after time T. Since V is exponentially distributed with parameter λ, we have $a(x) = \int_0^{x-T} v\lambda e^{-\lambda v}\, dv + (x - T)e^{-\lambda(x-T)} = \frac{1}{\lambda}[1 - e^{-\lambda(x-T)}]$. By $P(V \leq x - T) = 1 - e^{-\lambda(x-T)}$ for $x > T$, the expected cost incurred in one cycle is $c_0 F(T) + \int_T^\infty \left[c_1\left(1 - e^{-\lambda(x-T)}\right) + c_0 e^{-\lambda(x-T)}\right] f(x)\, dx$, where $F(x) = P(X \leq x)$. The long-run average cost per unit time is the ratio of E(cost incurred in one cycle) and E(length of a cycle).

9.33 For fixed clearing time $T > 0$, the stochastic process describing the number of messages in the buffer regenerates itself each time the buffer is cleared. The clearing epochs are regeneration epochs because of the memoryless property of the Poisson process. The expected length of a cycle is T. The expected cost incurred in the first cycle is $K + E[\sum_{n=1}^\infty h(S_n)] = K + \sum_{n=1}^\infty E[h(S_n)]$. Since $E[h(S_n)] = h \int_0^T (T - x) \lambda^n \frac{x^{n-1}}{(n-1)!} e^{-\lambda x}\, dx$, we get $\sum_{n=1}^\infty E[h(S_n)] = h \int_0^T (T - x)\lambda\, dx = \frac{1}{2}\lambda T^2$. Thus, the average cost per unit time is $(K + \frac{1}{2}\lambda T^2)/T$. This expression is minimal for $T = \sqrt{2K/(h\lambda)}$.

9.35 In view of the interpretation of an Erlang-distributed interoccurrence time as the sum of r independent phases, imagine that phases are completed according to a Poisson process with rate α, where the completion of each rth phase marks the occurrence of an event. Then $P(N(t) \leq k)$

can be interpreted as the probability that at most $(k+1)r - 1$ phases are completed up to time t.

9.37 Fix $s \geq 1$. Imagine that a reward of 1 is earned for each customer belonging to a batch of size s. Then the expected reward earned for a batch is sp_s. The expected number of customers in a batch is μ. Thus, the average reward per customer is $\frac{sp_s}{\mu}$, which intuitively explains the result. A formal proof is easily given by using the strong law of large numbers.

Chapter 10

10.1 Let X_n be the number of type-1 particles in compartment A after the nth transfer. The process $\{X_n\}$ is a Markov chain with state space $I = \{0, 1, \ldots, r\}$. The one-step transition probabilities are $p_{i,i-1} = i^2/r^2$, $p_{ii} = 2i(r-i)/r^2$, $p_{i,i+1} = (r-i)^2/r^2$, and $p_{ij} = 0$ otherwise.

10.3 Take as state the largest outcome in the last roll. Let X_n be the state after the nth roll, with the convention $X_0 = 0$. The process $\{X_n\}$ is a Markov chain with state space $I = \{0, 1, \ldots, 6\}$. The one-step transition probabilities are $p_{0k} = \frac{1}{6}$ for $1 \leq k \leq 6$ and $p_{jk} = \left(\frac{k}{6}\right)^j - \left(\frac{k-1}{6}\right)^j$ for $1 \leq j, k \leq 6$, using the relation $P(Y = k) = P(Y \leq k) - P(Y \leq k - 1)$.

10.5 Let's say that the system is in state $(0,0)$ if both machines are good, in state $(0,k)$ if one of the machines is good and the other one is in revision with a remaining repair time of k days for $k = 1, 2$, and in state $(1,2)$ if both machines are in revision with remaining repair times of one day and two days. Letting X_n be the state of the system at the end of the nth day, the process $\{X_n\}$ is a Markov chain. The one-step transition probabilities are given by $p_{(0,0)(0,0)} = \frac{9}{10}$, $p_{(0,0)(0,2)} = \frac{1}{10}$, $p_{(0,1)(0,0)} = \frac{9}{10}$, $p_{(0,1)(0,2)} = \frac{1}{10}$, $p_{(0,2)(0,1)} = \frac{9}{10}$, $p_{(0,2)(1,2)} = \frac{1}{10}$, $p_{(1,2)(0,1)} = 1$, and $p_{vw} = 0$ otherwise.

10.7 The system is in state i if the channel holds i messages (including any message in transmission). If the system is in state i at the beginning of a time slot, then the buffer contains $\max(i - 1, 0)$ messages. Define X_n as the state of the system at the beginning of the nth time slot. The process $\{X_n\}$ is a Markov chain with state space $I = \{0, 1, \ldots, K+1\}$. In a similar way as in Example 10.4, the one-step transition probabilities are obtained. Let $a_k = e^{-\lambda}\lambda^k/k!$. Then $p_{0j} = a_j$ for $0 \leq j \leq K - 1$, $p_{0,K} = \sum_{k=K}^{\infty} a_k$, $p_{i,i-1} = (1-f)a_0$ for $1 \leq i \leq K$, $p_{K+1,K} = 1 - f$, $p_{ij} = (1-f)a_{j-i+1} + fa_{j-i}$ for $1 \leq i \leq j \leq K$, $p_{i,K+1} = 1 - \sum_{j=i-1}^{K} p_{ij}$ for $1 \leq i \leq K$, and $p_{ij} = 0$ otherwise.

10.9 Use a Markov chain with four states SS, SR, RS, and RR. The probability that it is sunny five days from now is $p^{(5)}_{RR,RS} + p^{(5)}_{RR,SS} = 0.7440$. The calculation of $\mathbf{P}^n$ for larger values of n reveals that the long-run probability of a sunny day is $0.6923 + 0.0989 = 0.7912$. The expected number of sunny days in the coming 14 days is $\sum_{t=1}^{14}(p^{(t)}_{RR,RS} + p^{(t)}_{RR,SS}) = 10.18$.

10.11 Consider a two-state Markov chain with states 0 and 1, where state 0 means that the last bit was received incorrectly and state 1 means that the last bit was received correctly. The one-step transition probabilities are given by $p_{00} = 0.9$, $p_{01} = 0.1$, $p_{10} = 0.001$, and $p_{11} = 0.999$. The expected number of incorrectly received bits is $\sum_{n=1}^{5,000} p^{(n)}_{10} = 49.417$.

10.13 Using $\mathrm{var}\left(\sum_{t=1}^{n} I_t\right) = \sum_{t=1}^{n} \mathrm{var}(I_t) + 2\sum_{t=1}^{n-1}\sum_{u=t+1}^{n} \mathrm{cov}(I_t, I_u)$, we get the result for $\sigma^2[V_{ij}(n)]$. The normal approximation to the sought probability is $1 - \Phi\left(\frac{240.5 - 217.294}{12.101}\right) = 0.0276$ (the simulated value is 0.0267).

10.15 Use a Markov chain with states $s = (i, k)$, where $i = 1$ if England has won the last match, $i = 2$ if England has lost the last match, $i = 3$ if the last match was a draw, and $k \in \{0, 1, 2, 3\}$ denotes the number of matches England has won so far. For states $s = (i, k)$ with $k = 0$, the one-step transition probabilities are $p_{(1,0)(1,1)} = 0.44$, $p_{(1,0)(2,0)} = 0.37$, $p_{(1,0)(3,0)} = 0.19$, $p_{(2,0)(1,1)} = 0.28$, $p_{(2,0)(2,0)} = 0.43$, $p_{(2,0)(3,0)} = 0.29$, $p_{(3,0)(1,1)} = 0.27$, $p_{(3,0)(2,0)} = 0.30$, and $p_{(3,0)(3,0)} = 0.43$. Similarly, the one-step probabilities for the states $(i, 1)$ and $(i, 2)$. The one-step transition probabilities for the states $(i, 3)$ are not relevant and may be taken as $p_{(i,3)(i,3)} = 1$. Let p_k denote the probability that England will win k matches of the next three matches when the last match was a draw. Then $p_k = p^{(3)}_{(3,0)(1,k)} + p^{(3)}_{(3,0)(2,k)} + p^{(3)}_{(3,0)(3,k)}$ for $0 \le k \le 3$ (use the fact that the second component of state $s = (i, k)$ cannot decrease). This leads to $p_0 = 0.3842$, $p_1 = 0.3671$, $p_2 = 0.1964$, and $p_3 = 0.0523$. As a sanity check, $\sum_{k=0}^{3} k p_k = 0.9167$, in agreement with the answer to Problem 10.14.

10.17 Use a Markov chain with the 11 states $1, 2, \ldots, 10$ and $10+$, where state i means that the particle is in position i and state $10+$ means that the particle is in a position beyond position 10. The state $10+$ is taken to be absorbing. The one-step transition probabilities are $p_{ij} = 0.5$ for $j = i - \lfloor i/2 \rfloor$ and $j = 2i - \lfloor i/2 \rfloor$. The other p_{ij} are zero. The sought probability is $p^{(25)}_{1,10+} = 0.4880$.

10.19 Use a Markov chain with four states 0, 1, 2, and 3, where state 0 means neither a total of 7 nor a total of 12 for the last roll, state 1 means a total of 7 for the last roll but not for the roll before, state 2 means a total of 7 for the last two rolls, and state 3 means a total of 12 for the last roll. The states 2 and 3 are absorbing with $p_{22} = p_{33} = 1$. Further, $p_{00} = p_{10} = \frac{29}{36}$, $p_{01} = p_{12} = \frac{6}{36}$, $p_{03} = p_{13} = \frac{1}{36}$, and $p_{ij} = 0$ otherwise. Let f_i be the probability of absorption in state 2 when the initial state is i. The sought probability f_0 is $\frac{6}{13}$, as follows by solving $f_0 = \frac{29}{36}f_0 + \frac{6}{36}f_1$ and $f_1 = \frac{29}{36}f_0 + \frac{6}{36}$.

10.21 Take a specific city (say, Venice) and a particular number (say, 53). Consider a Markov chain with state space $I = \{0, 1, \ldots, 182\}$, where state i indicates the number of draws since the particular number 53 appeared for the last time in the Venice lottery. The state 182 is taken as an absorbing state. The one-step transition probabilities for the other states of the Markov chain are $p_{i0} = \frac{5}{90}$ and $p_{i,i+1} = \frac{85}{90}$ for $i = 0, 1, \ldots, 182$. The probability that in the next 1,040 draws of the Venice lottery there is some window of 182 consecutive draws in which the number 53 does not appear can be calculated as $p_{0,182}^{(1,040)}$. This probability has the value $p = 0.00077541$. The five winning numbers in a draw of the lottery are not independent of each other, but the dependence is weak enough to give a good approximation for the probability that in the next 1,040 drawings of the Venice lottery there is not some number that does not occur during some window of 182 consecutive draws. This probability is approximated by $(1-p)^{90}$. The lottery takes place in 10 cities. Thus, the sought probability is approximately $1 - (1-p)^{900} = 0.5025$. This problem is another illustration of the fact that coincidences can nearly always be explained by probabilistic arguments!

10.23 Take a Markov chain with state space $I = \{(i,j) : i, j \geq 0, i+j \leq 25\}$, where state (i,j) means that i pictures are in the pool once and j pictures are in the pool twice or more. State $(0, 25)$ is absorbing. The other one-step transition probabilities are $p_{(i,j),(i,j)} = \frac{i}{25} \times \frac{j}{25}$, $p_{(i,j),(i+1,j)} = 2 \times \frac{25-i-j}{25} \times \frac{j}{25}$, $p_{(i,j),(i+2,j)} = \frac{25-i-j}{25} \times \frac{24-i-j}{25}$, $p_{(i,j),(i,j+1)} = \frac{i}{25} \times \frac{25-i-j}{25} + \frac{25-i-j}{25} \times \frac{i+1}{25}$, $p_{(i,j),(i-1,j+1)} = \frac{i}{25} \times \frac{j+1}{25} + \frac{j}{25} \times \frac{i}{25}$, $p_{(i,j),(i-2,j+2)} = \frac{i}{25} \times \frac{i-1}{25}$. The probability $P(N > n)$ is given by $1 - p_{(0,0),(0,25)}^{(n)}$. We find $E(N) = 71.4$ weeks by solving a set of linear equations.

10.25 Take a Markov chain with state space $I = \{0, 1, \ldots, 5\}$, where state i means that Joe's bankroll is $i \times 200$ dollars. The states 0 and 5 are absorbing. The other one-step transition probabilities are $p_{10} = p_{20} = p_{31} = p_{43} = \frac{19}{37}$, $p_{12} = p_{24} = p_{35} = p_{45} = \frac{18}{37}$, and $p_{ij} = 0$ otherwise. (a) The probability that Joe will place more than n bets is $1 - p_{4,0}^{(n)} - p_{4,5}^{(n)}$. This probability has the values 0.2637, 0.1283, 0.0320, and 0.0080 for $n = 2, 3, 5,$ and 7. (b) To find the probability of Joe reaching his goal, solve the four linear equations $f_1 = \frac{18}{37}f_2$, $f_2 = \frac{18}{37}f_4$, $f_3 = \frac{18}{37} + \frac{19}{37}f_1$, and $f_4 = \frac{18}{37} + \frac{19}{37}f_3$. The probability of Joe reaching his goal is $f_4 = 0.78531$. (c) Parameterize and let s_i be the expected value of the total amount you will stake in the remaining part of the game when the current state is i. To find s_4, solve the linear equations $s_1 = 200 + \frac{18}{37}s_2$, $s_2 = 400 + \frac{18}{37}s_4$, $s_3 = 400 + \frac{19}{37}s_2$, and $s_4 = 200 + \frac{19}{37}s_3$. This gives $s_4 = 543.37$ dollars. As a sanity check, the ratio of your expected loss and the expected amount staked during the game is indeed equal to the house advantage of $0.0270 per dollar staked (the expected loss is $0.21469 \times 800 - 0.78531 \times 200 = 14.69$ dollars). Whatever roulette system you use, in the long run you will lose on average 2.7 cents on each dollar staked!

10.27 Use a Markov chain with the states $i = 0, 1, \ldots, 8$, where state i means that the dragon has i heads. The states 0, 7, and 8 are absorbing. The one-step transition probabilities are $p_{i,i-1} = 0.7$, $p_{i,i+1} = 0.3p$, and $p_{i,i+2} = 0.3(1-p)$ for $1 \leq i \leq 6$. The win probabilities for $p = 0, 0.5$, and 1 are the absorption probabilities 0.6748, 0.8255, and 0.9688.

10.29 Use a Markov chain with the states $0, 1, \ldots, 5$, where state 0 corresponds to the start of the game, state 1 means that all five dice show a different value, and state i with $i \geq 2$ means that you have i dice of a kind. State 5 is an absorbing state. In state 1 you re-roll all five dice and in state i with $i \geq 2$ you leave the i dice of a kind and re-roll the other $5 - i$ dice. The matrix of one-step transition probabilities is

from /to	0	1	2	3	4	5
0	0	$\frac{120}{1296}$	$\frac{900}{1296}$	$\frac{250}{1296}$	$\frac{25}{1296}$	$\frac{1}{1296}$
1	0	$\frac{120}{1296}$	$\frac{900}{1296}$	$\frac{250}{1296}$	$\frac{25}{1296}$	$\frac{1}{1296}$
2	0	0	$\frac{120}{216}$	$\frac{80}{216}$	$\frac{15}{216}$	$\frac{1}{216}$
3	0	0	0	$\frac{25}{36}$	$\frac{10}{36}$	$\frac{1}{36}$
4	0	0	0	0	$\frac{5}{6}$	$\frac{1}{6}$
5	0	0	0	0	0	1

The probability of getting Yahtzee within three rolls is $p_{05}^{(3)} = 0.04603$.

10.31 You may use a Markov chain with the eight states $(0, 0, 0), \ldots, (1, 1, 1)$, where 0 means an empty glass and 1 means a filled glass. However, for reasons of symmetry, a Markov chain with four states $i = 0, 1, 2$, and 3 suffices, where state i means that there are i filled glasses. State 0 is absorbing with $p_{00} = 1$. The other one-step transition probabilities are $p_{10} = \frac{1}{3}$, $p_{12} = \frac{2}{3}$, $p_{21} = \frac{2}{3}$, $p_{23} = \frac{1}{3}$, and $p_{32} = 1$. By solving the linear equations $\mu_3 = 1 + \mu_2$, $\mu_2 = 1 + \frac{2}{3}\mu_1 + \frac{1}{3}\mu_3$, and $\mu_1 = 1 + \frac{2}{3}\mu_2$, we find $E(N) = 10$. The probability $P(N > n) = 1 - p_{30}^{(n)}$ has the values 0.6049, 0.3660, 0.1722, 0.1042, and 0.0490 for $n = 5, 10, 15, 20$, and 25.

10.33 Use a Markov chain with the eight states $0, 1, \ldots, 6$ and -1, where state 0 is the starting state, state i with $1 \leq i \leq 6$ means that the outcome of the last roll is i, and state -1 means that the outcome of the last roll is less than the outcome of the roll preceding it. State -1 is absorbing with $p_{-1,-1} = 1$. To answer the first question, take the one-step transition probabilities $p_{0j} = \frac{1}{6}$ for $1 \leq j \leq 6$, $p_{ii} = \cdots = p_{i6} = \frac{1}{6}$, and $p_{i,-1} = \frac{i-1}{6}$ for $1 \leq i \leq 6$. The probability that each of the last three rolls is at least as large as the roll preceding it equals $1 - p_{0,-1}^{(4)} = 0.0972$. In a similar way, the probability that each of the last three rolls is larger than the roll preceding it is 0.0116.

10.35 Adjust the Markov matrix $\mathbf{P}$ in Problem 10.8 to the matrix $\mathbf{Q}$ by replacing the first row of $\mathbf{P}$ by $(1, 0, 0, 0)$, that is, state 1 is made absorbing. Next, calculate the matrix products $\mathbf{Q}^4$ and $\mathbf{Q}^5$. The probability that the car will be rented out more than five times before it returns to location 1 is $1 - q_{41}^{(5)} = 0.7574$ if the car is currently at location 4 and $p_{12}(1 - q_{21}^{(4)}) + p_{13}(1 - q_{31}^{(4)}) + p_{14}(1 - q_{41}^{(4)}) = 0.1436$ if the car is currently at location 1.

10.37 In Example 10.3, the solution of the equilibrium equations $\pi_1 = 0.50\pi_1 + 0.50\pi_3$, $\pi_2 = 0.5\pi_1 + 0.5\pi_3$, and $\pi_3 = \pi_2$ together with $\pi_1 + \pi_2 + \pi_3 = 1$ is $\pi_1 = \pi_2 = \pi_3 = \frac{1}{3}$. The long-run proportion of time the professor has his license with him is $\pi_1 = \frac{1}{3}$. In Problem 10.2 the Markov chain has six states $1, 2, \ldots, 6$, where state 1/2/3 means that the professor is driving to the office and has his driver's license with him/at home/at the office and state 4/5/6 means that the professor is driving to home and has his driver's license with him/at the office/at home. The solution of the equilibrium equations $\pi_1 = 0.75\pi_4 + 0.75\pi_6$, $\pi_2 = 0.25\pi_4 + 0.25\pi_6$, $\pi_3 = \pi_5$, $\pi_4 = 0.50\pi_1 + 0.50\pi_3$, $\pi_5 = 0.50\pi_1 + 0.50\pi_3$, and $\pi_6 = \pi_2$ together with $\pi_1 + \cdots + \pi_6 = 1$ is $\pi_1 = \pi_3 = \pi_4 = \pi_5 = 0.2143$, $\pi_2 = \pi_6 = 0.0714$. The long-run proportion of time the professor has his license with him is $\pi_1 + \pi_4 = 0.4286$.

10.39 Use a four-state Markov chain with the one-step transition probabilities $p_{ii} = 1 - r_i$ and $p_{ij} = \frac{1}{3}r_i$ for $j \neq i$. The Markov chain is aperiodic. Therefore, the limiting probabilities exist and are given by the equilibrium probabilities. Solving the equilibrium equations $\pi_j = (1 - r_j)\pi_j + \sum_{i \neq j} \frac{1}{3}r_i\pi_i$ for $1 \leq j \leq 4$ together with $\pi_1 + \cdots + \pi_4 = 1$, we get $\pi_1 = 0.2817$, $\pi_2 = 0.1690$, $\pi_3 = 0.2113$, and $\pi_4 = 0.3380$.

10.41 State j means that compartment A contains j particles of type 1. An intuitive guess is that $\pi_j = \binom{r}{j}\binom{r}{r-j}/\binom{2r}{r}$ for $j = 0, 1, \ldots, r$. These π_j satisfy the equilibrium equations $\pi_j = \frac{(r-j+1)^2}{r^2}\pi_{j-1} + \frac{2j(r-j)}{r^2}\pi_j + \frac{(j+1)^2}{r^2}\pi_{j+1}$, as can be verified by substitution. Since the Markov chain has no two or more disjoint closed sets, its equilibrium distribution is uniquely determined.

10.43 A Markov chain with two states 0 and 1 suffices, where state 0 means that Linda and Bob are in different venues and state 1 means that they are in the same venue. The one-step transition probability $p_{01} = 2 \times 0.4 \times (0.6 \times \frac{1}{3}) + (0.6 \times \frac{2}{3}) \times (0.6 \times \frac{1}{3})$, where the first term refers to the probability that exactly one of the two persons does not change venue and the other person goes to the venue of that person, and the second term refers to the probability that both persons change venue and go to the same venue. By a similar argument, $p_{11} = 0.4 \times 0.4 + 0.6 \times (0.6 \times \frac{1}{3})$. This gives $p_{01} = 0.24$ and $p_{11} = 0.28$. Further, $p_{00} = 1 - p_{01}$ and $p_{10} = 1 - p_{11}$. Solving the equations $\pi_0 = p_{00}\pi_0 + p_{10}\pi_1$ and $\pi_0 + \pi_1 = 1$ gives $\pi_0 = \frac{18}{19}$ and $\pi_1 = \frac{1}{19}$. The long-run fraction of weekends that Linda and Bob visit the same venue is $\pi_1 = \frac{1}{19}$. The limiting probability that they visit the same venue two weekends in a row is $\pi_1 \times p_{11} = 0.0147$ (the Markov chain is aperiodic).

10.45 Since $\sum_{k=1}^{N} p_{kj} = 1$ for $1 \leq j \leq N$, it follows that $\pi_j = \frac{1}{N}$ for all j is a solution of the equilibrium equations $\pi_j = \sum_{k=1}^{N} \pi_k p_{kj}$ for $1 \leq j \leq N$. The Markov chain has no two or more disjoint closed sets and so the discrete uniform distribution is its unique equilibrium distribution.

10.47 The Markov chain is a regenerative stochastic process. The times at which the process visits state r are taken as regeneration epochs and so a cycle is the time interval between two successive visits to state r. The expected length of one cycle is the mean recurrence time $\mu_{rr} = 1/\pi_r$. Fix state j. Assume that a reward of 1 is earned each time the Markov chain visits state j. Then, by the renewal–reward theorem, the long-run average reward per unit time is γ_{jr}/μ_{rr}. Further, the long-run average reward per unit time is π_j, showing that $\pi_j = \gamma_{jr}\pi_r$ for any state j.

10.49 (a) This result follows immediately from the inequality $p_{ab}^{(n+m)} \geq p_{ac}^{(n)} p_{cb}^{(m)}$ for all states a, b and c.

(b) To prove the "if" part, assume to the contrary that C is not irreducible. Then there is a closed set $S \subset C$ with $S \neq C$. Choose any $i \in C$. Since S is closed, $C(i) \subseteq S$ and so $C(i) \neq C$, contradicting that i communicates with all states in C. The "only if" part follows by showing that $C(i) = C$ for any $i \in C$. To show this, it suffices to prove that $C(i)$ is closed. Assume to the contrary that $C(i)$ is not closed. Then there are states $a \in C(i)$ and $b \notin C(i)$ with $p_{ab} > 0$. Since $a \in C(i)$, there is an integer $n \geq 1$ such that $p_{ia}^{(n)} > 0$ and so $p_{ib}^{(n+1)} \geq p_{ia}^{(n)} p_{ab} > 0$, contradicting that $b \notin C(i)$.

(c) Since C is finite and closed, there must be some state $j \in C$ such that $P(X_n = j$ for infinitely many $n \mid X_0 = j) = 1$. Such a state is recurrent.

(d) Assume to the contrary that some transient state t can be reached from some recurrent state r, that is, $p_{rt}^{(m)} > 0$ for some $m \geq 1$. Then $P(X_n = r$ for infinitely many $n \mid X_0 = r) = 1$ implies that this relation also holds with r replaced by t, which would mean that state t is recurrent.

(e) By (c), there is some state $j \in C$ that is recurrent. Take any other state $i \in C$. By (b), the states i and j communicate and so there are integers $r, s \geq 1$ such that $p_{ij}^{(r)} > 0$ and $p_{ji}^{(s)} > 0$. It now follows from $p_{ii}^{(r+n+s)} \geq p_{ij}^{(r)} p_{jj}^{(n)} p_{ji}^{(s)}$ that $\sum_{n=1}^{\infty} p_{ii}^{(n)} \geq p_{ij}^{(r)} p_{ji}^{(s)} \sum_{n=1}^{\infty} p_{jj}^{(n)}$. Since $\sum_{n=1}^{\infty} p_{jj}^{(n)} = \infty$ for the recurrent state j, it follows that $\sum_{n=1}^{\infty} p_{ii}^{(n)} = \infty$, showing that state i is recurrent.

(f) Since $P(N_{ij} > r) \leq P(N_{ij} > r_i) = 1 - p_{ij}^{(r_i)} \leq 1 - \rho$ for all $i \in C$, we have $P(N_{ij} > kr) \leq (1-\rho)^k$ for $k \geq 1$. Then, by $E(N_{ij}) = \sum_{n=0}^{\infty} P(N_{ij} > n)$, we get $E(N_{ij}) = 1 + \sum_{k=1}^{\infty} \sum_{l=(r-1)k+1}^{rk} P(N_{ij} > l) \leq 1 + \sum_{k=1}^{\infty} r(1-\rho)^{k-1}$, which proves that $E(N_{ij}) < \infty$. *Note:* Let T be the set of transient states of a Markov chain having a single irreducible set C of states. Mimicking the foregoing proof shows that the expected number of transitions until reaching the set C is finite for any starting state $i \in T$.

(g) Denote by d_k the period of any state $k \in C$. Choose $i, j \in C$ with $i \neq j$. By (b), there are integers $v, w \geq 1$ such that $p_{ij}^{(v)} > 0$ and $p_{ji}^{(w)} > 0$. Then $p_{ii}^{(v+w)} > 0$, and so $v + w$ is divisible by d_i. Let $n \geq 1$ be any integer with $p_{jj}^{(n)} > 0$. Then $p_{ii}^{(v+n+w)} > 0$ and so $v + n + w$ is divisible by d_i. Thus, n is divisible by d_i and so $d_i \leq d_j$. For reasons of symmetry, $d_j \leq d_i$, showing that $d_i = d_j$.

10.51 The long-run average stock on hand at the end of the week equals $\sum_{j=0}^{S} j\pi_j = 4.387$. The long-run average ordering frequency is $\sum_{j=0}^{s-1} \pi_j = 0.5005$. Let $L(j) = \sum_{k=j+1}^{\infty}(k-j)e^{-\lambda}\lambda^k/k!$ denote the expected amount of demand lost in the coming week if the current stock on hand just after review is j. By Rule 10.7, the long-run average amount of demand lost per week is $L(S)\sum_{j=0}^{s-1}\pi_j + \sum_{j=s}^{S} L(j)\pi_j = 0.0938$

10.53 A circuit board is said to have status 0 if it has failed and is said to have status i if it functions and has age i weeks. Let X_n be the state of the system at the beginning of the nth week just before any replacement, where state (i,j) means that one of the circuit boards has status i and the other has status j. The process $\{X_n\}$ is a Markov chain with state space $I = \{(i,j) : 0 \leq i \leq j \leq 6\}$. The one-step probabilities can be expressed in terms of the failure probabilities r_i. For example, for $0 \leq i < j \leq 5$, $p_{(i,j),(i+1,j+1)} = (1-r_i)(1-r_j)$, $p_{(i,j),(0,i+1)} = (1-r_i)r_j$, $p_{(i,j),(0,j+1)} = r_i(1-r_j)$, $p_{(i,j),(0,0)} = r_i r_j$, and $p_{(i,j),(v,w)} = 0$ otherwise. The long-run proportion of time the device operates properly is $1 - \pi_{(0,0)} = 0.9814$. By Rule 10.7, the long-run average weekly cost is $750\pi_{(0,0)} + 200[\pi_{(0,0)} + \pi_{(6,6)} + \pi_{(0,6)}] + 100\sum_{j=1}^{5}[\pi_{(0,j)} + \pi_{(j,6)}] = 52.46$ dollars.

10.55 Let X_n be the number of tokens in the buffer at the beginning of the nth time slot just before a new token arrives. Then $\{X_n\}$ is a Markov chain with state space $I = \{0, 1, \ldots, M\}$. Put for abbreviation $a_k = e^{-\lambda}\lambda^k/k!$. The one-step transition probabilities are $p_{jk} = a_{j+1-k}$ for $0 \leq j < M$ and $1 \leq k \leq j+1$, $p_{j0} = 1 - \sum_{k=1}^{j+1} p_{jk}$ for $0 \leq k < M$, $p_{Mk} = a_{M-k}$ for $1 \leq k \leq M$, and $p_{M0} = 1 - \sum_{k=1}^{M} p_{Mk}$. Let $\{\pi_j\}$ be the equilibrium distribution of the Markov chain. By Rule 10.7, the long-run average number of packets admitted in one time slot is $\sum_{j=0}^{M} c(j)\pi_j$, where $c(j) = \sum_{k=0}^{j} k a_k + (j+1)\left(1-\sum_{k=0}^{j} a_k\right)$ and $c(M) = \sum_{k=0}^{M-1} k a_k + M\left(1-\sum_{k=0}^{M-1} a_k\right)$.

10.57 The Markov matrix is doubly stochastic and irreducible. Therefore its unique equilibrium distribution is the uniform distribution, that is, $\pi_i = 0.25$ for all i, see Problem 10.45. Also, the Markov matrix has the property that $p_{jk} = p_{kj}$ for all j,k. This gives that $\pi_j p_{jk} = \pi_k p_{kj}$ for all j,k, showing that the Markov chain is reversible.

10.59 Since $\pi_i p_{ij} = \pi_j p_{ji}$ boils down to $\pi_i/\sum_k w_{ik} = \pi_j/\sum_k w_{jk}$, the detailed balance equations are satisfied by $\pi_i = \sum_k w_{ik}/\sum_j \sum_k w_{jk}$. Thus, the Markov chain is reversible and has the π_i as its unique equilibrium distribution. The equilibrium probabilities for the mouse problem are $\pi_1 = \pi_5 = \pi_{11} = \pi_{15} = \frac{2}{44}$, $\pi_2 = \pi_3 = \pi_4 = \pi_6 = \pi_{10} = \pi_{12} = \pi_{13} = \pi_{14} = \frac{3}{44}$, and $\pi_7 = \pi_8 = \pi_9 = \frac{4}{44}$ (take $w_{ij} = 1$ if the rooms i and j are connected by a door). The mean recurrence time from state i to itself is $\mu_{ii} = 1/\pi_i$.

10.61 Applying 100,000 iterations of the Metropolis–Hastings algorithm with random-walk sampling, we found for $a = 0.02, 0.2, 1$, and 5 the average values 97.3%, 70.9%, 34%, and 4.5% for the acceptance probability.

Further, the simulation results clearly showed that a high acceptance probability does not necessarily guarantee a good mixing of the state. We found experimentally that $a = 0.6$ gives an excellent mixing with an average acceptance probability of about 49.9% (based on 100,000 iterations). For the choice $a = 0.6$, the estimates 1.652 and 1.428 for the expected value and the standard deviation of X_1 were obtained after 100,000 iterations (the exact values are 1.6488 and 1.4294).

10.63 The univariate conditional densities $\pi_1(x_1 | x_2)$ and $\pi_2(x_2 | x_1)$ are given by the $N(7(1 + x_2^2)^{-1}/2, (1 + x_2^2)^{-1})$ density and the $N(7(1 + x_1^2)^{-1}/2, (1 + x_1^2)^{-1})$ density. The estimates 1.6495 and 1.4285 are found for $E(X_1)$ and $\sigma(X_1)$ after one million runs.

Chapter 11

11.1 Let state i mean that i units are in working condition, where $i = 0, 1, 2$. Let $X(t)$ be the state at time t. The process $\{X(t)\}$ is a continuous-time Markov chain with transition rates $q_{01} = \mu$, $q_{10} = \lambda$, $q_{12} = \mu$, and $q_{21} = \lambda + \eta$.

11.3 Let $X(t)$ denote the number of customers in the barbershop at time t. The process $\{X(t)\}$ is a continuous-time Markov chain with state space $I = \{0, 1, \ldots, 7\}$. The transition rates are $q_{i,i+1} = \lambda(1 - b(i))$ for $0 \le i \le 6$, $q_{10} = \mu$, $q_{i,i-1} = 2\mu$ for $2 \le i \le 7$, and the other $q_{ij} = 0$.

11.5 Let state $(i, 0)$ mean that i passengers are waiting at the stand and no sheroot is present ($0 \le i \le 7$), and let state $(i, 1)$ mean that i passengers are waiting at the stand and a sheroot is present ($0 \le i \le 6$). Let $X(t)$ be the state at time t. The process $\{X(t)\}$ is a continuous-time Markov chain with transition rates $q_{(i,0),(i+1,0)} = \lambda$ and $q_{(i,0),(i,1)} = \mu$ for $0 \le i \le 6$, $q_{(7,0),(0,0)} = \mu$, $q_{(i,1),(i+1,1)} = \lambda$ for $0 \le i \le 5$, and $q_{(6,1),(0,0)} = \lambda$.

11.7 Let $X_1(t)$ be the number of messages in the system at time t, and let $X_2(t)$ be 1 if the gate is not closed at time t and 0 otherwise. The process $\{(X_1(t), X_2(t))\}$ is a continuous-time Markov chain with state space $I = \{(0, 1), \ldots, (R - 1, 1)\} \cup \{(r + 1, 0), \ldots, (R, 0)\}$. The transition rates are $q_{(i,1),(i+1,1)} = \lambda$ for $0 \le i \le R - 2$, $q_{(R-1,1),(R,0)} = \lambda$, $q_{(i,1),(i-1,1)} = \mu$ for $1 \le i \le R - 1$, $q_{(i,0),(i-1,0)} = \mu$ for $r + 2 \le i \le R$, and $q_{(r+1,0),(r,1)} = \mu$.

11.9 Define the state as the number of machines in repair. In states $i = 0$ and 1 there are 10 machines in operation and $2 - i$ machines on standby, while in state $i \ge 2$ there $12 - i$ machines in operation and no machines on standby. Let $X(t)$ be the state at time t. The process $\{X(t)\}$ is a continuous-time Markov chain with state space $I = \{0, 1, \ldots, 12\}$. The transition rates are $q_{i,i+1} = 10\lambda$ for $i = 0, 1$, $q_{i,i+1} = (12 - i)\lambda$ for $2 \le i \le 11$, and $q_{i,i-1} = i\mu$ for $1 \le i \le 12$.

11.11 Using Problem 11.10 and Example 11.1 (continued) in Section 11.2, the answer is $\int_0^T \left[\frac{\mu_1}{\mu_1+\mu_2} - \frac{\mu_1}{\mu_1+\mu_2} e^{-(\mu_1+\mu_2)t}\right] dt = \frac{\mu_1 T}{\mu_1+\mu_2} - \frac{\mu_1}{(\mu_1+\mu_2)^2} (1 - e^{-(\mu_1+\mu_2)T})$.

11.13 (a) Let $N(t)$ be the number of state changes in $(0, t)$. The sought probability $P(N(T) = 1 \mid X(0) = a, X(T) = b)$ is $P(X(T) = b, N(T) = 1 \mid X(0) = a)$ divided by $P(X(T) = b \mid X(0) = a) = p_{ab}(T)$. Let $v_i = \sum_{j \ne i} q_{ij}$ and $p_{ij} = q_{ij}/v_i$. Then, by the law of conditional probability, $P(X(T) = b, N(T) = 1 \mid X(0) = a)$ is given by $\int_0^T e^{-v_b(T-t)} v_a e^{-v_a t} p_{ab} \, dt$. Thus, the sought probability is $q_{ab}(e^{-v_b T} - e^{-v_a T})/[(v_a - v_b)p_{ab}(T)]$ if $v_a \ne v_b$ and $q_{ab} T e^{-v_a T}/p_{ab}(T)$ if $v_a = v_b$.
(b) Since $P(I(t) = j \mid X(0) = a, X(T) = b) = p_{aj}(t)p_{jb}(T-t)/p_{ab}(T)$, the expected amount of time that the process will stay in state j during $(0, T)$ is given by $E\left(\int_0^T I(t) dt \mid X(0) = a, X(T) = b\right) = \frac{1}{p_{ab}(T)} \int_0^T p_{aj}(t) p_{jb}(T-t) \, dt$.

11.15 The number of messages in the buffer is described by a continuous-time Markov chain with state space $I = \{0, 1, \ldots, 10\}$ and transition rates $q_{i,i-1} = i\mu$ for $1 \le i \le 10$ and $q_{i,i+1} = \lambda$ for $0 \le i \le 9$. Using the alternative construction of a continuous-time Markov chain, $f_i = \frac{i\mu}{\lambda+i\mu} f_{i-1} + \frac{\lambda}{\lambda+i\mu} f_{i+1}$ for $1 \le i \le 9$, where $f_0 = 0$ and $f_{10} = 1$.

11.17 (a) Let $N(t)$ be the number of state transitions in $(0, t)$ and X_n be the state after the nth state transition. The random variable $N(t)$ is Poisson distributed with expected value vt and the embedded process $\{X_n\}$ is a discrete-time Markov chain. By the law of conditional probability, $P(J(t) = j \mid J(0) = i)$ is $\sum_{n=0}^{\infty} P(J(t) = j \mid J(0) = i, N(t) = n) P(N(t) = n)$ and so $P(J(t) = j \mid J(0) = i)$ equals $\sum_{n=0}^{\infty} P(X_n = j \mid X_0 = i) e^{-vt} \frac{(vt)^n}{n!}$, which gives the desired result. (b) The process $\{J(t)\}$ is a continuous-time Markov chain. Since a finite-state continuous-time Markov chain is uniquely determined by its transition rates, it suffices to verify that $\{J(t)\}$ has the q_{ij} as its transition rates. Using the expression for $p_{ij}(t) = P(J(t) = j \mid J(0) = i)$ in part (a) and noting that $r_{ij}^{(0)} = 0$ for $j \ne i$, it follows from $\lim_{t \to 0} p_{ij}(t)/t = \lim_{t \to 0} \frac{1}{t} \sum_{n=0}^{\infty} r_{ij}^{(n)} e^{-vt} \frac{(vt)^n}{n!}$ that $\lim_{t \to 0} p_{ij}(t)/t = r_{ij} v = q_{ij}$ for any $j \ne i$, as was to be proved.

11.19 This problem can be analyzed by a continuous-time Markov chain with nine states. The states and the transition rates are given in the figure. The sleep mode requires two states, because it takes an exponentially distributed time until the system is converted into the sleep mode and

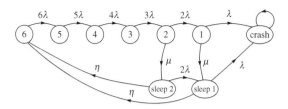

another gyroscope may fail during this time. The state labeled as the crash state is taken as an absorbing state. Using a numerical code for linear differential equations, the values 0.000504 and 0.3901 are found for the sought probabilities.

11.21 The principle "rate out of state i = rate into state i" gives the balance equations $\mu p_0 = \lambda p_1$, $(\lambda + \mu)p_1 = \mu p_0 + (\lambda + \eta)p_2$, and $(\lambda + \eta)p_2 = \mu p_1$. Together with the normalization equation $p_0 + p_1 + p_2 = 1$, these equations can be solved (one of the balance equations can be omitted). The long-run fraction of time the system is down is given by $p_0 = \frac{\lambda}{\lambda + \mu + \mu^2/(\lambda+\eta)}$. The probability p_0 is 0.0129 when $\lambda = 0.1$, $\eta = 0.05$, and $\mu = 1$.

11.23 There are five states. Let state $(0,0)$ mean that both stations are free, state $(0, 1)$ that station 1 is free and station 2 is busy, state $(1, 0)$ that station 1 is busy and station 2 is free, state $(1, 1)$ that both stations are busy, and state $(b, 1)$ that station 1 is blocked and station 2 is busy. Using

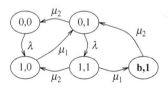

the transition-rate diagram, we get the balance equations $\lambda p(0,0) = \mu_2 p(0,1)$, $(\mu_2+\lambda)p(0,1) = \mu_1 p(1,0)+\mu_2 p(b,1)$, $\mu_1 p(1,0) = \lambda p(0,0)+\mu_2 p(1,1)$, $(\mu_1+\mu_2)p(1,1) = \lambda p(0,1)$, and $\mu_2 p(b,1) = \mu_1 p(1,1)$. The long-run fraction of time station 1 is blocked is $p(b, 1)$. The long-run fraction of items that are rejected is $p(1,0) + p(1,1) + p(b,1)$, by the property of Poisson arrivals see time averages.

11.25 State $(i, 0)$ means that i passengers are waiting at the stand and no sheroot is present, and state $(i, 1)$ means that i passengers are waiting at the stand and a sheroot is present. Using the transition-rate diagram, we get the balance equations $(\lambda + \mu)p(0,0) = \mu p(7,0) + \lambda p(6,1)$, $(\lambda + \mu)p(i,0) = \lambda p(i-1,0)$ for $1 \le i \le 6$, $\mu p(7,0) = \lambda p(6,0)$,

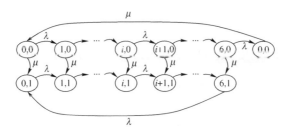

$\lambda p(0,1) = \mu p(0,0)$, $\lambda p(0,1) = \mu p(0,0)$, and $\lambda p(i,1) = \mu p(i,0) + \lambda p(i-1,1)$ for $1 \le i \le 5$. The long-run average number of waiting passengers is $\sum_{i=1}^{6} i[p(i,0) + p(i,1)] + 7p(7,0)$. The long-run fraction of potential passengers who go elsewhere is $p(7,0)$.

11.27 The state of the system is defined as (i,k) when $i \ge 1$ cars are in the gasoline station and the service time of the car in service is in phase k for $k = 1, 2$. State 0 means that the gasoline station is empty. Let $X(t)$ be the state at time t. Then $\{X(t)\}$ is a continuous-time Markov chain. Letting $\lambda = \frac{1}{6}$ and $\beta = \frac{1}{2}$, the balance equations are $\lambda p_0 = \beta p_{1,2}$, $(\lambda + \beta)p_{i,1} = \lambda p_{i-1,1} + \beta p_{i+1,2}$, and $(\lambda + \beta)p_{i,2} = \lambda p_{i-1,2} + \beta p_{i,1}$ for $1 \le i \le 3$, $\beta p_{4,1} = \lambda p_{3,1}$, and $\beta p_{4,2} = \beta p_{4,1}$, where $p_{0,1} = p_0$ and $p_{0,2} = 0$. Solving these equations, we get the following answers. (a) The fraction of time the pump is occupied is $1 - p_0 = 0.6319$. (b) The average number of cars waiting in queue is $L_q = \sum_{i=1}^{4}(i-1) \times (p_{i,1} + p_{i,2}) = 1.652$. (c) The fraction of cars not entering the station is $P_{rej} = p_{4,1} + p_{4,2} = 0.0529$, by the property of Poisson arrivals see time averages. (d) The average waiting time in a queue of a car entering the station is $W_q = \sum_{i=1}^{3} \frac{p_{i,1}}{1-P_{rej}}\left(i \times \frac{2}{\mu}\right) + \sum_{i=1}^{3} \frac{p_{i,2}}{1-P_{rej}}\left(\frac{1}{\mu} + (i-1) \times \frac{2}{\mu}\right) = 1.043$ (the probability that a car entering the station finds upon arrival state (i,k) is $p_{i,k}/(1-P_{rej})$). Note: $L_q = \lambda(1-P_{rej})W_q$. This relation is generally valid for finite-capacity queues, as can be explained with the renewal-reward theorem.

11.29 (a) In each of the problems the continuous-time Markov chain has a state space of the form $\{0, 1, \ldots, N\}$ and transition rates q_{jk} that are zero if $|j - k| > 1$. Equating the rate at which the system leaves the set of states $\{j, j+1, \ldots N\}$ to the rate at which the system enters this set gives that $p_j q_{j,j-1} = p_{j-1} q_{j-1,j}$ for all $j = 1, \ldots N$, showing that the Markov chain is reversible. (b) By the reversibility of the Markov process $\{X(t)\}$, $p_j q_{jk} = p_k q_{kj}$ for all $j, k \in A$ with $j \ne k$. This gives that $\overline{p}_j \overline{q}_{jk} = \overline{p}_k \overline{q}_{kj}$ for all $j, k \in A$ with $j \ne k$. Summing this equality over $k \in A$, we get $\overline{p}_j \sum_{k \ne j} \overline{q}_{jk} = \sum_{k \ne j} \overline{p}_k \overline{q}_{kj}$ for all $j \in A$. In other words, the $\overline{p}_j$ satisfy the balance equations of the continuous-time Markov chain $\{\overline{X}(t)\}$ and the normalization equation $\sum_{j\in A} \bar{p}_j = 1$. The solution of these equations is unique. This proves that the $\bar{p}_j$ are the limiting probabilities of the Markov process $\{\overline{X}(t)\}$.

11.31 Let T_i be the sojourn time in state i. Then $P(T_i \leq t + \Delta t \mid T_i > t) = (\mu_i \Delta t) \times (1 - a_i) + o(\Delta t)$ for Δt small, showing that T_i is exponentially distributed with mean $1/[(1 - a_i)\mu_i]$. Let $X(t) = i$ if a type-i part is processed at time t. Then $\{X(t)\}$ is a continuous-time Markov chain with state space $I = \{1, 2\}$ and transition rates $q_{12} = \mu_1(1 - a_1)$ and $q_{21} = \mu_2(1 - a_2)$. The equilibrium probabilities are $p_1 = \mu_2(1 - a_2)/[\mu_1(1 - a_1) + \mu_2(1 - a_2)]$ and $p_2 = \mu_1(1 - a_1)/[\mu_1(1 - a_1) + \mu_2(1 - a_2)]$. The average number of type-i parts processed per unit time is $\mu_i p_i$ for $i = 1, 2$.

11.33 Let state 0 mean that both stations are idle, state 1 mean that only station 1 is occupied, state 2 mean that only station 2 is occupied, and state 3 mean that both stations are occupied. Let $X(t)$ be the state at time t. Then the process $\{X(t)\}$ is a continuous-time Markov chain with transition rates $q_{01} = \lambda$, $q_{10} = \mu_1$, $q_{13} = \lambda$, $q_{21} = \mu_2$, $q_{23} = \lambda$, $q_{31} = \mu_2$, and $q_{32} = \mu_1$. The solution of the equilibrium equations is $p_0 = 0.4387$, $p_1 = 0.2494$, $p_2 = 0.1327$, and $p_3 = 0.1791$. The long-run fraction of time both stations are occupied is $p_3 = 0.1791$. By the property Poisson arrivals see time averages, this probability also gives the long-run fraction of items that are rejected. *Note:* the loss probability is 0.0467 in Problem 10.52, showing that the loss probability is very sensitive to the distributional form of the arrival process.

11.35 This problem can be solved through the Erlang loss model. The customers are the containers and the lots on the yard are the servers. A capacity for 18 containers is required. Then the loss probability is 0.0071. By the insensitivity property of the Erlang loss model, the answer is the same when the holding time of a customer has a uniform distribution.

11.37 This inventory model is a special case of the Erlang loss model. Identify the number of outstanding orders with the number of busy servers. The limiting distribution of the stock on hand is given by $r_j = \gamma(\lambda L)^{S-j}/(S-j)!$ for $0 \leq j \leq S$, where $\gamma = 1/\sum_{k=0}^{S}(\lambda L)^k/k!$. The average stock on hand is $\sum_{j=0}^{S} jr_j$. The fraction of lost demand is r_0.

11.39 The process describing the inventory position is a continuous-time Markov chain with state space $I = \{s+1, \ldots, s+Q\}$ and transition rates $q_{i,i-1} = \lambda$ for $s+2 \leq i \leq s+Q$ and $q_{s+1,s+Q} = \lambda$. Its limiting probabilities are $p_i = \frac{1}{Q}$ for $s+1 \leq i \leq s+Q$. Let $p_i(t)$ be the probability that the inventory position at time t is i and $r_k(t)$ be the probability that the stock on hand at time t is k for $0 \le k \le s + Q$. For any $t > L$, the stock on hand minus the amount back-ordered at time $t + L$ equals the inventory position at time t minus the total demand in $(t, t+L]$. Hence, $r_k(t) = \sum_{i=k}^{s+Q} e^{-\lambda L} \frac{(\lambda L)^i}{(i-k)!} p_i(t)$ for $1 \le k \le s + Q$ and $r_0(t) = \sum_{i=s+1}^{s+Q} p_i(t) \sum_{l=i}^{\infty} e^{-\lambda L} \frac{(\lambda L)^l}{l!}$. Noting that $\lim_{t \to \infty} p_i(t) = \frac{1}{Q}$, the limiting distribution of the stock on hand follows.

11.41 The $M/G/\infty$ model with an arrival rate of three customers per minute and a constant service time of 3 minutes applies. The sought probability is $\sum_{k>15} e^{-9} 9^k/k! = 0.0220$.

11.43 Define the state of the system as the number of taxis waiting at the stand. It is immediate from the state diagram that the $M/M/1$ queueing model with arrival rate $\lambda = \frac{7}{60}$ and service rate $\mu = \frac{10}{60}$ applies to the number of taxis waiting at the stand. Thus, the long-run average number of waiting taxis is $\frac{0.7}{1-0.7} = 2\frac{1}{3}$ and the long-run proportion of served customers is $1 - 0.3 = 0.7$.

11.45 This problem is an application of the infinite-server queueing model from Example 11.5. The steady-state probability that more than seven oil tankers are on the way to Rotterdam is insensitive to the shape of the sailing-time distribution and is given by the Poisson probability $\sum_{j=8}^{\infty} e^{-4} 4^j/j! = 0.0511$.

11.47 The process describing the number of service requests present is a birth-and-death process with transition rates $\lambda_i = \lambda$, $\mu_i = i\mu$ for $1 \le i \le s - 1$, and $\mu_i = s\mu + (i-s)\theta$ for $i \ge s$. The limiting probabilities p_j can be obtained recursively from $j\mu p_j = \lambda p_{j-1}$ for $1 \le j \le s$ and $(s\mu + (j-s)\theta) p_j = \lambda p_{j-1}$ for $j > s$. The long-run fraction of balking callers is $\sum_{j=s+1}^{\infty} (j-s)\theta p_j/\lambda$.

11.49 Let $X(t)$ be the number of busy channels at time t. The process $\{X(t)\}$ is a continuous-time Markov chain with transition rates $q_{i,i-1} = i\mu$ and $q_{i,i+1} = (M-i)\alpha$. The equilibrium probabilities p_j satisfy the recursive equations $j\mu p_j = (M-j+1)\alpha p_{j-1}$ for $1 \le j \le c$. The solution is the truncated binomial distribution $p_j = \binom{M}{j} p^j (1-p)^{M-j} / \sum_{k=0}^{c} \binom{M}{k} p^k (1-p)^{M-k}$ for $0 \le j \le c$, where $p = (1/\mu)/(1/\mu + 1/\alpha)$. The long-run average number of service requests generated per unit time when i service channels are busy is $(M-i)\alpha p_i$. Hence, the long-run fraction of service requests that are lost is $(M-c)\alpha p_c / \sum_{i=0}^{c} (M-i)\alpha p_i$. *Note*: The Engset loss model has the property that the limiting probabilities p_j are insensitive to the specific form of the service-time distribution and uses only its expected value.

11.51 (a) Let $X(t)$ be the stock on hand at time t. The process $\{X(t)\}$ is a continuous-time Markov chain with state space $I = \{1, 2, \ldots, R-1\}$. The transition rates are $q_{1Q} = \lambda$, $q_{i,i-1} = \lambda$ for $2 \leq i \leq R-1$, $q_{R-1,Q} = \mu$, $q_{i,i+1} = \mu$ for $1 \leq i \leq R-2$, and the other $q_{ij} = 0$. (b) The average stock on hand is $\sum_{i=1}^{R-1} i p_i$, where the p_i are the limiting probabilities of the Markov chain. (c) The average number of stock replenishments per unit time is λp_1 and the average number of stock reductions per unit time is μp_{R-1}.

Index

absorbing state, 363
acceptance–rejection method, 273, 387
accessibility, 362
addition rule, 29
almost-sure convergence, 331
aperiodicity, 376
array method, 457
axioms of probability, 5

balance equations, 377, 420
    detailed, 390, 432
balls and bins model, 64, 102, 293, 326
Bayes' rule, 67, 295
Bayes, T., 78
Bayesian credible interval, 298
Bayesian inference, 77, 294
Benford density, 201
Bernoulli distribution, 118
Bernoulli trial, 118
Bernoulli, J., 1, 329
Bernstein's inequality, 327
beta distribution, 179
binomial coefficients, 440
binomial distribution, 119, 305
    normal approximation, 185
birth-and-death process, 426
birthday problem, 28, 41, 308
birthday-coverage problem, 102, 131
bivariate normal distribution, 239, 318
Boole's inequality, 25
Borel–Cantelli lemma, 24, 54
boy–girl problem, 76
branching process, 311
Buffon's needle problem, 17

Cardano, G., 120
Cauchy density, 225, 229, 242, 313

Cauchy–Schwarz inequality, 236
central limit theorem, 184, 318
    multidimensional, 250
chain rule, 48
Chapman–Kolmogorov equations, 357
Chebyshev's inequality, 321
Chebyshev, P. L., 322
Chernoff bounds, 321
chi-square test, 257
chi-squared distribution, 194, 223, 317
Chow–Robbins game, 309
Chuck-a-Luck, 121
closed set, 373
coincidences, 29, 133, 138, 519
combination, 9, 440
complement rule, 27
complementary event, 25
complementary set, 444
compound experiment, 19
compound Poisson distribution, 306
conditional expectation, 276
    law of, 277
conditional probability, 43
    law of, 54, 269
conditional probability density, 264
conditional probability mass function, 262
confidence interval, 197, 458
conjugate prior, 297
continuity property, 23
continuous random variable, 148
continuous-time Markov chain, 403
    equilibrium equations, 420
    first-passage time probabilities, 416
    irreducible, 420
    limiting probabilities, 420
    Markovian property, 404

reversible, 432
time-dependent probabilities, 414
transition rates, 404
convergence in distribution, 332
convergence in mean square, 332
convergence in probability, 331
convergence with probability one, 23
convolution formula, 116, 221, 304
correlation coefficient, 234
countable, 5, 445
countably infinite, 4, 445
coupon-collector problem, 40, 102, 131, 292, 363
covariance, 233
craps, 62, 366
cumulative distribution function, 149

De Méré, C., 133
De Moivre, A., 116, 185, 302
De Morgan's laws, 444
discrete random variable, 86
discrete-time Markov chain, 351
   equilibrium equations, 377
   equilibrium probabilities, 377, 378
   Markovian property, 351
   reversible, 390
   time-dependent probabilities, 357
disjoint events, 6
distribution function, 149
doubling strategy, 96
drunkard's walk, 117, 190, 243, 349
   in dimension three, 246
   in dimension two, 243

Ehrenfest model, 352, 378
empty set, 444
Engset loss model, 437
entropy, 205
equilibrium distribution, 375
equilibrium equations, 377, 420
equilibrium excess distribution, 346
Erlang delay formula, 429
Erlang delay model, 428
Erlang distribution, 170, 178, 317
Erlang loss formula, 425
Erlang loss model, 425
Euler, L., 302, 450
event, 4, 6
excess life, 343
expectation, see expected value

expected value, 90
   continuous random variable, 156
   discrete random variable, 90
   of sum, 99
exponential distribution, 167, 203, 223
exponential function, 447
extinction probability, 312

failure rate, 406
failure-rate function, 202
fair game, 96
Fermat, P., 1, 119
first-step analysis, 58, 308
flow-rate-equation method, 422
fundamental principle of counting, 438

Galilei, G., 13
gambler's ruin problem, 57, 286
gamma distribution, 177, 316
gamma function, 177
Gauss, C. F., 181
Gaussian distribution, 181
generating function, 302
geometric distribution, 141
geometric probability model, 15
geometric series, 449
Gibbs sampler, 399
Gosset, W., 196

harmonic series, 449
hit-or-miss method, 387
hundred prisoners problem, 36
Huygens, C., 1, 89
hypergeometric distribution, 135

importance sampling, 389
inclusion–exclusion rule, 36, 104
independent events, 52
independent random variables, 113, 221
indicator random variable, 102
insensitivity, 425, 428
interchangeable random variables, 136
intersection of events, 25
intersection of sets, 443
inventory systems, 434
inverse-transformation method, 198
irreducible, 373, 384, 420

Jacobian, 228
Jensen's inequality, 107
joint probability density, 212
joint probability mass function, 210

Kelly betting, 194, 335
Kolmogorov differential equations, 415
Kolmogorov, A. N., 290, 329, 446
Kruskal count, 461

Laplace density, 318, 496
Laplace model, 7
Laplace, P. S., 2, 78, 185
law of the iterated logarithm, 333
law of the unconscious statistician, 106
Leibniz, G. W., 13
likelihood ratio, 68
linear predictor, 237, 282
Little's formula, 430
lognormal distribution, 193, 316

marginal density, 219
Markov chain Monte Carlo, 386
Markov chains, 348, 403
Markov's inequality, 321
Markov, A. A., 321, 348, 382
Markovian property, 350
Marsaglia's polar method, 230
matching problems, 37, 132
mean, *see* expected value
mean residual lifetime, 204
median, 143, 158, 182
memoryless property, 168
Metropolis–Hastings algorithm, 393
minimum error predictor, 282
mixed random variable, 104
moment-generating function, 314
Montmort, P. R., 132
multinomial distribution, 124
multiplication rule, 47
multivariate normal distribution, 248
  simulating from, 250
mutually exclusive events, 6

natural logarithm function, 447, 450
negative binomial distribution, 142, 305
newsboy problem, 110
Newton–Pepys problem, 122
normal density, 229
normal distribution, 180, 222, 315
  bivariate, 239
  multivariate, 248

$o(h)$, 405
odds, 67
odds form of Bayes' rule, 67

one-stage-look-ahead rule, 92
optimal stopping, 92, 287, 289
order statistics, 231

p-value, 81
page-ranking algorithm, 380
Pareto density, 164
Pascal, B., 1, 89, 119
Pearson, K., 258
Penney Ante game, 371
percentile, 182
periodicity, 376
permutation, 439
permutation cycle, 106
Poisson arrivals see time averages, 423
Poisson distribution, 109, 117, 124, 304, 306
  normal approximation, 185
Poisson heuristic, 130
Poisson process, 170, 232, 347, 407
Poisson, S. D., 124
Poissonization trick, 290
positive semi-definite, 237
posterior probability, 67, 79, 295
principle of maximum entropy, 205
prior probability, 67, 79, 295
probability density function, 149
probability mass function, 86
probability measure, 5
probability space, 6
problem of points, 119

queues, 425, 427, 428, 436

random number, 147, 165, 452
random permutation, 106, 456
random point, 16, 217, 451
random process, 339
random variable, 86
random walk, 57, 65, 243
random-number generator, 452
rare event, 168
rate, 404
Rayleigh distribution, 246
records, 306
recurrent state, 381, 384
regeneration cycle, 339
regression curve, 281
regression to the mean, 238, 283
reliability function, 202
reliability modeling, 169, 179, 283, 341, 411
renewal equation, 342

renewal function, 342
renewal process, 342
renewal–reward process, 339
renewal–reward theorem, 340
reversible, 390, 432

sample mean, 117, 458
sample space, 3
sample variance, 117, 458
set theory, 443
sigma algebra, 445
Simpson trial, 71
simulation, 267, 273, 386, 387, 451
  accuracy, 457, 459
square-root law, 116
St. Petersburg game, 96
standard deviation, 107
standard normal distribution function, 181
statistical equilibrium, 375
Stein–Chen identity, 134
Stirling's approximation, 10
stochastic optimization, 287
stochastic process, 339
strong birthday problem, 135
strong law of large numbers, 23, 328
Student-t distribution, 195, 231
subjective probability, 7, 79, 295

substitution rule, 106, 226
success runs, 58, 308, 364

tabu probability, 368
transformation rule, 228
transient state, 381, 384
transition probabilities, 352, 357
transition rate diagram, 422
transition rates, 405
triangular distribution, 166

uncorrelatedness, 234
uncountable, 5, 445
uniform distribution, 111, 140, 147, 165, 225
uniformization method, 419
union bound, 25
union of events, 25
union of sets, 6, 443

variance, 107, 160

waiting-time paradox, 343
Wald's equation, 118
weak law of large numbers, 323
Weibull distribution, 178, 203
Wright–Fisher model, 367

Made in the USA
Middletown, DE
18 December 2019